Accuracy Verification Methods

Computational Methods in Applied Sciences

Volume 32

Series Editor

E. Oñate
International Center for Numerical Methods in Engineering (CIMNE)
Technical University of Catalonia (UPC)
Edificio C-1, Campus Norte UPC
Gran Capitán, s/n
08034 Barcelona, Spain
onate@cimne.upc.edu
url: http://www.cimne.com

For further volumes:
www.springer.com/series/6899

Olli Mali • Pekka Neittaanmäki • Sergey Repin

Accuracy Verification Methods

Theory and Algorithms

Springer

Olli Mali
Department of Mathematical Information Technology
University of Jyväskylä
Jyväskylä, Finland

Pekka Neittaanmäki
Department of Mathematical Information Technology
University of Jyväskylä
Jyväskylä, Finland

Sergey Repin
Steklov Institute of Mathematics
Russian Academy of Sciences
St. Petersburg, Russia
and
University of Jyväskylä
Jyväskylä, Finland

ISSN 1871-3033 Computational Methods in Applied Sciences
ISBN 978-94-024-0498-2 ISBN 978-94-007-7581-7 (eBook)
DOI 10.1007/978-94-007-7581-7
Springer Dordrecht Heidelberg New York London

Printed on acid-free paper

Springer is part of Springer Science+Business Media (www.springer.com)

Preface

Nowadays, mathematical and numerical modeling has become an essential component of the general scientific process. Ever since the 1960s, numerical analysis and scientific computation have made up the most rapidly growing part of mathematics. One of the challenging problems in this area is the creation of fully reliable computer simulation methods, which could become an adequate complement to experimental sciences. This book aims to give an overview of mathematical methods and computer technologies focused on reliable verification of computed solutions and present recently developed methods. We hope that it will be useful for an audience much larger than just advanced specialists in numerical analysis and computer simulation methods. In actuality, the book can be used in three different ways.

For engineers and specialists in natural sciences interested in quantitative analysis of mathematical models, it is best to concentrate on algorithms and prescriptions, which explain how to measure the accuracy of a numerical solution. In Chap. 2, we discuss various error indicators, which are used in mesh adaptive numerical algorithms in order to achieve proper restructuring (refinement) of the computational mesh (or changing the set of trial functions). We suggest a unified approach to this question and discuss different error indicators. Chapter 3 is concerned with the question: "how can guaranteed and computable bounds of errors associated with approximations of differential equations be derived?". We tried to explain this in simple terms without a deep excursion into the mathematical background. In other words, the reader whose main purpose is to use the results (estimates) will find the corresponding detailed recommendations. Certainly, they are given for a limited amount of typical problems. Other cases can be found in the literature cited or require additional analysis (in the latter case, a good understanding of the mathematical theory is necessary).

For advanced specialists interested in the development of new error estimation methods, Chaps. 3–5 are the most interesting. Here, we discuss mathematical technologies that provide guaranteed error control and applications to analysis of problems with uncertain data. These chapters essentially use materials exposed in the books P. Neittaanmäki and S. Repin [NR04] and S. Repin [Rep08] (in [NR04] the reader can find a complete set of a posteriori error estimation theory generated by the variational duality approach and [Rep08] is mainly devoted to the method using transformations of integral identities, which define generalized solutions of boundary value problems). We recommend them for further study of the mathematical theory of a posteriori error estimation. However, in this book (unlike the abovementioned publications) we pay more attention to computational aspects and try to supply the reader with practical prescriptions. Chapter 5 is devoted to a special but important topic: analysis of effects caused by indeterminacy (incomplete knowledge) of problem data. It contains many new results. We show that studying problems with incompletely known data leads to conceptions and methods, which differ from those used in "classical" error analysis. In particular, they lead to the notion of an *a priori limited accuracy*, which leads to a new perspective on quantitative analysis of mathematical models. Chapter 5 and Sects. 4.1.2 and 4.1.3 (related to beams) use materials of the Ph.D. thesis of O. Mali [Mal11]. The material exposed in Chaps. 4 and 5 may be especially interesting for specialists in computational mechanics interested in finding bounds of the accuracy generated by approximation errors and data indeterminacy.

The entire book (maybe with the exception of Chaps. 4–6) can also be considered as a *textbook* for undergraduate and postgraduate students studying applied mathematics and mathematics of computations. For these reasons, we append three chapters (Appendices A, B, and C), in which basic mathematical knowledge is summarized. These chapters present a concise lecture course "Numerical analysis of differential equations" (which has been developed by the authors for graduate and undergraduate students of the University of Jyväskylä). It discusses the main methods used for quantitative analysis of partial differential equations. Chapters 2 and 3 are also written in the textbook style. Here, we have used materials from lecture courses on a posteriori error estimation methods that have been delivered to undergraduate and postgraduate students by S. Repin in Jyväskylä, Radon Institute of Computational and Applied Mathematics in Linz, Helsinki University of Technology, and University of Saarbrucken.

We would like to express our gratitude to the University of Jyväskylä and to the Academy of Finland for their support.

We are especially grateful to I. Anjam and S. Matculevich for contributions to the material exposed in the book, discussions, and proofreading and to M.-L. Rantalainen for her help in preparing the electronic version of our book.

Many materials related to theoretical justification and practical implementation of new a posteriori error estimation methods are results of joint research exposed in joint publications with our colleagues, which are referred to in the respective parts of the book. We express sincere gratitude to all of them for the cooperation and interesting discussions. Finally, we would like to thank Springer-Verlag publishing group for the friendly cooperation.

Jyväskylä, Finland
2013

Olli Mali
Pekka Neittaanmäki
Sergey Repin

Contents

Notation

$\hookrightarrow$	compact embedding
$\mapsto$	continuous embedding
$:=$	equals by definition
$\neq$	not equal
$\equiv$	logical equivalence
$\forall$	for all
$\\|\cdot\\|_{\mathrm{div}}$	norm in space $H(\Omega, \mathrm{div})$
$\\|\cdot\\|_{\mathrm{Div}}$	norm in space $H(\Omega, \mathrm{Div})$
$\\|\cdot\\|_{\mathrm{curl}}$	norm in space $H(\Omega, \mathrm{curl})$
$\\|\cdot\\|_{m,p,\Omega}$	norm in space $W^{m,p}$
$\\|\cdot\\|_{\infty,\Omega}$	norm in $L^\infty(\Omega)$
$\\|\cdot\\|$	norm in $L^2(\Omega)$
$\\|w\\|_A$	$(\int_\Omega Aw \cdot w\mathrm{d}x)^{1/2}$
$\\|w\\|_{A^{-1}}$	$(\int_\Omega A^{-1}w \cdot w\mathrm{d}x)^{1/2}$
$\\|\cdot\\|_\alpha$	norm in $L^\alpha(\Omega)$
$\\|\cdot\\|_{\alpha,\omega}$	norm in $L^\alpha(\omega)$
$\\|\\|\cdot\\|\\|$	energy norm
$\\|[(\cdot,\cdot)]\\|$	combined primal-dual norm
$\|[\boldsymbol{w}]\|_{(\gamma,\delta)}$	$(\int_\Omega(\gamma\|\mathrm{curl}\ \boldsymbol{w}\|^2 + \delta\|\boldsymbol{w}\|^2)\mathrm{d}x)^{1/2}$
$\|\cdot\|$	norm of a vector
$A_\circ$, $\mathcal{A}_\circ$	mean data operator
$\wp(T)$	aspect ratio of a simplex T
$\Box a \Box$	$\sum_{k=1}^N a_k$, $a \in \mathbf{B}^N$
$a \cdot b$	scalar product of vectors
$a \times b$	vector product of vectors
$a \otimes b$	diad product of vectors
$[\cdot]$	jump, difference of left-hand side limit and right-hand side limit
$\mathbf{B}^N$	set of boolean vectors
$\mathring{C}^\infty(\Omega)$	space of all infinitely differentiable functions with compact supports in Ω
$C^k(\Omega)$	spaces of k-times differentiable scalar-valued functions

$C^k(\Omega, \mathbb{R}^d)$	spaces of k-times differentiable vector-valued functions
$\overset{\circ}{C}^k(\Omega)$	subspace of $C^k(\Omega)$ that contains functions vanishing on the boundary
$\operatorname{cond} A$	condition number of A
curl	rotor of a vector-valued function
$D^\alpha v$	derivative of order $\lvert\alpha\rvert$
Δ	Laplace operator; $\Delta v = \operatorname{div} \nabla v$
$\mathcal{D}$	set of admissible data
$D_\circ$	mean data
$\operatorname{diam} \Omega$	diameter of the set Ω
div	divergence of a vector-valued function $\operatorname{div} v = \sum_{i=1,d} v_{i,i}$
Div	divergence of a tensor-valued function $(\operatorname{Div} \tau)_j = \sum_{i=1,d} \tau_{ij,i}$
$\operatorname{dist}(x, \omega)$	distance between x and ω
E_{ls}	edge of a simplex
e	error
$e_{\max}$	maximal (worst case) error
$e_{\min}$	minimal (best case) error
$\mathbb{E}$	error indicator
$f_{,i}$	partial derivative with respect to i-th coordinate
$\{\lvert g \rvert\}_S$	mean value of g on S
$\widetilde{g}_\omega$	$g - \{\lvert g \rvert\}_\omega$
$\Gamma, \Gamma_1, \Gamma_2$	boundary of Ω and its parts
H	Hilbert space
$H^m(\Omega)$	Sobolev space of square summable functions with square summable derivatives up to the order m
$H^{-1}(\Omega)$	space dual to $\overset{\circ}{H}{}^1(\Omega)$
$\overset{\circ}{H}{}^m(\Omega)$	subset of $H^m(\Omega)$ formed by the functions vanishing on Γ
$H(\Omega, \mathrm{curl})$	subspace of $L^2(\Omega, \mathbb{R}^d)$ that contains vector-valued functions with square-summable rotor
$H(\Omega, \mathrm{div})$	subspace of $L^2(\Omega, \mathbb{R}^d)$ that contains vector-valued functions with square-summable divergence
$H(\Omega, \mathrm{Div})$	subspace of $L^2(\Omega, \mathbb{R}^{d \times d})$ that contains tensor-valued functions with square-summable divergence
I_{eff}	efficiency index
$\ker \ell$	kernel of linear functional ℓ
ℓ	linear functional
ℓ_v, ℓ_{u_h}	residual functional
$\lvert\!\lvert\!\lvert \ell \rvert\!\rvert\!\rvert$	norm of ℓ
$\overset{\circ}{L}{}^2(\Omega)$	set of square-summable functions with zero mean
$L^2(\Omega, \mathbb{M}^{d \times d})$	space of tensor-valued functions with components that are square summable in Ω
$L^2(\Omega, \mathbb{R}^d)$	space of vector-valued functions with components that are square summable in Ω
$L^\alpha(\omega)$	space of functions integrable with power α over ω

$L^\infty(\Omega)$	space of functions uniformly bounded almost everywhere
$\mathbb{M}$	marker
$\overline{\mathsf{M}}$	majorant, subindex refers to a problem
$\underline{\mathsf{M}}$	minorant, subindex refers to a problem
$\mathbb{M}^{d\times d}$	space of real $d \times d$ matrixes
$\mathbb{M}_s^{d\times d}$	space of symmetric real $d \times d$ matrixes
∇	gradient of a scalar-valued function $\nabla\phi = (\phi_{,1}, \dots, \phi_{,d})$
$\lvert\omega\rvert$	Lebesgue measure of a set ω
Ω	open bounded connected set in $\mathbb{R}^d$ with Lipschitz continuous boundary
$\varpi(T_k)$, $\varpi(E_{ks})$	patches associated with simplex T_k and edge E_{ks}
π_h	interpolation operator
$P^k(\Omega)$	set of polynomial functions defined in $\Omega \subset \mathbb{R}^d$
$\mathbb{R}^d$	space of real d-vectors
$\mathbf{R}(\Omega)$	space of rigid deflections
r	radius of the solution set
$\bar{r}$	normalized radius of the solution set
$\lvert\sigma\rvert$	norm of tensor
$\sigma : \varepsilon$	scalar product of tensors
$(\sigma, \tau)_{\rm Div}$	scalar product in space $H(\Omega, \mathrm{Div})$
$\mathcal{S}$	solution mapping
$\mathcal{S}(\mathcal{D})$	solution set
T_k	simplex
$\lvert T_k\rvert$	area of a simplex
$\tau^{\rm D}$	deviator of τ
$\operatorname{tr}\tau$	trace of τ
$(u, v)_{\rm div}$	scalar product in space $H(\Omega, \mathrm{div})$
$\mathrm{S}(\Omega)$	set of solenoidal fields
$\overset{\circ}{\mathrm{S}}$	the subset of functions from $\mathrm{S}(\Omega)$ vanishing on the boundary
$W^{m,p}(\Omega)$	space of functions summable with power p the generalized derivatives of which up to order m belong to L^p
$\sharp$	number of elements (integer)

Chapter 1
Errors Arising in Computer Simulation Methods

Abstract The goal of this introductory chapter is to discuss in general terms different classes of errors arising in computer simulation methods and to direct the reader to the chapters and sections of the book where these errors are analyzed. Moreover, we describe the error estimation methodology applied in this book.

1.1 General Scheme

Mathematical modeling and computer simulation allows us to perform virtual experiments without costly physical equipment, construct predictions based on our assumptions, investigate events from the past, investigate prototypes of industrial objects, etc. However, without proper understanding and estimation of errors generated during the modeling process, there is a risk of drawing wrong conclusions based on unreliable numerical results.

The modeling process consists of several stages. First, physical (or biological, financial, etc.) reality is described using mathematical relations, which generate the respective mathematical model. Then, we obtain a mathematical problem, which in general terms is as follows: Find $u \in V$ such that

$$\mathcal{L}u = f,$$

where the set V, the operator $\mathcal{L}$, and the source term f are defined in accordance with the features of the problem. A mathematical model always represents an "abridged" version of a physical object, so that the error of the mathematical model is always greater than zero. This error is the difference between u and the corresponding physical function, which we denote by ε_1.

Approximation error arises when continual (differential) models are replaced by a finite dimensional (discrete) problem: Find $u_h \in V_h$ such that

$$\mathcal{L}_h u_h = f_h,$$

where $u_h \in V_h \subset V$ and h is the mesh size parameter. A certain norm of the difference between u and u_h is the approximation error ε_2.

Numerical errors arise because finite dimensional problems are also solved approximately, using numerical integration, iteration procedures, and operations performed with a limited amount of digital numbers. For this reason, instead of u_h we obtain $\widetilde{u}_h$. A norm of $u_h - \widetilde{u}_h$ is denoted by ε_3 (see Fig. 1.1).

O. Mali et al., *Accuracy Verification Methods*,
Computational Methods in Applied Sciences 32, DOI 10.1007/978-94-007-7581-7_1,

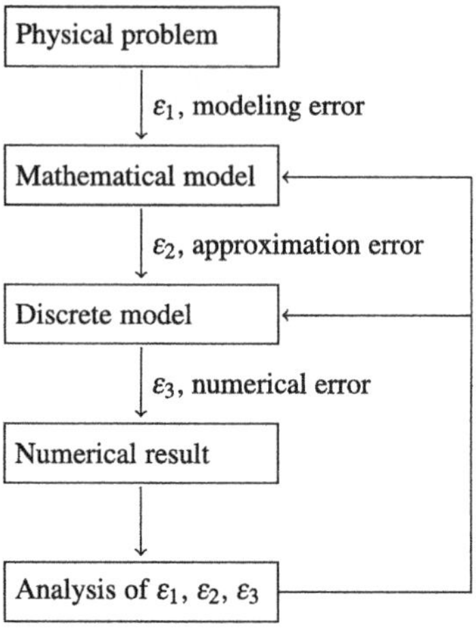

Fig. 1.1 Errors in mathematical modeling

This book is devoted to the *quantitative* estimation of these errors. This is necessary not only for the sake of reliability, but also because it provides information that can be used to improve mathematical and/or discrete models. Chapters 3 and 4 consider guaranteed and computable estimates of errors encompassed in $\widetilde{u}_h$. We explain how to construct computable upper and lower bounds of errors. We emphasize that the error estimation functionals (minorants and majorants) do not depend on the applied discrete model (in particular they do not depend on the parameter h) and are valid for any approximation in V. In particular, we do not assume that u_h is the exact solution of the corresponding discrete problem (so that it may not satisfy the Galerkin orthogonality property). Thus, they measure the numerical error ε_3 as well and satisfy the following general relation:

$$\underline{\mathsf{M}}(\widetilde{u}_h) \leq \varepsilon_2 + \varepsilon_3 \leq \overline{\mathsf{M}}(\widetilde{u}_h).$$

The benefits of these estimates are as follows:

- Estimates are always guaranteed (not only in the asymptotic sense as $h \to 0$).
- Computation of the estimates can be performed in many different ways and depending on the circumstances, we can balance between the computational cost and the accuracy of the estimates. For linear problems, we can prove that two-sided error estimates can be computed with any desired accuracy (i.e., the estimates have no gaps).
- Finally, the estimates depend explicitly on the problem data, providing us with an efficient mathematical tool to investigate modeling errors and errors generated by the incompletely know data.

Now, we discuss different types of errors in more detail and refer to particular parts of the book in which the corresponding error estimates are considered.

1.2 Errors of Mathematical Models

Let U be a physical quantity that characterizes some phenomenon or process. In mathematical modeling, it is described by a certain mathematical model, the exact solution of which is u (depending on a model u may be a vector, a function, or a collection of functions dependent on spatial variables and time). Then, the quantity

$$\varepsilon_1 = |U - u|$$

represents the error generated by the mathematical model (henceforth, we call it the *modeling error*). Here the symbol $|\cdot|$ is understood in a broad sense: it may denote a suitable norm in the corresponding functional space or some special (e.g., local) quantity used to estimate the difference between the results of physical experiments and the numbers generated by a mathematical model.

A mathematical model always represents an "abridged" version of a physical object, so that ε_1 is always greater than zero.

The evaluation of modeling errors is one of the most difficult problems in quantitative analysis of mathematical models, which in the vast majority of cases is yet to be solved.

Typical sources of modeling errors are the following:

- A mathematical model neglects some of the really existing effects (physical relations, dependencies, influences).
- Physical data involved in the mathematical model are defined with limited precision.
- The problem is solved using a simplified geometric description or dimension reduction.

In this book, we first of all focus our attention on the second case, i.e., on errors generated by incomplete knowledge of data, which are studied in Chap. 5. In Sect. 6.6 the reader will also find an overview of results related to modeling errors of other types.

1.3 Approximation Errors

Approximation errors arise when continual (differential) models are replaced by finite dimensional (discrete) models.

Usually finite dimensional problems are created with the help of meshes, which cover the corresponding domain (set in $\mathbb{R}^d$). Let $\mathcal{T}_h$ be such a mesh, and let h denote the "character" (mean) distance between neighboring elements of the mesh. By u_h we denote an approximate solution computed on $\mathcal{T}_h$. Obviously, u_h encompasses the *approximation error*

$$\varepsilon_2 = |u - u_h|.$$

In modern scientific computing, the major attention is usually paid to ε_2. Classical a priori error analysis aims to construct estimates of the type

$$\varepsilon_2 \leq Ch^k \| f \|,$$

which indicate that the solution provided by the discrete model converges to the one of the exact model by $O(h^k)$ convergence rate as $h \to 0$. Thus, the method to obtain the discrete model is well justified. In Appendix C, we shortly discuss a priori asymptotic methods with the paradigm of elliptic partial differential equations. This topic is well studied, and a priori error estimates qualified in terms of mesh size parameter(s) have been derived for many problems. The goal of the chapter is to explain the main principles of a priori error analysis in the context of conforming variational methods and mixed approximations of PDEs. The reader interested in further investigation of the subject is provided with necessary references.

A posteriori indicators of approximation errors are considered in Chap. 2, where we present a unified outlook on this question, which leads to a clear classification of different error indication methods.

Guaranteed bounds of approximation errors can be computed with the help of methods discussed in Chaps. 3 and 4. Some generalizations to nonlinear problems are presented in Chap. 6, which ends with a list of challenging problems arising within the framework of fully reliable modeling conception.

1.4 Numerical Errors

Numerical errors arise because finite dimensional problems are solved approximately, using numerical integration, iteration procedures, and operations performed with a limited amount of digital numbers.

In general, computers cannot perform elementary mathematical operations absolutely accurately, so that instead of u_h we obtain $\widetilde{u}_h$. The quantity

$$\varepsilon_3 = |u_h - \widetilde{u}_h|$$

shows the *error of numerical operations* performed by a concrete computer. Typically, this error includes errors arising in iteration processes, errors of numerical integration, and roundoff errors. Errors associated with various quadrature (cubature) formulas are well studied (we direct the reader to, e.g., [PTVF07]). Errors of iteration processes are considered in Sect. 6.7 and roundoff errors are briefly discussed in Sect. 6.8.

It remains to comment on errors caused by defects (bugs) in numerical codes. They create a special class of errors arising in large codes with complicated logical structures. Certainly, rough errors in a code usually lead to evident discrepancies and are easily detectable. However, experienced numerical analysts know that in some cases code bugs may produce relatively small effects (which are not easy to recognize in a particular numerical test) and much bigger effects in other cases (which may seriously corrupt results of expensive numerical experiments). The latter situation is rather dangerous because may lead to misleading conclusions. In principle, two different methods can be used to discover defects in codes. The first method suggests the theory of algorithms in the mathematical logic and theoretical computing (see, e.g., [CLRS01, GK90, Knu97]). Another method follows from a posteriori error estimation theory. In particular, estimates considered in Chaps. 3 and 4 include such type errors (if they indeed exist). If the corresponding error majorant does not decrease and shows relatively big errors even for fine meshes, then this fact may indicate that the algorithm is not quite correct and generates solutions containing some essential and nondecreasing errors.

Chapter 2
Indicators of Errors for Approximate Solutions of Differential Equations

Abstract Error indicators play an important role in mesh-adaptive numerical algorithms, which currently dominate in mathematical and numerical modeling of various models in physics, chemistry, biology, economics, and other sciences. Their goal is to present a comparative measure of errors related to different parts of the computational domain, which could suggest a reasonable way of improving the finite dimensional space used to compute the approximate solution. An "ideal" error indicator must possess several properties: efficiency, computability, and universality. In other words, it must correctly reproduce the distribution of errors, be indeed computable, and be applicable to a wide set of approximations. In practice, it is very difficult to satisfy all these requirements simultaneously so that different error indicators are focused on different aims and stress some properties at the sacrifice of others. We discuss the mathematical origins and algorithmic implementation of the most frequently used error indicators. Our goal is twofold: to discuss the main types of error indicators, which have already gained high popularity in numerical practice, and to suggest a unified conception, which covers practically all methods used in error indication.

For differential equations, we discuss indicators of two types. Indicators of the first type show the distribution of errors in the whole computational domain. Another group of indicators is focused on the so-called goal-oriented error functionals typically associated with some subdomains ("zones of interest"), where the accuracy of an approximate solution is especially important. Usually, the indicators of the latter type use solutions of adjoint boundary value problems. We discuss some new forms of these indicators, which do not exploit extra regularity of solutions and special properties of respective approximations (such as, e.g., superconvergence). Indicators that follow from a posteriori error majorants of the functional type are discussed in Chap. 3.

2.1 Error Indicators and Adaptive Numerical Methods

Adaptive numerical methods are based on the conception that efficient approximations should be constructed by means of a sequence of consequently refined finite dimensional spaces $\{V_k\}$, $k = 1, 2, \dots$ such that the amount of linearly independent trial functions in V_{k+1} is larger than in V_k (i.e., $\dim V_{k+1} > \dim V_k$). Typically,

O. Mali et al., *Accuracy Verification Methods*,
Computational Methods in Applied Sciences 32, DOI 10.1007/978-94-007-7581-7_2,

the structure of these spaces is a priori unknown. Within the framework of the adaptive modeling conception, the generation of V_{k+1} is based upon the information encompassed in the approximation u_k associated with V_k. For this reason, it is necessary to have computable quantities that furnish information on the error e presented in terms of a certain error measure (e.g., in terms of the energy norm). Such quantities are called *Error Indicators*. Throughout the book, we denote them by the symbol $\mathbb{E}$ (which is generated by the initial letters E and I). Error indicators play an important role in mesh-adaptive numerical algorithms, which follow the formal scheme

$$V_1 \overset{\mathbb{E}(u_1)}{\longrightarrow} V_2 \overset{\mathbb{E}(u_2)}{\longrightarrow} \cdots V_k \overset{\mathbb{E}(u_k)}{\longrightarrow} V_{k+1}.$$

A "good" error indicator must be easily computable and must correctly reproduce the distribution of errors. It is also desirable that an indicator be applicable to a wide set of approximations and imply quantities that provide a realistic presentation on the overall (global) error. In practice, it is very difficult to satisfy all these requirements simultaneously, so that different error indicators are focused on different aims and stress some properties at the expense of the others.

In this chapter, we discuss the general principles of error indication and examples of error indicators with the paradigm of finite element approximations of elliptic partial differential equations.

2.1.1 Error Indicators for FEM Solutions

Let $T_s, s = 1, 2, \ldots, N$ be elements (subdomains) associated with the mesh $\mathfrak{T}_h$ (with characteristic size h), and let u_h be an approximate solution computed on this mesh. Henceforth, the corresponding finite dimensional space is denoted by V_h, so that $u_h \in V_h$. Then, the true error is $e = u - u_h$. Denote by $m_s(e)$ the value of the error measure m associated with T_s. Usually, the error measure $m_s(e)$ is defined as a certain integral of $u - u_h$ related to T_s. For example, local error measures of approximate solutions to linear elliptic problems are often presented by the integrals

$$\left(\int_{T_s} |u - u_h|^2 \,\mathrm{d}x\right)^{1/2} \quad \text{or} \quad \left(\int_{T_s} \left|\nabla(u - u_h)\right|^2 \mathrm{d}x\right)^{1/2}.$$

The components of the vector

$$\mathbf{m}(e) = \left\{m_1(e), m_2(e), \ldots, m_N(e)\right\}$$

are nonnegative numbers, which may be rather different.

If the overall error encompassed in u_h is too big, then a new approximate solution should be computed on a new (refined) mesh $\mathfrak{T}_{h_{\mathrm{ref}}}$. Comparative analysis of $m_s(e)$ suggests where to add new degrees of freedom (new trial functions). However, in real life computations the vector $\mathbf{m}(e)$ is not known and, therefore, an error indicator

$\mathbb{E}(u_h)$ is used. The corresponding approximate values of errors $\mathbb{E}_s$ associated with the elements form the vector

$$\mathbb{E}(u_h) = \{\mathbb{E}_1, \mathbb{E}_2, \ldots, \mathbb{E}_N\},$$

which is used instead of $\mathbf{m}(e)$. If the vector $\mathbb{E}(u_h)$ is close to $\mathbf{m}(e)$, i.e.,

$$\mathbf{m}(e) \approx \mathbb{E}(u_h), \tag{2.1}$$

then a new mesh $\mathfrak{T}_{h_{\mathrm{ref}}}$ can be efficiently constructed on the basis of comparative analysis of $\mathbb{E}_s$. However, the fact that the adaptive procedure is efficient depends on how accurately the condition (2.1) is satisfied and how efficiently the information encompassed in $\mathbb{E}(u_h)$ is used to improve approximations.

2.1.2 Accuracy of Error Indicators

Certainly, the condition (2.1) looks vague unless a formal definition of the sign $\approx$ is given. Despite the huge amount of publications focused on error indication, to the best of our knowledge no commonly used definition has yet been accepted. Different authors may claim (explicitly or implicitly) different things, so the words "good error indicator" may take on a variety of meanings.

Below we suggest definitions, which can be used for a reasonable qualification of error indicators. They define "strong" and "weak" meanings of $\approx$, respectively.

Definition 2.1 The indicator $\mathbb{E}(u_h)$ is ε-accurate on the mesh $\mathfrak{T}_h$ if

$$\mathcal{M}\big(\mathbb{E}(u_h)\big) := \frac{|\mathbf{m}(e) - \mathbb{E}(u_h)|}{|\mathbf{m}(e)|} \leq \varepsilon. \tag{2.2}$$

The value of $\mathcal{M}(\mathbb{E}(u_h))$ is the strongest quantitative measure of the accuracy of $\mathbb{E}(u_h)$.

This definition imposes strong requirements on $\mathbb{E}(u_h)$. Indeed, (2.2) guarantees that inaccuracies in the error distribution computed by $\mathbb{E}(u_h)$ are much smaller (provided that ε is a small number) than the overall error. Therefore, an indicator should be regarded as "accurate" if it meets (2.2) with relatively coarse ε.

From (2.2) it follows that the so-called efficiency index

$$I_{\mathrm{eff}}\big(\mathbb{E}(u_h)\big) := \frac{|\mathbb{E}(u_h)|}{|\mathbf{m}(e)|} \leq 1 + \mathcal{M}\big(\mathbb{E}(u_h)\big) \tag{2.3}$$

is close to 1, which means that $|\mathbb{E}(u_h)|$ provides a good evaluation of the overall error $|\mathbf{m}(e)|$.

The efficiency of $\mathbb{E}(u_h)$ may be different for different meshes and approximate solutions. It is desirable that the indicator is accurate for a sufficiently wide class

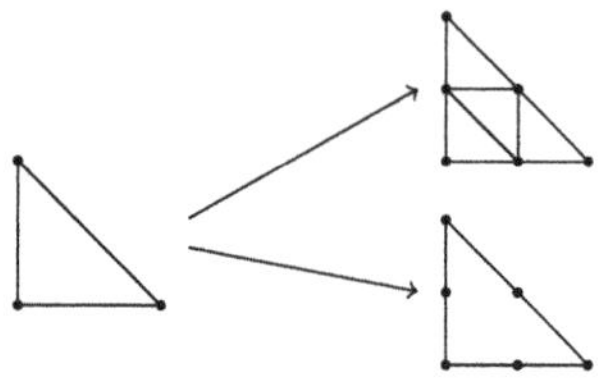

Fig. 2.1 Typical h-refinement and p-refinement

of approximations and meshes. The wider the class of approximations served by an indicator, the better it is from the computational point of view.

The majority of indicators suggested for finite element approximations are applicable only to Galerkin approximations (or to approximations that are very close to Galerkin solutions). Properties of the mesh used are also very important, and theoretical estimates of the quality of error indicators usually involve constants dependent on the aspect ratio of finite elements.

2.1.2.1 Marking Procedures

In adaptive finite element schemes, subsequent approximations are often constructed on nested meshes, where a refined mesh is obtained by "splitting" elements (h-refinement) or by increasing the amount and order of basis functions (p-refinement) of the current mesh. In Fig. 2.1, we depict typical refinement strategies for a linear triangular element, the degrees of freedom of which are function values at nodes. A detailed discussion on refinement methods can be found in, e.g., [BGP89, Dem07]. Alternative procedures intended to increase the set of basis functions lead to nonconforming methods (cf. Appendix B).

Typical adaptive schemes consists of solving the problem several times on a sequence of improving subspaces. In this type of practice, error indicators are used together with a certain *marker* that marks elements (subdomains) where errors are excessively high. A new subspace $V_{h_{\rm ref}}$ is constructed in such a way that these errors are diminished.

Let $\mathbf{B}$ denote the Boolean set $\{0, 1\}$ (we can assign the meaning "NO" to 0 and "YES" to 1). By $\mathbf{B}^N$ we denote the set of Boolean valued arrays (associated with one-, two- or multidimensional meshes) of total length N. If $\mathbf{b} = \{\flat_1, \flat_2, \ldots, \flat_N\} \in \mathbf{B}^N$, then $\flat_s \in \mathbf{B}$ for any $s = 1, 2, \ldots, N$. It is assumed that in the new mesh the elements (subdomains) marked by 1 should be refined, while those marked by 0 should be preserved (see Fig. 2.2). Note that the refined mesh in Fig. 2.2 contains the so-called "hanging nodes". In order to avoid them it is often necessary to refine also some neighboring subdomains marked by 0.

Remark 2.1 Modern mesh adaptation algorithms often make coarsening of a mesh in subdomains where local errors are insignificant (see, e.g., [BNP10, BS12, KM10,

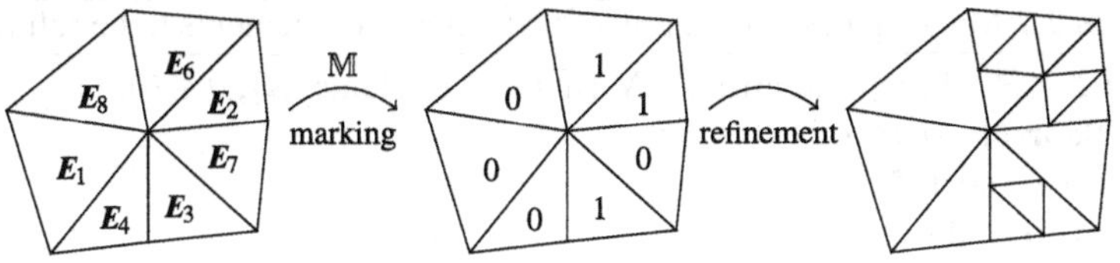

Fig. 2.2 Marking procedure and a refined mesh

Algorithm 2.1 Marking based on comparison with the average value

---

Input: $\mathbb{E}(u_h) \in \mathbb{R}^N$ {vector of errors indicated by $\mathbb{E}$}, N {number of elements}
$\widetilde{\mathbb{E}} = \frac{1}{N}\sum_{i=1}^{N} \mathbb{E}_i$ {Averaged value of the error on mesh elements}
for $i = 1 : N$ **do**
  if $\mathbb{E}_i \geq \widetilde{\mathbb{E}}$ **then**
    $\flat_i = 1$
  else
    $\flat_i = 0$
  end if
end for
Output: **b** {Marking of elements}

PPB12, Rhe80, SDW$^+$10, SMGG12] and the references cited therein). In this case, elements of $\mathbf{B}^N$ may attain three values: $\{-1, 0, 1\}$. The elements marked by -1 should be further aggregated in larger blocks.

From the mathematical point of view, marking is an operation performed by a special operator.

Definition 2.2 Marker $\mathbb{M}$ is a mapping (operator) acting from the set $\mathbb{R}_+^N$ (which contains estimated values of local errors) to the set $\mathbf{B}^N$.

Different markers generate different selection procedures, which are applied to the array of errors evaluated by an indicator $\mathbb{E}(u_h)$ in order to obtain a boolean array **b**. Further refinement is performed with the help of data encompassed in **b**.

Example 2.1 Algorithm 2.1 determines the simplest marker, which classifies the components of **e** into two groups by comparing with the average value.

Example 2.2 As before, $\mathbb{E}(u_h)$ is a vector with nonnegative components containing indicated errors and $\theta \in (0, 1)$ is a parameter (which determines the percentage of refined elements). Algorithm 2.2 ranks the values of $\mathbb{E}_i$ (from minimal to maximal

Algorithm 2.2 Marking based on a predefined amount of elements to be refined

Input: $\mathbb{E} \in \mathbb{R}^N$ {vector of errors}, N {number of elements}, $\theta \in (0, 1)$
$i_{cut} =$ **floor**$((1-\theta)N)$
$\{\mathbb{E}_{\text{sorted}}, \mathbf{I}\} =$ **sort**$(\mathbb{E})$
for $i = 1$ **to** N **do**
  if $i < i_{cut}$ **then**
    $\flat(\mathbf{I}(i)) = 1$
  else
    $\flat(\mathbf{I}(i)) = 0$
  end if
end for
Output: **b** {Marking of elements}

values) and assigns 1 to the largest θN values. All other elements are marked by 0. In the formal description of the algorithm, we use a "sorting procedure" **sort**, which input is the array $\mathbb{E}$ and output is the array $\mathbb{E}_{\text{sorted}}$ containing local errors sorted in the descending order (i.e., $\mathbb{E}_{\text{sorted}}(j) \geq \mathbb{E}_{\text{sorted}}(j+1)$), and the array **I**, which contains natural numbers (indexes of sorted elements) in the original vector, i.e., for any $j = 1, 2, \ldots, N$, $\mathbb{E}_{\text{sorted}}(j) = \mathbb{E}(\mathbf{I}(j))$. Algoritmization of such a procedure is a technical task, which we are not focused on. The procedure **floor**(z) selects the largest integer not greater than z.

Example 2.3 In the literature related to adaptive procedures, a selection method called the "bulk criterion" is often used. In it, we select by a certain method a set of elements for which the summed indicated error is greater than some "bulk" of the total indicated error (one of the first papers related to this method is [Dör96]; see also [BCH09]). Algorithm 2.3 forms the subset of elements which contains the highest indicated errors. The process stops when the error accumulated on previous steps exceeds the "bulk" level. This is sometimes referred to a "greedy" algorithm.

In order to demonstrate the performance of the above-discussed marking procedures, we consider the following diffusion problem:

$$-\Delta u = 1, \quad \text{in } \Omega := (0,1)^2 \setminus \big([0.5, 1] \times [0, 0.5]\big),$$
$$u = 0, \quad \text{on } \Gamma.$$

We compute u_h by the finite element method using piecewise affine approximations (Courant elements), and use the indicator $\mathbb{E}(u_h)$ generated by the gradient-averaging method (see Sect. 2.2.2.1). We apply both Algorithms 2.1 and 2.2. In Fig. 2.3 the mesh and elements marked by a certain method are depicted (above) and the histogram of indicated errors and the marked elements are presented (below). In general, Algorithms 2.1 and 2.2 may suggest to refine rather different amount of elements.

Algorithm 2.3 Marking based on the bulk criterion

Input: $\boldsymbol{E}(u_h)$ {vector of errors}
$\theta \in (0, 1)$ {bulk factor}
$\{\boldsymbol{E}_{\text{sorted}}, \mathbf{I}\} = \mathbf{sort}(\boldsymbol{E})$
$\boldsymbol{E}_{tot} = \sum_{i=1}^{N} \boldsymbol{E}_i$ {total error}
$\boldsymbol{E}_{\text{bulk}} = \theta\, \boldsymbol{E}_{tot}$ {value of the "bulk" error}
$i = 1$
$\boldsymbol{E}_{tmp} = 0$ {temporary value of accumulated error}
while $\boldsymbol{E}_{\text{bulk}} \geq \boldsymbol{E}_{tmp}$ **do**
  $\flat(\mathbf{I}(i)) = 1$
  $\boldsymbol{E}_{tmp} = \boldsymbol{E}_{tmp} + \boldsymbol{E}_{\text{sorted}}(i)$
  $i = i + 1$
end while
Output: $\mathbf{b}$ {Marking of elements}

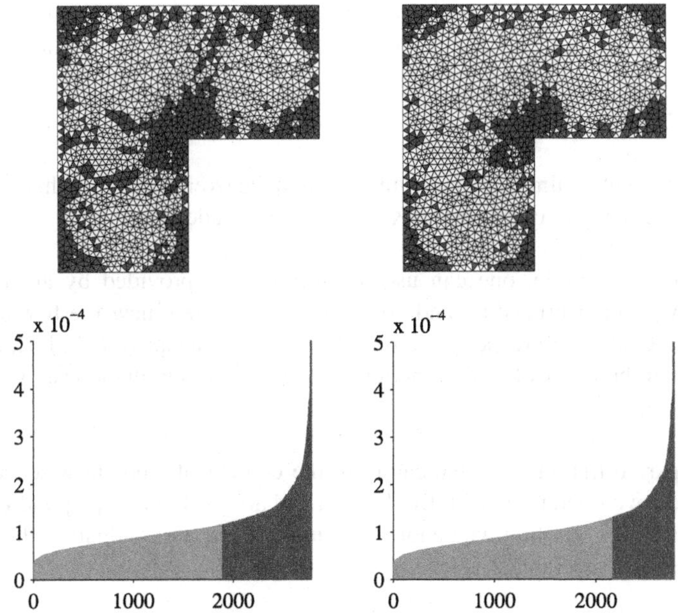

Fig. 2.3 Marking by Algorithms 2.1 (*left*) and 2.2 (*right*), marked elements ($\flat_i = 1$) are *colored darker*. *Above* are meshes and *below* the histograms of element-wise errors

Remark 2.2 We note that the marking of elements with the highest errors makes sense only if the errors differ significantly. If they have close values, then any ranking is not really motivated. For example, consider an almost uniform error distribution and two markings presented in Fig. 2.4. It is obvious that in this the case refining only the shadowed elements mesh is a rather disputable strategy because

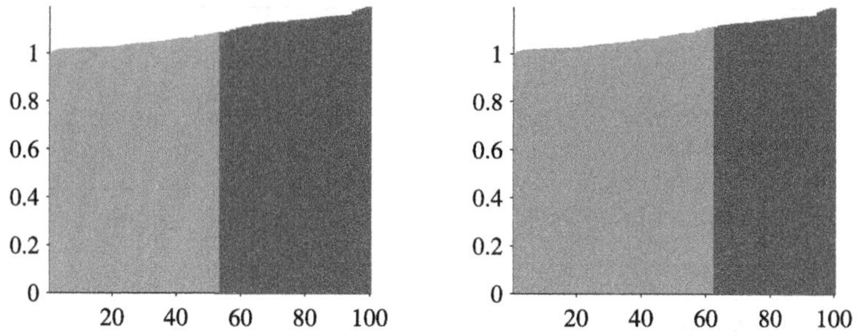

Fig. 2.4 Algorithms 2.1 (*left*) and 2.2 (*right*) are applied to mark elements of almost uniform error distribution, elements to be refined are *darker*

Table 2.1 Logical operation $\equiv$ in Definition 2.3

a	b	$a \equiv b$
0	0	1
1	0	0
0	1	0
1	1	1

$$\begin{aligned} \mathbb{M}(\mathbf{m}(e)) &= [1 \quad 0 \quad 0 \quad 0 \quad 1 \quad 1 \quad 0 \quad 0 \quad 1 \quad 1] \\ \mathbb{M}(\boldsymbol{E})(u_h) &= [0 \quad 1 \quad 1 \quad 0 \quad 1 \quad 1 \quad 1 \quad 0 \quad 1 \quad 0] \\ (\mathbb{M}(\mathbf{m}(e)) \equiv \mathbb{M}(\boldsymbol{E})(u_h)) &= [0 \quad 0 \quad 0 \quad 1 \quad 1 \quad 1 \quad 0 \quad 1 \quad 1 \quad 0] \end{aligned}$$

every element makes almost equal contribution to the overall error. In this situation, the uniform refinement of all elements would be more adequate.

Remark 2.3 In principle, one can use the information provided by an indicator without any ranking procedure and construct a completely new mesh where element sizes are related to respective errors. Moreover, in adaptive hp-FEM, the element size and the order of basis functions can be varied simultaneously (see, e.g., [AS99, Dem07]).

To compare different error indicators in the context of element-wise marking, we introduce two operations with Boolean valued arrays. Let $\mathbf{a} = \{a_i\}$ and $\mathbf{b} = \{b_i\}$ be elements of $\mathbf{B}^N$. By $[\![\mathbf{a}]\!]$ we denote the sum $\sum_{i=1}^{N} a_i$ and $\equiv$ denotes the logical equivalence rule (see Table 2.1, left).

Definition 2.3 An indicator $\boldsymbol{E}(u_h)$ is ε-accurate on the mesh $\mathfrak{T}_h$ with respect to the marker $\mathbb{M}$ if

$$\mathcal{M}\big(\boldsymbol{E}(u_h), \mathbb{M}\big) := 1 - \frac{[\![\mathbb{M}(\mathbf{m}(e)) \equiv \mathbb{M}(\boldsymbol{E}(u_h))]\!]}{N} \leq \varepsilon. \tag{2.4}$$

This definition is illustrated by Table 2.1 (right). We see that the operation $\equiv$ counts the cases in which markings based on the true error measure and on $\boldsymbol{E}(u_h)$ coincide. In the array of $N = 10$ elements the number of coincides is 5. Hence, in this example $\mathcal{M}(\boldsymbol{E}(u_h), \mathbb{M}) = 1 - \frac{5}{10} = 0.5$. This quantity shows that the indicator

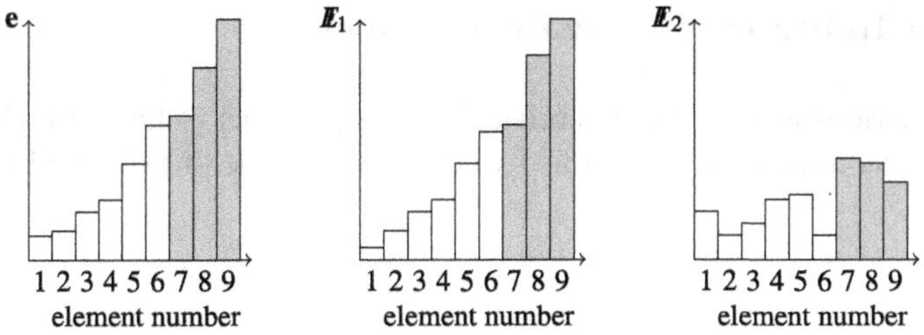

Fig. 2.5 True error distribution **e** for a set of nine elements and local errors generated by two indicators $\mathbb{E}_1$ and $\mathbb{E}_2$

is unacceptably coarse. If in another example we have an array containing, e.g., 10000 elements and the number of inconveniences (with respect to $\mathbb{M}(\mathbf{m}(e))$) is 8, then $\mathcal{M}(\mathbb{E}(u_h), \mathbb{M}) = 1 - \frac{9992}{10000} = 0.0008$. This shows the high accuracy of the indicator.

It is easy to see that the accuracy measure $\mathcal{M}(\mathbb{E}(u_h), \mathbb{M})$ is much weaker than the measure introduced in Definition 2.1. For example, in Fig. 2.5 we depict the exact distribution of local errors (left) and two distributions generated by two indicators (which are rather different). However, a marker designed to select three elements with the highest errors would select the same elements (shadowed). This example shows the difference between the accuracy measures (2.2) and (2.4). We see that the indicator $\mathbb{E}_2$ may be accurate in the sense of (2.4), but do not provide a true idea of the values of errors. This situation is quite typical. Often error indicators are based on heuristic argumentation and have no mathematical justification (in the best case they can be justified only in the above weak sense). Nevertheless, numerical analysts and engineers use them. Customarily they motivate this by saying that in some tests performed with the help of a marking procedure the indicator manages to properly mark the elements. In general, these arguments are not convincing because there is no guarantee that similar results will be obtained in other computations.

If $\mathbb{E}(u_h)$ is not accurate in the strong sense (i.e., it does not show actual values of the error), then the quality of marking may be good for one marker (mesh) and quite bad for another. Therefore, we believe that the indicators suggested for reliable numerical experiments should satisfy Definition 2.1.

It is clear that direct accuracy verification for an error indicator can be performed only in test examples where the exact solutions are known (so that we can find e). In other cases, the validity of an indicator is usually motivated by some indirect arguments (e.g., by those based on a priori regularity and asymptotic analysis). Some of the most popular motivations are considered below.

2.2 Error Indicators for the Energy Norm

To present various error indicators related to energy norms of linear elliptic equations within the framework of a unified scheme, we consider the classical Poisson's problem

$$-\Delta u = f \quad \text{in } \Omega, \tag{2.5}$$

$$u = 0 \quad \text{on } \Gamma, \tag{2.6}$$

where Ω is an open bounded connected subset in $\mathbb{R}^d$ with Lipschitz continuous boundary Γ and $f \in L^2(\Omega)$.

The generalized solution (see Sect. B.1) satisfies the relation

$$\int_\Omega \nabla u \cdot \nabla w \, dx = \int_\Omega f w \, dx, \qquad \forall w \in V_0 := \overset{\circ}{H}{}^1(\Omega). \tag{2.7}$$

Let $v \in V_0$ be an approximation of u. We are interested in evaluation of the global error norm $\|\nabla e\| = \|\nabla(u - v)\|$ and local errors $m_s(e)$ associated with subdomains (elements).

Note that

$$\sup_{w \in V_0} \left\{ \int_\Omega (\nabla(u - v) \cdot \nabla w) \, dx - \frac{1}{2} \|\nabla w\|^2 \right\}$$
$$\leq \sup_{\tau \in L^2(\Omega, \mathbb{R}^d)} \int_\Omega \left(\nabla(u - v) \cdot \tau - \frac{1}{2} |\tau|^2 \right) dx = \frac{1}{2} \|\nabla(u - v)\|^2.$$

On the other hand,

$$\sup_{w \in V_0} \int_\Omega \left(\nabla(u - v) \cdot \nabla w - \frac{1}{2} |\nabla w|^2 \right) dx \geq \frac{1}{2} \|\nabla(u - v)\|^2.$$

Thus,

$$\|\nabla(u - v)\|^2 = \sup_{w \in V_0} \int_\Omega (2\nabla(u - v) \cdot \nabla w - |\nabla w|^2) \, dx$$
$$= \sup_{w \in V_0} \left\{ -\|\nabla w\|^2 - 2 \int_\Omega (\nabla v \cdot \nabla w - f w) \, dx \right\},$$

and we conclude that

$$\|\nabla(u - v)\|^2 = \sup_{w \in V_0} \left\{ -\|\nabla w\|^2 - 2\ell_v(w) \right\}, \tag{2.8}$$

where

$$\ell_v(w) := \int_\Omega (\nabla v \cdot \nabla w - f w)\,dx$$

is the *residual functional*. This relation serves as a basis for various error estimation methods.

It is easy to show that the variational problem on the right-hand side of (2.8) has a unique solution and this solution is $w = u - v$. Indeed,

$$\ell_v(u - v) = \int_\Omega \big(\nabla v \cdot \nabla(u - v) - \nabla u \cdot \nabla(u - v)\big)\,dx = -\big\|\nabla(u - v)\big\|^2,$$

and we see that the right-hand side coincides with the left-hand one. Hence, (2.8) implies

$$\big|\ell_v(u - v)\big| = \big\|\nabla(u - v)\big\|^2. \tag{2.9}$$

We can use (2.9) to indicate the error $\|\nabla(u - v)\|$ and classify the following three principal ways:

A: Estimate $\ell_v(u - v)$ in (2.8) from the above, and use the computable part(s) of the estimate as error indicator(s).
B: Replace ℓ_v in (2.8) by a close functional, which leads to a directly computable estimator.
C: Solve the problem (2.8) numerically.

Below we discuss several error indicators, which are based on the approaches (A), (B), or (C).

2.2.1 Error Indicators Based on Interpolation Estimates

Error estimators of this type can be referred to the group (A). They originate from the papers [BR78b, BR78a]. In the literature, they are often called "residual type a posteriori error estimators". Various modifications and advanced forms have been discussed in numerous publications (see, e.g., [AO92, AO00, BS01, BWS11, Car99, CV99, DR98, EJ88, JH92, Ver96]). Let the approximate solution $v = u_h$ be the Galerkin approximation computed on $V_{0h} \subset V_0$, i.e.,

$$\int_\Omega \nabla u_h \cdot \nabla w_h\,dx = \int_\Omega f w_h\,dx, \quad \forall w_h \in V_{0h}. \tag{2.10}$$

With the help of (2.10), we can deduce an upper bound of the residual functional and suggest error indicators associated with computable parts of the estimate.

We represent the residual functional in the form

$$\ell_{u_h}(w) = \int_\Omega \big(\nabla u_h \cdot \nabla (w - \pi_h w) - f(w - \pi_h w)\big)\,dx,$$

where $\pi_h : V_0 \to V_{0h}$ denotes an interpolation operator. Assume that Ω consists of subdomains (e.g., simplexes T_k, which form the mesh $\mathfrak{T}_h$), and u_h is sufficiently regular on each subdomain. Then, we integrate by parts and obtain

$$\ell_{u_h}(w) = \sum_{k=1}^{N} \int_{T_k} (\Delta u_h + f)(\pi_h w - w)\,dx + \sum_{\substack{l,s=1\\ l>s}}^{N} \int_{E_{ls}} [\nabla u_h \cdot n_{ls}](w - \pi_h w)\,ds, \tag{2.11}$$

where [ ] denotes the jump, E_{ls} is the common boundary (edge) of T_l and T_s (boundary edges do not have this term), and n_{ls} is the unit normal vector to E_{ls} outward to T_l if $l > s$ (we recall that the integral over E_{ls} is assumed to be equal to zero if the elements l and s have no common edge).

It is easy to see that

$$\int_{T_k} (\Delta u_h + f)(\pi_h w - w)\,dx \le \|\Delta u_h + f\|_{T_k} \|w - \pi_h w\|_{T_k},$$

$$\int_{E_{ls}} \left[\frac{\partial u_h}{\partial n}\right](w - \pi_h w)\,ds \le \left\|\left[\frac{\partial u_h}{\partial n}\right]\right\|_{E_{ls}} \|w - \pi_h w\|_{E_{ls}}.$$

Now, we need to bound $\|w - \pi_h w\|_{T_k}$ and $\|w - \pi_h w\|_{E_{ls}}$ by $\|\nabla w\|$, i.e., we need *interpolation estimates* associated with the operator π_h. The derivation of such estimates is more difficult than for the operator Π_h considered in Sect. C.2. It is clear that the estimates must rely on geometrical features of T_k and properties of V_{0h}. In the case of piecewise affine continuous approximations, a polygonal $\Omega \subset \mathbb{R}^2$, and a simplicial mesh, the corresponding interpolation operator $\pi_h : H^1(\Omega) \to V_{0h}$ has been studied in [Clé75].

Let $v \in \overset{\circ}{H}{}^1(\Omega)$ and X_j be an inner node of the triangulation $\mathfrak{T}_h$. We define the set

$$\omega_j := \{x \in T_t \mid X_j \in \overline{T}_t, t = 1, 2, \ldots, N\},$$

which contains all the elements having common node X_j. Define $p_j(x) \in P^1(\omega_j)$ by the relation

$$\int_{\omega_j} (v - p_j) q\,dx = 0, \quad \forall q \in P^1(\omega_j). \tag{2.12}$$

This definition means that p_j is the L^2-projection of v on ω_j. Now, π_h is defined by setting

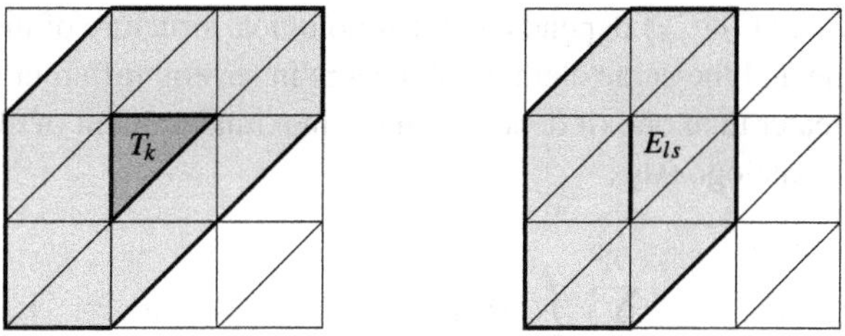

Fig. 2.6 The sets $\varpi(T_k)$ and $\varpi(E_{ls})$ on a regular mesh

$$\pi_h v(X_j) = p(X_j), \quad \forall X_j \in \operatorname{int}\Omega, \tag{2.13}$$

$$\pi_h v(X_j) = 0, \quad \forall X_j \in \Gamma. \tag{2.14}$$

This mapping is linear, continuous, and is subject to the relations (see, e.g., [Ver96])

$$\|v - \pi_h v\|_{2,T_k} \le C_{1k}^{int} \operatorname{diam} T_k \|v\|_{1,2,\varpi(T_k)}, \tag{2.15}$$

$$\|v - \pi_h v\|_{2,E_{ls}} \le C_{2ls}^{int} |E_{ls}|^{1/2} \|v\|_{1,2,\varpi(E_{ls})}, \tag{2.16}$$

where the sets (patches) associated with T_k and E_{ls} are defined as follows:

$$\varpi(T_k) := \{x \in \overline{T}_t \mid \overline{T}_t \cap \overline{T}_k \ne \emptyset, t = 1, 2, \dots, N\},$$

$$\varpi(E_{ls}) := \{x \in \overline{T}_t \mid \overline{T}_t \cap \overline{E}_{ls} \ne \emptyset, t = 1, 2, \dots, N\}.$$

See Fig. 2.6 for a clarifying illustration.

The constants C_{1k}^{int} and C_{2ls}^{int} depend on the structure of the mesh, and the factors $\operatorname{diam}(T_k)$ and $|E_{ls}|^{1/2}$ depend on the mesh size parameter h. We have

$$\begin{aligned}
&\sum_{k=1}^{N} \int_{T_k} (\Delta u_h + f)(w - \pi_h w)\,\mathrm{d}x \\
&\quad \le \sum_{k=1}^{N} \|\Delta u_h + f\|_{2,T_k} \|w - \pi_h w\|_{2,T_k} \\
&\quad \le \sum_{k=1}^{N} \|\Delta u_h + f\|_{2,T_k} C_{1k}^{int} \operatorname{diam} T_k \|w\|_{1,2,\varpi(T_k)} \\
&\quad \le \left(\sum_{k=1}^{N} (C_{1k}^{int})^2 (\operatorname{diam} T_k)^2 \|\Delta u_h + f\|_{2,T_k}^2 \right)^{1/2} \sqrt{\iota_T(w)},
\end{aligned} \tag{2.17}$$

where $\iota_T(w) = \sum_{k=1}^{N} \|w\|_{1,2,\varpi(T_k)}^2$. It is easy to see that

$$\iota_T(w) \le C_T^2(\mathfrak{T}_h) \|w\|_{1,2,\Omega}^2, \tag{2.18}$$

where $C_T(\mathfrak{T}_h)$ depends on the topological structure of the mesh. We note that since one and the same element T_k occurs in several different patches ϖ, the constant is greater than one (it depends on the maximal amount of elements in a patch).

Analogously,

$$\begin{aligned}
&\sum_{\substack{l,s=1\\ l>s}}^{N} \int_{E_{ls}} [\nabla u_h \cdot n_{ls}](w - \pi_h w)\,\mathrm{d}s \\
&\quad\le \sum_{\substack{l,s=1\\ l>s}}^{N} \left\| [\nabla u_h \cdot n_{ls}] \right\|_{2,E_{ls}} C^{int}_{2ls} |E_{ls}|^{1/2} \|w\|_{1,2,\varpi(E_{ls})} \\
&\quad\le \left(\sum_{\substack{l,s=1\\ l>s}}^{N} (C^{int}_{2ls})^2 |E_{ls}| \left\| [\nabla u_h \cdot n_{ls}] \right\|^2_{2,E_{ls}} \right)^{1/2} \sqrt{\iota_E(w)},
\end{aligned} \tag{2.19}$$

where

$$\iota_E(w) = \sum_{\substack{l,s=1\\ l>s}}^{N} \|w\|^2_{1,2,\varpi(E_{ls})}.$$

We have

$$\iota_E(w) \le C_E^2(\mathfrak{T}_h) \|w\|^2_{1,2,\Omega}, \tag{2.20}$$

where $C_E(\mathfrak{T}_h)$ also depends on the mesh.

By (2.17) and (2.19), we find that

$$\begin{aligned}
\left|\ell_{u_h}(w)\right| \le \Bigg(&C_T \left(\sum_{k=1}^{N} (C^{int}_{1k})^2 (\operatorname{diam} T_k)^2 \|\Delta u_h + f\|^2_{2,T_k} \right)^{1/2} \\
&+ C_E \left(\sum_{\substack{l,s=1\\ l>s}}^{m} (C^{int}_{2ls})^2 |E_{ls}| \left\| [\nabla u_h \cdot n_{ls}] \right\|^2_{2,E_{ls}} \right)^{1/2} \Bigg) \|w\|_{1,2,\Omega}.
\end{aligned} \tag{2.21}$$

Let $C = \max\{C_T, C_E\} \sqrt{1 + C^2_{F\Omega}}$. Then,

$$\left|\ell_{u_h}(w)\right| \le C\,\mathbb{E}(u_h) \|\nabla w\|, \tag{2.22}$$

where

$$\boldsymbol{E}(u_h) = \left(\sum_{k=1}^{N} (C_{1k}^{int})^2 (\text{diam}\, T_k)^2 \|\Delta u_h + f\|_{2,T_k}^2 \right)^{1/2} + \left(\sum_{\substack{l,s=1 \\ l>s}}^{N} (C_{2ls}^{int})^2 |E_{ls}| \big\| [\nabla u_h \cdot n_{ls}] \big\|_{2,E_{ls}}^2 \right)^{1/2}.$$

By (2.8), we obtain

$$\big\|\nabla(u - u_h)\big\|^2 \leq \sup_{w \in V_0} \left\{ -\|\nabla w\|^2 + 2C\boldsymbol{E}(u_h)\|\nabla w\| \right\} \leq C^2 \boldsymbol{E}^2(u_h).$$

Hence,

$$\big\|\nabla(u - u_h)\big\| \leq C\boldsymbol{E}(u_h). \tag{2.23}$$

We can represent this estimate in a slightly different form

$$\big\|\nabla(u - u_h)\big\| \leq \hat{C}\hat{\boldsymbol{E}}(u_h), \tag{2.24}$$

where the indicator

$$\hat{\boldsymbol{E}}(u_h) = \left(\sum_{k=1}^{N} (C_{1k}^{int})^2 (\text{diam}\, T_k)^2 \|\Delta u_h + f\|_{2,T_k}^2 + \sum_{\substack{l,s=1 \\ l>s}}^{N} (C_{2ls}^{int})^2 |E_{ls}| \big\| [\nabla u_h \cdot n_{ls}] \big\|_{2,E_{ls}}^2 \right)^{1/2}$$

is a sum of locally defined quantities.

It is worth outlining that in the process of deriving (2.23) and (2.24), we several times considerably overestimated the right-hand side, so that the equality sign in (2.8) and (2.11) is irretrievably lost. For this reason, the estimates obtained with the help of the above mathematical arguments may overestimate the error even if we manage to find and use sharp values of the interpolation constants C_{1k}^{int} and C_{2ls}^{int}. However, the latter task is not easy (especially for nonuniform meshes, which arise in the process of mesh adaptation). Indeed, to find C_{1k}^{int} we must solve the problem

$$\sup_{w \in V_0} \frac{\|w - \pi_h w\|_{2,\varpi(T_k)}}{\|w\|_{1,2,\varpi(T_k)}}, \tag{2.25}$$

which is an infinite dimensional problem. In some publications, it is suggested to find the constant approximately (e.g., by using a finite dimensional space formed by low order polynomial functions w). In this case, the true value of sup in (2.25) may be not achieved and, therefore, the overall estimate looses reliability. Moreover, solving a large number of local problems (2.25) (even for finite dimensional spaces) requires considerable numerical efforts. The corresponding computational

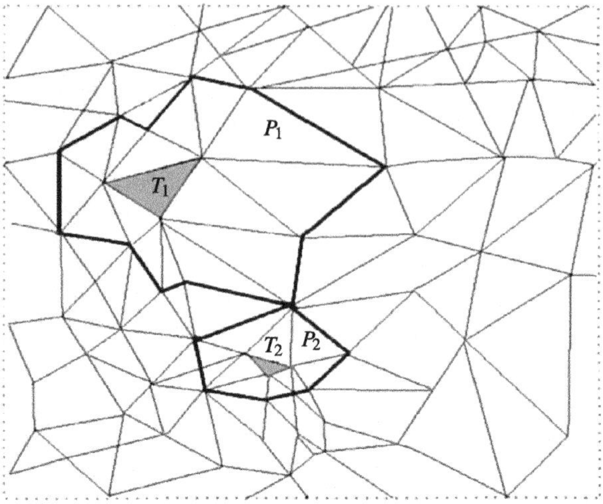

Fig. 2.7 Two patches of a nonuniform mesh

expenditures must be taken into account. After each mesh refinement, new constants associated with patches of the new mesh must be recomputed. Patches of highly nonuniform meshes may contain a different number of elements and complicated geometry (especially in 3D). For example, in Fig. 2.7 bold lines show boundaries of two patches P_1 and P_2 associated with two elements T_1 and T_2 of a nonuniform plane mesh. In real life computations, adaptive methods may generate meshes with much higher irregularities than those depicted in Fig. 2.7. In the case of highly irregular mesh, it is impossible to compute all the constants within the framework of a certain unified procedure similar to that we use for the constants in $H^2 \to C^0$ interpolation estimates, which can be fairly easily evaluated by interpolation estimates on the basic (etalon) simplex (see Sect. C.2). Thus, sharp computations of all the constants C_{1k}^{int} and C_{2ls}^{int} for thousands of different patches lead to high computational expenditures.

In view of these reasons, getting realistic and guaranteed error bounds with the help of (2.23) and (2.24) is rather challenging even for relatively simple elliptic equations (see, e.g., [CF00a], where these questions are systematically studied with the paradigm of boundary value problems in L-shaped domains).

A true meaning of the indicator $\hat{\mathbb{E}}$ is that it suggests easily computable quantities associated with elements, which can be used as error indicators. The standard argument for this is as follows. Assume that we use a quasi-uniform mesh. Then, we may assume that all (or almost all) constants C_{1k}^{int} have approximately the same value, and can be replaced by a single constant C_1^{int}. If the constants C_{2ls}^{int} are also replaced by a single constant C_2^{int}, then (2.21) implies an estimate

$$\hat{\mathbb{E}}(u_h) \approx \hat{C}\left(\sum_{k=1}^{N}\eta^2(T_k)\right)^{1/2}, \tag{2.26}$$

where

$$\eta^2(T_k) = \left(C_1^{int}\right)^2 (\operatorname{diam} T_k)^2 \|\Delta u_h + f\|^2_{2,T_k} + \frac{(C_2^{int})^2}{2} \sum_{E_{ls} \in \overline{T}_k} |E_{ls}| \left\| [\nabla u_h \cdot n_{ls}] \right\|^2_{2,E_{ls}}. \tag{2.27}$$

The multiplier 1/2 arises in the second term because any interior edge is common for two elements.

Remark 2.4 Sometimes only the last term containing jumps is used as an efficient error indicator (in many cases it dominates, see, e.g., [CV99]).

2.2.2 *Error Indicators Based on Approximation of the Error Functional*

Assume that the functional ℓ_v in (2.8) can be efficiently approximated by another functional, i.e., $\ell_v \simeq \widetilde{\ell}_v$, and, moreover, for the new functional we have the estimate

$$\left|\widetilde{\ell}_v(w)\right| \leq Q(v) \|\nabla w\|, \tag{2.28}$$

where $Q(v)$ is a computable nonnegative functional. Then, (cf. (2.8))

$$\begin{aligned} \left\|\nabla(u-v)\right\|^2 &= \sup_{w \in V_0} \left\{-\|\nabla w\|^2 - 2\ell_v(w)\right\} \simeq \sup_{w \in V_0} \left\{-\|\nabla w\|^2 - 2\widetilde{\ell}_v(w)\right\} \\ &\leq \sup_{w \in V_0} \left\{-\|\nabla w\|^2 + 2Q(v)\|\nabla w\|\right\} = Q^2(v). \end{aligned} \tag{2.29}$$

This relation shows the general idea of generating indicators of the group (B) and motivates the indicator $Q(v)$. Certainly, the quality of such an error indicator depends on the closeness of ℓ_v and $\widetilde{\ell}_v$.[1] The functional $\widetilde{\ell}_v$ can be constructed by a certain post-processing procedure.

Post-processing is a computational procedure that adjusts computed data to some a priori knowledge on properties of the exact solution. This procedure should be fairly simple, being compared with the expenditures required for computing the approximate solution.

Below, we describe several post-processing procedures.

[1] In general, the functionals must be close in the sense of $H^{-1}(\Omega)$.

2.2.2.1 Averaging of Gradients (Fluxes)

Gradient averaging procedures are often used to post-process gradients (fluxes, stresses) computed by finite element approximations of elliptic boundary value problems. Among first publications in this direction we mention the papers [ZZ87, ZZ88], which generated an interest in gradient recovery methods. Similar methods were investigated in numerous publications (see, e.g., [AO92, BC02, BR93, BS01, HTW02, Ver96, Wan00, WY02, ZBZ98, ZN05]). Mathematical justifications of the error indicators obtained in this way follow from the *superconvergence* phenomenon (see, e.g., [KN84, KNS98, Wah95]). Superconvergence arises on regular (quasiregular) meshes and, in simple terms, means that some components of approximate solutions obtained by inexpensive post-processing procedures converge to the corresponding components of the exact solution with a rate higher than the rate that can be predicted by standard a priori estimates. One of the most widely known results justified by superconvergence claims that a relatively simple averaging of ∇u_h yields a vector-valued function, which approximates ∇u much better than ∇u_h. Assume that in our problem this phenomenon takes place, and the gradient ∇u can be successfully represented by $G_h(\nabla u_h)$, where G_h is a certain post-processing operator. Then,

$$\int_\Omega (\nabla u_h \cdot \nabla w - f w)\,\mathrm{d}x \simeq \int_\Omega Z(u_h) \cdot \nabla w\,\mathrm{d}x,$$

where $Z(u_h) := \nabla u_h - G_h(\nabla u_h)$ (and (2.28) holds if we set $Q(u_h) = \|Z(u_h)\|$).
We recall (2.8) and deduce the relation

$$\|\nabla e\|^2 \simeq \sup_{w \in V_0} \left\{ -\|\nabla w\|^2 - 2\int_\Omega Z(u_h) \cdot \nabla w\,\mathrm{d}x \right\} \leq \left\| Z(u_h) \right\|^2,$$

which means that

$$\|\nabla e\| \simeq \left\| Z(u_h) \right\|.$$

This relation suggests the idea to use the function $Z(u_h)$ as an error indicator and set

$$\mathbf{E}_s(u_h) = \left\| Z(u_h) \right\|_{T_s}.$$

So far we did not define particular forms of the operator G_h, which can be constructed by many different methods. Some of them are discussed below. At this point, we only note that

Various post-processing procedures (averaging, smoothing, regularization) lead to various error indicators.

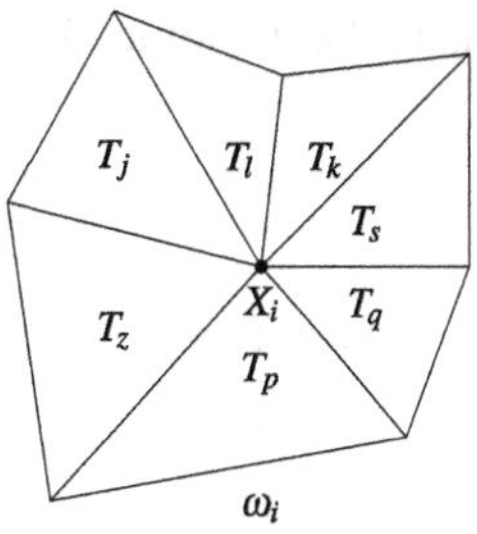

Fig. 2.8 A patch ω_i associated with the node X_i. $I_{\omega_i} = \{s, j, k, p, l, q, z\}$

2.2.2.2 Averaging of Fluxes in H^1

In the majority of cases, post-processing is performed by local averaging procedures. Consider the patch ω_i associated with the node X_i (see Fig. 2.8)

$$\overline{\omega}_i = \bigcup_{j \in I_{\omega_i}} \overline{T}_j,$$

where I_{ω_i} contains indexes of simplexes in ω_i.

Define $\mathbf{g}^{(i)}$ as the vector-valued function in $P^k(\omega_i, \mathbb{R}^d)$ solving the minimization problem:

$$\inf_{\mathbf{g} \in P^k(\omega_i, \mathbb{R}^d)} \int_{\omega_i} |\mathbf{g} - \nabla u_h|^2 \, \mathrm{d}x. \tag{2.30}$$

Using $\mathbf{g}^{(i)}$, we can define values of an averaged gradient at the node X_i.

Consider the simplest case $k = 0$ and assume that u_h is a piecewise affine continuous function. Then, the components of ∇u_h are constants on T_j. We denote them by $(\nabla u_h)_j$ and find $\mathbf{g}^{(i)} \in P^0(\omega_i, \mathbb{R}^d)$ such that

$$\int_{\omega_i} \left|\mathbf{g}^{(i)} - \nabla u_h\right|^2 \mathrm{d}x = \inf_{\mathbf{g} \in P^0(\omega_i, \mathbb{R}^d)} \int_{\omega_i} |\mathbf{g} - \nabla u_h|^2 \, \mathrm{d}x. \tag{2.31}$$

It is easy to see that

$$\mathbf{g}^{(i)} = \sum_{j \in I_{\omega_i}} \frac{|T_j|}{|\omega_i|} (\nabla u_h)_j. \tag{2.32}$$

We set $\mathrm{G}(\nabla u_h)(X_i) = \mathbf{g}^{(i)}$. Repeat this procedure for all nodes and define the vector-valued function $y_{\mathrm{G}} := \mathrm{G}(\nabla u_h)$ by the piecewise affine extrapolation of these values. This vector-valued function belongs to H^1 and in many cases approximates ∇u much better then the original (numerical) flux ∇u_h. This fact is justified by the *superconvergence phenomenon* (see, e.g., [KNS98, Wah95]).

Various averaging formulas of this type are represented in the form

$$\mathbf{g}^{(i)} = \sum_{j \in I_{\omega_i}} \lambda_j (\nabla u_h)_j, \quad \sum_{j \in I_{\omega_i}} \lambda_j = 1, \tag{2.33}$$

where the quantities λ_j are weight factors. In (2.32), we set

$$\lambda_j = \frac{|T_j|}{|\omega_i|}.$$

If the mesh is regular and all the quantities $|T_{ij}|$ are equal, then (2.32) reads

$$\mathbf{g}^{(i)} = \frac{1}{M} \sum_{j \in I_{\omega_i}} (\nabla u_h)_j, \tag{2.34}$$

where M is the number of elements in ω_i. For internal nodes, the factors λ_{ij} may also be defined by the rule

$$\lambda_j = \frac{|\gamma_j|}{2\pi},$$

where $|\gamma_j|$ is the radian measure of the angle of T_j associated with the node X_i. However, if a node belongs to the boundary, then it is better to choose special weights. Their values depend on the mesh and on the boundary type (see, e.g., [HK87]).

Another way of defining $\mathbf{g}^{(i)}$ is to solve the problem

$$\inf_{g \in \mathbb{P}^k(\omega_i, \mathbb{R}^d)} \sum_{s=1}^{m_i} \left| g(x_s) - \nabla u_h(x_s) \right|^2,$$

where the points $x_s \in \overline{\omega}_i$ are so-called *superconvergent* points (see, e.g., [KN87, KNS98]).

If $k = 0$, then by similar arguments we obtain

$$\mathbf{g}^{(i)} = \frac{1}{m_i} \sum_{s=1}^{m_i} \nabla u_h(x_s). \tag{2.35}$$

As in the previous case, we define the vector-valued function $\mathrm{G}_h(\nabla u_h)$ by the piecewise affine extrapolation of these values.

2.2.2.3 Averaging of Fluxes in $H(\Omega, \mathrm{div})$

Post-processing operators for fluxes can be based on Raviart–Thomas elements of the lowest order (see, e.g., in [BF86, RT91]). The corresponding averaging operator G_{RT} generates an averaged flux in the space $H(\Omega, \mathrm{div})$ by averaging normal components of fluxes. Since the true flux belongs to this space (provided that $f \in L^2(\Omega)$), this way of averaging is quite natural.

Consider a patch formed by two elements T_i and T_j having a common edge E_{ij} (see Fig. 2.9). If u_h is constructed by P^1-approximations, then $(\nabla u_h)|_{T_i}$ and

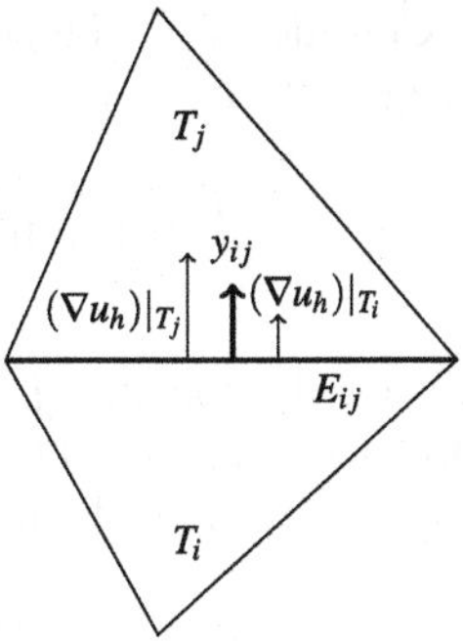

Fig. 2.9 Patch related to E_{ij} and averaged flux y_{ij}

$(\nabla u_h)|_{T_j}$ are constant vectors. In general, their normal components on E_{ij} are different. We define the (common) normal flux on E_{ij} as follows:

$$(y \cdot n_{ij})|_{E_{ij}} = \left(\kappa_{ij}(\nabla u_h)|_{T_i} + (1 - \kappa_{ij})(\nabla u_h)|_{T_j}\right) \cdot n_{ij},$$

where $\kappa_{ij} \in (0, 1)$ is the weight factor associated with E_{ij}. In the simplest case, $\kappa_{ij} = 1/2$. Another option (which takes into account the sizes of the elements) is

$$\kappa_{ij} = \frac{|T_i|}{|T_i| + |T_j|}.$$

For the boundary edges, we use the only one existing flux. Thus, three normal fluxes on three sides of each element are determined. The field inside the element is obtained by the standard RT^0-extension of normal fluxes. As a result, we have an averaged flux

$$y_{\mathrm{RT}} = \mathrm{G}_{\mathrm{RT}}(\nabla u_h) \in H(\Omega, \mathrm{div}).$$

Similar averaging procedures can be constructed in the case of 3D approximations, e.g., by averaging normal fluxes over the faces of a tetrahedron.

2.2.2.4 Averaging of Fluxes with Partial Equilibration

Since the exact flux p must satisfy the equilibrium (balance) equation $\operatorname{div} p + f = 0$, it is sensible to post-process it in such a way that the residual of this equation is minimal (e.g., in the integral sense). There are methods that produce equilibrated (or almost equilibrated) fluxes (see, e.g., [AO00, Bra07, LL83]). Sometimes these methods are rather sophisticated and use solutions of local Neumann type problems on patches. We have no space to properly discuss them here more systematically and, therefore, refer the reader to the above-mentioned and many other publications cited therein.

We conclude by describing a simple relaxation type algorithm, which allows to quasi-equilibrate y_{RT}.

Consider two neighboring elements with common edge E_{ij}. Our goal is to select the quantity $\gamma_{ij} = y \cdot n_{ij}$ in such a way that

$$\int_{T_i} \left((\operatorname{div} y)|_{T_i} + f\right)^2 \mathrm{d}x + \int_{T_j} \left((\operatorname{div} y)|_{T_j} + f\right)^2 \mathrm{d}x \to \min .$$

We use the identity $\int_{T_i} \operatorname{div} y \,\mathrm{d}x = \int_{\partial T_i} y \cdot n_i \,\mathrm{d}x$ and the fact that $(\operatorname{div} y)|_{T_i}$ and $(\operatorname{div} y)|_{T_j}$ are constant on T_i and T_j, respectively. Then, the corresponding value of γ_{ij} is explicitly defined by the relation (see [Rep08])

$$\gamma_{ij} = \frac{\mu_j |T_i| - \mu_i |T_j| + |T_i||T_j|(\{f\}_{T_j} - \{f\}_{T_i})}{|E_{ij}|(|T_i| + |T_j|)}, \tag{2.36}$$

where $\{f\}_{T_j}$ is the mean value of f on T_j.

Using the same idea, we recompute normal fluxes for all edges. At each step of this procedure the value of $\| \operatorname{div} y + f \|_{\Omega}^2$ decreases. After several cycles of minimization we obtain a vector-valued field, which is equilibrated much better than the original one.

2.2.2.5 Global Averaging

In many cases, an efficient averaging operator is obtained if local minimization problems on patches are replaced by a global problem (this method may generate essential computational expenditures). Consider the following problem: Find $\bar{\mathbf{g}}_h$ in a certain (global) set $U_h(\Omega)$, which minimizes the quantity $\sum_i \int_{T_i} |\mathbf{g}_h - \nabla u_h|^2 \mathrm{d}x$ among all $\mathbf{g}_h \in U_h(\Omega)$. Very often $\bar{\mathbf{g}}_h$ is a better image of ∇u than the functions obtained by local procedures. Moreover, mathematical justifications of the methods based on global averaging procedures can be performed under weaker assumptions, which makes them applicable to a wider class of problems (see, e.g., [CB02, CF00b, HTW02]).

2.2.2.6 Averaging by Least Squares Surface Fitting

In [Wan00], it was suggested a different recovery procedure, which is efficient for problems with sufficiently smooth solutions. The analysis is based on the representation

$$u - Q_\tau u_h = (u - Q_\tau u) + Q_\tau (u - u_h), \tag{2.37}$$

where u is the exact solution of a linear elliptic problem, u_h is the Galerkin approximation computed on a mesh $\mathfrak{T}_h$, and Q_τ is the L^2-projection operator on the finite dimensional space constructed on a mesh $\mathfrak{T}_\tau$ with the help of piecewise polynomial functions of the order $r \geq 0$. The key estimate is

$$\| Q_\tau u - Q_\tau u_h \| \leq C h^{s-1+\alpha \min\{0,2-s\}} \| u - u_h \|_{H^1}, \tag{2.38}$$

where $\alpha \in (0, 1)$ is a parameter that connects h and τ in the way $\tau = h^\alpha$. The original problem is assumed to be H^s-regular with $1 \le s \le k+1$, and k is the degree of polynomials used in the Galerkin approximation. From (2.38) it follows that

$$\|u - Q_\tau u_h\| \le C h^{\beta(h,\tau,r,k)} \tag{2.39}$$

provided that $u \in H^{k+1}(\Omega) \cap H^{r+1}(\Omega_0) \cap V_0$. In (2.39), the rate β depends on h, τ, r, and k, and is greater than 2, provided that u is regular enough, and the space V_τ is selected appropriately (i.e., it is sufficiently rich). The constant C depends on the norm of u. Concrete values of the convergence rate for various k, r, and α are presented in the paper [Wan00].

2.2.2.7 Error Indicators Based on Solutions of Local Subproblems

The splitting of the error functional $\ell_{u_h}(w)$ into a number of functionals (defined by solutions of local subproblems (see, e.g., [Ain98, AO00] and further developments in [AR10]) generates another class of error indicators, which can be assigned to the group (B). Below we present a sketch of the underlying ideas. For a consequent study, we address the reader to the above-cited literature and many other publications cited therein.

Let Ω be a union of nonoverlapping domains (elements) Ω_i, $i = 1, 2, \ldots, N$. Denote the common edge of Ω_i and Ω_j by Γ_{ij} and $\Gamma_{0i} := \partial\Omega_i \cap \Gamma$ and assume that for each Ω_i we know a function u_i such that

$$\ell_{u_h}(w) = \sum_{i=1}^{N} \int_{\Omega_i} \nabla u_i \cdot \nabla w \, dx. \tag{2.40}$$

Consider a function $\bar{u} : \Omega \to \mathbb{R}$ that coincides with $u_i(x)$ if $x \in \Omega_i$. Assume that the functions u_i preserve continuity on the boundaries Γ_{ij} and the function $\bar{u}(x)$ belongs to $H^1(\Omega)$. Then, (2.40) reads

$$\int_\Omega \nabla(\bar{u} + u_h) \cdot \nabla w \, dx = \int_\Omega f w \, dx, \quad \forall w \in V_0(\Omega). \tag{2.41}$$

The relation (2.41) means that $u = u_i + u_h$ on Ω_i. Therefore, $u_i = u - u_h$, and we know the errors.

One way to determine u_i is to use solutions of local subproblems with Neumann (or Dirichlet–Neumann) type boundary conditions. For each Ω_i we solve the following problem: Find $u_i \in H^1(\Omega_i)$ such that $u_i = 0$ on Γ_{i0} and

$$\int_{\Omega_i} \nabla u_i \cdot \nabla w \, dx \cong \int_{\Omega_i} f w \, dx + \sum_{j=1}^{N} \int_{\Gamma_{ij}} \zeta_{ij} g w \, ds - \int_{\Omega_i} \nabla u_h \cdot \nabla w \, dx, \quad \forall w \in V_0(\Omega_i), \tag{2.42}$$

where g is a reconstruction of ∇u_h (in the simplest case this reconstruction can be performed by averaging of $\nabla u_h \cdot n_{ij}$ associated with two neighboring elements). The space $V_0(\Omega_i)$ is defined as follows. If $\Gamma_{0i} \neq \emptyset$, then $V_0(\Omega_i)$ is a subspace of $H^1(\Omega_i)$, which contains the functions vanishing on Γ_{i0}. If $\Gamma_{0i} = \emptyset$, then the local problem is considered with Neumann conditions and $V_0(\Omega_i)$ is the subspace of $H^1(\Omega_i)$ containing functions with zero mean. The weight ζ_{ij} is equal to zero if $i = j$. If $i > j$, then it is equal to 1, and $\zeta_{ij} = -1$ in the opposite case. It is easy to see that each internal boundary Γ_{ij} generates two integrals with equal absolute values and opposite signs. Therefore, the sum of all integrals does not contain such terms, and we obtain

$$\sum_{i=1}^{N} \int_{\Omega_i} \nabla u_i \cdot \nabla w \, dx = \int_{\Omega} f w \, dx - \int_{\Omega} \nabla u_h \cdot \nabla w \, dx$$
$$= \ell_{u_h}(w), \quad \forall w \in V_0(\Omega), \tag{2.43}$$

which shows that the relation (2.40) holds.

This simple procedure may contain certain technical difficulties. One of them is that for internal domains the function g (which defines the Neumann type boundary conditions of the local subproblems) cannot be taken arbitrarily. This follows from the fact that the Neumann problem may be unsolvable if the external data do not satisfy an additional condition. For the problem (2.42) this condition is as follows:

$$\int_{\Omega_i} f \, dx + \sum_{j=1}^{N} \int_{\Gamma_{ij}} \zeta_{ij} g \, ds = 0. \tag{2.44}$$

Therefore, a special equilibration procedure that transforms g in order to satisfy (2.44) on each element is required. After that, exact solutions u_i of local problems must be found. Except special cases, this problem cannot be solved exactly and, therefore, instead of u_i some approximations $\widetilde{u}_i$ of local solutions are often used. Then, $\ell_{u_h}(w)$ is replaced by a directly computable functional

$$\widetilde{\ell}_{u_h}(w) = \sum_{i=1}^{N} \int_{\Omega_i} \nabla \widetilde{u}_i \cdot \nabla w \, dx. \tag{2.45}$$

It generates the quantities $\mathbb{E}_i(u_h) = \|\nabla \widetilde{u}_i\|_{\Omega_i}$, which can be used to indicate local errors, and the quantity $|\mathbb{E}(u_h)| = (\sum_i (\mathbb{E}_i(u_h))^2)^{1/2}$, which serves as an indicator of the global error. Accuracy of such an estimate depends on the choice of g and on the accuracy of the computed approximations $\widetilde{u}_i$.

2.2.3 Error Indicators of the Runge Type

Consider again the case $v = u_h$, where u_h is the Galerkin approximation on $V_{0h} \subset V_0$. We can try to get an error indicator by solving the variational problem in (2.8) numerically using a certain finite dimensional subspace $V_{0h_{\text{ref}}}$ instead of V_0 (cf. (2.7)), i.e., by applying the relation

$$\|\nabla(u-u_h)\|^2 \geq \sup_{w\in V_{0h_{\text{ref}}}} \left\{-\|\nabla w\|^2 - 2\ell_{u_h}(w)\right\}. \tag{2.46}$$

Thus, in our classification, estimators of this group belong to the class (C). It should be noted that this procedure makes sense only if the space $V_{0h_{\text{ref}}}$ is essentially richer than V_{0h} (if $V_{0h_{\text{ref}}} = V_{0h}$ then $\ell_{u_h}(w) = 0$, for any $w \in V_{0h_{\text{ref}}}$ and, therefore, the value of sup in (2.46) is zero).

Assume that

$$V_{0h} \subset V_{0h_{\text{ref}}}, \qquad \dim V_{0h_{\text{ref}}} > \dim V_{0h}. \tag{2.47}$$

The function $w_{h_{\text{ref}}}$ maximizing the right-hand side of (2.46) satisfies the relation

$$\int_\Omega \nabla w_{h_{\text{ref}}} \cdot \nabla w \,\mathrm{d}x = \int_\Omega (fw - \nabla u_h \cdot \nabla w)\,\mathrm{d}x, \quad \forall w \in V_{0h_{\text{ref}}}, \tag{2.48}$$

which is equivalent to

$$\int_\Omega \nabla(w_{h_{\text{ref}}} + u_h)\cdot \nabla w \,\mathrm{d}x = \int_\Omega fw\,\mathrm{d}x, \quad \forall w \in V_{0h_{\text{ref}}}. \tag{2.49}$$

Hence, $u_{h_{\text{ref}}} = w_{h_{\text{ref}}} + u_h$, where $u_{h_{\text{ref}}}$ is the Galerkin solution on $V_{0h_{\text{ref}}}$. We have

$$\|\nabla(u-u_h)\|^2 \geq -\|\nabla(u_{h_{\text{ref}}} - u_h)\|^2 - 2\ell_{u_h}(u_{h_{\text{ref}}} - u_h).$$

Since

$$\begin{aligned}\ell_{u_h}(u_{h_{\text{ref}}} - u_h) &= \int_\Omega \big(\nabla u_h \cdot \nabla(u_{h_{\text{ref}}} - u_h) - f(u_{h_{\text{ref}}} - u_h)\big)\,\mathrm{d}x \\ &= \int_\Omega \big(\nabla u_h \cdot \nabla(u_{h_{\text{ref}}} - u_h) - \nabla u_{h_{\text{ref}}} \cdot \nabla(u_{h_{\text{ref}}} - u_h)\big)\,\mathrm{d}x \\ &= -\|\nabla(u_h - u_{h_{\text{ref}}})\|^2,\end{aligned}$$

we conclude that the quantity $\|\nabla(u_h - u_{h_{\text{ref}}})\|$ estimates $\|\nabla e\|$ from below. If $V_{0h_{\text{ref}}}$ is much wider than V_{0h}, then $\|\nabla(u_h - u_{h_{\text{ref}}})\|$ can be used to measure the global error, and the corresponding contributions $\boldsymbol{E}_s$ can be used for indication of element-wise errors. It is easy to see that this type error indicator always underestimates the error. In fact, it coincides with the indicator suggested by C. Runge at the beginning of the 20th century. In the simplest form, it reads as follows: *if the difference between two approximate solutions computed on a coarse mesh $\mathfrak{T}_h$ and on a certain refined mesh $\mathfrak{T}_{h_{\text{ref}}}$ (e.g., $h_{\text{ref}} = h/2$) has become small, then both $u_{h_{\text{ref}}}$ and u_h are probably close to the exact solution u.*

In other words, this rule suggests the use of global or local norms of $u_h - u_{h_{\text{ref}}}$ as error indicators. Henceforth, we denote it by $\boldsymbol{E}_{\text{Runge}}(u_h)$. This indicator is simple

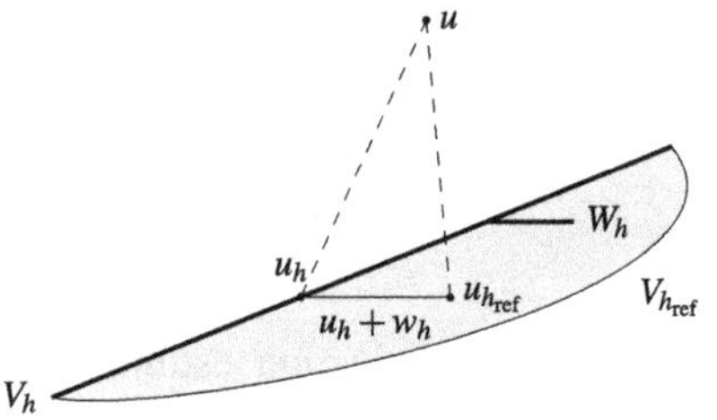

Fig. 2.10 The subspaces V_h, W_h, and $V_{h_{\rm ref}}$, the exact solution u and solutions u_h and $u_{h_{\rm ref}}$, from the respective subspaces

and looks very natural. For these reasons, it was easily accepted by engineers, who often consider it as a self-evident criterion. However, it is not difficult to find examples showing that this heuristic rule may be wrong. In particular, $\mathbb{E}_{\rm Runge}(u_h)$ may lead to misleading conclusions if the space V_h has been refined "improperly", i.e., if new (appended) trial functions do not really improve the approximation. In that case, u_h and $u_{h_{\rm ref}}$ may be quite close to each other but not close to u. We note that a correct form of the Runge's rule, which indeed provides guaranteed upper bounds of approximation errors, follows from error majorants of the functional type (see Sect. 3.6 of [Rep08] and Sect. 3.5.1 of this book).

Below we discuss *hierarchically based error indication methods*, where error indicators are constructed with the help of auxiliary problems on enriched finite dimensional subspaces (local or global) (see, e.g., [Ago02, DLY89, DMR91, DN02] and the references therein). Thus, in principle they invoke the same idea as does the Runge indicator, but in a more economical way.

Assume that the spaces V_h and $V_{h_{\rm ref}}$ are constructed in such a way that

$$V_{h_{\rm ref}} = V_h \oplus W_h.$$

In Fig. 2.10, we schematically depict the space V, the subspaces V_h, W_h, and $V_{h_{\rm ref}}$ and the corresponding approximate solutions u_h and $u_{h_{\rm ref}}$. It is easy to see that

$$\int_\Omega |\nabla(u-u_h)|^2\,{\rm d}x = \int_\Omega |\nabla(u-u_{h_{\rm ref}})|^2\,{\rm d}x + \int_\Omega |\nabla(u_h-u_{h_{\rm ref}})|^2\,{\rm d}x + 2\int_\Omega \nabla(u-u_{h_{\rm ref}})\cdot\nabla(u_{h_{\rm ref}}-u_h)\,{\rm d}x,$$

where

$$\int_\Omega \nabla(u-u_{h_{\rm ref}})\cdot\nabla(u_{h_{\rm ref}}-u_h)\,{\rm d}x = \int_\Omega f(u_{h_{\rm ref}}-u_h)\,{\rm d}x - \int_\Omega f u_{h_{\rm ref}}\,{\rm d}x + \int_\Omega f u_h\,{\rm d}x = 0.$$

Hence,

$$\|\nabla(u-u_h)\|^2 = \|\nabla(u-u_{h_{\rm ref}})\|^2 + \|\nabla(u_h-u_{h_{\rm ref}})\|^2 = \|\nabla(u-u_{h_{\rm ref}})\|^2 + \|\mathbb{E}_{\rm Runge}(u_h)\|^2.$$

Further analysis is based on the so-called *saturation assumption*

$$\|\nabla(u - u_{h_{\mathrm{ref}}})\| \le \lambda \|\nabla(u - u_h)\|, \quad \lambda \le 1, \tag{2.50}$$

which formalizes a rather natural condition: $u_{h_{\mathrm{ref}}}$ is closer to u than u_h. Usually, the space W_h is constructed by locally based approximations of higher order (e.g., by "bubble-functions"). In this case, the asymptotic relation $\lambda \sim h^q$ is often considered as a justification of the saturation property. However, in general, proving this inequality (with an explicit $\lambda < 1$) is a difficult task.

With the help of (2.50), we obtain

$$\left(1 - \lambda^2\right)\|\nabla(u - u_h)\|^2 = \|\boldsymbol{E}_{\mathrm{Runge}}(u_h)\|^2 \le \|\nabla(u - u_h)\|^2. \tag{2.51}$$

This inequality can be used for error control, provided that λ is known, but even in that case, the computation of $u_{h_{\mathrm{ref}}}$ may be too expensive. Since $V_{h_{\mathrm{ref}}}$ differs from V_h only by the orthogonal complement W_h, the difference $u_{h_{\mathrm{ref}}} - u_h = \widehat{w}_h$ belongs to this subspace. This fact suggests the idea to compute the correction function with the help of a subsidiary problem defined on W_h (instead of $V_{h_{\mathrm{ref}}}$). However, in general, the projection of $u_{h_{\mathrm{ref}}}$ onto V_h does not coincide with u_h and the true projection $\widehat{u}_h$ is unknown. Instead, an approximation of $u_{h_{\mathrm{ref}}}$ is sought in the form $u_h + w_h$, where w_h is defined as an element minimizing the distance from $u_h + \widetilde{w}_h$ to u, which leads to the problem

$$\inf_{w_h \in W_h} \frac{1}{2} \int_\Omega |\nabla(u - u_h - w_h)|^2 \,\mathrm{d}x.$$

It is easy to see that the latter problem is equivalent to

$$\inf_{w_h \in W_h} \left\{ \frac{1}{2}\|\nabla w_h\|^2 - \int_\Omega \nabla(u - u_h)\cdot\nabla w_h \,\mathrm{d}x \right\}$$

or

$$\inf_{w_h \in W_h} \left\{ \frac{1}{2}\|\nabla w_h\|^2 - \int_\Omega f w_h \,\mathrm{d}x + \int_\Omega \nabla u_h \cdot \nabla w_h \,\mathrm{d}x \right\}.$$

We arrive at the following problem: Find $\widetilde{w}_h \in W_h$ such that

$$\int_\Omega \nabla\widetilde{w}_h \cdot \nabla w_h \,\mathrm{d}x = \int_\Omega f w_h \,\mathrm{d}x - \int_\Omega \nabla u_h \cdot \nabla w_h \,\mathrm{d}x, \quad \forall w_h \in W_h. \tag{2.52}$$

The following questions rise: how large is the difference between $\widetilde{w}_h$ and $\widehat{w}_h$, and when $\widetilde{w}_h$ can be used instead of $\widehat{w}_h$ (we recall that $u_{h_{\mathrm{ref}}} = \widehat{u}_h + \widehat{w}_h$). To answer them, we first recall that u satisfies the integral relation

$$\int_\Omega \nabla u \cdot \nabla w \,\mathrm{d}x = \int_\Omega f w \,\mathrm{d}x, \quad \forall w \in V, \tag{2.53}$$

u_h and $u_{h_{\rm ref}}$ are Galerkin solutions, i.e.,

$$\int_\Omega \nabla u_h \cdot \nabla w_h \, dx = \int_\Omega f w_h \, dx, \qquad \forall w_h \in V_h,$$
$$\int_\Omega \nabla u_{h_{\rm ref}} \cdot \nabla w_{h_{\rm ref}} \, dx = \int_\Omega f w_{h_{\rm ref}} \, dx, \quad \forall w_{h_{\rm ref}} \in V_{h_{h_{\rm ref}}} \subset V,$$

and

$$\int_\Omega (\nabla u_{h_{\rm ref}} - u_h) \cdot \nabla w_h \, dx = 0, \quad \forall w_h \in V_h. \tag{2.54}$$

Also, we assume that the spaces V_h and W_h are such that the *strengthened Cauchy inequality*

$$\left| \int_\Omega \nabla v_h \cdot \nabla w_h \, dx \right| \leq \gamma \left(\int_\Omega \nabla v_h \cdot \nabla v_h \, dx \right)^{1/2} \left(\int_\Omega \nabla w_h \cdot \nabla w_h \, dx \right)^{1/2} \tag{2.55}$$

holds, where $\gamma \in (0, 1)$ is a constant independent of h. In this case,

$$\left\| \nabla (u - u_h) \right\| \leq C_{\lambda\gamma} \| \nabla \widetilde{w}_h \|. \tag{2.56}$$

To prove this fact, we argue as follows. By the Galerkin orthogonality (cf. (2.54)), we have

$$\int_\Omega \nabla (u_{h_{\rm ref}} - u_h) \cdot \nabla (\widehat{u}_h - u_h) \, dx = 0. \tag{2.57}$$

In view of (2.52),

$$\int_\Omega \nabla \widetilde{w}_h \cdot \nabla \widehat{w}_h \, dx = \int_\Omega f \widehat{w}_h \, dx - \int_\Omega \nabla u_h \cdot \nabla \widehat{w}_h \, dx = \int_\Omega \nabla (u_{h_{\rm ref}} - u_h) \cdot \nabla \widehat{w}_h \, dx,$$

whence

$$\int_\Omega \nabla (u_{h_{\rm ref}} - u_h - \widetilde{w}_h) \cdot (\nabla \widehat{w}_h) \, dx = 0. \tag{2.58}$$

From (2.57) and (2.58), we conclude that

$$\begin{aligned}
0 &= \int_\Omega \nabla (u_{h_{\rm ref}} - u_h - \widetilde{w}_h) \cdot \nabla \widehat{w}_h \, dx + \int_\Omega \nabla (u_{h_{\rm ref}} - u_h) \cdot \nabla (\widehat{u}_h - u_h) \, dx \\
&= \int_\Omega \nabla (u_{h_{\rm ref}} - u_h) \cdot \nabla (\widehat{w}_h + \widehat{u}_h - u_h) \, dx - \int_\Omega \nabla \widetilde{w}_h \cdot \nabla \widehat{w}_h \, dx \\
&= \| u_{h_{\rm ref}} - u_h \|^2 - \int_\Omega \nabla \widetilde{w}_h \cdot \nabla \widehat{w}_h \, dx.
\end{aligned}$$

Thus,

$$\left\| \nabla (u_{h_{\rm ref}} - u_h) \right\|^2 = \int_\Omega \nabla \widetilde{w}_h \cdot \nabla \widehat{w}_h \, dx. \tag{2.59}$$

Note that

$$\|\nabla(u_{h_{\mathrm{ref}}} - u_h)\|^2 = \|\nabla(u_{h_{\mathrm{ref}}} - \widehat{u}_h)\|^2 + \|\nabla(\widehat{u}_h - u_h)\|^2 + 2\int_\Omega \nabla(u_{h_{\mathrm{ref}}} - \widehat{u}_h) \cdot \nabla(\widehat{u}_h - u_h)\,\mathrm{d}x.$$

Here $\widehat{u}_h - u_h \in V_h$ and $u_{h_{\mathrm{ref}}} - \widehat{u}_h = \widehat{w}_h \in W_h$, so that we use (2.55) and obtain

$$\begin{aligned}\|\nabla(u_{h_{\mathrm{ref}}} - u_h)\|^2 &\geq \|\nabla\widehat{w}_h\|^2 + \|\nabla(\widehat{u}_h - u_h)\|^2 - 2\gamma\|\nabla\widehat{w}_h\|\,\|\nabla(\widehat{u}_h - u_h)\| \\ &\geq (1-\gamma^2)\|\nabla\widehat{w}_h\|^2.\end{aligned}$$

From this relation and (2.59), we find that

$$\|\nabla\widehat{w}_h\|^2 \leq \frac{1}{1-\gamma^2}\|\nabla(u_{h_{\mathrm{ref}}} - u_h)\|^2 = \frac{1}{1-\gamma^2}\int_\Omega \nabla\widetilde{w}_h \cdot \nabla\widehat{w}_h\,\mathrm{d}x. \tag{2.60}$$

Thus, we see that the true correction function $\widehat{w}_h$ is subject to $\widetilde{w}_h$:

$$\|\nabla\widehat{w}_h\| \leq \frac{1}{1-\gamma^2}\|\nabla\widetilde{w}_h\|. \tag{2.61}$$

Now, we recall that $\|\nabla(u - u_h)\|^2 = \|\nabla(u - u_{h_{\mathrm{ref}}})\|^2 + \|\nabla(u_h - u_{h_{\mathrm{ref}}})\|^2$ and use (2.59). We have

$$\begin{aligned}\|\nabla(u - u_h)\|^2 &= \|\nabla(u - u_{h_{\mathrm{ref}}})\|^2 + \int_\Omega \nabla\widetilde{w}_h \cdot \widehat{w}_h\,\mathrm{d}x \\ &\leq \lambda^2\|\nabla(u - u_h)\|^2 + \|\nabla\widetilde{w}_h\|\,\|\nabla\widehat{w}_h\| \\ &\leq \lambda^2\|\nabla(u - u_h)\|^2 + \frac{1}{1-\gamma^2}\|\nabla\widetilde{w}_h\|^2.\end{aligned}$$

From here, we conclude that

$$\|\nabla(u - u_h)\|^2 \leq \frac{1}{(1-\lambda^2)(1-\gamma^2)}\|\nabla\widetilde{w}_h\|^2, \tag{2.62}$$

which shows that $\|\nabla e\| \simeq \|\nabla\widetilde{w}_h\|$ and motivates using $\|\nabla\widetilde{w}_h\|$ as an error indicator.

2.3 Error Indicators for Goal-Oriented Quantities

Evaluation of approximation errors in terms of special "goal-oriented" quantities is very popular in engineering computations. A consequent exposition can be found in [BR03] and in numerous publications devoted to *goal-oriented* a posteriori error estimates and applications of them to various problems (see, e.g, [BR12, BR96, HRS00, KM10, MS09, OP01, PP98, Ran00, RV10, SO97, SRO07]). In this method, estimates are derived for the quantity $\langle \ell, u - u_h \rangle$, where ℓ is a given linear functional and u_h is a conforming approximation. In general, ℓ belongs to the dual energy space V_0^*. Typically, ℓ is focused on some special properties of approximate solutions. For example, if ℓ is an integral type functional (e.g., $\ell \in L^2(\Omega)$) localized in a certain subdomain $\omega \subset \Omega$, then $|\langle \ell, u - u_h \rangle|$ characterizes the quality of u_h in ω. A way of evaluating this quantity is based on the following idea, which we discuss with the example of the basic elliptic problem: Find $u \in V_0 := \overset{\circ}{H}{}^1(\Omega)$ such that

$$\int_\Omega A\nabla u \cdot \nabla w \, \mathrm{d}x = \int_\Omega f w \, \mathrm{d}x, \qquad \forall w \in V_0, \tag{2.63}$$

where A is a positive definite matrix with bounded coefficients.

Let $A^\star$ be the matrix adjoint to A and u_ℓ the solution of the respective *adjoint* problem

$$\int_\Omega A^\star \nabla u_\ell \cdot \nabla w \, \mathrm{d}x = \langle \ell, w \rangle, \qquad \forall w \in V_0. \tag{2.64}$$

From (2.63) and (2.64), it follows that

$$\langle \ell, u - u_h \rangle = \int_\Omega A^\star \nabla u_\ell \cdot \nabla (u - u_h) \, \mathrm{d}x \tag{2.65}$$

$$= \int_\Omega (f u_\ell - A\nabla u_h \cdot \nabla u_\ell) \, \mathrm{d}x =: I_\ell(u_\ell, u_h). \tag{2.66}$$

Hence, $\langle \ell, u - u_h \rangle$ is equal to the functional $I_\ell(u_\ell, u_h)$ and can be easily estimated, provided that u_ℓ is known (we note that finding u_ℓ amounts to solving another boundary value problem having the same complexity as (2.63)). In the majority of cases, u_ℓ is unknown and, therefore, it is replaced by an approximation $u_{\ell\tau}$ computed on an *adjoint mesh* $\mathcal{T}_\tau$ (which does not necessarily coincide with $\mathcal{T}_h$). Then, the non-computable quantity $I_\ell(u_\ell, u_h)$ is approximated by the computable quantity $I_\ell(u_{\ell\tau}, u_h)$.

If $u_{\ell\tau}$ is a sharp approximation of u_ℓ (in general, it should be sharper than u_h), then the quantity $|\mathbb{E}_\ell(u_{\ell\tau}, u_h)| := |I_\ell(u_{\ell\tau}, u_h)|$ serves as an indicator of the goal-oriented error $|\langle \ell, u - u_h \rangle|$. However, getting a sharp approximation of u_ℓ may lead to essential additional expenditures. In order to minimize them, one can apply different modifications (generalizations) of (2.65), which the reader can find in the publications mentioned at the beginning of Sect. 2.3.

2.3.1 Error Indicators Relying on the Superconvergence of Averaged Fluxes in the Primal and Adjoint Problems

Henceforth, for the sake of simplicity we assume that A is a symmetric matrix. We rewrite I_ℓ in the form

$$I_\ell(u_h, u_{\ell\tau}) = I_{\ell 1}(u_h, u_{\ell\tau}) + I_{\ell 2}(u_h, u_{\ell\tau}; u, u_\ell), \tag{2.67}$$

where

$$I_{\ell 1}(u_h, u_{\ell\tau}) := \int_\Omega (f u_{\ell\tau} - A\nabla u_h \cdot \nabla u_{\ell\tau})\,\mathrm{d}x$$

is a directly computable functional and

$$I_{\ell 2}(u_h, u_{\ell\tau}; u, u_\ell) := \int_\Omega A(\nabla u - \nabla u_h) \cdot (\nabla u_\ell - \nabla u_{\ell\tau})\,\mathrm{d}x$$

involves unknown u and u_ℓ, i.e., the exact solutions of (2.63) and (2.64), respectively. Note that if u_h is a Galerkin approximation and $\mathcal{T}_\tau$ coincides with $\mathcal{T}_h$, then $I_{\ell 1}(u_h, u_{\ell\tau}) = 0$.

Estimate (2.67) is a source of various indicators. One of them is based on the idea of replacing unknown fluxes

$$p := A\nabla u \quad \text{and} \quad p_\ell := \nabla u_\ell$$

by $G_h p_h$ and $G_\tau p_{\ell\tau}$, where $p_h := A\nabla u_h$, $p_{\ell\tau} := A\nabla u_{\ell\tau}$, and G_h and G_τ are some suitable averaging operators associated with the primal and adjoint meshes, respectively. In [KNR03, NR04], it is proved that under the standard assumptions (which guarantee superconvergence of averaged fluxes computed for the primal and adjoint problems) such a replacement generates errors of a higher order (with respect to h and τ). In view of this fact, the quantity

$$\mathbb{E}_\ell(u_h, u_{\ell\tau}) := I_{\ell 1}(u_h, u_{\ell\tau}) + \mathbb{E}_{\ell 2}(u_h, u_{\ell\tau}), \tag{2.68}$$

where

$$\mathbb{E}_{\ell 2}(u_h, u_{\ell\tau}) := \int_\Omega A^{-1}(G_h p_h - p_h) \cdot (G_\tau p_{\ell\tau} - p_{\ell\tau})\,\mathrm{d}x$$

is used instead of $I_\ell(u_h, u_{\ell\tau})$. However, such an indicator is justified only if both problems (primal and adjoint) are sufficiently regular, so that u_h and $u_{\ell\tau}$ possess superconvergent fluxes. This fact imposes rather obligatory conditions on $\mathcal{T}_\tau$, which may be difficult to satisfy. Typically, the mesh $\mathcal{T}_h$ generated by commonly used solvers is sufficiently regular (so that one can await the superconvergence of p_h, at least in the major part of Ω). For the adjoint mesh $\mathcal{T}_\tau$, such a regularity is difficult to guarantee. Indeed, this mesh should satisfy two conditions, which in fact contradict each other. On the one hand, $\dim V_\tau$ should not significantly exceed $\dim V_h$ (otherwise the adjoint problem is computationally much more expensive than the primal one). On the other hand, $\mathcal{T}_\tau$ should be "sufficiently dense" in the vicinity of ω. This observation motivates attempts at finding other error indicators which are not based on the superconvergence of adjoint fluxes.

2.3.2 Error Indicators Using the Superconvergence of Approximations in the Primal Problem

An error indicator that does not attract the superconvergence of averaged gradients in the adjoint problem was suggested in [NRT08]. The idea behind is to represent the term $I_{\ell 2}(u_h, u_{\ell\tau}; u, u_\ell)$ in a new form, namely:

$$\begin{aligned} I_{\ell 2}&(u_h, u_{\ell\tau}; u, u_\ell) \\ &:= \sum_{T_i \in \mathcal{T}_\tau} \int_{T_i} (\nabla u - \nabla u_h) \cdot (p_\ell - p_{\ell\tau})\,\mathrm{d}x \\ &= \sum_{T_i \in \mathcal{T}_\tau} \left(\int_{T_i} (u_h - u)\mathcal{R}(p_{\ell\tau})\,\mathrm{d}x + \int_{\partial T_i} (u - u_h)(p_\ell - p_{\ell\tau}) \cdot \nu_i\,\mathrm{d}s \right) \\ &= I_{\ell 21}(u_h, p_{\ell\tau}; u) + I_{\ell 22}(u_h, p_{\ell\tau}; u, p_\ell), \end{aligned}$$

where ν_i is a unit outward normal to ∂T_i and

$$\mathcal{R}(p_{\ell\tau}) := \operatorname{div} p_{\ell\tau} + \ell.$$

Since u, u_h, and p_ℓ are continuous on interelement boundaries, we find that

$$\begin{aligned} I_{\ell 22}(u_h, p_{\ell\tau}; u, p_\ell) &= \sum_{T_i \in \mathcal{T}_\tau} \int_{\partial T_i} (u - u_h)(p_\ell - p_{\ell\tau}) \cdot \nu_i\,\mathrm{d}s \\ &= \sum_{E_{ij} \in \mathcal{E}_\tau} \int_{E_{ij}} (u_h - u)[p_{\ell\tau} \cdot \nu_{ij}]_{E_{ij}}\,\mathrm{d}s. \end{aligned}$$

Here, $\mathcal{E}_\tau$ is the set of edges in the adjoint mesh, ν_{ij} is the unit normal to the edge E_{ij} (common for T_i and T_j), which is external to T_i if $i < j$. Since u_h and u satisfy the same Dirichlet boundary conditions, $\mathcal{E}_\tau$ contains only internal edges. In this functional, the exact solution of the adjoint problem is *completely excluded.* Therefore, the justification of the estimator is not connected with superconvergence in the adjoint problem, and we may hope that it is insensitive with respect to adjoint mesh structure. To obtain a computable error indicator, in [NRT08] the *superconvergent post-processing* of the function u_h (by the operator Q_τ; see (2.37)–(2.39)) and a *regularization* of the adjoint flux $p_{\ell\tau}$ (which eliminates the jumps $[p_{\ell\tau} \cdot \nu_{ij}]_{E_{ij}}$) were used. Below, the corresponding regularization operator is denoted by G_τ and the Wang projection operator by $\mathcal{W}$. In particular, such an operator can be constructed with the help of Hsieh–Clough–Tocher finite element approximations (see, e.g., [BH81, Cia78b]). Then, $I_{\ell 22} = 0$ and $I_{\ell 21}$ is replaced by

$$\boldsymbol{E}_{\ell 21}(u_h, u_{\ell\tau}; u) := \int_\Omega \big(u_h - \mathcal{W}(u_h)\big)\mathcal{R}\big(G_\tau(p_{\ell\tau})\big)\,\mathrm{d}x,$$

and we arrive at the indicator

$$\langle \ell, u - u_h \rangle \approx \mathbb{E}_\ell(u_h) := I_{\ell 1}(u_h, u_{\ell\tau}) + \int_\Omega \big(u_h - \mathcal{W}(u_h)\big)\mathcal{R}\big(G_\tau(p_{\ell\tau})\big)\,\mathrm{d}x. \quad (2.69)$$

Another representation of $I_{\ell 2}$ leads to a somewhat different error indicator. Let q be a vector-valued function in $H(\Omega, \mathrm{div})$. Then,

$$\begin{aligned} I_{\ell 2}&(u_h, u_{\ell\tau}; u, u_\ell) \\ &:= \int_\Omega (\nabla u - \nabla u_h)(p_\ell - p_{\ell\tau})\,\mathrm{d}x \\ &= \int_\Omega (\nabla u - \nabla u_h)(q - p_{\ell\tau})\,\mathrm{d}x + \int_\Omega (u - u_h)(\mathrm{div}\, q + \ell)\,\mathrm{d}x \\ &= \int_\Omega A^{-1}(p - p_h)\cdot(q - p_{\ell\tau})\,\mathrm{d}x + \int_\Omega (u - u_h)(\mathrm{div}\, q + \ell)\,\mathrm{d}x. \end{aligned} \quad (2.70)$$

In this relation, u_ℓ is excluded from the right-hand side without a regularization of $q_{\ell\tau}$. This relation implies an error indicator if one reconstructs p and u with the help of the recovery operators G_h and $\mathcal{W}$, respectively.

We have

$$\begin{aligned} \langle \ell, u - u_h \rangle \approx{}& \mathbb{E}_\ell(u_h, p_{\ell\tau}) \\ :={}& I_{\ell 1}(u_h, u_{\ell\tau}) \\ &+ \int_\Omega A^{-1}\big(G_h(p_h) - p_h\big)\cdot(q - p_{\ell\tau})\,\mathrm{d}x \\ &+ \int_\Omega \big(u_h - \mathcal{W}(u_h)\big)(\mathrm{div}\, q + \ell)\,\mathrm{d}x, \end{aligned} \quad (2.71)$$

where q is an arbitrary vector valued function. If q is equilibrated (or almost equilibrated), then the last term can be ignored and we obtain a simpler indicator

$$\mathbb{E}_\ell(u_h, p_{\ell\tau}) := I_{\ell 1}(u_h, u_{\ell\tau}) + \int_\Omega A^{-1}\big(G_h(p_h) - p_h\big)\cdot(q - p_{\ell\tau})\,\mathrm{d}x. \quad (2.72)$$

It is clear that properties of $\mathbb{E}_\ell(u_h, p_{\ell\tau})$ depend on superconvergence properties of averaged fluxes in the primal problem and on the difference between q and $p_{\ell\tau}$. Numerical examples and asymptotic exactness of the above-introduced indicators are discussed in [NRT08]. One of the examples is presented below.

Example 2.4 We consider the following elliptic type problem:

$$\Delta u + 1 = 0 \quad \text{in } \Omega, \qquad u = 0 \quad \text{on } \partial\Omega, \quad (2.73)$$

and define

$$\langle \ell, u - u_h \rangle = \int_\Omega \ell_\omega (u - u_h)\,\mathrm{d}x, \quad (2.74)$$

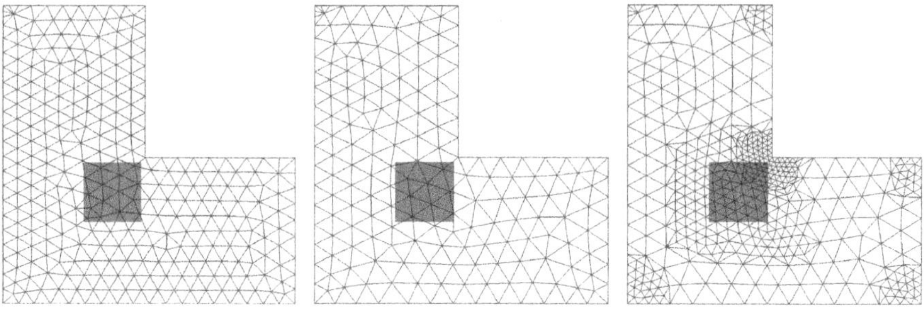

Fig. 2.11 The meshes $\mathcal{T}_1$ (315 nodes) (*left*), $\mathcal{T}_2$ (193 nodes) (*middle*), and $\mathcal{T}_3$ (451 nodes) (*right*) used in the test; the region of interest ω is *shadowed*

where

$$\ell_\omega(x) = \begin{cases} 1, & \text{if } x \in \omega \subset \Omega, \\ 0, & \text{otherwise.} \end{cases} \tag{2.75}$$

Both primal and adjoint problems are solved with the help of piecewise linear finite element approximations. As usual, the efficiency index is defined by the relation

$$i_{\text{eff}} := \frac{\mathbb{E}_\ell(u_h)}{|\langle \ell, u - u_h \rangle|}.$$

The primal problem is solved on the mesh $\mathcal{T}_1$ (see Fig. 2.11). It is known that the corresponding exact solution u has singularity in the re-entrant corner. The adjoint problem was solved on $\mathcal{T}_1$, on a rather coarse regular mesh $\mathcal{T}_2$ and on the mesh $\mathcal{T}_3$ adapted to the configuration of the domain ω (shadowed). Numerical results are summarized in Table 2.2, where we compare the indicators (2.68), (2.69), and (2.71). We see that error indicators based on (2.69) and (2.71) demonstrate better performance than (2.68). Other tests in [NRT08] for problems with regular and rather irregular solutions confirm advantages of (2.69) and especially of (2.71).

2.3.3 Error Indicators Based on Partial Equilibration of Fluxes in the Original Problem

First, we prove one principal result, which yields another (in a sense more convenient) form of the functional $I_{\ell 2}(u_h, u_{\ell\tau}; u, u_\ell)$.

Proposition 2.1 *The term $I_{\ell 2}(u_h, u_{\ell\tau}; u, u_\ell)$ is equal to the quantity*

$$\int_\Omega A^{-1}\left(\mathsf{P}_{Q_f}(p_h) - p_h\right) \cdot (\eta_\ell - A\nabla u_{\ell\tau})\,\mathrm{d}x := I_{\ell 2}(p_h, u_{\ell\tau}, \eta_\ell), \tag{2.76}$$

where η_ℓ is an arbitrary function in the set

Table 2.2 Efficiency of the estimators in Example 2.4

Indicator	N_{nod}	$\mathcal{T}_\tau$	$I_{\ell 1}$	$\mathbf{E}_{\ell 2}$	$\mathbf{E}_\ell$	i_{eff}
(2.68)	315	$\mathcal{T}_1$	0.00000	0.00264	0.00264	1.58
	193	$\mathcal{T}_2$	0.00119	0.00138	0.00257	1.54
	451	$\mathcal{T}_3$	0.00184	0.00040	0.00223	1.34

Indicator	N_{nod}	$\mathcal{T}_\tau$	$I_{\ell 1}$	$\mathbf{E}_{\ell 21}$	$\mathbf{E}_\ell$	i_{eff}
(2.69)	315	$\mathcal{T}_1$	0.00163	0.00051	0.00213	1.28
	193	$\mathcal{T}_2$	0.00189	0.00064	0.00253	1.51
	451	$\mathcal{T}_3$	0.00181	0.00013	0.00193	1.16

Indicator	N_{nod}	$\mathcal{T}_\tau$	$I_{\ell 1}$	$\mathbf{E}_{\ell 21}$	$\mathbf{E}_\ell$	i_{eff}
(2.71)	315	$\mathcal{T}_1$	0.00108	0.00055	0.00163	0.98
	193	$\mathcal{T}_2$	0.00126	0.00053	0.00179	1.07
	451	$\mathcal{T}_3$	0.00178	0.00000	0.00178	1.06

$$Q_\ell(\Omega) := \{q \in H(\Omega, \mathrm{div}) \mid \mathrm{div}\, q + \ell = 0\},$$

and the operator $P_{Q_f} : Q \to Q_f$ is defined by the relation

$$\left\| q - \mathsf{P}_{Q_f}(q) \right\|_{A^{-1}} \le \| q - q_f \|_{A^{-1}}, \quad \forall q_f \in Q_f. \tag{2.77}$$

Proof Let η_0 be a solenoidal vector-valued function. Then,

$$I_{\ell 2}(u_h, u_{\ell\tau}; u, u_\ell) = \int_\Omega (\nabla u - \nabla u_h) \cdot (A\nabla u_\ell + \eta_0 - A\nabla u_{\ell\tau})\, \mathrm{d}x.$$

Since $A\nabla u_\ell \in Q_\ell$, we conclude that

$$I_{\ell 2}(u_h, u_{\ell\tau}; u, u_\ell) = \int_\Omega A^{-1}(p - p_h) \cdot (\eta_\ell - A\nabla u_{\ell\tau})\, \mathrm{d}x,$$

where η_ℓ is an arbitrary element of Q_ℓ. From (2.77) with $q = p_h$, it follows that

$$\int_\Omega A^{-1}\left(p_h - \mathsf{P}_{Q_f}(p_h)\right) \cdot \eta_0\, \mathrm{d}x = 0, \quad \forall \eta_0 \in Q_0. \tag{2.78}$$

Since p and $\mathsf{P}_{Q_f}(p_h)$ belong to $Q_f(\Omega)$, we conclude that $(p - \mathsf{P}_{Q_f}(p_h)) \in Q_0$. In view of (2.78), we obtain

$$\begin{aligned} 0 &= \int_\Omega A^{-1}\left(p_h - \mathsf{P}_{Q_f}(p_h)\right) \cdot \left(p - \mathsf{P}_{Q_f}(p_h)\right) \mathrm{d}x \\ &= \int_\Omega A^{-1}\left(p_h - p + p - \mathsf{P}_{Q_f}(p_h)\right) \cdot \left(p - \mathsf{P}_{Q_f}(p_h)\right) \mathrm{d}x \end{aligned}$$

$$= \int_{\Omega} (\nabla u_h - \nabla u) \cdot (p - \mathsf{P}_{Q_f}(p_h))\, dx + \|p - \mathsf{P}_{Q_f}(p_h)\|^2_{A^{-1}}$$
$$= \|p - \mathsf{P}_{Q_f}(p_h)\|^2_{A^{-1}},$$

and the relation (2.76) follows. □

We note that the term $I_{\ell 2}(p_h, u_{\ell\tau}, \eta_\ell)$ does not contain the exact solution of the adjoint problem. The only difficulty in computing $I_{\ell 2}(p_h, u_{\ell\tau}, \eta_\ell)$ consists of the projection to Q_f. A computable error indicator arises if the exact projection $\mathsf{P}_{Q_f}(p_h)$ is replaced by an approximate $\widetilde{p}_h$ (which can be constructed with the help of a certain quasi-equilibration procedure). Then, we replace $I_{\ell 2}(p_h, u_{\ell\tau}, \eta_\ell)$ by the term

$$\mathbb{E}_{\ell 2}(p_h, \widetilde{p}_h, u_{\ell\tau}, \eta_\ell) := \int_{\Omega} A^{-1}(\widetilde{p}_h - p_h) \cdot (\eta_\ell - A\nabla u_{\ell\tau})\, dx \tag{2.79}$$

and find that

$$\langle \ell, u - u_h \rangle = I_{\ell 1}(u_h, u_{\ell\tau}) + \mathbb{E}_{\ell 2}(p_h, \widetilde{p}_h, u_{\ell\tau}, \eta_\ell) + \mathcal{R}(p_h, \widetilde{p}_h, u_{\ell\tau}, \eta_\ell), \tag{2.80}$$

where the first two terms are explicitly computable and the remainder term is defined by the relation

$$\mathcal{R}(p_h, \widetilde{p}_h, u_{\ell\tau}, \eta_\ell) := \int_{\Omega} A^{-1}\big(\mathsf{P}_{Q_f}(p_h) - \widetilde{p}_h\big) \cdot (\eta_\ell - A\nabla u_{\ell\tau})\, dx.$$

An upper bound of this term can be explicitly evaluated.

Proposition 2.2 *The remainder term is subject to the estimate*

$$\big|\mathcal{R}(p_h, \widetilde{p}_h, u_{\ell\tau}, \eta_\ell)\big|$$
$$\leq \left(\|p_h - \widetilde{p}_h\|_{A^{-1}} + \frac{C_{F\Omega}}{c_1} \|\operatorname{div} \widetilde{p}_h + f\| \right) \|\eta_\ell - A\nabla u_{\ell\tau}\|_{A^{-1}} := \mu_{h\tau}. \tag{2.81}$$

Proof We have

$$\big|\mathcal{R}(p_h, \widetilde{p}_h, u_{\ell\tau}, \eta_\ell)\big| \leq \big\|\mathsf{P}_{Q_f}(p_h) - \widetilde{p}_h\big\|_{A^{-1}} \|\eta_\ell - A\nabla u_{\ell\tau}\|_{A^{-1}}.$$

It is easy to see that

$$\big\|\mathsf{P}_{Q_f}(\widetilde{p}_h) - \mathsf{P}_{Q_f}(p_h)\big\|_{A^{-1}} \leq \|\widetilde{p}_h - p_h\|_{A^{-1}}.$$

This fact follows from the relation

$$\int_{\Omega} A^{-1}\big(p_h - \widetilde{p}_h - \mathsf{P}_{Q_f}(p_h) + \mathsf{P}_{Q_f}(\widetilde{p}_h)\big) \cdot \eta_0\, dx = 0, \quad \forall \eta_0 \in Q_0,$$

Fig. 2.12 Actual domains Ω and sample domains Ω_e

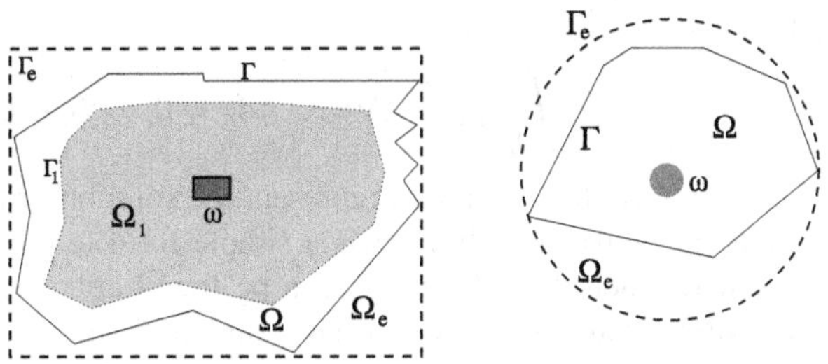

if we set $\eta_0 = \mathsf{P}_{Q_f}(\widetilde{p}_h) - \mathsf{P}_{Q_f}(p_h) \in Q_0$. Hence,

$$\big\|\mathsf{P}_{Q_f}(p_h) - \widetilde{p}_h\big\|_{A^{-1}} \le \|p_h - \widetilde{p}_h\|_{A^{-1}} + \big\|\mathsf{P}_{Q_f}(\widetilde{p}_h) - \widetilde{p}_h\big\|_{A^{-1}}.$$

Since

$$\big\|\mathsf{P}_{Q_f}(\widetilde{p}_h) - \widetilde{p}_h\big\|_{A^{-1}} = \inf_{q_f \in Q_f} \|\widetilde{p}_h - q_f\|_{A^{-1}} \le \frac{C_{F\Omega}}{c_1}\|\operatorname{div}\widetilde{p}_h + f\|,$$

we arrive at (2.81). □

Remark 2.5 From (2.80) and (2.81), it follows that

$$\begin{aligned} &I_{\ell 1}(u_h, u_{\ell\tau}) + \mathbb{E}_{\ell 2}(p_h, \widetilde{p}_h, u_{\ell\tau}, \eta_\ell) - \mu_{h\tau} \\ &\quad \le \langle \ell, u - u_h\rangle \le I_{\ell 1}(u_h, u_{\ell\tau}) + \mathbb{E}_{\ell 2}(p_h, \widetilde{p}_h, u_{\ell\tau}, \eta_\ell) + \mu_{h\tau}, \end{aligned}$$

which yields guaranteed error bounds. Certainly these bounds are sensible only if the quantity $\mu_{h\tau}$ is small compared to the first two terms. Since $\mu_{h\tau}$ is directly computable, this requirement can be verified in practical computations.

Finally, we discuss a particular form of the above-introduced error indicator based on solutions of specially constructed *sample problems*. In (2.80), the function $u_{\ell\tau}$ can be replaced by any conforming approximation v_ℓ of u_ℓ (in the derivation of this relation the Galerkin orthogonality of $u_{\ell\tau}$ was not used). Therefore,

$$\langle \ell, u - u_h\rangle = I_{\ell 1}(u_h, v_\ell) + \mathbb{E}_{\ell 2}(p_h, \widetilde{p}_h, v_\ell, \eta_\ell) + \mathcal{R}(p_h, \widetilde{p}_h, v_\ell, \eta_\ell). \qquad (2.82)$$

A way of constructing v_ℓ and η_ℓ is to use the exact solution of an adjoint problem for a close domain Ω_e having a simple geometric form. In Fig. 2.12 (left), this domain is presented by a dashed rectangular and ω is the domain (zone) of interest, in which ℓ is nonzero. In Fig. 2.12 (right), this domain is a circle. In the simplest form, the idea of the method is as follows (see [NR09] for more details). Consider the problem (2.63) with the boundary condition $u_0 = 0$. Let $\Omega \subset \Omega_e$. Assume that we know the functions $p_e \in H(\Omega_e, \operatorname{div})$ and $u_e \in V_0(\Omega_e)$ such that

$$\int_{\Omega_e} p_e \cdot \nabla w \,\mathrm{d}x = \int_{\Omega_e} \ell w \,\mathrm{d}x, \quad \forall w \in V_0(\Omega_e), \qquad (2.83)$$

and

$$\int_{\Omega_e} (p_e - A\nabla u_e) \cdot \eta \, dx = 0, \quad \forall \eta \in Q(\Omega_e). \tag{2.84}$$

It is easy to see that u_e and p_e represent the solution of the adjoint problem in Ω_e and the respective flux. If Ω_e has a simple form (e.g., it is a rectangular, a cube or a sphere) then these functions can be found either analytically or numerically with a high accuracy (since Ω_e has a simple form, sharp approximations can be constructed with the help of, e.g., spectral methods or other methods adapted to such type domains).

Let ϕ be a continuous function such that

$$\phi = 0 \quad \text{on } \Gamma, \qquad 0 \leq \phi(x) \leq 1 \quad \text{in } \Omega,$$

$$\phi(x) = 1 \quad \text{in } \Omega_1, \qquad \nabla\phi \in L^\infty(\Omega, \mathbb{R}^d).$$

Set $\eta_\ell = p_e$ and $v_\ell = \phi u_e$. Since $\phi u_e \in V_0(\Omega)$, we can use it in the indicator. Then, $A\nabla v_\ell = \phi A\nabla u_e + u_e A\nabla\phi$, $\eta_\ell = A\nabla u_e$ and the remainder term has the following form:

$$\mathcal{R}(p_h, \widetilde{p}_h, v_\ell, \eta_\ell) := \int_{\Omega\setminus\Omega_1} A^{-1}\big(\mathsf{P}_{Q_f}(p_h) - \widetilde{p}_h\big) \cdot \big((1-\phi)p_e - u_e A\nabla\phi\big) \, dx.$$

If the flux $\widetilde{p}_h$ is almost equilibrated in the boundary strip $\Omega \setminus \Omega_1$, then the remainder term is very small so that the two first computable terms in (2.82) dominate and represent the major part of $\langle \ell, u - u_h \rangle$. Therefore, the quality of the error indicator

$$\langle \ell, u - u_h \rangle \approx \mathbb{E}_\ell(u_h, \widetilde{p}_h, v_\ell, \phi, \Omega_e) := I_{\ell 1}(u_h, v_\ell) + \mathbb{E}_{\ell 2}(p_h, \widetilde{p}_h, v_\ell, \eta_\ell)$$

depends mainly on the equilibration properties of $\widetilde{p}_h$ in the boundary strip.

Chapter 3
Guaranteed Error Bounds I

Abstract In this chapter, we discuss foundations of new error control methods developed during the last 10–12 years. First, we consider the simplest boundary value problems generated by ordinary differential equations and show that proper transformations of the corresponding integral identity yield a guaranteed bound of the difference between the exact solution and any conforming approximation. Subsequently, this method is extended to partial differential equations of the elliptic type.

Our goal is not only to explain how fully reliable error bounds are derived but also discuss their main properties, which are as follows:

- the estimates are guaranteed,
- they do not contain mesh-dependent constants, and
- the estimates are valid for any conforming approximation of a problem.

The theory provides a way of creating new error estimation algorithms. First, we present them with the paradigm of the stationary diffusion problem. In subsequent sections, the error control techniques and step-by-step algorithms are discussed for several main classes of linear elliptic problems.

3.1 Ordinary Differential Equations

We begin with the boundary value problem

$$-\big(a(x)u'(x)\big)' + b(x)u(x) = f(x) \quad x \in \Omega := (\xi_1, \xi_2), \tag{3.1}$$

$$u(\xi_1) = u_1, \tag{3.2}$$

$$u(\xi_2) = u_2, \tag{3.3}$$

generated by the Sturm–Liouville operator with bounded coefficients a and b satisfying the conditions

$$a(x) \geq a_0 > 0, \qquad b(x) \geq 0, \quad \text{and} \quad f \in L^2(\Omega).$$

Problem (3.1)–(3.3) is one of the most simple boundary value problems, which can be solved by different numerical methods. Let

$$v \in H^1(\Omega), \quad v(\xi_1) = u_1, \qquad v(\xi_2) = u_2$$

O. Mali et al., *Accuracy Verification Methods*,
Computational Methods in Applied Sciences 32, DOI 10.1007/978-94-007-7581-7_3,

be a function computed by some method. Our goal is to obtain a guaranteed (fully reliable) estimate of $u - v$ in terms of the energy norm

$$|||u - v|||^2 := \int_{\xi_1}^{\xi_2} \left(a\left(u' - v'\right)^2 + b(u - v)^2\right) \mathrm{d}x.$$

A function $u \in H^1(\Omega)$ satisfying the boundary conditions is a generalized solution of (3.1)–(3.3) if it meets the relation

$$\int_{\xi_1}^{\xi_2} \left(au'w' + buw\right) \mathrm{d}x = \int_{\xi_1}^{\xi_2} fw \, \mathrm{d}x \tag{3.4}$$

for any trial function $w \in V_0$, where V_0 contains the functions from $H^1(\Omega)$ vanishing at ξ_1 and ξ_2 (see Sect. B.1).

3.1.1 Derivation of Guaranteed Error Bounds

In order to deduce a computable upper bound of $u - v$, we rewrite (3.4) in the equivalent form

$$\int_{\xi_1}^{\xi_2} \left(a(u - v)'w' + b(u - v)w\right) \mathrm{d}x = \int_{\xi_1}^{\xi_2} \left(fw - av'w' - bvw\right) \mathrm{d}x. \tag{3.5}$$

Let $y(x)$ be an arbitrary function in $H^1(\Omega)$. Since

$$\int_{\xi_1}^{\xi_2} (yw)' \, \mathrm{d}x = (yw)\Big|_{\xi_1}^{\xi_2} = 0, \quad \forall w \in V_0, \tag{3.6}$$

we rewrite the right-hand side of (3.5) as follows:

$$\int_{\xi_1}^{\xi_2} \left(fw - av'w' - bvw\right) \mathrm{d}x = \int_{\xi_1}^{\xi_2} \left((f + y' - bv)w - \left(av' - y\right)w'\right) \mathrm{d}x.$$

Now, we set $w = u - v$ and obtain

$$|||u - v|||^2 = \int_{\xi_1}^{\xi_2} \left((f + y' - bv)(u - v) - \left(av' - y\right)(u - v)'\right) \mathrm{d}x. \tag{3.7}$$

If b is strictly positive, then

$$\int_{\xi_1}^{\xi_2} (f + y' - bv)(u - v) \, \mathrm{d}x \le \left(\int_{\xi_1}^{\xi_2} \frac{1}{b} |f + y' - bv|^2 \, \mathrm{d}x\right)^{1/2} \left(\int_{\xi_1}^{\xi_2} b|u - v|^2 \, \mathrm{d}x\right)^{1/2}.$$

Analogously,

$$\int_{\xi_1}^{\xi_2} (av' - y)(u - v)' \,\mathrm{d}x \le \left(\int_{\xi_1}^{\xi_2} \frac{1}{a} |av' - y|^2 \,\mathrm{d}x \right)^{1/2} \left(\int_{\xi_1}^{\xi_2} a|u' - v'|^2 \,\mathrm{d}x \right)^{1/2}.$$

By means of the algebraic inequality $\lambda_1\delta_1 + \lambda_2\delta_2 \le \sqrt{\lambda_1^2 + \lambda_2^2}\sqrt{\delta_1^2 + \delta_2^2}$, we find that

$$|||u - v|||^2 = \int_{\xi_1}^{\xi_2} \big((f + y' - bv)(u - v) - a(v' - y)(u - v)' \big) \,\mathrm{d}x \tag{3.8}$$

$$\le \left(\int_{\xi_1}^{\xi_2} \left(\frac{1}{b} |f + y' - bv|^2 + \frac{1}{a} |y - av'|^2 \right) \mathrm{d}x \right)^{1/2} |||u - v|||. \tag{3.9}$$

Thus, we arrive at the estimate

$$|||u - v|||^2 \le \int_{\xi_1}^{\xi_2} \left(\frac{1}{b} |f + y' - bv|^2 + \frac{1}{a} |y - av'|^2 \right) \mathrm{d}x =: \overline{\mathsf{M}}_1^2(v, y), \tag{3.10}$$

the right-hand side of which is a nonnegative functional depending on v and problem data. It presents a guaranteed upper bound of the error and *does not involve* u. Henceforth, such type functionals are called a posteriori estimates of functional type, or *error majorants*.

The method used in the process of deriving (3.10) is based on the idea to *split* the residual functional by means of the integration by parts formula, which involves a "free function" y. Originally, this method was introduced in [Rep01b]. In subsequent sections, we show that it can be extended to a wide spectrum of boundary value problems.

Since y is an arbitrary function in $H^1(\Omega)$, we find that

$$|||u - v||| \le \inf_{y \in H^1(\Omega)} \overline{\mathsf{M}}_1(v, y). \tag{3.11}$$

It is easy to show that (3.11) holds as equality. Indeed, if we set $y = au'$, then

$$\frac{1}{b} |f + y' - bv|^2 = \frac{1}{b} |b(u - v)|^2, \qquad \frac{1}{a} |y|^2 + a|v'|^2 - 2yv' = a(u' - v')^2$$

and

$$\overline{\mathsf{M}}_1(v, y) = \left(\int_{\xi_1}^{\xi_2} \big(b|u - v|^2 + a(u' - v')^2 \big) \,\mathrm{d}x \right)^{1/2} = |||u - v|||.$$

Another estimate follows from (3.7) if we apply the simplest Friedrichs inequality

$$\int_{\xi_1}^{\xi_2} |w|^2 \, dx \leq C_F \int_{\xi_1}^{\xi_2} a|w'|^2 \, dx, \tag{3.12}$$

where w is a function in $H^1((\xi_1, \xi_2))$ such that $w(\xi_1) = w(\xi_2) = 0$ and

$$C_F \leq \bar{C}_F := \frac{\xi_2 - \xi_1}{\pi} \operatorname*{ess\,sup}_{x \in \Omega} a(x).$$

Then,

$$\int_{\xi_1}^{\xi_2} (f + y' - bv)(u - v) \leq \bar{C}_F \| f + y' - bv \| \, |||u - v|||,$$

and we find that

$$|||u - v||| \leq \left(\int_{\xi_1}^{\xi_2} \frac{1}{a} |y - av'|^2 \, dx \right)^{1/2} + \bar{C}_F \| f + y' - bv \| =: \overline{\mathsf{M}}_2(v, y).$$

Moreover,

$$|||u - v||| \leq \inf_{y \in H^1(\Omega)} \overline{\mathsf{M}}_2(v, y). \tag{3.13}$$

It is easy to see that $\overline{\mathsf{M}}_1(v, y)$ and $\overline{\mathsf{M}}_2(v, y)$ vanishes if and only if the functions y and v are such that

$$y = av' \tag{3.14}$$

and

$$y' - bv + f = 0. \tag{3.15}$$

Since v satisfies the boundary conditions, these two relations imply that v coincides with u and y with au'.

The majorants $\overline{\mathsf{M}}_1(v, y)$ and $\overline{\mathsf{M}}_2(v, y)$ provide the guaranteed upper bounds of the overall error $|||u - v|||$. They are nonnegative functionals, which depend on the problem data (a, f, Ω), approximate solution v, and a function y, which can be considered as an approximation of au'. We emphasize that y is completely at our disposal, and the majorants provide the guaranteed upper bound with any y.

In practical computations, we can use both majorants and select the best estimate. However, since $\overline{\mathsf{M}}_2(v, y)$ does not contain b^{-1}, it is more convenient to use it if b attains small (or zero) values. A method to derive more efficient (advanced) forms of the majorants is discussed in the next section with the paradigm of a boundary value problem generated by a partial differential equation.

3.1.2 Computation of Error Bounds

Assume that approximate solution is a piecewise affine continuous function defined by nodal values on a regular mesh with N intervals and $h = \frac{\xi_2 - \xi_1}{N}$. Such type approximation can be viewed as the simplest finite element approximation (cf. Sect. B.4.3). We denote it by v_h, and the corresponding finite dimensional space by V_h. From the computational point of view, it is convenient to slightly modify the estimate. We square $\overline{\mathsf{M}}_2(v_h, y)$, use (A.4), and obtain

$$|||u - v_h|||^2 \le \int_{\xi_1}^{\xi_2} \left(\frac{1+\beta}{a} |y - a v_h'|^2 + \bar{C}_F \left(1 + \frac{1}{\beta}\right) |f + y' - b(x) v_h|^2 \right) \mathrm{d}x, \quad (3.16)$$

where β is an arbitrary positive number and y is an arbitrary differentiable function. We denote the right-hand side of (3.16) by $\overline{\mathsf{M}}_2^2(v_h, y, \beta)$ and two parts of it by

$$\overline{\mathsf{M}}_2^{\mathrm{D}}(v_h, y) := \left(\int_{\xi_1}^{\xi_2} \frac{1}{a} |y - a v_h'|^2 \,\mathrm{d}x \right)^{1/2}$$

and

$$\overline{\mathsf{M}}_2^{\mathrm{Eq}}(v_h, y) := \left(\int_{\xi_1}^{\xi_2} |f + y' - b v_h|^2 \,\mathrm{d}x \right)^{1/2}.$$

Clearly, they are related to violations of (3.14) and (3.15), respectively. We note that (3.14) is the simplest form of the duality relation (cf. (A.44)) and (3.15) is a simple equilibrium (balance) equation.

The right-hand side of (3.16) contains only known functions, namely, v_h, a, and b are given and y and β are in our disposal (they are changeable). We outline that no special conditions are imposed on v_h, so that the estimate can be applied to any function v_h regardless of the way used to construct it. However, getting a good upper bound needs a rational selection of the "free" function y and "free" parameter β. The latter task is easy: if y is given, then the best β is easy to find by the relation

$$\beta_{\mathrm{opt}}(v_h, y) := \bar{C}_F \frac{\overline{\mathsf{M}}_2^{\mathrm{Eq}}(v_h, y)}{\overline{\mathsf{M}}_2^{\mathrm{D}}(v_h, y)} \quad (3.17)$$

provided that the numerator does not equal to zero (if it is zero, then the majorant contains only one term with the factor 1 instead of $1 + 1/\beta$).

To construct a suitable y_h, we can use different methods. One of the most simple is as follows. Using v_h, we compute a rough approximation of the flux au' presented

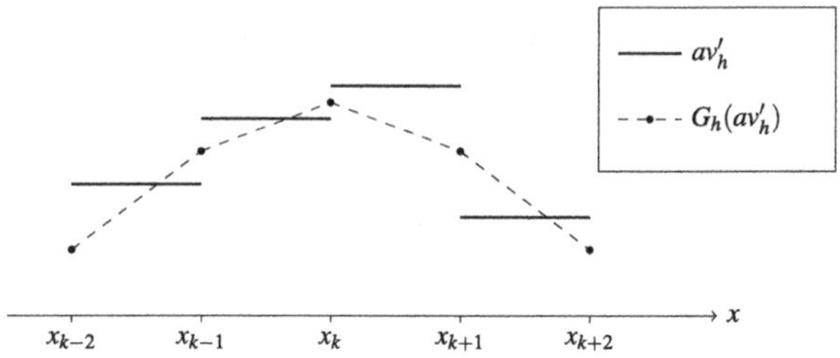

Fig. 3.1 Piecewise constant function av'_h and averaged $G_h(av'_h)$

by a piecewise constant function $q_h := \{|a|\}_k v'_h$ (where $\{|a|\}_k$ is the mean value of a on the interval number k). This function is non-differentiable and therefore we cannot use it in the majorant. However, we can construct another function $y_h^{(0)} = G_h q_h$, where the operator G_h averages values of q_h at the nodal points (where q_h has jumps), assigns these mean values to the nodal points, and extends to the internal points of the intervals by the affine extension. The functions av'_h and $G_h(av'_h)$ are depicted in Fig. 3.1. Since $y_h^{(0)} = G_h q_h$ is differentiable, we compute $\overline{\mathsf{M}}_2^{\mathrm{Eq}}(v_h, y_h^{(0)})$ and $\overline{\mathsf{M}}_2^{\mathrm{D}}(v_h, y_h^{(0)})$. Then, we select $\beta^{(0)}$ in accordance with (3.17) and compute the majorant $\overline{\mathsf{M}}_2(v_h, y_h^{(0)}, \beta^{(0)})$, which provides a coarse (but guaranteed) upper bound of the error. Further improvements of the majorant can be performed by different iteration procedures, which minimize the majorant with respect to y.

In Algorithm 3.1, this part is presented in the simplest form (as a cycle with m steps). The minimization method is not specified. In fact many different methods can be used for this relatively simple quadratic minimization problem (from direct minimization methods solving the problem approximately to multigrid type solvers of linear systems able to get the exact minimizer over some predefined subspace). In general, the choice of a particular method depends on preferences of a computer analyst and on the quality of error bounds one wishes to obtain.

Example 3.1 Consider the problem (3.1)–(3.2), where $a(x) = 1$, $b(x) = 0$, $f(x) = 2$, $\xi_1 = 0$, $\xi_2 = 1$, and $u_1 = 0u_2 = 0$. In this case, $C_F = 1/\pi^2$, and the exact solution and the flux are known:

$$u = -x(x-1) \quad \text{and} \quad p = -2x + 1.$$

We generate approximation v by interpolating a perturbed exact solution. Then, we apply Algorithm 3.1, where the minimization of the majorant with respect to y (y belongs to piecewise linear functions) is performed by the function `fminunc` from the optimization toolbox of Matlab. In Table 3.1, we present the estimates obtained by minimizing y for k iteration steps. The efficiency index of the majorant is defined by the formula

$$I_{\mathrm{eff}} := \frac{\overline{\mathsf{M}}_2(v, y, \beta)}{|||u - v|||}.$$

For comparison, on the bottom line we also present the values obtained by substituting the exact flux p to the majorant.

Algorithm 3.1 Estimation of approximation errors

Input: $v_h \in V_h$ {piecewise affine approximation defined on a certain mesh},
m {number of iterations}
$i = 0$
$y_h^{(0)} = G_h(av_h')$ {averaged approximation of the flux}
Compute $\overline{\mathrm{M}}_2^{\mathrm{D}}(v_h, y_h^{(0)})$ and $\overline{\mathrm{M}}_2^{\mathrm{Eq}}(v_h, y_h^{(0)})$
Compute $\beta^{(0)} := \beta_{\mathrm{opt}}(v_h, y^{(0)})$ by (3.17).
Compute $\overline{\mathrm{M}}_2(v_h, y_h^{(0)}, \beta^{(0)})$ {Coarse upper bound of the error}
while $i \le m$ **do**
Find $y_h^{(i+1)} \in Y_h$ such that $\overline{\mathrm{M}}_2(v_h, y_h^{(i+1)}, \beta^{(i)}) < \overline{\mathrm{M}}_2(v_h, y_h^{(i)}, \beta^{(i)})$
{exact or approximate minimization of the majorant with $\beta^{(i)}$}
Compute $\overline{\mathrm{M}}_2^{\mathrm{D}}(v_h, y_h^{(i+1)})$ and $\overline{\mathrm{M}}_2^{\mathrm{Eq}}(v_h, y_h^{(i+1)})$
Compute $\beta^{(i+1)} := \beta_{\mathrm{opt}}(v_h, y^{(i+1)})$ by (3.17).
Compute $\overline{\mathrm{M}}_2(v_h, y_h^{(i+1)}, \beta^{(i+1)})$ {error majorant on the iteration $i+1$}
$i = i + 1$
end while
Output: $y^{(m)}$ {an approximation of the flux}
$\overline{\mathrm{M}}_2(v_h, y_h^{(m)}, \beta^{(m)})$ {error majorant computed after m iterations}
end

Table 3.1 Application of Algorithm 3.1

k	I_{eff}	$\overline{\mathrm{M}}_2(v_h, y^{(k)}, \beta^{(k)})$	$\overline{\mathrm{M}}_2^{\mathrm{D}}(v_h, y^{(k)})$	$\overline{\mathrm{M}}_2^{\mathrm{Eq}}(v_h, y^{(k)})$	$\beta^{(k)}$
0	3.9757	0.059433	0.005926	0.15635	2.6733
5	1.9800	0.029598	0.005946	0.06434	1.0962
10	1.8935	0.028305	0.005994	0.06024	1.0182
20	1.6970	0.025369	0.006780	0.04833	0.7223
40	1.4406	0.021535	0.009472	0.02816	0.3012
80	1.1316	0.016916	0.014131	0.00537	0.0385
150	1.0017	0.014975	0.014952	0.00004	0.0003

I_{eff}	$\overline{\mathrm{M}}_2(v_h, p, 0)$	$\overline{\mathrm{M}}_2^{\mathrm{D}}(v_h, p)$	$\overline{\mathrm{M}}_2^{\mathrm{Eq}}(v_h, p)$	β
1.0000	0.014949	0.014949	8.5047e-015	5.7644e-014

As we can see from Table 3.1, the majorant provides a guaranteed upper bound. Moreover, the bound becomes sharper if we invest more to the computation of a suitable y. It is easy to see that the parts of the majorant (as well as the whole majorant) converges toward the values obtained by substituting p (actually, one can show that p is the exact minimizer of the majorant).

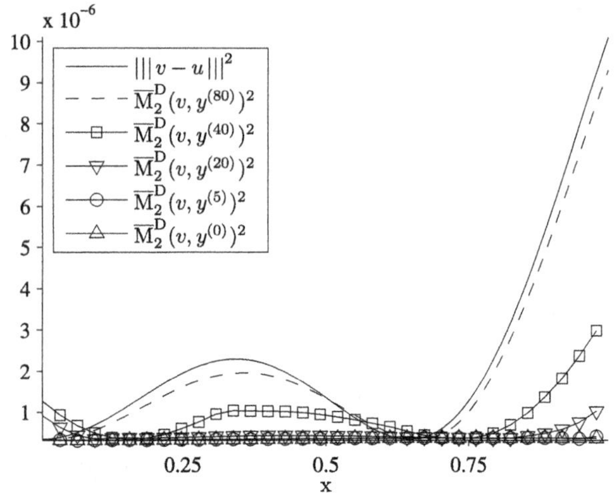

Fig. 3.2 Interval-wise values of integrals $\overline{\mathsf{M}}_2^{\mathrm{D}}(v, y^{(k)})$ after k iterations steps and the respective interval-wise values of the exact error distribution $|||v - u|||^2$

The selection of y in $\overline{\mathsf{M}}(v, y)$ is at our disposal. There is a wide spectrum of methods to obtain an approximation of y. For example, one possibility is to compute the minimizer of the majorant directly, from the necessary conditions. This approach is discussed in detail in Sect. 3.3.1 in the framework of partial differential equations. The general rule is that the more computational power you invest in computing y, the sharper is the value of the majorant.

Finally, we note that numerical efficiency of the error majorant for boundary value problems generated by ordinary differential equations was firstly tested in [Rep99b]. The dependence of the effectiveness of a posteriori estimation of an approximate solution of an elliptic boundary value problem on the input data and the algorithm parameters has been systematically studied in [BMP09].

Example 3.2 Several error indication methods have already been discussed in Chap. 2. Here and in Sect. 3.4 we show that the majorant can be used for the same purpose. As Table 3.1 shows, for a "good" y, $\overline{\mathsf{M}}_2^{\mathrm{Eq}} \approx 0$ and $\overline{\mathsf{M}}_2^{\mathrm{D}}$ accurately represents the error. This motivates us to use $\overline{\mathsf{M}}_2^{\mathrm{D}}$ computed on each interval in order to obtain an error indicator and identify the intervals where the error $|||u - v|||$ is large compared with other errors.

In Fig. 3.2, we depict the exact interval-wise error distribution and the error distribution indicated by $\overline{\mathsf{M}}_2^{\mathrm{D}}$. In Table 3.2 we measure the performance of the error indicator $\boldsymbol{E}^{(k)} := \overline{\mathsf{M}}_2^{D}(v, y(k))$ in the strong sense (with respect to Definition 2.1) and by applying various marking procedures and Definition 2.3. $\mathbb{M}_1$ is based on comparison with the average indicated error (Algorithm 2.1), $\mathbb{M}_2$ marks 40 % of the intervals where the error is largest (Algorithm 2.2), and $\mathbb{M}_3$ selects intervals of highest indicated error, until the selected elements contain 40 % of the "error mass" (Algorithm 2.3). It is easy to see that after a sufficient number of iterations, the indicated error distribution approaches the exact one. In other words, the indicator becomes sharp in the sense of Definition 2.1.

Table 3.2 Accuracy of error indicator $\mathbb{E}^{(k)}$ measured in the strong sense and with respect to various markers

k	$\mathcal{M}(\mathbb{E}^{(k)})$	$\mathcal{M}(\mathbb{E}^{(k)}, \mathbb{M}_1)$	$\mathcal{M}(\mathbb{E}^{(k)}, \mathbb{M}_2)$	$\mathcal{M}(\mathbb{E}^{(k)}, \mathbb{M}_3)$
0	0.9247	0.4040	0.4040	0.3232
5	0.9229	0.3838	0.4040	0.3232
10	0.9200	0.3737	0.4040	0.3232
20	0.8777	0.2020	0.3434	0.1818
40	0.6901	0.2424	0.2424	0.0808
80	0.1012	0.0707	0.0404	0.0101
150	0.0011	0	0	0

However, weaker measures strongly depend on the marker. They may be rather small, even tough the error distribution computed by $\mathbb{E}$ differs considerably from the exact error. On the other hand, if the strong measure is small, than all weak measures are also small. This observation confirms our conclusion that reliable error indication should be based on indicators which are able to produce realistic measurements of errors in the strong sense.

3.2 Partial Differential Equations

Now, we extend the method presented in Sect. 3.1 to linear partial differential equations. Consider the system

$$-\operatorname{div} p + \varrho^2 u = f \qquad \text{in } \Omega, \tag{3.18}$$

$$p = A\nabla u \qquad \text{in } \Omega, \tag{3.19}$$

$$u = u_0 \qquad \text{on } \Gamma, \tag{3.20}$$

where A is a symmetric matrix satisfying the condition $Az \cdot z \geq c_1 |z|^2$ for any $z \in \mathbb{R}^d$ and ϱ is a nonnegative function of x (this system is often used as a model of a stationary reaction diffusion process, where A is the diffusion matrix and ρ describes a reaction).

The generalized solution $u \in V_0 + u_0$ of (3.18)–(3.20) is defined by the integral identity

$$\int_\Omega (A\nabla u \cdot \nabla w + \varrho^2 u w)\,\mathrm{d}x = \int_\Omega f w\,\mathrm{d}x, \qquad w \in V_0. \tag{3.21}$$

It minimizes the functional

$$I(w) = \int_\Omega \left(\frac{1}{2} A\nabla w \cdot \nabla w + \frac{\varrho^2}{2} |w|^2 - f w \right) \mathrm{d}x \tag{3.22}$$

on the set $V_0 + u_0$.

3.2.1 Maximal Deviation from the Exact Solution

Let $v \in V_0 + u_0$. By (3.21) we deduce the relation

$$\int_\Omega A\big(\nabla(u-v)\cdot\nabla w + \varrho^2(u-v)w\big)\,dx = \int_\Omega \big(fw - \varrho^2 vw - \nabla v\cdot\nabla w\big)\,dx,$$

which holds for any $w \in V_0$. Since w vanishes at the boundary, we rewrite this relation as follows:

$$\int_\Omega \big(A\nabla(u-v)\cdot\nabla w + \varrho^2(u-v)w\big)\,dx$$
$$= \int_\Omega \big((f - \varrho^2 v + \operatorname{div} y)w + (y - A\nabla v)\cdot\nabla w\big)\,dx, \tag{3.23}$$

where $y \in H(\Omega, \operatorname{div})$. It is easy to see that

$$\int_\Omega (y - A\nabla v)\cdot\nabla w\,dx \le \|y - A\nabla v\|_{A^{-1}}\|\nabla w\|_A, \tag{3.24}$$

$$\int_\Omega \big(f - \varrho^2 v + \operatorname{div} y\big)\cdot w\,dx \le \left\|\frac{1}{\varrho}\big(f - \varrho^2 v + \operatorname{div} y\big)\right\|\|\varrho w\|, \tag{3.25}$$

$$\int_\Omega \big(f - \varrho^2 v + \operatorname{div} y\big)\cdot w\,dx \le \mathrm{C}\|f - \varrho^2 v + \operatorname{div} y\|\,\|\nabla w\|_A, \tag{3.26}$$

where

$$\|y\|_A^2 := \int_\Omega Ay\cdot y\,dx \quad \text{and} \quad \|y\|_{A^{-1}}^2 := \int_\Omega A^{-1}y\cdot y\,dx$$

are the norms equivalent to the natural norm of the space $Q(\Omega) := L^2(\Omega, \mathbb{R}^d)$ and C is a constant in the inequality

$$\|w\| \le \mathrm{C}\|\nabla w\|_A, \quad \forall w \in V_0. \tag{3.27}$$

In view of (A.28), we find that $\mathrm{C} \le c_1^{-1/2} C_{F\Omega}$ (thus, if the constant $C_{F\Omega}$ or a computable upper bound of it are known, then the constant C is easily computable).

By (3.23), (3.24), and (3.25), we deduce the estimate

$$|\!|\!|u - v|\!|\!| \le \overline{\mathrm{M}}_1(v, y)$$
$$:= \left(\|A\nabla v - y\|_{A^{-1}}^2 + \left\|\frac{1}{\varrho}\big(f - \varrho^2 v + \operatorname{div} y\big)\right\|^2\right)^{1/2}, \tag{3.28}$$

where

$$|||w|||^2 := \|\nabla w\|_A^2 + \|\varrho w\|^2$$

is the energy norm related to the problem.

From (3.23), (3.24), and (3.26), we obtain another estimate

$$|||u - v||| \leq \overline{\mathsf{M}}_2(v, y) := \|A\nabla v - y\|_{A^{-1}} + \mathrm{C}\big\|f - \varrho^2 v + \operatorname{div} y\big\|. \quad (3.29)$$

Note that the majorants $\overline{\mathsf{M}}_1(v, y)$ and $\overline{\mathsf{M}}_2(v, y)$ are generalizations of the majorants $\overline{\mathsf{M}}_1(v, y)$ and $\overline{\mathsf{M}}_2(v, y)$ derived in the previous section. It is easy to show that

$$\inf_{y \in H(\Omega, \operatorname{div})} \overline{\mathsf{M}}_1(v, y) = |||u - v|||.$$

This fact follows from the relation

$$\overline{\mathsf{M}}_1^2(v, p) = \big\|A\nabla(v - u)\big\|_{A^{-1}}^2 + \left\|\frac{1}{\varrho}\big(f - \varrho^2 v + \operatorname{div} p\big)\right\|^2$$

$$= \big\|\nabla(v - u)\big\|_A^2 + \int_\Omega \varrho(u - v)^2\,\mathrm{d}x = |||u - v|||^2,$$

where $p = A\nabla u$.

However, $\overline{\mathsf{M}}_1(v, y)$ has an essential drawback: if ϱ is small, then the second term has a large multiplier, which makes the whole estimate sensitive with respect to the residual

$$\mathcal{R}(v, y) := f - \varrho^2 v + \operatorname{div} y.$$

In those problems where ϱ has small (or zero) values in one part of Ω and large in the other one, the majorant $\overline{\mathsf{M}}_1(v, y)$ may lead to a considerable overestimation of the error. In an opposite case, $\overline{\mathsf{M}}_2(v, y)$ is robust with respect to small ϱ but it may have an inherent gap between the left and right-hand sides of (3.29).

In order to overcome the above difficulties and obtain an estimate that possesses positive features of the above estimates, we apply another modus operandi for the deviation of an upper bound of $\|u - v\|$ suggested in [RS06].

Let us represent the first integral on the right-hand side of (3.23) as follows

$$\int_\Omega \mathcal{R}(v, y) w\,\mathrm{d}x = \int_\Omega \alpha \mathcal{R}(v, y) w\,\mathrm{d}x + \int_\Omega (1 - \alpha)\mathcal{R}(v, y) w\,\mathrm{d}x,$$

where $\alpha \in L^\infty_{[0,1]}(\Omega) := \{\alpha \in L^\infty(\Omega) \mid 0 \leq \alpha(x) \leq 1\}$ is a weight function at our disposal. Then, we have

$$\left|\int_\Omega \mathcal{R}(v, y) w\,\mathrm{d}x\right| \leq \left\|\frac{\alpha}{\varrho}\mathcal{R}(v, y)\right\| \|\varrho w\| + \mathrm{C}\big\|(1 - \alpha)\mathcal{R}(v, y)\big\| \|\nabla w\|_A.$$

By setting $w = u - v$ we arrive at the estimate

$$|||u - v|||^2 \leq \left(\mathrm{C}\left\|(1-\alpha)\mathcal{R}(v, y)\right\| + \|A\nabla v - y\|_{A^{-1}}\right)^2 + \left\|\frac{\alpha}{\varrho}\mathcal{R}(v, y)\right\|^2. \quad (3.30)$$

We denote the right-hand side of (3.30) by $\overline{\mathsf{M}}^2_\alpha(v, y)$. It is easy to see that (3.28) and (3.29) are particular cases of (3.30).

Squaring both parts of (3.25) and (A.6) yields

$$\begin{aligned} |||u - v|||^2 \leq{}& \mathrm{C}(1+\beta)\left\|(1-\alpha)\mathcal{R}(v, y)\right\|^2 \\ &+ \frac{1+\beta}{\beta}\|A\nabla v - y\|^2_{A^{-1}} + \left\|\frac{\alpha}{\varrho}\mathcal{R}(v, y)\right\|^2, \end{aligned} \quad (3.31)$$

where β is an arbitrary positive number.

Minimization of the right-hand side of (3.31) with respect to α is reduced to the following auxiliary variational problem: find $\widehat{\alpha} \in L^\infty_{[0,1]}(\Omega)$ such that

$$\Upsilon(\widehat{\alpha}) = \inf_{\alpha \in L^\infty_{[0,1]}(\Omega)} \Upsilon(\alpha), \quad (3.32)$$

where

$$\Upsilon(\alpha) := \int_\Omega \left(\alpha^2 P(x) + (1-\alpha)^2 Q(x)\right) \mathrm{d}x,$$

and P and Q are nonnegative integrable functions, which do not vanish simultaneously. It is easy to find that for almost all x

$$\widehat{\alpha}(x) = \frac{Q}{P+Q} \in [0, 1], \quad \Upsilon(\widehat{\alpha}) = \frac{PQ}{P+Q}.$$

In our case, $P = \rho^{-2}\mathcal{R}^2(v, y)$ and $Q = \mathrm{C}^2(1+\beta)\mathcal{R}^2(v, y)$. Therefore, we obtain

$$\begin{aligned} |||u - v|||^2 \leq{}& \int_\Omega \frac{\mathrm{C}^2(1+\beta)}{\mathrm{C}^2\varrho^2(1+\beta)+1}\mathcal{R}^2(v, y)\,\mathrm{d}x + \frac{1+\beta}{\beta}\|A\nabla v - y\|^2_{A^{-1}} \\ :={}& \overline{\mathsf{M}}^2_{\widehat{\alpha}}(v, y, \beta). \end{aligned} \quad (3.33)$$

The majorant $\overline{\mathsf{M}}^2_{\widehat{\alpha}}(v, y, \beta)$ is robust with respect to small values of ρ and, at the same time, it remains sharp. To prove the latter fact, we note that

$$\overline{\mathsf{M}}^2(v, p, \beta) = \int_\Omega \left(\frac{\mathrm{C}^2(1+\beta)}{\mathrm{C}^2\varrho^2(1+\beta)+1}\varrho^4(v-u)^2 + \frac{1+\beta}{\beta}|\nabla(v-u)|^2\right)\mathrm{d}x$$

and the majorant tends to the exact error norm if $\beta \to +\infty$. Therefore,

$$\inf_{\substack{y \in H(\Omega,\mathrm{div}),\\ \beta>0}} \overline{\mathsf{M}}^2(v, y, \beta) \leq \inf_{\beta>0} \overline{\mathsf{M}}^2_{\widetilde{\alpha}}(v, p, \beta) = |||u - v|||^2,$$

and we see that the estimate (3.33) has no "gap". The structure of the first term of (3.33) is such that it is not sensitive with respect to small values of ϱ.

It is easy to see that for any $\beta > 0$ the majorant vanishes if and only if

$$y = A\nabla v \quad \text{and} \quad \mathcal{R}(v, y) = 0.$$

Since v satisfies the boundary conditions, these relations mean that $v = u$. Hence, we arrive at the following conclusion.

$\overline{\mathsf{M}}_{\widetilde{\alpha}}(v, p, \beta)$ is a guaranteed upper bound of the distance between u and a function $v \in V_0 + u_0$. It vanishes if and only if v coincides with u and y coincides with the exact flux $p = A\nabla u$. Different y and β lead to different upper bounds. By selecting these parameters, we can find an upper bound of the error arbitrarily close to the exact value of the error.

3.2.2 Minimal Deviation from the Exact Solution

Now, we wish to find a guaranteed lower bound of the distance between the exact solution and a given function. In many cases, such type estimates also contain important information. They allow us to verify the efficiency of error majorants and are important in analysis of modeling errors and errors generated by indeterminate data. In this section, we consider a way of deriving such error estimates.

The simplest method of deriving error minorants is based on the relation (B.72), which holds for problems with quadratic energy functionals. In our case, this functional is $J(w) = \frac{1}{2}|||w|||^2 - \int_\Omega f w \, \mathrm{d}x$, and it is easy to show that

$$\frac{1}{2}|||u - v|||^2 = J(v) - J(u), \tag{3.34}$$

where v is an arbitrary function in $V_0 + u_0$. Since u is the minimizer of I, we know that $J(u) \leq J(v + w)$ for any $w \in V_0$. Therefore,

$$\begin{aligned}\frac{1}{2}|||u - v|||^2 &\geq J(v) - J(v + w)\\ &= \frac{1}{2}\|v\|_A^2 + \frac{\rho}{2}\|v\|^2 - \int_\Omega f v \, \mathrm{d}x\end{aligned}$$

$$
-\frac{1}{2}\|v+w\|_A^2 - \frac{\rho}{2}\|v+w\|^2 + \int_\Omega f(v+w)\,\mathrm{d}x
$$

$$
= -\frac{1}{2}\|w\|_A^2 - \frac{\rho}{2}\|w\|^2 - \int_\Omega (A\nabla v\cdot\nabla w + \rho v w)\,\mathrm{d}x + \int_\Omega f w\,\mathrm{d}x.
$$

Hence, we find that for any $w \in V_0$,

$$
|||u-v|||^2 \geq -2\int_\Omega (A\nabla v\cdot\nabla w + \rho v w)\,\mathrm{d}x + 2\int_\Omega f w\,\mathrm{d}x - |||w|||^2
$$

$$
=: \underline{\mathsf{M}}^2(v,w). \tag{3.35}
$$

If we set $w = u - v$, then

$$
\int_\Omega f w\,\mathrm{d}x = \int_\Omega \big(A\nabla u\cdot\nabla(u-v) + \rho u(u-v)\big)\,\mathrm{d}x,
$$

and (3.35) holds as the equality. This fact means that the minorant $\underline{\mathsf{M}}(v,w)$ is theoretically sharp, and for any v there exists w such that the minorant coincides with the exact error.

Error minorant $\underline{\mathsf{M}}^2(v,w)$ shows a guaranteed lower bound of the error for any function $w \in V_0$. It involves only known data and, therefore, is fully computable. By a proper selection of w, we can find a lower bound arbitrarily close to the exact error.

3.2.3 Particular Cases

If $\varrho = 0$, then the problem (3.18)–(3.20) has the form

$$
-\operatorname{div} A\nabla u = f \quad \text{in } \Omega, \tag{3.36}
$$

$$
u = u_0 \quad \text{on } \Gamma, \tag{3.37}
$$

and estimates (3.29) and (3.33) have simplified forms

$$
\big\|\nabla(u-v)\big\|_A \leq \|A\nabla v - y\|_{A^{-1}} + \mathrm{C}\|f + \operatorname{div} y\| := \overline{\mathsf{M}}_{\operatorname{div} A\nabla}(v,y) \tag{3.38}
$$

and

$$\|\nabla(u-v)\|_A^2 \le C^2(1+\beta)\|f+\operatorname{div} y\|^2 + \frac{1+\beta}{\beta}\|A\nabla v - y\|_{A^{-1}}^2$$
$$:= \overline{\mathsf{M}}^2_{\operatorname{div} A\nabla}(v, y, \beta), \tag{3.39}$$

respectively. For the problem (3.36)–(3.37), the minorant has the form

$$\|\nabla(u-v)\|_A^2 \ge 2\int_\Omega (fw - A\nabla v \cdot \nabla w)\,dx - |||w|||^2 := \underline{\mathsf{M}}^2_{\operatorname{div} A\nabla}(v, w). \tag{3.40}$$

If $A = I$, then we arrive at the Poisson problem

$$\Delta u + f = 0 \quad \text{in } \Omega, \tag{3.41}$$

$$u = u_0 \quad \text{on } \Gamma, \tag{3.42}$$

for which two-sided bounds of the error are presented by the relations

$$\underline{\mathsf{M}}^2_\Delta(v, w) \le \|\nabla(u-v)\|^2 \le \overline{\mathsf{M}}^2_\Delta(v, y, \beta), \quad \forall w \in V_0,\ y \in H(\Omega, \operatorname{div}), \tag{3.43}$$

where β is an arbitrary positive number,

$$\underline{\mathsf{M}}^2_\Delta(v, w) := \int_\Omega \left(2fw - 2\nabla v \cdot \nabla w - |\nabla w|^2\right) dx, \tag{3.44}$$

and

$$\overline{\mathsf{M}}^2_\Delta(v, y, \beta) := C^2_{F\Omega}(1+\beta)\|f+\operatorname{div} y\|^2 + \frac{1+\beta}{\beta}\|\nabla v - y\|_A^2. \tag{3.45}$$

Also, we can use the estimate

$$\|\nabla(u-v)\|_A \le \overline{\mathsf{M}}_\Delta(v, y) := C_{F\Omega}\|f+\operatorname{div} y\| + \|\nabla v - y\|. \tag{3.46}$$

It is easy to prove that the exact lower bound of the majorant $\overline{\mathsf{M}}_\Delta(v, y)$ (and of $\overline{\mathsf{M}}^2_\Delta(v, y, \beta)$) with respect to y is attained on a subspace of $H(\Omega, \operatorname{div})$. Indeed, for any $v \in V_0$ (and any $\beta > 0$) the majorant is convex, continuous, and coercive on $H(\Omega, \operatorname{div})$. By known results in the calculus of variations (e.g., see Theorem B.5), we conclude that a minimizer $\bar{y}(v)$ exists. Since $\overline{\mathsf{M}}^2_\Delta(v, y, \beta)$ is a quadratic functional, the corresponding minimizer $\bar{y}$ (which depends on v and β) is unique.

Lemma 3.1 *Let $\bar{y}$ be such that*

$$\overline{\mathsf{M}}_\Delta(v, \bar{y}) = \inf_{y \in H(\Omega, \operatorname{div})} \overline{\mathsf{M}}_\Delta(v, y). \tag{3.47}$$

There exists $\bar{w} \in V_0$ such that $\bar{y} = \nabla \bar{w}$.

Proof For any y_0 from the set of solenoidal fields $\mathrm{S}(\Omega)$, we have

$$\|\nabla v - \bar{y}\| + C_{F\Omega}\|\operatorname{div}\bar{y} + f\| \leq \|\nabla v - y_0 - \bar{y}\| + C_{F\Omega}\|\operatorname{div}\bar{y} + f\|.$$

From the above, we conclude that for any y_0,

$$\int_\Omega \bar{y}\cdot y_0\,\mathrm{d}x + \frac{1}{2}\|y_0\|^2 \geq 0.$$

This inequality holds if and only if

$$\int_\Omega \bar{y}\cdot y_0\,\mathrm{d}x = 0, \quad \forall y_0 \in \mathrm{S}(\Omega). \tag{3.48}$$

Recall that $\bar{y} \in L^2(\Omega, \mathbb{R}^d)$ admits the decomposition $\bar{y} = \nabla\bar{w} + \tau_0$, where $\bar{w} \in V_0$ and τ_0 is a solenoidal field. Set $y_0 = \tau_0$. From (3.48), it follows that $\|\tau_0\| = 0$. Thus, $\bar{y} = \nabla\bar{w}$. □

Remark 3.1 If Y_k is a sequence of finite dimensional subspaces of $H(\Omega, \operatorname{div})$, which is limit dense in this space, then a sequence of the corresponding minimizers $\{y_k\}$ converges to the exact flux (see [Rep08, RSS03]). In view of this fact, the integrand of $\overline{\mathrm{M}}^2_\Delta(v, y, \beta)$ can be used as an error indicator.

3.2.4 Problems with Mixed Boundary Conditions

Diffusion problems are often considered with the mixed boundary conditions. Consider the problem

$$-\operatorname{div}(A\nabla u) = f \quad \text{in } \Omega, \tag{3.49}$$

$$u = u_0 \quad \text{on } \Gamma_D, \tag{3.50}$$

$$n\cdot A\nabla u + \kappa(x)u = F \quad \text{on } \Gamma_N, \tag{3.51}$$

where $\kappa(x) \geq 0$ and $F \in L^2(\Gamma_N)$. For this case, estimates of the deviation from u can be derived from the integral identity

$$\int_\Omega \left(A\nabla u\cdot\nabla w + \varrho^2 uw\right)\mathrm{d}x + \int_{\Gamma_N}\kappa uw\,\mathrm{d}s$$
$$= \int_\Omega fw\,\mathrm{d}x + \int_{\Gamma_N} Fw\,\mathrm{d}x, \quad \forall w \in V_0 := \left\{w \in H^1(\Omega) \mid w = 0 \text{ on } \Gamma_D\right\} \tag{3.52}$$

by the method discussed in the previous section (see also Sect. 4.2 in [Rep08]).

Let $v \in V_0 + u_0$. Then, (3.52) infers the relation

$$\int_\Omega \big(A\nabla(u-v)\cdot\nabla w + \varrho^2(u-v)w\big)\,dx + \int_{\Gamma_N} \kappa(u-v)w\,ds$$
$$= \int_\Omega \big(fw - \varrho^2 vw - A\nabla v\cdot\nabla w\big)\,dx - \int_{\Gamma_N} (\kappa v - F)w\,ds$$
$$= \int_\Omega \mathcal{R}(v,y)w\,dx + \int_\Omega (y - A\nabla v)\cdot\nabla w\,dx - \int_{\Gamma_N} (y\cdot n + \kappa v - F)w\,ds, \tag{3.53}$$

where $\mathcal{R}(v,y) = f - \varrho^2 v + \operatorname{div} y$. Set $w = u - v$, then (3.53) implies the estimate

$$|||u-v||| \le \left(\|A\nabla v - y\|^2_{A^{-1}} + \left\|\frac{\mathcal{R}(v,y)}{\varrho}\right\|^2 + \left\|\frac{y\cdot n + \kappa v - F}{\sqrt{\kappa}}\right\|^2_{\Gamma_N} \right)^{1/2}, \tag{3.54}$$

where

$$|||w|||^2 := \|\nabla w\|^2_A + \|\varrho w\|^2 + \|\sqrt{\kappa} w\|^2_{\Gamma_N}. \tag{3.55}$$

The right-hand side of (3.54) presents the simplest error majorant for the problem with mixed Dirichlét–Robin (or Dirichlet–Neumann) boundary conditions. However, if κ and/or ϱ vanish (or attain very small values), then (3.54) cannot be applied. To avoid this drawback, we apply the estimates

$$\int_\Omega \mathcal{R}(v,y)w\,dx \le \frac{C_{F\Omega}}{c_1}\|\mathcal{R}(v,y)\|\,\|\nabla w\|_A,$$

and

$$\int_{\Gamma_N} (y\cdot n + \kappa v - F)w\,ds \le \frac{C_{tr}}{c_1}\|y\cdot n + \kappa v - F\|\,\|\nabla w\|_A,$$

where C_F and C_{tr} come from the inequalities (see Sects. A.2.2 and A.3.2)

$$\|w\| \le C_{F\Omega}\|\nabla w\|, \quad \forall w \in V_0, \tag{3.56}$$

and

$$\|w\|_{\Gamma_N} \le C_{tr}\|\nabla w\|, \quad \forall w \in V_0. \tag{3.57}$$

Then, we obtain another upper bound:

$$\big|[u-v]\big| \le |||A\nabla v - y|||_* + \frac{C_{F\Omega}}{c_1}\big\|\mathcal{R}(v,y)\big\| + \frac{C_{tr}}{c_1}\|y\cdot n + \kappa v - F\|_{\Gamma_N}, \tag{3.58}$$

which is not sensitive with respect to small values of ϱ and κ.

Remark 3.2 By combining the methods used for the derivation of (3.54) and (3.58), one can deduce a more general estimate (an analog of (3.30)) valid for the case of mixed Dirichlét–Robin boundary condition.

3.2.5 Estimates of Global Constants Entering the Majorant

We note that the constants $C_{F\Omega}$ and C_{tr} in (3.56) and (3.57) (or suitable upper bounds of them) must be known. If $\Gamma = \Gamma_N$, then an upper bound of $C_{F\Omega}$ is easy to find (e.g., by taking the lowest eigenvalue of the operator Δ in the rectangular domain encompassing Ω; cf. (A.30)).

In general, this problem is equivalent to finding a lower bound of the minimal eigenvalue associated with the corresponding differential operator. It is well-known that upper bounds of the eigenvalues can be computed fairly easily with the help of Rayleigh quotients. However, the problem of finding explicitly computable lower bounds for the minimal eigenvalue of a selfadjoint differential operator in an arbitrary domain is a complicated problem which still awaits a complete solution. Several methods have been suggested to find the lower bound of the minimal positive eigenvalue. One group of methods, (see, e.g., [FW60, Gou57]) is mainly based on various extensions of the set of admissible functions. If the smallest eigenvalue of the extended problem is known (or computable), then we obtain a certain lower bound.

Another group of methods uses the so-called positive (positone) solutions and the following statement (see, e.g., [KC67]).

Theorem 3.1 *Let $Lu = -\operatorname{div} A\nabla u + a_0 u$ be a uniformly elliptic operator with continuously differentiable coefficients and boundary conditions $\alpha u + \beta \nabla u \cdot n = 0$ on Γ, where n is the unit outward normal to Γ. Assume that we have a positive function $\rho(x) \in C(\Omega)$ and a function $\phi(x) \in C^2(\Omega)$ such that $B\phi = 0$ on Γ (where B is the operator of boundary conditions) and*

$$L\phi - \lambda\rho\phi > 0 \quad \text{in } \Omega. \tag{3.59}$$

Then, $\phi(x) > 0$ if and only if $\lambda < \lambda_1$, where λ_1 is the minimal eigenvalue of the problem

$$L\phi - \lambda\rho\phi > 0 \quad \text{in } \Omega, \tag{3.60}$$

$$B\phi = 0 \quad \text{on } \Gamma. \tag{3.61}$$

This theorem opens a way of finding guaranteed minorants of eigenvalues. However, in practice this way is difficult to realize because it is necessary to construct a function in C^2 which simultaneously satisfies (3.59) and the strict positivity condition. For nonconvex domains with piecewise smooth boundaries this may be a very complicated task.

Another method was suggested in [KS78, KS84], where it was shown that if λ_* and u_* are approximations of an eigenvalue and an eigenfunction and w is the exact solution of the problem

$$\Delta w = \Delta u_* + \lambda_* u_* \quad \text{in } \Omega \tag{3.62}$$

with the boundary condition $w = u_*$ of Γ, then

$$\min_l \left| \frac{\lambda_l - \lambda_*}{\lambda_l} \right| \leq \frac{\|w\|_\Omega^2}{\|u_*\|_\Omega^2}. \tag{3.63}$$

Practical application of this method also leads to difficult problems. First, we need the exact solution of the auxiliary problem (3.62). Certainly, there is an obvious idea to bypass this difficulty, using an approximate solution $\widetilde{w}$ of (3.62) instead of w. Since the error $\|\nabla(\widetilde{w} - w)\|$ is controlled by the error majorant (3.45), we can easily obtain a certain lower bound of the eigenvalue. However, there is another condition: we must guarantee that λ_1 is the eigenvalue closest to λ_*. For problems in geometrically complicated (e.g., multiconnected) domains, this condition can hardly be guaranteed, so that the corresponding estimates cannot be considered as fully reliable.

Below we discuss a simple practical algorithm by which a suitable value of the constant $C_{F\Omega}$ can be found numerically with the help of standard minimization methods developed for convex functionals. Certainly, this method also does not generate fully reliable bounds of the constant, but in the vast majority of cases gives approximate bounds, the accuracy of which is quite sufficient for engineering computations.

To find a bound of $C_{F\Omega}$ we consider the functional

$$G_\mu(w) := \int_\Omega \left(|\nabla w|^2 - \mu|w|^2\right) dx.$$

It is clear that if $\mu < \lambda_1$, where $\lambda_1 = 1/C_{F\Omega}^2$, then the functional $G_\mu(w)$ is coercive and nonnegative on V_0. This fact suggests a way of finding approximate values of $C_{F\Omega}$ by minimizing $G_\mu(w)$ on some subspace(s) of V_{0h} (the latter subspaces can be constructed by, e.g., finite element or spectral approximations). In the process of minimization, we can obtain either a sequence of positive numbers (which tends to zero) or a sequence that tends to $-\infty$. In the latter case, on some step G_μ attains negative values which shows that $\mu > \mu_1$.

The algorithm starts with some small $\mu = \mu_0$ and minimizes G_μ on V_{0h} (e.g., with the help of a gradient or relaxation type method). If the minimal value is zero, then we increase μ with the step μ_{inc}. If minimization generates negative values of the functional, then the value of μ_0 must be diminished. We increase the value of μ with smaller steps until the minimization process generates a negative value again. The previous value of μ is taken as an approximation of λ_1. The initial value μ_0 can be taken as $\kappa\widetilde{\lambda}_1$, where $0 < \kappa \ll 1$ and $\widetilde{\lambda}_1$ is a coarse estimate of the first eigenvalue (which is computed, e.g., with the help of Rayleigh quotients). This simple algorithm usually generates quite good values of the constant $C_{F\Omega}$ (which is inverse to $\sqrt{\mu}$). Similar constants arising in functional inequalities associated with other differential operators can be approximately evaluated by the same method. Also, we can use well-known numerical methods for eigenvalues supplied with error indicators and mesh adaptation (see, e.g., [HWZ10]).

Recently, two new methods have been developed. In [KR13, Kuz09], guaranteed lower bounds for eigenvalues are derived within the framework of a domain decomposition method using overlapping domains. A different method based on integration by parts relations for adjoint differential operators is presented in [Rep12].

3.2.6 Error Majorants Based on Poincaré Inequalities

Now we consider another method, which allows us to deduce fully computable and guaranteed error bounds. It uses Poincaré type inequalities for subdomains. For the sake of simplicity, we discuss it in application to the problem

$$\operatorname{div} A\nabla u + f = 0 \quad \text{in } \Omega, \tag{3.64}$$

$$u = u_0 \quad \text{on } \Gamma_D, \tag{3.65}$$

$$\nabla u \cdot n = F \quad \text{on } \Gamma_N. \tag{3.66}$$

A more detailed discussion of this error estimation method and examples related to convection-diffusion, elasticity, and general linear elliptic problems is presented in [Rep08] (see Sects. 3.5.3, 4.3.3, and 7.1.2).

Similarly to previous cases, we use the relation

$$\begin{aligned}\int_\Omega A\nabla(u-v)\cdot\nabla w\,\mathrm{d}x &= \int_\Omega (fw - A\nabla v\cdot\nabla w)\,\mathrm{d}x + \int_{\Gamma_N} Fw\,\mathrm{d}s\\ &= \int_\Omega (\operatorname{div} y + f)w\,\mathrm{d}x + \int_\Omega (y - A\nabla v)\cdot\nabla w\,\mathrm{d}x\\ &\quad - \int_{\Gamma_N} (y\cdot n - F)w\,\mathrm{d}s,\end{aligned} \tag{3.67}$$

which follows from the integral identity and contains a "free" function $y \in H(\Omega, \operatorname{div})$. Let y be selected such that $y \cdot n = F$ on Γ_N (usually this condition is easy to satisfy). Consider the first term on the right-hand side of (3.67). Assume that Ω is decomposed into a set $\mathcal{T}$ of subdomains Ω_i (in particular, Ω_i may coincide with finite elements) with Lipschitz continuous boundaries (see Fig. 3.3), i.e.,

$$\overline{\Omega} = \bigcup_{i=1,\dots,N} \overline{\Omega}_i, \quad \text{and} \quad \Omega_i \cap \Omega_j = \emptyset, \quad \text{if } i \neq j.$$

It is not difficult to see that

$$\begin{aligned}\int_\Omega (\operatorname{div} y + f)w\,\mathrm{d}x &= \sum_{i=1}^N \int_{\Omega_i} (\operatorname{div} y + f)w\,\mathrm{d}x\\ &= \sum_{i=1}^N \Bigg(\int_{\Omega_i} \big(\operatorname{div} y + f - \{\!|\operatorname{div} y + f|\!\}_{\Omega_i}\big) w\,\mathrm{d}x\\ &\quad + \{\!|\operatorname{div} y + f|\!\}_{\Omega_i} \int_{\Omega_i} w\,\mathrm{d}x \Bigg).\end{aligned} \tag{3.68}$$

If we impose the conditions

$$\{\!|\operatorname{div} y + f|\!\}_{\Omega_i} = 0, \quad \forall i = 1, 2, \dots, N, \tag{3.69}$$

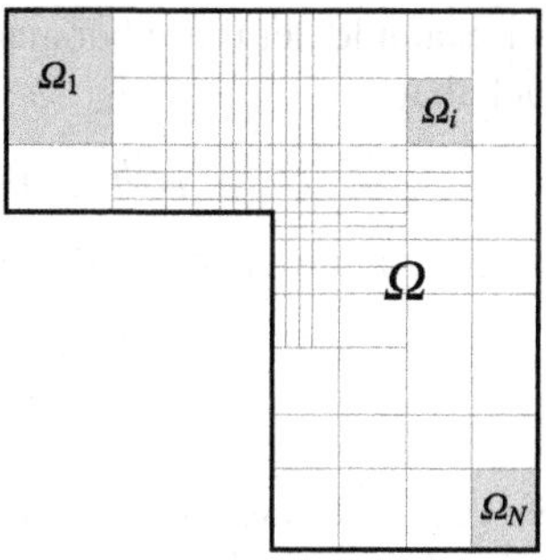

Fig. 3.3 Decomposition of Ω into subdomains

then

$$\int_\Omega (\operatorname{div} y + f) w \, \mathrm{d}x = \sum_{i=1}^{N} \int_{\Omega_i} (\operatorname{div} y + f) w \, \mathrm{d}x$$
$$= \sum_{i=1}^{N} \int_{\Omega_i} (\operatorname{div} y + f)\big(w - \{\!|w|\!\}_{\Omega_i}\big) \, \mathrm{d}x. \tag{3.70}$$

We recall that by (A.19),

$$\big\| w - \{\!|w|\!\}_{\Omega_i} \big\|_{\Omega_i} \leq C_{P\Omega_i} \|\nabla w\|_{\Omega_i}. \tag{3.71}$$

From (3.70) and (3.71), we deduce the estimate

$$\int_\Omega (\operatorname{div} y + f) w \, \mathrm{d}x \leq \sum_{i=1}^{N} C_{P\Omega_i} \|\operatorname{div} y + f\|_{\Omega_i} \|\nabla w\|_{\Omega_i}$$
$$\leq \left(\sqrt{\sum_{i=1}^{N} \|\operatorname{div} y + f\|_{\Omega_i}^2 C_{P\Omega_i}^2} \right) \|\nabla w\|. \tag{3.72}$$

The term $\int_\Omega (y - A\nabla v) \cdot \nabla w \, \mathrm{d}x$ is estimated by (3.24). We set $w = u - v$, use (3.67) and obtain

$$\big\|\nabla(u - v)\big\|_A \leq \|A\nabla v - y\|_{A^{-1}} + \frac{1}{\sqrt{c_1}} \sqrt{\sum_{i=1}^{N} C_{P\Omega_i}^2 \|\operatorname{div} y + f\|_{\Omega_i}^2}. \tag{3.73}$$

Instead of the constant $C_{F\Omega}$, this estimate involves constants $C_{P\Omega_i}$ associated with the subdomains Ω_i.

Consider a special but important case, where Ω_i are convex. Then, $C_{P\Omega_i}$ is estimated from the above by diam $\Omega_i \pi^{-1}$ and we obtain

$$\big\|\nabla(u - v)\big\|_A \leq \|A\nabla v - y\|_{A^{-1}} + \frac{1}{\sqrt{c_1}} \sqrt{\sum_{i=1}^{N} \frac{(\operatorname{diam} \Omega_i)^2}{\pi^2} \|\operatorname{div} y + f\|_{\Omega_i}^2}. \tag{3.74}$$

For example, let the subdomains be associated with a simplicial decomposition $\mathcal{T}_h$ such that

$$\mu_1 h \leq \operatorname{diam} T_i \leq \mu_2 h, \quad \forall i = 1, 2, \ldots, N, \tag{3.75}$$

for any simplex $T_i \in \mathcal{T}_h$. Define the constant

$$C_{\max P} = \max_i \{C_{P\Omega_i}\} = \frac{\mu_2 h}{\pi}.$$

For regular triangulations, the constant $C_{\max P}$ is of the same order as all other constants $C_{P\Omega_i}$, so that without a significant overestimation we can replace all these constants by $C_{\max P}$. Then, we arrive at the error majorant

$$\|\nabla(u - v)\|_A \leq \|A\nabla v - y\|_{A^{-1}} + \frac{\mu_2 h}{\pi\sqrt{c_1}} \|\operatorname{div} y + f\|. \tag{3.76}$$

We recall that such type estimates provide guaranteed upper bounds provided that the conditions (3.69) are satisfied together with the Neumann condition $y \cdot n = F$ on Γ_N. If N is not large, then this integral balancing of the flux can be performed fairly easily.

However, if $\Omega_i = T_i$, where T_i are finite elements, then N may be very large. In this case, exact satisfaction of all conditions (3.69) may generate a technical problem. For relatively simple elliptic problem (3.64)–(3.66), these difficulties can be overcame within the framework of the dual mixed method (see below). However, in general the satisfaction of a large amount of integral type conditions may be an obstacle. One way of solving it is discussed below.

Assume (for the sake of simplicity only) that $\Gamma = \Gamma_D$. Then, the desired flux y can be constructed by the dual mixed method (see Sect. B.4.4.2), which gives a pair of functions $(\widehat{u}_h, \widehat{p}_h) \in \widehat{V}_h \times \widehat{Q}_{Fh}$, satisfying the system

$$\int_\Omega \left(A^{-1}\widehat{p}_h \cdot \widehat{q}_h + \widehat{u}_h \operatorname{div}\widehat{q}_h\right) \mathrm{d}x = g(u_0, \widehat{q}_h), \quad \forall \widehat{q}_h \in \widehat{Q}_{0h}, \tag{3.77}$$

$$\int_\Omega (\operatorname{div}\widehat{p}_h + f)\widehat{v}_h \,\mathrm{d}x = 0, \qquad \forall \widehat{v}_h \in \widehat{V}_h. \tag{3.78}$$

Here V_h contains piecewise constant functions ($p_h = const$ on T_h), and Q_h is constructed with the help of Raviart–Thomas elements of the lowest order. We assume that the finite element mesh satisfies (3.75). From (3.78) we find that

$$\int_{T_h} (\operatorname{div}\widehat{p}_h + f)\,\mathrm{d}x = 0, \quad \forall T_h.$$

This means that $\widehat{p}_h$ satisfies (3.69) (if we identify Ω_i with elements). We use $\widehat{p}_h$ instead of y, apply (3.76) and deduce the estimate

$$\|\nabla(u - v)\|_A \leq \|A\nabla v - \widehat{p}_h\|_{A^{-1}} + \frac{\mu_2 h}{\pi} \|\operatorname{div}\widehat{p}_h + f\|, \tag{3.79}$$

which is valid for any conforming approximation $v \in V_0 + u_0$. Obviously, $\widehat{u}_h$ cannot be considered as a conforming approximation. Let $G_h : \widehat{V}_h \to V_0 + u_0$ be a smoothing operator. It is easy to construct such an operator using, e.g., the methods discussed in the context of gradient-averaging. Then, we obtain the estimate

$$\left\| \nabla(u - G_h \widehat{u}_h) \right\|_A \le \| A \nabla G_h \widehat{u}_h - \widehat{p}_h \|_{A^{-1}} + \frac{\mu_2 h}{\pi} \| \operatorname{div} \widehat{p}_h + f \|, \tag{3.80}$$

which gives a computable upper bound for the dual mixed approximation.

3.2.7 Estimates with Partially Equilibrated Fluxes

If we manage to find a vector-valued function y_f such that $\operatorname{div} y_f + f = 0$ and y_f is close (in L^2) to the exact flux p, then the estimates (3.39), (3.58), and (3.74) are reduced to the hypercircle estimate

$$\left\| \nabla(u - v) \right\|_A \le \| A \nabla v - y_f \|_{A^{-1}}. \tag{3.81}$$

However, even in this rather simple elliptic problem, getting an exactly equilibrated flux close to the exact one may not be a simple task. In more difficult problems (linear elasticity, models with convection and diffusion, nonlinear elliptic equations) such requirements are too demanding to be satisfied in real life computations. Nevertheless, various procedures have been invented in order to construct partially equilibrated (balanced) fields (see, e.g., [LL83]). Being used in (3.81) they lead to error indicators (which are often very efficient). Such procedures can be easily used in error majorants, e.g., in (3.39), where they (unlike (3.81)) result in guaranteed upper bounds of errors. In this section, we briefly discuss estimates of this type.

Assume that we have a vector-valued function $y_{\bar{f}}$ such that

$$\operatorname{div} y_{\bar{f}} + \bar{f} = 0, \tag{3.82}$$

where $\bar{f}$ is close to f in L^2-norm. In this case, we say that $y_{\bar{f}}$ is almost (or partially) equilibrated. Set $y = y_{\bar{f}} + \tau_0$, where $\tau_0 \in \mathrm{S}(\Omega)$. Then, $\operatorname{div} y + \bar{f} = 0$. Using y in (3.39), we arrive at the estimate

$$\left\| \nabla(u - v) \right\|_A \le \| A \nabla v - \tau_0 - y_{\bar{f}} \|_{A^{-1}} + \mathrm{C} \| f - \bar{f} \|. \tag{3.83}$$

In particular, if we represent τ_0 in the form $\tau_0 = \operatorname{curl} \eta$ (where η is an arbitrary vector-valued function in $H(\Omega, \operatorname{curl})$), then we obtain

$$\left\| \nabla(u - v) \right\|_A \le \| \tau - \operatorname{curl} \eta \|_{A^{-1}} + \mathrm{C} \| f - \bar{f} \|, \tag{3.84}$$

where $\tau = \nabla v - y_{\bar{f}}$ is the given vector-valued function. This estimate has practical sense, provided that $\| f - \bar{f} \|$ is significantly smaller than the tolerance level accepted for approximations. Then, finding a sharp upper bound is reduced to the problem

$$\min_{\eta \in H(\Omega,\mathrm{curl})} \|\tau - \mathrm{curl}\,\eta\|_{A^{-1}}, \tag{3.85}$$

which can be solved numerically using a suitable finite dimensional subspace for η of $H(\Omega, \mathrm{curl})$.

A modification of (3.74) follows similar ideas. Let $\bar{f}$ denote the function defined by mean values of f on the subdomains Ω_i, i.e., $f_i(x) = \{\!|f|\!\}_{\Omega_i}$ if $x \in \Omega_i$. Assume that we have $y_{\bar{f}}$ satisfying (3.82) and such that

$$\{\!|\mathrm{div}\, y_{\bar{f}} + \bar{f}|\!\}_{\Omega_i} = 0, \quad i = 1, 2, \dots, N. \tag{3.86}$$

Then,

$$\big\|\nabla(u - v)\big\|_A \le \|\nabla v - \tau_0 - y_{\bar{f}}\|_{A^{-1}} + e(\bar{f}), \tag{3.87}$$

where

$$e(\bar{f}) := \frac{1}{\sqrt{c_1}} \Big(\sum_i C_P^2(\Omega_i) \|f - f_i\|^2_{\Omega_i}\Big)^{1/2}.$$

Since the term $e(\bar{f})$ is easily computable, minimization of the error majorant is reduced to (3.85). In particular, we can apply this estimate to approximations of u and p computed by the dual mixed method (see Sect. B.4.4).

It is easy to show that $e(\bar{f})$ represents the error generated by local averaging of f. Indeed, let $\bar{u}$ be the exact solution of the problem with $\bar{f}$. Then,

$$\int_\Omega A\nabla(u - \bar{u}) \cdot \nabla w \,\mathrm{d}x = \int_\Omega (fw - A\nabla\bar{u} \cdot \nabla w)\,\mathrm{d}x = \int_\Omega (f - \widehat{f})w \,\mathrm{d}x.$$

Hence,

$$\big\|\nabla(u - \bar{u})\big\|_A^2 = \sum_{i=1}^N \int_{\Omega_i} (f - f_i)(u - \bar{u})\,\mathrm{d}x.$$

Since

$$\int_{\Omega_i} (f - f_i)\,\mathrm{d}x = 0, \quad \forall i = 1, 2, \dots, N,$$

we find that

$$\sum_{i=1}^N \int_{\Omega_i} (f - f_i)(u - \bar{u})\,\mathrm{d}x = \sum_{i=1}^N \int_{\Omega_i} (f - f_i)(u - \bar{u} - c_i)\,\mathrm{d}x,$$

where c_i are arbitrary constants. We set $c_i = \{\!|u - \bar{u}|\!\}_{\Omega_i}$ and apply the estimate (A.23), which yields

$$\|u - \bar{u} - c_i\|_{\Omega_i} \leq C_{P\Omega_i} \left\|\nabla(u - \bar{u})\right\|_{\Omega_i} \leq \frac{C_{P\Omega_i}}{\sqrt{c_1}} \left\|\nabla(u - \bar{u})\right\|_{A,\Omega_i}.$$

Then,

$$\begin{aligned}\left\|\nabla(u - \bar{u})\right\|_A^2 &\leq \frac{1}{c_1} \sum_{i=1}^{N} C_{P\Omega_i} \|f - f_i\|_{\Omega_i} \left\|\nabla(u - v)\right\|_{A,\Omega_i} \\ &\leq \frac{1}{c_1} \left(\sum_{i=1}^{N} C_{P\Omega_i}^2 \|f - f_i\|_{\Omega_i}^2 \right)^{1/2} \left\|\nabla(u - v)\right\|_A\end{aligned}$$

and, therefore,

$$\left\|\nabla(u - \bar{u})\right\|_A^2 \leq e(\bar{f}). \tag{3.88}$$

If all the subdomains are convex, then $e(\bar{f})$ is an easily computable quantity, which can be computed a priori (in principle, we can view $e(\bar{f})$ as a modeling error generated by simplification of f). Depending on the desired accuracy ε, we may have two different situations.

If $\varepsilon \gg e(\bar{f})$, then the boundary value problem with $\bar{f}$ can be efficiently used instead of the original one. In the context of finite element approximations, the value of $e(\bar{f})$ on a particular mesh is easy to compute. Since this quantity is proportional to h, we can always detect when the mesh is so fine that we can ignore local oscillations of f within the accepted tolerance level.

On the other hand, if $e(\bar{f})$ is of the order ε or larger, then the estimate (3.87) cannot estimate the error efficiently because it contains an irremovable gap (which value is unacceptably large). In this case, we advise the use of the basic estimates (3.38) or (3.74).

3.3 Error Control Algorithms

Error majorants $\overline{\mathrm{M}}_1(v, y)$, $\overline{\mathrm{M}}_2(v, y)$, $\overline{\mathrm{M}}_{\hat{\alpha}}(v, y)$, $\overline{\mathrm{M}}_{\mathrm{div}\,A\nabla}(v, y)$ and analogous majorants for other elliptic boundary value problems can be used in two different ways:

(a) Finding sharp error bounds by minimization of the majorant.
(b) Getting a preliminary (coarse) error bound with minimal expenditures.

If we really need a profound investigation of errors encompassed in the numerical solution (in terms of global and local norms), then it is necessary to select the method (a). In this case, error control is reduced to a quadratic type minimization problem, the efficient solution of which requires special methods (some of them are discussed below).

If we need only preliminary knowledge on the accuracy of a computed solution, then we may apply computationally light procedures and obtain coarse (but guaranteed) error bounds within the framework of (b). Below we discuss both ways.

3.3.1 Global Minimization of the Majorant

The sharpest estimates of the error can be obtained if the majorant is minimized with respect to y over a certain subspace $Y_\tau \subset H(\Omega, \mathrm{div})$. We discuss this method with the paradigm of the Poisson problem (3.64)–(3.66). In general, Y_τ may be constructed using a mesh $\mathfrak{T}_\tau$ that differs from $\mathfrak{T}_h$. Then,

$$\|\nabla(u - u_h)\| \leq \inf_{y_\tau \in Y_\tau} \left\{ \|\nabla u_h - y_\tau\| + C_{F\Omega} \| \operatorname{div} y_\tau + f \| \right\}.$$

The wider Y_τ is, the sharper the upper bound obtained. A detailed discussion of the minimization methods and numerical results can be found in [FNR03, NR04, Rep99b, RSS03] and some other publications cited therein. We emphasize that the motivation to spend resources for the minimization with respect to y is not solely to find a sharp upper bound for the error, but to obtain a good approximation of the flux itself. Typically, reconstructions of the flux obtained by this method are close to the best possible on a given mesh (generating the subspace Y_τ).

If we intend to define y_τ by minimization of the majorant, then it is preferable to represent the problem in the quadratic form:

$$\min_{\beta > 0} \min_{y_\tau \in Y_\tau} \overline{\mathsf{M}}^2_\Delta(v, y_\tau; \beta),$$

where

$$\overline{\mathsf{M}}^2_\Delta(v, y_\tau; \beta) = (1 + \beta)\|\nabla v - y_\tau\|^2 + \left(1 + \frac{1}{\beta}\right) C^2_{F\Omega} \| \operatorname{div} y_\tau + f \|^2.$$

We recall that two terms of the error majorant are related to the decomposed form of the equation. The first part evaluates violations of the relation (3.18) where $A = I$. We denote it

$$\overline{\mathsf{M}}^{\mathrm{D}}_\Delta := \|y - \nabla v\|^2. \tag{3.89}$$

The second part represents the error in the balance equation, i.e., (3.18), where $\varrho = 0$. We denote this part by

$$\overline{\mathsf{M}}^{\mathrm{Eq}}_\Delta(y_\tau) := \| \operatorname{div} y_\tau + f \|^2. \tag{3.90}$$

Therefore, in the process of minimization we also obtain information about these two physically meaningful parts of the error.

If the majorant is minimized with respect to the positive scalar β, then the minimum is attained at

$$\beta_{\min} := \left(\frac{C^2_{F\Omega} \| f + \operatorname{div} y \|^2}{\| y - \nabla v \|^2} \right)^{1/2}.$$

If β is fixed, then the necessary condition for the minimizer y can be computed as follows. Consider the variation of $\overline{\mathsf{M}}_\Delta$, namely,

$$\overline{\mathsf{M}}^2_\Delta(v, y + t\mu; \beta) = \left(1 + \frac{1}{\beta}\right) C^2_{F\Omega} \| f + \operatorname{div} y + t \operatorname{div} \mu \|^2_{\mathcal{V}} + (1+\beta) \| y + t\mu - \nabla v \|^2,$$

where $\mu \in H(\operatorname{div}, \Omega)$. It is easy to see that

$$\frac{1}{2} \frac{\mathrm{d}\overline{\mathsf{M}}^2_\Delta(v, y + t\mu; \beta)}{\mathrm{d}t} = \left(1 + \frac{1}{\beta}\right) C^2_{F\Omega} \int_\Omega (f + \operatorname{div} y + t \operatorname{div} \mu) \operatorname{div} \mu \, \mathrm{d}x$$
$$+ (1+\beta) \int_\Omega (y + t\mu - \nabla v) \cdot \mu \, \mathrm{d}x$$

and the condition

$$\left. \frac{\mathrm{d}\overline{\mathsf{M}}^2_\Delta(v, y + t\mu; \beta)}{\mathrm{d}t} \right|_{t=0} = 0$$

means that

$$C^2_{F\Omega} \int_\Omega \operatorname{div} y \operatorname{div} \mu \, \mathrm{d}x + \beta \int_\Omega y \cdot \mu \, \mathrm{d}x$$
$$= -C^2_{F\Omega} \int_\Omega f \operatorname{div} \mu \, \mathrm{d}x + \beta \int_\Omega \nabla v \cdot \mu \, \mathrm{d}x. \tag{3.91}$$

Assume that $y \in \operatorname{span}\{\phi^1, \phi^2, \ldots, \phi^N\} =: Y_\tau \subset H(\operatorname{div}, \Omega)$, i.e., $y = \sum_{i=1}^N \gamma_i \phi^i$. The condition (3.91) leads to a system of linear equations,

$$\sum_{i=1}^N \gamma_i \left(C^2_{F\Omega} \int_\Omega \operatorname{div} \phi^i \operatorname{div} \phi^j \, \mathrm{d}x + \beta \int_\Omega \phi^i \cdot \phi^j \, \mathrm{d}x \right)$$
$$= -C^2_{F\Omega} \int_\Omega f \operatorname{div} \phi^j \, \mathrm{d}x + \beta \int_\Omega \nabla v \cdot \phi^j \, \mathrm{d}x, \quad j = \{1, \ldots, N\}.$$

Let

$$\{S_{ij}\}_{i,j=1}^N = \int_\Omega \operatorname{div} \phi^i \operatorname{div} \phi^j \, \mathrm{d}x, \qquad \{K_{ij}\}_{i,j=1}^N = \int_\Omega \phi^i \cdot \phi^j \, \mathrm{d}x, \tag{3.92}$$

$$\{z_j\}_{j=1}^N = -\int_\Omega f \operatorname{div} \phi^j \, \mathrm{d}x, \quad \text{and} \quad \{g_j\}_{j=1}^N = \int_\Omega \nabla v \cdot \phi^j \, \mathrm{d}x. \tag{3.93}$$

Then, the system can be written in the matrix form

$$\left(\frac{C_{F\Omega}^2}{c_1}S+\beta K\right)\gamma=\frac{C_{F\Omega}^2}{c_1}z+\beta g, \tag{3.94}$$

where γ is the vector of the unknown coefficients. We define γ and compute the corresponding value of the majorant.

$$\overline{\mathsf{M}}_{\Delta}^2(v,y,\beta)=\left(1+\frac{1}{\beta}\right)C_{F\Omega}^2\big(\gamma^T S\gamma-2\gamma^T z+\|f\|^2\big)$$
$$+(1+\beta)\big(\gamma^T K\gamma-2\gamma^T g+\|\nabla v\|^2\big).$$

These observations motivate Algorithm 3.2, in which the majorant is minimized by means of an iteration procedure. It begins with assigning β a certain value (e.g., one). Then, the majorant is minimized with respect to y (which amounts to solving (3.94)). Using this solution, we recompute both parts of the majorant and find new β by minimizing the majorant with respect to this parameter. Then, the process is repeated. In the algorithm, the amount of iteration steps is limited by the number $I_{\max}$. In practice, other stopping criteria can be used (e.g., iterations are terminated if relative changes between the values of the majorant on two consequent steps become insignificant or if the value of the majorant has become smaller than some predefined tolerance level). Numerical experiments show that usually five or six iteration steps are quite enough to obtain a very good approximation of the minimizer $y\in Y_\tau$. Moreover, the required number of iterations is independent of N.

We recall that $Y_\tau\subset H(\Omega,\mathrm{div})$ can be constructed by standard piecewise affine approximations of vector-valued functions. This is well motivated if $\mathcal{T}_\tau$ coincides with the mesh $\mathcal{T}_h$ and v is a finite element approximation computed on this mesh.

Of course, any conforming subspaces of $H(\Omega,\mathrm{div})$ and V_0 can be used. One can use, e.g., higher order polynomials, local mesh refinements, etc. Moreover, all the standard methods used to improve the quality of approximation and to accelerate the solution process can be applied to (3.91), e.g., domain decomposition, multigrid methods (see [Val09]), and isogeometric elements (NURBS, Non-Uniform Rational B-spline; see [KT13]).

Example 3.3 Consider the model problem

$$-\mathrm{div}(A\nabla u)=f \quad \text{in } \Omega, \tag{3.95}$$

$$u=0 \quad \text{on } \Gamma, \tag{3.96}$$

where

$$\Omega:=\big((-1,1)\times(0,1)\big)\setminus\big([-0.5,0.5]\times[0,0.5]\big). \tag{3.97}$$

We set $A=I$ and consider the problem with the solution

$$u(x_1,x_2)=x_2(x_1-1)(x_1+1)(x_1-0.5)(x_1+0.5)(x_2-1)(x_2-0.5).$$

Algorithm 3.2 Global minimization of the majorant

Input: v {Approximate solution}
ϕ_i {Basis functions}
$I_{\max}$ {Number of iteration steps}

Compute $\|f\|^2$ and $\|\nabla v\|^2$.
Assemble matrices S and K, and vectors z and g as in (3.92)–(3.93)
$\beta_1 = 1$
for $k = 1$ **to** $I_{\max}$ **do**
Solve the system

$$\left(\frac{C^2_{F\Omega}}{c_1} S + \beta_k K\right) \gamma_{k+1} = \frac{C^2_{F\Omega}}{c_1} z + \beta_k g.$$

Compute two parts of the majorant:

$$\overline{\mathsf{M}}^{\mathrm{Eq}}_{\Delta} = \gamma^T_{k+1} S \gamma_{k+1} - 2y^T_{k+1} z + \|f\|^2$$

$$\overline{\mathsf{M}}^{\mathrm{D}}_{\Delta} = \gamma^T_{k+1} K \gamma_{k+1} - 2y^T_{k+1} g + \|\nabla v\|^2$$

Compute new value of β:

$$\beta_{k+1} = \left(\frac{C^2_{F\Omega} \overline{\mathsf{M}}^{\mathrm{Eq}}_{\Delta}}{\overline{\mathsf{M}}^{\mathrm{D}}_{\Delta}}\right)^{1/2}$$

end for
$\overline{\mathsf{M}}^2_{\Delta} = (1 + \frac{1}{\beta}) C^2_F \overline{\mathsf{M}}^{\mathrm{Eq}}_{\Delta} + (1 + \beta) \overline{\mathsf{M}}^{\mathrm{D}}_{\Delta}$ {Compute the majorant}
$y = \sum_{i=1}^N \gamma_i \phi_i$ {Find the flux}

Output: $\overline{\mathsf{M}}^2_{\Delta}$ {Guaranteed upper bound of the error}
y {Reconstruction of the flux}

The approximation v is computed on an initial mesh, which generates the space $V^0_0 \subset V_0$ of piecewise affine functions (i.e., the space generated by first order Courant elements) with $\dim(V^0_0) = 280$. Spaces $V^k_0 \subset V_0$ are generated by mesh refinements. The subspaces $Y^k \subset H(\Omega, \mathrm{div})$ are created by means of the lowest order Raviart–Thomas elements using the same mesh as for V^k_0.

The resulting upper bound is

$$\overline{\mathsf{M}}_{\mathrm{div}\, A\nabla}\big(v, y^k_{\mathrm{glo}}\big) = \inf_{y \in Y^k} \overline{\mathsf{M}}_{\mathrm{div}\, A\nabla}(v, y),$$

and the lower bound is computed in accordance with the relation

$$|\!|\!| u - v |\!|\!|^2 \geq 2\big(J(v) - J(w)\big) =: \underline{\mathsf{M}}^2_{\mathrm{div}\, A\nabla}(v, w), \quad \forall w \in V_0.$$

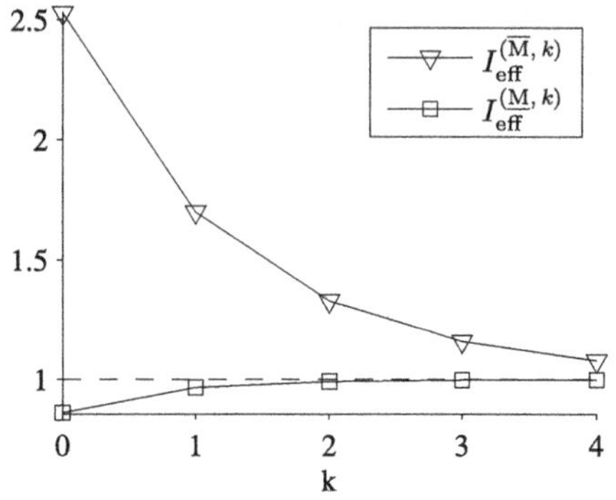

Fig. 3.4 Efficiency of the majorant and minorant, when y and w are found in consequently refined subspaces

The approximate solution of the original problem on V_0^{k+1} we denote by w_{glo}^{k+1}. In Fig. 3.4 the corresponding efficiency indexes

$$I_{\text{eff}}^{(\overline{\mathsf{M}},k)} := \frac{\overline{\mathsf{M}}_{\text{div}\,A\nabla}(v, y_{\text{glo}}^k)}{|||u-v|||} \quad \text{and} \quad I_{\text{eff}}^{(\underline{\mathsf{M}},k)} := \frac{\underline{\mathsf{M}}_{\text{div}\,A\nabla}(v, w_{\text{glo}}^{k+1})}{|||u-v|||}$$

are depicted. It is easy to see that both of them tend to one, as the subspaces for y and w are improved.

Other details are shown in Table 3.3. This example illustrates the "sharpness property" of the majorant and minorant, which can be proved theoretically (see, e.g., [NR04]). In other words, it shows that the majorant (minorant) converges to the exact error from the above (below), if we increase computational efforts.

3.3.2 Getting an Error Bound by Local Procedures

Let $V_{0h} \subset V_0$ be a finite dimensional space constructed with the help of finite element approximations. For example, V_{0h} may contain piecewise affine finite element approximations generated by the triangulation $\mathfrak{T}_h$. Assume that $v_h \in V_{0h}$ is an approximate solution of the problem $\Delta u + f = 0$ with the boundary condition $u = u_0$.

Table 3.3 Refinement of subspaces of y and w

	k				
	0	1	2	3	4
$\dim(V_0^{k+1})$	1047	4045	15897	63025	250977
$\dim(Y^k)$	767	2998	11852	47128	187952
$\overline{\mathsf{M}}_{\text{div}\,A\nabla}(v, y_{\text{glo}}^k)$	0.0229	0.0154	0.0120	0.0105	0.0097
$\underline{\mathsf{M}}_{\text{div}\,A\nabla}(v, w_{\text{glo}}^{k+1})$	0.0078	0.0087	0.0090	0.0090	0.0090
$I_{\text{eff}}^{(\overline{\mathsf{M}},k)}$	2.5282	1.6982	1.3283	1.1586	1.0779
$I_{\text{eff}}^{(\underline{\mathsf{M}},k)}$	0.8621	0.9671	0.9918	0.9980	0.9995

In particular, v_h may coincide with the *Galerkin approximation* $u_h \in V_{0h} + u_0$ defined by the relation

$$\int_\Omega \nabla u_h \cdot \nabla w_h \, dx = \int_\Omega f w_h \, dx, \quad \forall w_h \in V_{0h}. \tag{3.98}$$

Also, it may be any other approximation which differs from u_h owing to the presence of a numerical integration, roundoff, or other errors. Using v_h, we find a rough approximation of the flux

$$p_h := \nabla v_h \in L^2(\Omega, \mathbb{R}^d). \tag{3.99}$$

Generally, p_h does not belong to $H(\Omega, \mathrm{div})$, and we cannot directly substitute $y = p_h$ in the majorant $\overline{\mathrm{M}}_\Delta(u_h, y)$. For this reason, it is necessary to regularize p_h by a post-processing operator $\mathrm{G}_h : L^2(\Omega, \mathbb{R}^d) \to H(\Omega, \mathrm{div})$. After that, we obtain a vector-valued function $\mathrm{G}_h p_h$, which yields an easily computable estimate

$$\|\nabla(u - u_h)\| \le \|\nabla u_h - \mathrm{G}_h p_h\| + C_{F\Omega} \| \mathrm{div}\, \mathrm{G}_h p_h + f \|. \tag{3.100}$$

The quality of the upper bound given by (3.100) depends on the properties of the post-processing operator used. In Sect. 2.2.2, we have discussed main classes of post-processing (gradient-averaging) operators. Any of them can be applied to improve p_h. In particular, we recommend gradient-averaging based on low order Raviart–Thomas elements.

If the value of the term $\| \mathrm{div}\, \mathrm{G}_{\mathrm{RT}} p_h + f \|$ is too large (in comparison with the term $\|\nabla u_h - \mathrm{G}_{\mathrm{RT}} p_h\|$), then we can apply partial minimization of the majorant in order to reduce it (e.g., with the help of a relaxation procedure, which uses normal fluxes on edges as free parameters). However, in general, substituting a post-processed gradient does not give a very accurate upper bound. Numerical experiments have shown that if $\mathrm{G}_h p_h$ is constructed with the help of simple patch-averaging on the same mesh, then the upper bound given by the right-hand side of (3.100) is rather coarse. More sophisticated averaging procedures or additional post-processing usually lead to better estimate.

Example 3.4 We consider the problem (3.95)–(3.97). Let $v \in V_0^h \subset V_0$ be an approximation computed by piecewise affine Courant type elements. The corresponding numerical flux

$$y_0 := A \nabla v$$

is a piecewise constant vector-valued function. One way is to use the patch-averaging procedure (see Sect. 2.2.2.1) and compute

$$y_G := \mathrm{G}_h y_0.$$

Another option is to use edge-averaging and Raviart–Thomas elements (see Sect. 2.2.2.3) and compute

$$y_{\mathrm{RT}}^0 := \mathrm{G}_{\mathrm{RT}} y_0.$$

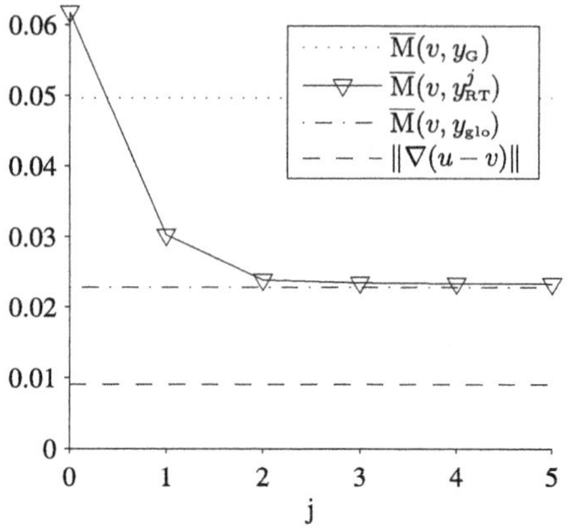

Fig. 3.5 Development of the majorant during the iteration of approximate flux

We apply the iterative quasi-equilibration procedure, denoted by the operator P_{RT} (one cycle of local equilibration treating patches related to every edge). The procedure is described in Sect. 2.2.2.4. Subsequent iterations produce functions

$$y^j_{RT} := P^j_{RT} y^0_{RT}, \quad j = \{1, 2, 3, \ldots\}. \tag{3.101}$$

Each approximate flux y generates the respective upper bound $\overline{M}_{\mathrm{div}\,A\nabla}(v, y)$. The resulting upper bounds are compared with the globally minimized value

$$\overline{M}_{\mathrm{div}\,A\nabla}(v, y_{\mathrm{glo}}) = \inf_{y \in Y^h} \overline{M}_{\mathrm{div}\,A\nabla}(v, y) \tag{3.102}$$

and the exact error in Fig. 3.5. After several iterations of local minimization we can obtain practically the same value as by global minimization of the majorant.

In Table 3.4, the efficiency indexes of the upper bounds are depicted, i.e.,

$$I^{(j)}_{\mathrm{eff}} := \frac{\overline{M}_{\mathrm{div}\,A\nabla}(v, y^j_{RT})}{\|\nabla(u-v)\|} \quad \text{and} \quad I^{(\mathrm{glo})}_{\mathrm{eff}} := \frac{\overline{M}_{\mathrm{div}\,A\nabla}(v, y_{\mathrm{glo}})}{\|\nabla(u-v)\|}.$$

In these tests, approximate solutions are piecewise affine finite element approximations with 280, 26118, and 53383 degrees of freedom, respectively. In all of these cases, after sufficient minimization cycles the globally minimized majorant and the one improved by the local iteration cycles have almost the same value.

Table 3.4 Efficiency indexes of majorant for different approximations

# nodes	Iteration, $I^{(j)}_{\mathrm{eff}}$										$I^{(\mathrm{glo})}_{\mathrm{eff}}$
	1	2	3	4	5	7	9	14	19	31	
280	6.83	3.34	2.64	2.59	2.59	2.58	2.58	2.58	2.58	2.58	2.53
26118	43.2	12.2	5.07	3.53	3.05	2.74	2.64	2.57	2.55	2.53	2.49
53383	60.8	16.9	6.62	4.26	3.49	2.94	2.75	2.61	2.57	2.54	2.49

Moreover, the efficiency index of the globally minimized majorant does not depend on the size of the mesh generating V_0^h. Instead, it depends on the relation between spaces V_0^h and RT_0 and the problem considered.

Remark 3.3 We note that the method of minimizing the upper bound presented here is only one of many options. In particular, all other post-processing methods discussed in this book (as well as in others) can be used together with error majorants in order to generate computable and reliable bounds of approximation errors.

3.4 Indicators Based on Error Majorants

Guaranteed upper bounds of errors considered in previous sections imply new error indicators that can be used in marking and mesh refinement procedures. We discuss this subject with the paradigm of the problem (3.64)–(3.66). We are interested in approximation of the error function $\mathcal{E}(x) := A\nabla(u-v)\cdot\nabla(u-v)$. The majorant

$$\overline{\mathsf{M}}_{\operatorname{div} A}(v, y) := \|A\nabla v - y\|_{A^{-1}} + \mathrm{C}\|f + \operatorname{div} y\|$$

indeed suggests a way of finding a function close to $\mathcal{E}(x)$. It is based on the following argumentation. Let y_τ be a vector-valued function found by minimization of $\overline{\mathsf{M}}_{\operatorname{div} A}(v, y)$ with respect to y on a certain finite dimensional space Y_τ. An efficient minimization procedure leads to a situation in which the first term of the majorant dominates and contains the major part of the error. Then, it is natural to use the function

$$\boldsymbol{E}_{\overline{\mathsf{M}}}(v, y_\tau) = A\nabla v\cdot\nabla v + A^{-1}y_\tau\cdot y_\tau - 2\nabla v\cdot y_\tau \tag{3.103}$$

as an indicator of $\mathcal{E}(x)$. Since

$$\begin{aligned}\mathcal{E}(x) - \boldsymbol{E}_{\overline{\mathsf{M}}}(v, y_\tau) &= A\nabla u\cdot\nabla u - A^{-1}y_\tau\cdot y_\tau + 2\nabla v\cdot(y_\tau - A\nabla u)\\ &= A^{-1}p\cdot p - A^{-1}y_\tau\cdot y_\tau + 2\nabla v\cdot(y_\tau - p),\end{aligned} \tag{3.104}$$

we see that the indicator $\boldsymbol{E}_{\overline{\mathsf{M}}}(v, y_\tau)$ is close to $\mathcal{E}(x)$ if y_τ is close to p.

In the classification of Chap. 2, this indicator belongs to the group (B) because it is generated by the relation

$$\|\nabla(u-v)\|_A^2 = \sup_{w\in V_0}\left\{-\|\nabla w\|_A^2 - 2\ell_v(w)\right\}, \tag{3.105}$$

where

$$\ell_v(w) = \int_\Omega (A\nabla v\cdot\nabla w - fw)\,\mathrm{d}x = \int_\Omega A\nabla(v-u)\cdot\nabla w\,\mathrm{d}x.$$

Since

$$|\ell_v(w)| \le \left(\|y_\tau - A\nabla v\|_{A^{-1}} + \mathrm{C}\|\operatorname{div} y_\tau + f\|\right)\|\nabla w\|_A$$

and we assume that the first term dominates, it is reasonable to use (3.103).

Let $v = u_h$, where u_h is a finite element approximation computed on $\mathfrak{T}_h$. Assume that $\{y_{\tau_k}\}$ is a sequence of fluxes computed by minimization of $\overline{\mathsf{M}}_{\operatorname{div} A}(v, y)$ on expanding spaces $\{Y_{\tau_k}\}$, which are limit dense in $H(\Omega, \operatorname{div})$. In this case,

$$\overline{\mathsf{M}}_{\operatorname{div} A}(v, y_{\tau_k}) \to \|\nabla(u - v)\|_A. \tag{3.106}$$

Hence, the sequence $\{y_{\tau_k}\}$ is bounded in $H(\Omega, \operatorname{div})$, and a weak limit $\widetilde{y}$ of this sequence exists. Since $\overline{\mathsf{M}}_{\operatorname{div} A}(u_h, y)$ is convex and continuous with respect to y, we know that

$$\begin{aligned} \|\nabla(u - u_h)\|_A &= \lim_{k\to+\infty} \overline{\mathsf{M}}_{\operatorname{div} A}(u_h, y_{\tau_k}) \geq \overline{\mathsf{M}}_{\operatorname{div} A}(u_h, \widetilde{y}) \\ &= \|\nabla u_h - \widetilde{y}\|_{A^{-1}} + \mathrm{C}\|\operatorname{div}\widetilde{y} + f\| \geq \|\nabla(u - u_h)\|_A. \end{aligned} \tag{3.107}$$

Thus, we conclude that

$$\|\nabla u_h - \widetilde{y}\|_{A^{-1}} + \mathrm{C}\|\operatorname{div}\widetilde{y} + f\| = \|\nabla(u - u_h)\|_A$$

and, therefore, $\widetilde{y}$ minimizes the functional $\overline{\mathsf{M}}_{\operatorname{div} A}(u_h, y)$. Using the same arguments as in [Rep08], we conclude that y_{τ_k} tends to p. Then, (3.104) shows that $\boldsymbol{E}_{\overline{\mathsf{M}}}(v, y_{\tau_k})$ is close to $\mathcal{E}(x)$. Note that y_{τ_k} tends to p, so that the second term of the majorant decreases and tends to zero. The first term remains finite and tends to the exact error. Therefore, we have an easily verifiable criterion of that $\boldsymbol{E}_{\overline{\mathsf{M}}}$ is indeed close to $\mathcal{E}(x)$, namely:

If further minimization of $\overline{\mathsf{M}}_{\operatorname{div} A}(u_h, y)$ with respect to y does not essentially decrease the majorant and the term $\|y_\tau - A\nabla v\|_{A^{-1}}$ is much larger than the second term of the majorant, then $\boldsymbol{E}_{\overline{\mathsf{M}}}(v, y_\tau)$ is close to the function $\mathcal{E}(x)$ and shows the distribution of local (element-wise) errors.

The indicator $\boldsymbol{E}_{\overline{\mathsf{M}}}$ was verified in numerous tests for diffusion models, linear elasticity, viscous flow problems, and problems related to the Maxwell equation (where certain analogs of $\boldsymbol{E}_{\overline{\mathsf{M}}}(v, y_\tau)$ were used). Experiments (see, e.g., [AMM+09, FNR02, FNR03, GNR06, Gor07, Rep99b]) have confirmed its efficiency and stability with respect to approximations of different types. Some of these results are discussed in Chap. 4.

Example 3.5 Below we present results of several numerical tests for the problem (3.95)–(3.97), where $f = 1$ and the coefficients are strongly discontinuous, namely,

$$A = \begin{bmatrix} 1 & 0 \\ 0 & 10 \end{bmatrix} \quad \text{in } \Omega_1 \quad \text{and} \quad A = \begin{bmatrix} 5 & 0 \\ 0 & 1 \end{bmatrix} \quad \text{in } \Omega_2,$$

where the subdomains Ω_1 and Ω_2 are depicted in Fig. 3.6. Approximate solutions

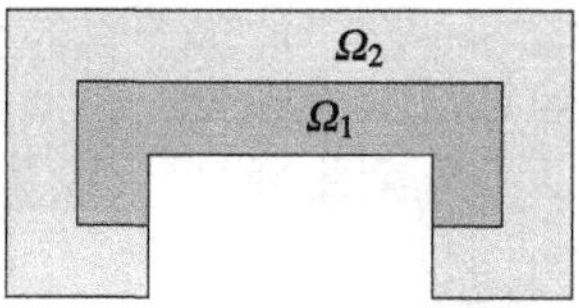

Fig. 3.6 Domain Ω

were computed by linear Courant type elements. In order to compare errors obtained by different error indicators with the true error, we precomputed the corresponding reference solutions using second order Courant type elements on a very fine mesh with 196608 elements. In the tests, we compare true error distributions with distributions computed by the following error indicators:

- $\mathbb{E}_{\overline{\mathrm{M}}}(v, y_{\mathrm{G}})$, where y_{G} is obtained by a commonly used gradient-averaging procedure on patches associated with nodes (see Sect. 2.2.2.2).
- $\mathbb{E}_{\overline{\mathrm{M}}}(v, y^0_{\mathrm{RT}})$, where y^0_{RT} is obtained by edge-wise averaging of normal fluxes (see Sect. 2.2.2.3).
- $\mathbb{E}_{\overline{\mathrm{M}}}(v, y^j_{\mathrm{RT}})$, where y^j_{RT} is obtained from y^0_{RT} by means of the iterative quasi-equilibration procedure (see (3.101) and Sect. 2.2.2.4).
- $\mathbb{E}_{\overline{\mathrm{M}}}(v, y_{\mathrm{glo}})$, where y_{glo} is obtained by global minimization of the majorant (see Algorithm 3.2 and (3.102)).
- $\mathbb{E}(\eta_{RF})$ (full residual type indicator), where element-wise error contribution is (see (2.27))

$$\eta_{RF,T} := \left(h_T^2 \| f_T \|_T^2 + \frac{1}{2} \sum_{E \in \mathcal{E}_h(T)/\mathcal{E}_{h,\partial\Omega}} |E| \left\| [n_E \cdot A\nabla u_h]_E \right\|_E^2 \right)^{1/2}. \tag{3.108}$$

- $\mathbb{E}(\eta_{RJ})$ (residual type indicator containing only jump terms), where elementwise error contribution is (see (2.27) and Remark 2.4)

$$\eta_{RJ,T} := \left(\frac{1}{2} \sum_{E \in \mathcal{E}_h(T)/\mathcal{E}_{h,\partial\Omega}} |E| \left\| [n_E \cdot A\nabla u_h]_E \right\|_E^2 \right)^{1/2}. \tag{3.109}$$

In Fig. 3.7, the true error distribution and indicated element-wise error distributions are depicted for a finite element approximation computed on a regular mesh with $N = 3072$ elements. We see that all indicators manage to locate errors associated with corner singularities and the points where the line of discontinuity of diffusion coefficients intersects with the boundary (we note that the necessity of mesh adaptation in this area is clear a priori). However, the values of $\mathbb{E}(\eta_{RF})$ and $\mathbb{E}(\eta_{RJ})$ are substantially larger. This is also seen on histograms in Fig. 3.8, which provide another view on these results. Here, all element-wise errors are ranked in the decreasing order in accordance with the true error distribution. Thus, the very first (left) vertical bar corresponds to the element with the largest error (the number of which is 1) and the very last one to the element with the smallest error (the number of which is N). Then, the order of elements exposed along the horizontal axis is fixed and all other distributions are presented in the same order. It is clear that if $\mathbb{E}$

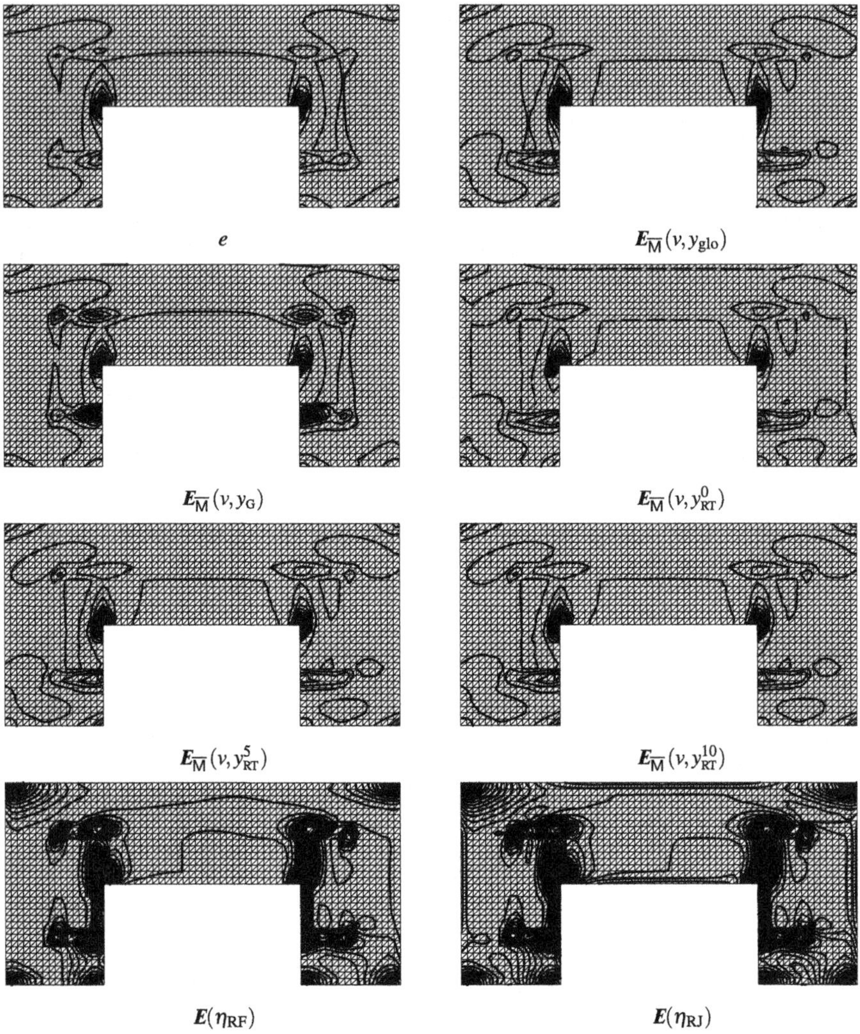

Fig. 3.7 Contour lines of true and indicated error distributions for the approximation computed on a regular mesh with 3072 elements

is accurate in the strong sense (and can be called fully reliable, see Definition 2.1), then the corresponding histogram must resemble the histogram generated by the true error. We see that not all indicators meet this condition. Similar tests have been made using finer meshes with 12288 and 49152 elements. They generate approximations with 7 % and 4 % of relative error, respectively. The corresponding histograms of the indicated errors on meshes are depicted in Figs. 3.9 and 3.10.

In Table 3.5, Table 3.6, and Table 3.7, we measure accuracy of indicators. We use the accuracy measure in Definition 2.1. Also, the accuracy of error indicators in the

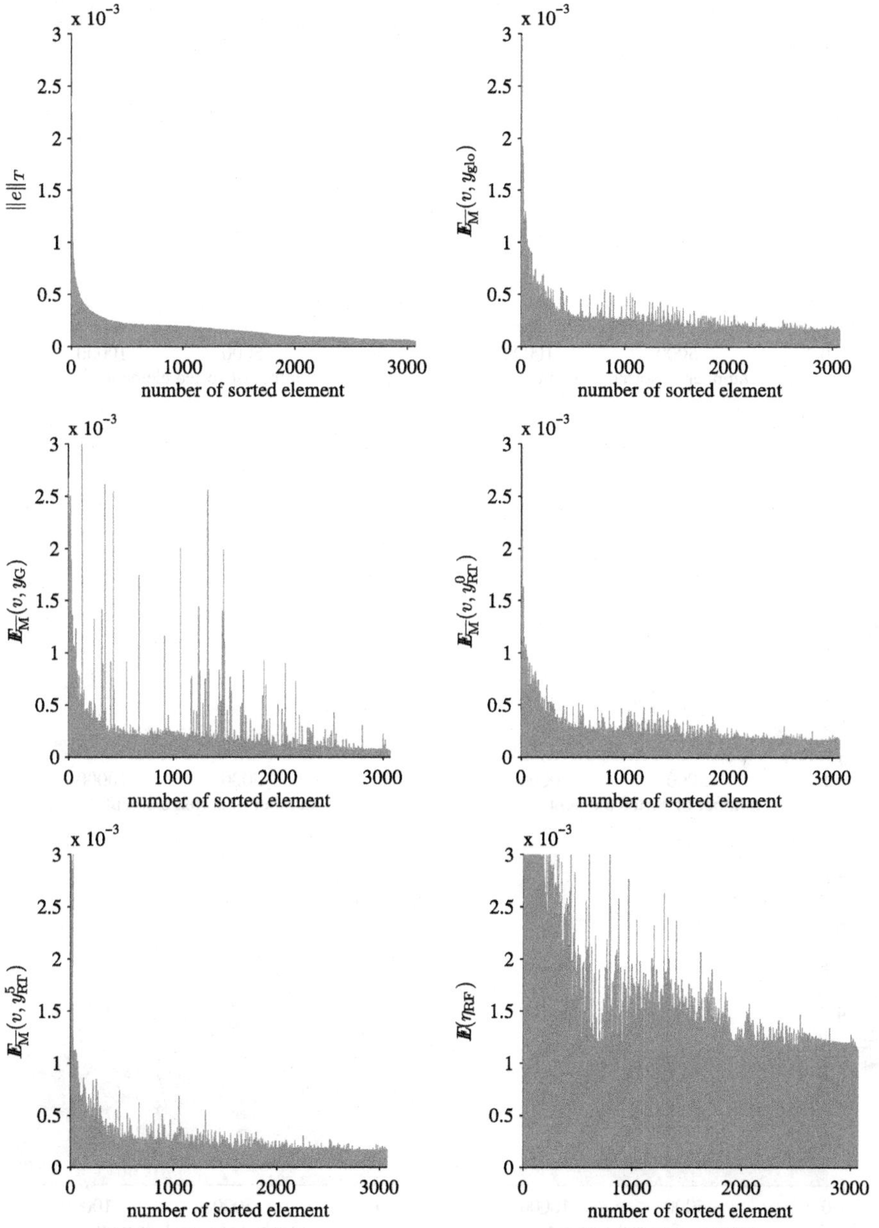

Fig. 3.8 Histograms of true and indicated error distributions for the approximation computed on a regular mesh with 3072 elements

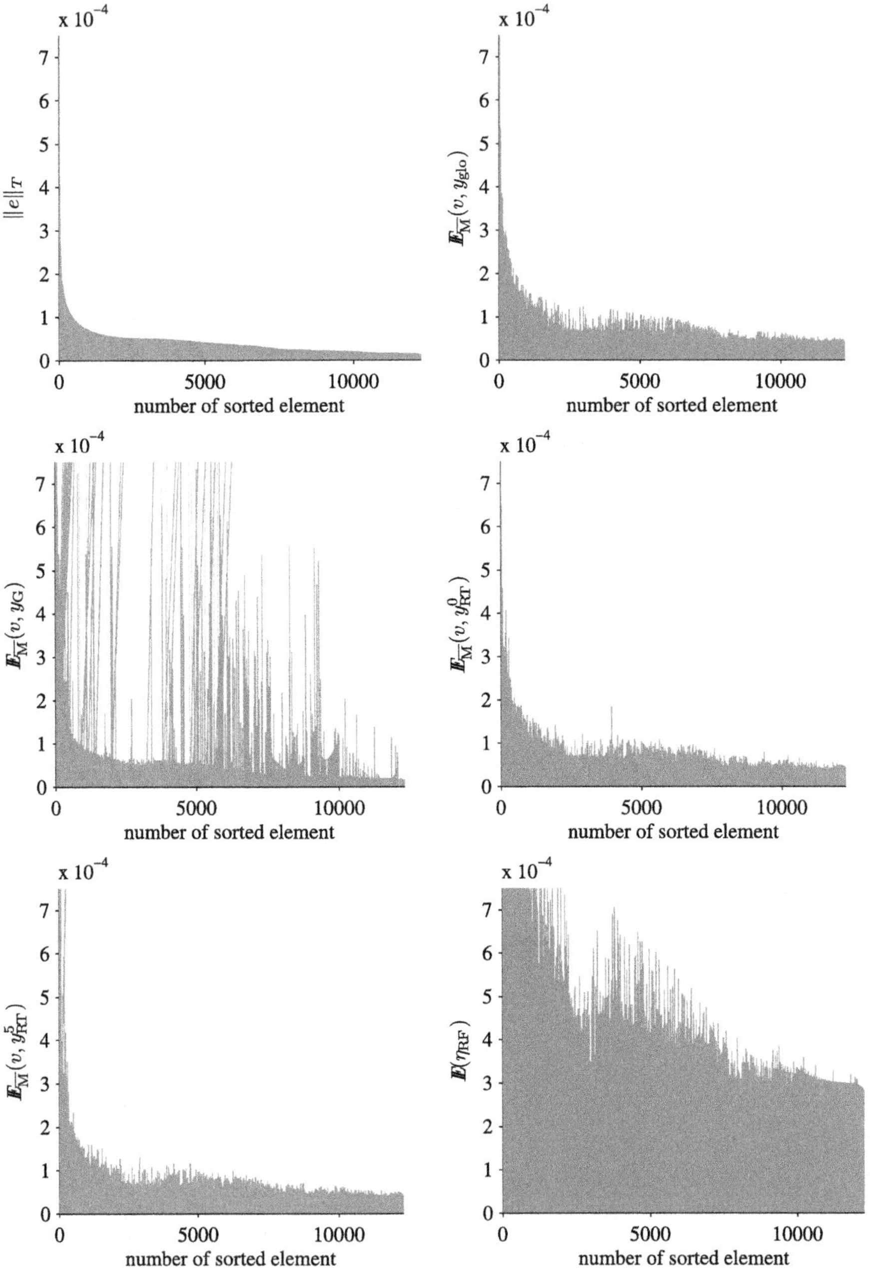

Fig. 3.9 Histograms of true and indicated error distributions for the approximation computed on a regular mesh with 12288 elements

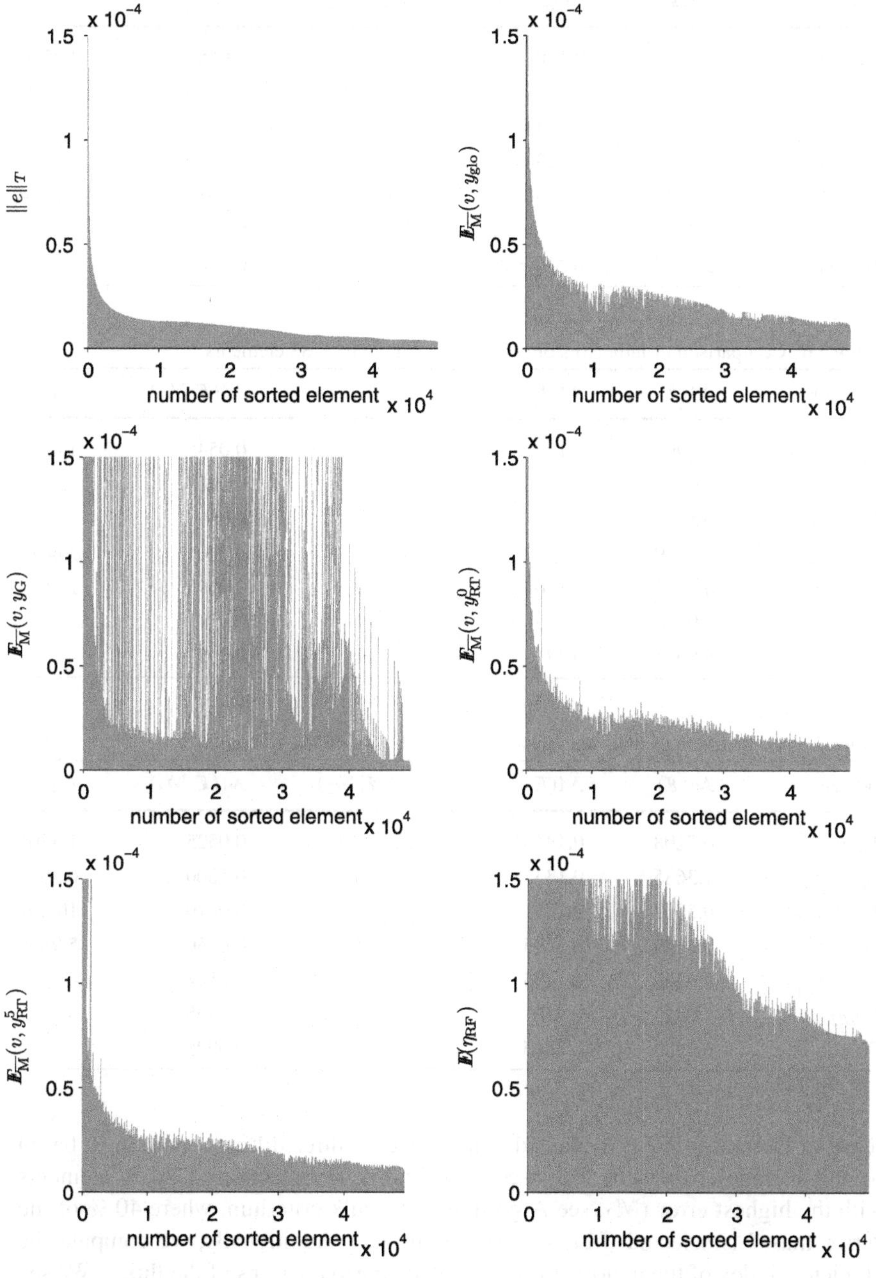

Fig. 3.10 Histograms of true and indicated error distributions for approximation computed on a regular mesh with 49152 elements

Table 3.5 Comparison of indicators on a regular mesh with 3072 elements

Indicator	$\mathcal{M}(\mathbb{E})$	$\mathcal{M}(\mathbb{E}, \mathbb{M}_1)$	$\mathcal{M}(\mathbb{E}, \mathbb{M}_2)$	$\mathcal{M}(\mathbb{E}, \mathbb{M}_3)$	I_{eff}
$\mathbb{E}_{\overline{\mathrm{M}}}(v, y_{\text{glo}})$	0.4988	0.1204	0.0703	0.0654	1.4220
$\mathbb{E}_{\overline{\mathrm{M}}}(v, y_{\mathrm{G}})$	0.6877	0.1156	0.1029	0.1110	16.351
$\mathbb{E}_{\overline{\mathrm{M}}}(v, y^0_{\mathrm{RT}})$	0.5534	0.1243	0.0957	0.0846	24.443
$\mathbb{E}_{\overline{\mathrm{M}}}(v, y^5_{\mathrm{RT}})$	0.5487	0.1234	0.0755	0.0700	2.3728
$\mathbb{E}_{\overline{\mathrm{M}}}(v, y^{10}_{\mathrm{RT}})$	0.5643	0.1250	0.0742	0.0687	2.0144
$\mathbb{E}(\eta_{RF})$	6.9200	0.2692	0.2617	0.1634	–
$\mathbb{E}(\eta_{RJ})$	5.5587	0.2767	0.2617	0.1104	–

Table 3.6 Comparison of indicators on a regular mesh with 12288 elements

Indicator	$\mathcal{M}(\mathbb{E})$	$\mathcal{M}(\mathbb{E}, \mathbb{M}_1)$	$\mathcal{M}(\mathbb{E}, \mathbb{M}_2)$	$\mathcal{M}(\mathbb{E}, \mathbb{M}_3)$	I_{eff}
$\mathbb{E}_{\overline{\mathrm{M}}}(v, y_{\text{glo}})$	0.4994	0.1281	0.0672	0.0545	1.4275
$\mathbb{E}_{\overline{\mathrm{M}}}(v, y_{\mathrm{G}})$	1.0027	0.1192	0.0685	0.0987	32.556
$\mathbb{E}_{\overline{\mathrm{M}}}(v, y^0_{\mathrm{RT}})$	0.5617	0.1245	0.0788	0.0692	48.364
$\mathbb{E}_{\overline{\mathrm{M}}}(v, y^5_{\mathrm{RT}})$	0.5650	0.1303	0.0675	0.0601	3.4817
$\mathbb{E}_{\overline{\mathrm{M}}}(v, y^{10}_{\mathrm{RT}})$	0.5833	0.1305	0.0669	0.0595	2.6653
$\mathbb{E}(\eta_{RF})$	6.9584	0.2636	0.2614	0.1515	–
$\mathbb{E}(\eta_{RJ})$	5.8981	0.2719	0.2614	0.0977	–

Table 3.7 Comparison of indicators on a regular mesh with 49152 elements

Indicator	$\mathcal{M}(\mathbb{E})$	$\mathcal{M}(\mathbb{E}, \mathbb{M}_1)$	$\mathcal{M}(\mathbb{E}, \mathbb{M}_2)$	$\mathcal{M}(\mathbb{E}, \mathbb{M}_3)$	I_{eff}
$\mathbb{E}_{\overline{\mathrm{M}}}(v, y_{\text{glo}})$	0.5208	0.1313	0.0653	0.0525	1.4501
$\mathbb{E}_{\overline{\mathrm{M}}}(v, y_{\mathrm{G}})$	1.3685	0.1337	0.0406	0.1000	68.656
$\mathbb{E}_{\overline{\mathrm{M}}}(v, y^0_{\mathrm{RT}})$	0.5807	0.1251	0.0671	0.0610	102.01
$\mathbb{E}_{\overline{\mathrm{M}}}(v, y^5_{\mathrm{RT}})$	0.6059	0.1285	0.0622	0.0550	5.9855
$\mathbb{E}_{\overline{\mathrm{M}}}(v, y^{10}_{\mathrm{RT}})$	0.6280	0.1295	0.0620	0.0544	4.1468
$\mathbb{E}(\eta_{RF})$	7.0463	0.2581	0.2623	0.1465	–
$\mathbb{E}(\eta_{RJ})$	6.2373	0.2665	0.2623	0.0925	–

sense of Definition 2.3 is evaluated with respect to three different markings: based on the average error value ($\mathbb{M}_1$, see Algorithm 2.1); selection of 30 % elements with the highest error ($\mathbb{M}_2$, see Algorithm 2.2); bulk criterium, where 40 % of the "error mass" is selected ($\mathbb{M}_3$, see Algorithm 2.3). Additionally, we compute the efficiency index of the majorant for computed approximations of the flux. We see that an indicator can be accurate in a weak sense with respect to a certain marker but inaccurate in the strong sense. However, in this case it might be much less accurate with respect to another marker.

3.5 Applications to Adaptive Methods

Adaptive strategies are aimed to generate sequences of meshes which provide sufficiently accurate approximations with the minimal amount of unknowns (degrees of freedom). The success depends on the applied error indication method, marking procedure, geometry of the domain, and on the differential problem itself. Moreover, the computational cost of producing the "optimal" mesh should be considered also. This is a special and important problem studied by many authors (see, e.g, [AS99, BR03, DLY89, Dör96, EJ88, JH92, JS95, PP98, Ran02, Ran00, Rhe80, RV12, SO97, SRO07, Ver96, ZBZ98, ZZ88]).

In this section, we discuss applications of various error control methods to adaptive numerical schemes. Our goal is to present algorithms which not only provide efficient adaptation but also generate approximations with a guaranteed a priori given accuracy.

3.5.1 Runge's Type Estimate

The simplest way of applying the above-discussed error estimates to mesh-adaptive numerical schemes is the following. Let $u_{h_1}, u_{h_2}, \ldots, u_{h_k}, \ldots$ be a sequence of approximations on consequently refined meshes $\mathcal{T}_{h_k}$. Compute $p_{h_k} := \nabla u_{h_k}$ and average it by an averaging operator G_{h_k} acting on $\mathcal{T}_{h_k}$. Then, the accuracy of the approximation $u_{h_{k-1}}$ can be measured by the estimate

$$\left\|\nabla(u - u_{h_{k-1}})\right\| \leq \|\nabla u_{h_{k-1}} - \mathrm{G}_{h_k} p_{h_k}\| + C_{F\Omega}\| \operatorname{div} \mathrm{G}_{h_k} p_{h_k} + f\|. \quad (3.110)$$

This estimate involves approximate solutions computed on two consequent meshes $\mathcal{T}_{h_{k-1}}$ and $\mathcal{T}_{h_k}$. Thus, it follows the same strategy as the Runge's indicator. Unlike the latter indicator, the majorant is mathematically justified and provides a guaranteed upper bound for any pair of consequent meshes. In Algorithm 3.3, we apply (3.110) to construct an adaptive method on nested meshes.

3.5.2 Getting Approximations with Guaranteed Accuracy by an Adaptive Numerical Algorithm

First, we describe a general numerical scheme, which exploits a certain solver and two-sided error estimates in order to obtain an approximate solution of a boundary value problem with an a priori given accuracy $\mathfrak{e}$.

Assume that $\{V_k\} \subset V$ is a sequence of finite dimensional subspaces which are limit dense in V (see Definition B.3) and let v_k be the Galerkin approximation associated with V_k. If the original problem and its discrete analogs are well-posed, then the respective sequence of approximate solutions $\{v_k\}$ tends to the exact solution u as k tends to infinity.

Algorithm 3.3 Adaptive method based on the correct form of the Runge's rule

Input: v {Approximate solution}
$I_{\max}$ {Number of refinements, $I_{\max} \geq 2$}
$\mathcal{T}_{h_1}$ {Initial mesh}
ε {Tolerance}
Compute u_{h_1} on mesh $\mathcal{T}_{h_1}$.
$\boldsymbol{E}_1 = \nabla u_{h_1} - \mathrm{G}_{h_1} \nabla u_{h_1}$
$\overline{\mathsf{M}}_{\mathrm{Runge}_1} = \|\nabla u_{h_1} - \mathrm{G}_{h_1} \nabla u_{h_1}\| + C_{F\Omega} \|\operatorname{div} \mathrm{G}_{h_1} \nabla u_{h_1} + f\|.$
$k = 1$
while $\overline{\mathsf{M}}_{\mathrm{Runge}_k} > \varepsilon$ **and** $k \leq I_{\max}$ **do**
Refine $\mathcal{T}_{h_k}$ to $\mathcal{T}_{h_{k+1}}$ by means of the indicator $\boldsymbol{E}_k$.
Compute $u_{h_{k+1}}$ on mesh $\mathcal{T}_{h_{k+1}}$.
$\overline{\mathsf{M}}_{\mathrm{Runge}_k} = \|\nabla u_{h_k} - \mathrm{G}_{h_{k+1}} \nabla u_{h_{k+1}}\| + C_{F\Omega} \|\operatorname{div} \mathrm{G}_{h_{k+1}} \nabla u_{h_{k+1}} + f\|.$
$\boldsymbol{E}_{k+1} = \|\nabla u_{h_k} - \mathrm{G}_{h_{k+1}} \nabla u_{h_{k+1}}\|$
$k = k + 1$
end while
Output: $u_{h_{k-1}}$ {Approximate solution}
$\overline{\mathsf{M}}_{\mathrm{Runge}_{k-1}}$ {Error estimate}

For each approximation, we compute $\overline{\mathsf{M}}_k$ and use the stopping criteria

$$\frac{\overline{\mathsf{M}}_k}{|||v^k|||} \leq \mathfrak{e}. \tag{3.111}$$

If $\mathfrak{e}$ exceeds the normalized majorant normalized by the energy norm, then the desired accuracy is achieved. In a schematic form, this procedure is presented by Algorithm 3.4.

Since $v_k \to u$ and the majorant is sharp, Algorithm 3.4 ends with finding a proper approximation v_k. However, any particular computer has a certain limited power, so that any problem can be solved only if $\mathfrak{e} \geq \mathfrak{e}_0$, where $\mathfrak{e}_0$ depends on the problem, computer and numerical method used.

Certainly, Algorithm 3.4 is rather schematic and can be viewed only as a skeleton of reliable numerical algorithms to be used in practice. Such type algorithms should include numerous improvements focused first of all on accelerating the process of computations. For example, in intermediate steps it may be efficient to make refinements with the help of simple indicators and perform a sharp computation of the majorant on some selected steps. Moreover, computation of the majorant can be accelerated by using iterative local procedures, which we have discussed.

It could happen that computations must be terminated by time limitations. In this case, the very last value of the majorant shows the best accuracy achieved, which gives an idea of the required power of the computer to be used for finding an approximate solution with the tolerance $\mathfrak{e}$.

Finally, it seems worthwhile to add one more remark. Finding sharp lower and upper bounds requires solving variational problems, so that one may ask about the sensitivity of the algorithm with respect to the inaccuracy of their solutions. To clar-

Algorithm 3.4 Computing approximate solution of a BVP with a guaranteed accuracy

Input: $\mathfrak{e}$ {Tolerance}
V_1 {Initial space}
Solve problem $\mathcal{P}$ on the space V_1 and find v_1.
Select a space Y_1 for the dual variable.
Compute $\overline{\mathsf{M}}_k = \min_{y \in Y_1} \overline{\mathsf{M}}_{\mathcal{P}}(v_1, y)$ {Minimization by Algorithm 3.2}
$k = 1$.
while $\mathfrak{e} < \frac{\overline{\mathsf{M}}_k}{|||v_k|||}$ **do**
 $k = k + 1$.
 Refine space V_k from V_{k-1} using $\boldsymbol{E}_{\overline{\mathsf{M}}_{k-1}}$ as an error indicator.
 Solve problem $\mathcal{P}$ on the space V_k and find v_k.
 Select Y_k.
 Compute $\overline{\mathsf{M}}_k = \min_{y \in Y_k} \overline{\mathsf{M}}_{\mathcal{P}}(v_k, y)$.
end while
Output: v_k {Approximate solution}
y_k {Approximate flux, minimizer of $\overline{\mathsf{M}}_{\mathcal{P}}(v_k, y)$}
$\overline{\mathsf{M}}_k$ {Error bound}

ify this point, we recall that $\overline{\mathsf{M}}(v_k, y, \beta)$ provides an upper bound of the error *for any* $y \in Y$ and $\underline{\mathsf{M}}(v_k, w)$ provides a lower bound *for any* $w \in V$. For this reason, exact solutions of these variational problems are, in general, not required. For example, it may occur that on some stage of the minimization procedure the value of $\frac{\overline{\mathsf{M}}_k}{|||v_k|||}$ becomes less than $\mathfrak{e}$. Then, computations may be terminated even if y^* does not minimize $\overline{\mathsf{M}}_k$.

3.6 Combined (Primal-Dual) Error Norms and the Majorant

In modern numerical technologies (such as, e.g., mixed finite element methods discussed in Sect. B.4.4), approximations are generated for both primal and dual components of the solution. From the physical point of view, this approach is well motivated. Consider, for example, the stationary diffusion problem in Ω. The corresponding mathematical model consists of two physical relations

$$-\operatorname{div} p = f \in L^2(\Omega), \tag{3.112}$$

$$p = A\nabla u \tag{3.113}$$

supplied with boundary conditions $u = u_0$ on Γ_D and $p \cdot n = 0$ on Γ_N. These relations include the functions u and p, which have a clear physical meaning (e.g., the temperature and the heat flux). Similar formulations arise in the elasticity theory, where the corresponding pair of functions presents the displacement and stress, respectively (see Sect. 4.1.6).

Within the framework of this conception, we should measure the deviation of the (computed) pair (v, y) from the exact solution (u, p). In this case, it is natural to use combined primal-dual norms. To this end, different equivalent norms associated with the space $H^1(\Omega) \times H(\Omega, \mathrm{div})$ can be used. For this purpose, we select the norm

$$\big\|[(u,p)-(v,y)]\big\| := \big\|\nabla(u-v)\big\|_A + \|p-y\|_{A^{-1}} + \frac{C_{F\Omega}}{c_1}\big\|\mathrm{div}(p-y)\big\|, \quad (3.114)$$

the last term of which can be equivalently rewritten as $\frac{C_{F\Omega}}{c_1}\|\,\mathrm{div}\, y + f\,\|$. Obviously, it represents the deviation of y from the space of equilibrated fluxes, Q_f. There are other equivalent norms, e.g.,

$$\big\|[(u,p)-(v,y)]\big\|_{(2)} := \sqrt{\big\|\nabla(u-v)\big\|_A^2 + \|p-y\|_{A^{-1}}^2 + \big\|\mathrm{div}(p-y)\big\|^2}.$$

We can use other weights for terms related to different components of the error norm and consider other equivalent norms.

In [RS05, RSS07], it was shown that the error majorant

$$\overline{\mathsf{M}}_{\mathrm{div}\, A\nabla}(v,y) = \|A\nabla v - y\|_{A^{-1}} + \frac{C_{F\Omega}}{c_1}\|\,\mathrm{div}\, y + f\,\|$$

is equivalent to the deviation measured in the combined norm (3.114). Moreover,

$$\overline{\mathsf{M}}_{\mathrm{div}\, A\nabla}(v,y) \leq \big\|[(u,p)-(v,y)]\big\| \leq 3\overline{\mathsf{M}}_{\mathrm{div}\, A\nabla}(v,y). \quad (3.115)$$

Indeed, by (3.112), (3.113), and (3.38) we see that

$$\begin{aligned}
\overline{\mathsf{M}}_{\mathrm{div}\, A\nabla}(v,y) &= \|A\nabla v - A\nabla u + A\nabla u - y\|_{A^{-1}} + \frac{C_{F\Omega}}{c_1}\|\,\mathrm{div}\, y + f\,\| \\
&\leq \big\|A\nabla(v-u)\big\|_{A^{-1}} + \|p-y\|_{A^{-1}} + \frac{C_{F\Omega}}{c_1}\|\,\mathrm{div}\, y - \mathrm{div}\, p\| \\
&= \big\|[(u,p)-(v,y)]\big\|.
\end{aligned}$$

On the other hand

$$\begin{aligned}
\big\|[(u,p)-(v,y)]\big\| &= \big\|\nabla(v-u)\big\|_A + \|p - A\nabla v + A\nabla v - y\|_{A^{-1}} \\
&\quad + \frac{C_{F\Omega}}{c_1}\|\,\mathrm{div}\, y - \mathrm{div}\, p\| \\
&\leq \big\|\nabla(v-u)\big\|_A + \big\|A\nabla(v-u)\big\|_{A^{-1}} + \|A\nabla v - y\|_{A^{-1}} \\
&\quad + \frac{C_{F\Omega}}{c_1}\|\,\mathrm{div}\, y - \mathrm{div}\, p\|
\end{aligned}$$

$$= 2\left\| \nabla(v-u) \right\|_A + \|A\nabla v - y\|_{A^{-1}} + \frac{C_{F\Omega}}{c_1}\|\operatorname{div} y + f\|$$

$$\leq 3\overline{\mathsf{M}}_{\operatorname{div} A\nabla}(v, y). \tag{3.116}$$

The relation (3.115) states that the corresponding efficiency index

$$I_{\text{eff}} := \frac{3\overline{\mathsf{M}}_{\operatorname{div} A\nabla}(v, y)}{|||[(u, p) - (v, y)]|||}$$

is always within the interval $[1, 3]$!

This fact shows that the majorant is a reliable and efficient measure of the error in combined primal-dual norms for any pair $(v, q) \in (V_0 + u_0) \times H(\Omega, \operatorname{div})$. In [Rep08], similar equivalence results have been proved for other problems related to partial differential equations of the divergent form (e.g., for the convection-diffusion, elasticity, Stokes, and other models).

Another important conclusion that comes out of this analysis is that efficient reconstruction of the flux in the space $H(\Omega, \operatorname{div})$ is equivalent to the minimization of $\overline{\mathsf{M}}_{\operatorname{div} A\nabla}(v, y)$. Indeed, the best reconstruction in $H(\Omega, \operatorname{div})$ must minimize the difference $y - p$ in the norm of this space. In view of (3.115), this minimization is equivalent to minimization of $\overline{\mathsf{M}}_{\operatorname{div} A\nabla}(v, y)$.

Remark 3.4 In many numerical methods (e.g., in the classical variational difference method or in the finite difference method) the major efforts are focused on finding v with minimal error in $H^1(\Omega)$, the corresponding y being subsequently computed by the relation (3.113). Often this approach provides a poor approximation of the flux. To overcome this drawback, one can use the post-processing methods discussed in Sect. 2.2.2.3.

Example 3.6 We consider the test problem (3.95)–(3.97) and discuss error estimation in terms of the combined norm (3.114).

Again, we compute a sequence of successive approximations $v_k \in V_k$, where V_k is a space generated by linear nodal triangular elements. Y_k is the space formed by Raviart–Thomas elements on the same mesh. We compare two ways to compute approximate flux: y_{G}^k (see Sect. 2.2.2.2) obtained by nodal gradient-averaging on a mesh generating V_k, and y_{glo}^k (see (3.102)) obtained by global minimization of the majorant on the same mesh.

Values of the majorants $\overline{\mathsf{M}}_{\operatorname{div} A}(v_k, y_{\text{G}}^k)$ and $\overline{\mathsf{M}}_{\operatorname{div} A}(v_k, y_{\text{glo}}^k)$ are computed on a sequence of uniformly refined meshes. The exact deviation of (v_k, y_{G}^k) and (v_k, y_{glo}^k) from (u, p) and the respective error bounds found by the majorants are depicted in Fig. 3.11 and Table 3.8, where the efficiency indexes

$$I_{\text{eff},\oplus}^{(\text{G},k)} = \frac{3\overline{\mathsf{M}}_{\operatorname{div} A}(v^k, y_{\text{G}}^k)}{|||[(u, p) - (v^k, y_{\text{G}}^k)]|||} \quad \text{and} \quad I_{\text{eff},\ominus}^{(\text{G},k)} = \frac{\overline{\mathsf{M}}_{\operatorname{div} A}(v^k, y_{\text{G}}^k)}{|||[(u, p) - (v^k, y_{\text{G}}^k)]|||}$$

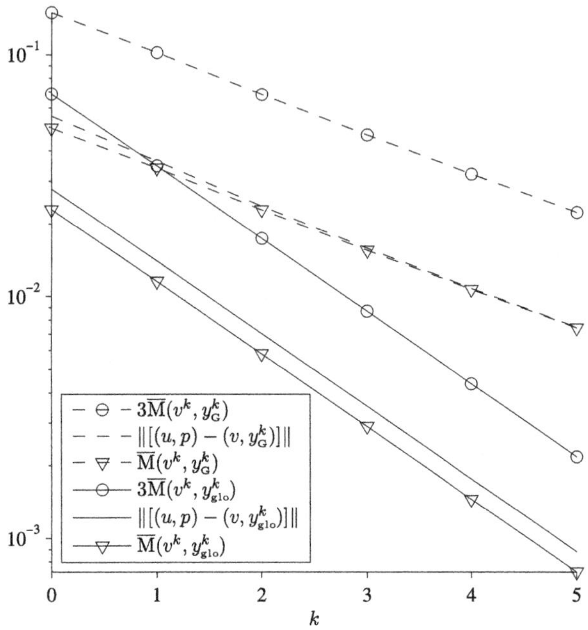

Fig. 3.11 Exact error in the combined norm (3.114) and two-sided bounds (3.115) computed on a sequence of uniformly refined meshes

are presented.

As one could predict, y^k_{glo} is a much better approximation of the flux than y^k_G. This is easy to observe if we compare the fourth and ninth lines of Table 3.8. For $k = 1$, the accuracy of the pair (v_k, y^k_G) is two times less than that of the pair

Table 3.8 Error control in combined norm

	k					
	1	2	3	4	5	6
$\dim(V^k_0)$	280	1047	4045	15897	63025	250977
$\dim(Y^k)$	767	2998	11852	47128	187952	750688
$3\overline{\mathsf{M}}_{\operatorname{div} A}(v^k, y^k_G)$	0.1489	0.1020	0.0684	0.0466	0.0321	0.0223
$\|[(u,p)-(v^k, y^k_G)]\|$	0.0557	0.0363	0.0237	0.0158	0.0108	0.0075
$\overline{\mathsf{M}}_{\operatorname{div} A}(v^k, y^k_G)$	0.0496	0.0340	0.0228	0.0155	0.0107	0.0074
$I^{(G,k)}_{\mathrm{eff},\oplus}$	2.6733	2.8093	2.8926	2.9424	2.9699	2.9846
$I^{(G,k)}_{\mathrm{eff},\ominus}$	0.8911	0.9364	0.9642	0.9808	0.9900	0.9949
$3\overline{\mathsf{M}}_{\operatorname{div} A}(v^k, y^k_{glo})$	0.0686	0.0347	0.0174	0.0087	0.0044	0.0022
$\|[(u,p)-(v^k, y^k_{glo})]\|$	0.0278	0.0140	0.0070	0.0035	0.0018	0.0009
$\overline{\mathsf{M}}_{\operatorname{div} A}(v^k, y^k_{glo})$	0.0229	0.0116	0.0058	0.0029	0.0015	0.0007
$I^{(glo,k)}_{\mathrm{eff},\oplus}$	2.4638	2.4698	2.4713	2.4716	2.4717	2.4717
$I^{(glo,k)}_{\mathrm{eff},\ominus}$	0.8213	0.8233	0.8238	0.8239	0.8239	0.8239

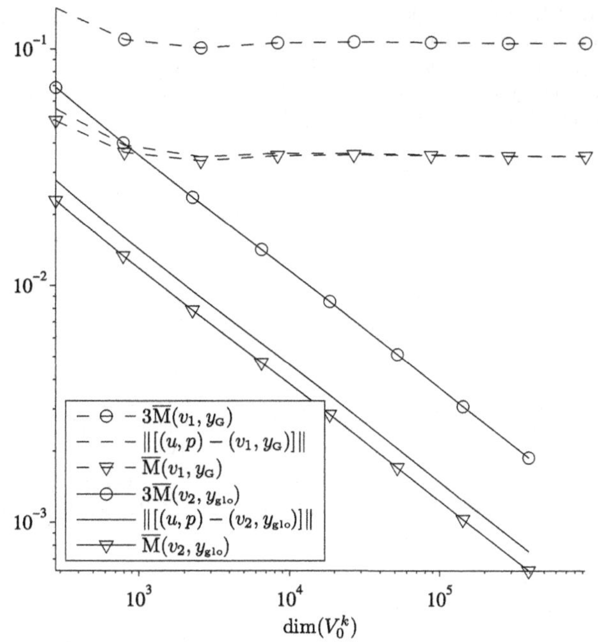

Fig. 3.12 Exact error in the combined norm (3.114) and two-sided bounds (3.115) computed on a sequence of adaptively refined meshes

(v_k, y^k_{glo}). For $k = 5$, the norm is about five times more accurate, and for $k = 6$ this difference is even larger. It is worth noting that although the error related to (v^k, y^k_{G}) is higher the efficiency of the respective bounds generated by the majorant does not deteriorate. This was expected, due to the equivalence relation (3.115).

We repeat the experiment by using an adaptive method. The entire majorant is used as an error indicator and the bulk marking with $\theta = 0.5$ is applied. We use the above-discussed methods to reconstruct the flux. The resulting error bounds together with the exact error in the combined norm are depicted in Fig. 3.12.

We observe that for non-uniform meshes generated by consequent local refinements, the method of generating the approximate flux by means of G_h is not adequate if we wish to have a good reconstruction in $H(\Omega, \operatorname{div})$. Figure 3.12 shows how the majorant captures defects of the flux generated by the gradient-averaging scheme. It provides guaranteed error bounds in the sense of the combined norm in all cases due to the equivalence (3.115).

All components of the exact error are depicted in Fig. 3.13, which clearly indicates that the flux y_{G} computed by simple averaging fails to satisfy the equilibrium condition. In contrast, the flux y_{glo} obtained by global minimization of the majorant behaves correctly.

This phenomenon is rather typical and can be seen in many other examples. It implies conclusions related to mesh adaptation procedures. Namely, if we wish to create a sequence of meshes such that the corresponding flux reconstructions are well-balanced, then using error indicators based on simple averaging is not the best way. In Fig. 3.13, v_1 denotes the finite element solution computed on a sequence

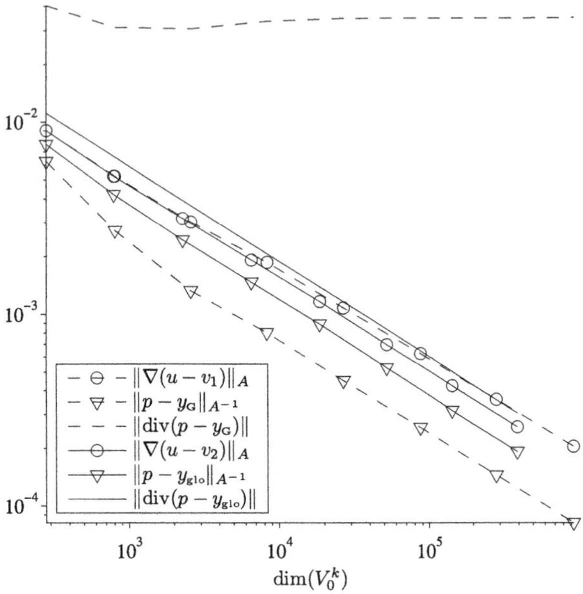

Fig. 3.13 Error components of approximation pairs (v_1, y_G) and (v_2, y_{glo}) computed on a sequence of adaptively refined meshes

of meshes generated by the averaging indicator, and v_2 denotes solutions computed on meshes constructed by means of the majorant. We see that the fluxes computed within the framework of the first adaptation method converge in L^2, but do not converge in $H(\Omega, \mathrm{div})$. The second method generates fluxes which satisfy the equation of the balance with increasing accuracy. We note that instead of y_{glo} a simpler reconstruction can be used. For example, we project y_G to the RT-space and perform several rounds of iteration steps described in Sect. 2.2.2.4. This method generates fluxes with approximation properties close to those y_{glo} has.

Chapter 4
Guaranteed Error Bounds II

Abstract In Chap. 3, we discussed the main ideas of fully reliable error control methods and the corresponding numerical algorithms with the paradigm of simple elliptic type problems. This chapter is intended to show a deep connection between a posteriori estimates of the functional type and physical relations generating the problem. Also, the goal of this chapter is to consider a wider set of problems arising in various applications and explain things in terms of computational mechanics. For this purpose, we begin with a simple class of mechanical problems (straight beams) and after that consider curvilinear beams and more complicated models of continuum mechanics (linear elasticity, viscous fluids, Maxwell type problem). At the end of the chapter we consider a generalized mathematical model, which includes almost all earlier discussed problems as particular cases.

4.1 Linear Elasticity

4.1.1 Introduction

We start an overview of models describing elastic bodies with models of elastic beams. They lead to ordinary differential equations and allow us to discuss the main ideas in the most transparent form.

A domain can be considered as a beam, if its long and slender, i.e., if it's length in one dimension is substantially larger than in two others (see Fig. 4.1). The simplest beam model studies only an axial deformation of the beam and includes only loads acting parallel to the beam. We use it to demonstrate close relations between the fundamental physical laws generating the problem and components of the respective error majorant.

We assume that the cross section perpendicular to the load does no deform, i.e., the beam elongates only with respect to the x-axis. The displacement (elongation) of the beam is denoted by function u.

The linearized relation between the strain and elongation is simple,

$$\varepsilon = u'. \tag{4.1}$$

O. Mali et al., *Accuracy Verification Methods*,
Computational Methods in Applied Sciences 32, DOI 10.1007/978-94-007-7581-7_4,

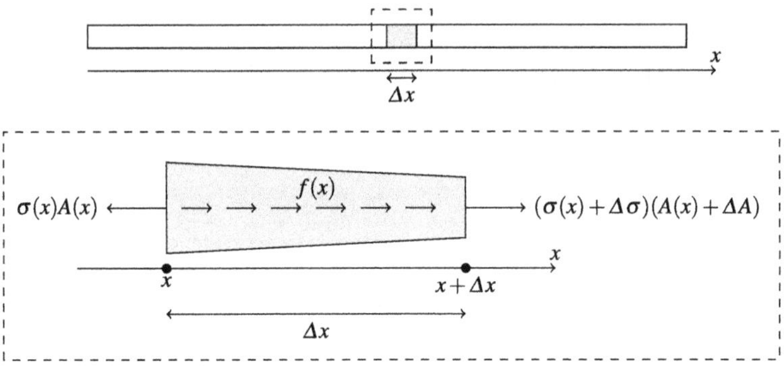

Fig. 4.1 Axially distributed load and stresses on an interval of length Δx of an elastic beam

Strain describes a "density" or "intensity" of the deformation. Equation (4.1) is the *kinematic* relation between the strain and displacement.

Here we assume that the normal stress σ is uniformly distributed over the cross section, and its resultant over the cross section (of area A) is σA. Consider a beam under an axially distributed load f. In Fig. 4.1, we depict the interval $[x, x+\Delta x]$, and assume that over it f, σ, and A change linearly, i.e, $f(x+\Delta x) = f(x) + \Delta f$, $\sigma(x+\Delta x) = \sigma(x) + \Delta\sigma$, and $A(x+\Delta x) = A(x) + \Delta A$. The equation of equilibrium for the beam section is

$$\int_x^{x+\Delta x} f(t)\,\mathrm{d}t + \big(\sigma(x) + \Delta\sigma\big)\big(A(x) + \Delta A\big) - \sigma(x)A(x) = 0.$$

Taking only linear terms, we arrive at the incremental relation $\Delta(\sigma A) + f(x)\Delta x = 0$, which yields the differential equation

$$-(A\sigma)' = f. \tag{4.2}$$

The material behavior of the beam is described by the Hooke's law

$$\sigma = E\varepsilon, \tag{4.3}$$

where E is the Young's modulus (Elastic modulus) of the material of the beam.

We note that (4.1), (4.2), and (4.3) are the basic relations defining the problem. Combining them yields the differential equation

$$-\big(EAu'\big)' = f, \tag{4.4}$$

which together with the boundary conditions generates a boundary value problem. The boundary conditions define the displacement u or axial stress EAu' (or their linear combination) on both ends of the beam. In particular, $u = 0$ means that end is fixed, and $EAu' = 0$ means that the end is not subjected to load. The boundary conditions must be defined in such a manner that the beam is statically determined, i.e., no rigid body motion can occur.

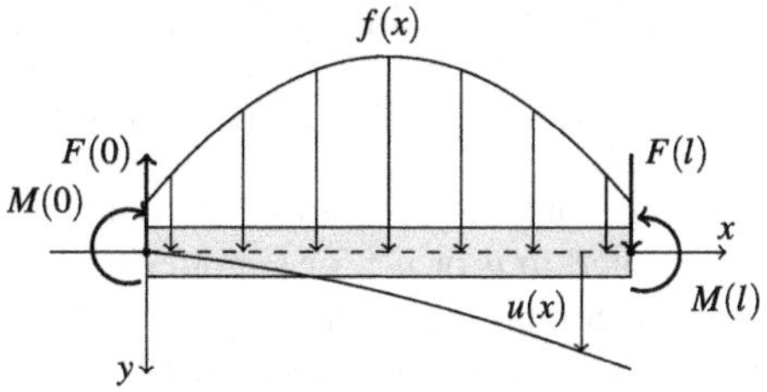

Fig. 4.2 Loads and displacement of the Euler–Bernoulli beam

It is easy to see that (4.4) (together with the boundary conditions $u(0) = 0$ and $u(l) = 0$, i.e., both ends of the beam are fixed) is a particular form of (3.1). Using the results of Chap. 3 we obtain the estimate

$$|||u - v||| \leq \left(\int_0^l \frac{1}{EA}(y - EAv')^2 \,\mathrm{d}x\right)^{1/2} + C\left(\int_0^l (f + y')^2 \,\mathrm{d}x\right)^{1/2}, \qquad (4.5)$$

where the energy norm is defined as

$$|||w||| = \left(\int_0^l EA(u')^2 \,\mathrm{d}x\right)^{1/2},$$

and the constant C satisfies the inequality $\|w\| \leq C\|w'\|$ for all w, which satisfy the boundary conditions and possess a square summable derivative.

Two terms of the majorant (4.5) penalize violations of two fundamental relations (4.1) and (4.2), respectively. Moreover, the inequality holds as the equality if $y := EAu'$, i.e., y is the normal stress.

4.1.2 Euler–Bernoulli Beam

The Euler–Bernoulli beam model is one of the most used beam model in engineering. Model examples can be found in numerous textbooks (see, e.g., [TG51, TY45]). The model describes vertical deflection of the beam (see Fig. 4.2) under a given load f. On both ends of the beam, a bending moment M or shear force F may be given.

The energy of the beam is presented by the functional

$$J(w) := \int_0^l \left(\frac{1}{2}EI(w'')^2 - fw\right) \mathrm{d}x - Fw\Big|_0^l + Mw'\Big|_0^l,$$

where E is Young's modulus and I is the second moment of the cross section, i.e., $I := \int_A y^2 \mathrm{d}A$, where y is the distance from the centroid axis.

The solution u minimizes the energy, i.e.,

$$J(u) \leq J(w), \quad \forall w \in V_0,$$

where V_0 contains functions satisfying the kinematic boundary conditions (i.e., those related to u or u'; for simplicity we consider beams for which the kinematic boundary conditions are homogeneous.). Since the energy functional contains second derivatives, we must restrict the set of admissible functions and assume they have second generalized derivatives (see Sect. A.2.1). Formally, this means that

$$V_0 := \{ w \in H^2((0,l)) \mid w \text{ satisfies the kinematic boundary conditions} \}.$$

A beam can be supported in multiple ways, generating many different boundary conditions. The boundary conditions should be such that the beam is statically determined, i.e., the boundary conditions do not allow any rigid body motion to occur.

It is not difficult to see that the function u minimizing the energy must satisfy the relation

$$\int_0^l EIu''w''\,\mathrm{d}x = \int_0^l fw\,\mathrm{d}x + Fw\Big|_0^l - Mw'\Big|_0^l, \quad \forall w \in V_0. \tag{4.6}$$

If the solution has derivatives up to the fourth order, then we obtain the classical form of the Euler–Bernoulli beam problem, i.e.,

$$\left(EIu''\right)'' = f \tag{4.7}$$

and the respective boundary conditions. This equation can be decomposed into two physically motivated relations

$$-M'' = f \tag{4.8}$$

$$-EIu'' = M, \tag{4.9}$$

where (4.8) is the equilibrium relation of the beam and (4.9) is a linearized form of the law that relates the curvature and the bending moment. Relation (4.9) is based on several assumptions concerning the deformation: (a) cross sections of the beam remain perpendicular to the neutral axis, i.e., there is no "twisting" of the cross section, (b) beam is in the state of pure bending, i.e., the deformation occurs due to the bending moment of the beam and the effects of shear force are neglected, and (c) the elongation of the beam is neglected. Moreover, we define the space of admissible bending moments

$$Q_0 := \{ y \in H^2((0,l)) \mid y \text{ satisfies natural boundary conditions} \},$$

i.e., $y = M$ and $y' = F$ at the endpoints of the beam.

Theorem 4.1 *Let $v \in V_0$, then*

$$|||u - v|||^2 \leq \overline{\mathsf{M}}^2_{\mathrm{BE}}(v, y, \beta), \quad \forall y \in Q_0, \beta > 0,$$

where

$$\overline{\mathrm{M}}^2_{\mathrm{BE}}(v, y, \beta) := (1+\beta)\left\| \frac{1}{\sqrt{EI}}(y + EIv'') \right\|^2 + \frac{1+\beta}{\beta}\frac{C_{\mathrm{KL}}}{\alpha}\|y'' + f\|^2,$$

where $\alpha := \min_{x\in[0,l]} E(x)I(x)$, C_{KL} *is from* (4.20), *and the energy norm for the problem is*

$$|||w|||^2 := \int_0^l EI(w'')^2\,\mathrm{d}x.$$

Proof Note that by integration by parts formulae,

$$\int_0^l y''w\,\mathrm{d}x = \int_0^l yw''\,\mathrm{d}x + y'w\Big|_0^l - yw'\Big|_0^l, \quad \forall y, w \in H^2((0,l)). \tag{4.10}$$

From (4.6) and (4.10),

$$\begin{aligned}\int_0^l EI(u'' - v'')w''\,\mathrm{d}x = &-\int_0^l EIv''w''\,\mathrm{d}x + \int_0^l fw\,\mathrm{d}x + Fw\Big|_0^l - Mw'\Big|_0^l \\ &+ \int_0^l y''w\,\mathrm{d}x - \int_0^l yw''\,\mathrm{d}x - y'w\Big|_0^l + yw'\Big|_0^l.\end{aligned}$$

We can reorganize it as follows:

$$\begin{aligned}\int_0^l EI(u'' - v'')w''\,\mathrm{d}x = &-\int_0^l (y + EIv'')w''\,\mathrm{d}x + \int_0^l (y'' + f)w\,\mathrm{d}x \\ &+ (F - y')w\Big|_0^l - (M - y)w'\Big|_0^l.\end{aligned}$$

Assume that $y \in Q_0$. Since $w \in V_0$ the terms on the second line vanish. Then,

$$\int_0^l EI(u'' - v'')w''\,\mathrm{d}x \le \left\| \frac{1}{\sqrt{EI}}(y + EIv'') \right\| \|\sqrt{EI}w''\| + \|y'' + f\|\,\|w\|.$$

By (4.20) we find that

$$\int_0^l EI(u'' - v'')w''\,\mathrm{d}x \le \left(\left\| \frac{1}{\sqrt{EI}}(y + EIv'') \right\| + \frac{C_{\mathrm{KL}}}{\alpha}\|y'' + f\| \right) \|\sqrt{EI}w''\|,$$

where $\alpha := \min_{x\in[0,l]} EI$. Substituting $w := u - v$ and dividing by $|||u - v|||$ yields the estimate

$$|||u - v||| \le \left\| \frac{1}{\sqrt{EI}}(y + EIv'') \right\| + \frac{C_{\mathrm{KL}}}{\alpha}\|y'' + f\|.$$

Squaring both sides and applying (A.5) leads to the statement. □

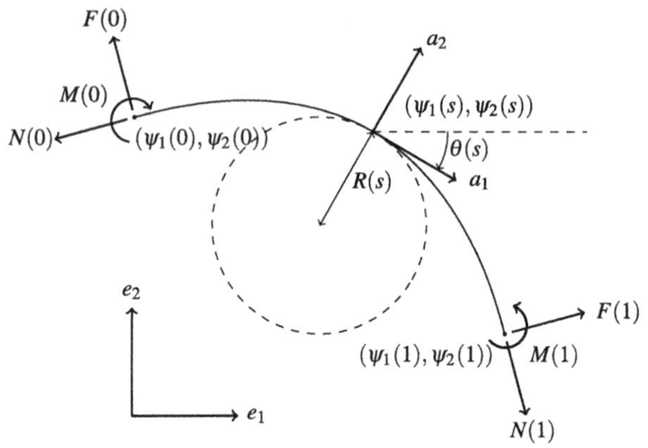

Fig. 4.3 Kirchhoff–Love arch

We see that the majorant $\overline{\mathsf{M}}_{\mathrm{BE}}$ has the same structure as the majorants we have studied before. It consists of the terms, which are measures of violations of the relations (4.8) and (4.9).

Remark 4.1 The Reissner–Mindlin plate (see Sect. 4.1.5) can be viewed as a two-dimensional analog of the Timoshenko beam. We address the reader interested in Timoshenko beams to the material exposed in that section.

4.1.3 The Kirchhoff–Love Arch Model

We consider a plane arch with a constant cross section and assume that the character size of it is small compared to the length of the arch. Following [Cia78a], the arch and all related functions are presented in a parameterized form. The $\psi : [0, 1] \to \mathbb{R}^2$ is a smooth parameterized non-self-intersecting curve that defines the shape of the arch. The displacement vector $u = (u_1, u_2)$ and the exterior load $f = (f_1, f_2)$ are presented in a local coordinate system (a_1, a_2), which varies along the arch. Here a_1 is the tangential and a_2 is the normal direction. The angle between the horizontal axis and a_1 is denoted as θ. On both ends of the beam, other external loads may occur. N denotes the normal force, F is the shear force and M is the bending moment (see Fig. 4.3).

A systematic exposition of the classical beam theory can be found, e.g., in [TG51, TY45] and a more advanced one in [NST06, TV05], where regularity requirements for ψ are substantially relaxed.

The constitutive relation of the curved beam is

$$\begin{cases} EA(u_1' - cu_2) = p_1, \\ EI(cu_1 + u_2')' = p_2, \end{cases} \tag{4.11}$$

Table 4.1 Boundary conditions of the Kirchhoff–Love arch

Kinematic	Natural
u_1 (tangential disp.)	N (tangential stress)
u_2 (normal disp.)	F (shear force)
u_2' (rotation)	M (bending moment)

where $c : [0, 1] \to \mathbb{R}$,

$$c(s) := \frac{\psi_2''(s)\psi_1'(s) - \psi_1''(s)\psi_2'(s)}{(\psi_1(s)'^2 + \psi_2(s)^2)^{3/2}} \tag{4.12}$$

is the curvature of the arch, p_1 is the tangential stress, and p_2 is the bending moment. E is the Young's modulus of the material, A is the area of the cross section, and I is the second moment of inertia of the cross section.

The equilibrium conditions are presented by the equations:

$$\begin{cases} -p_1' - cp_2' = f_1, \\ -cp_1 + p_2'' = f_2. \end{cases} \tag{4.13}$$

These relations present the physical laws governing the beam problem. For the straight beam ($c = 0$), (4.11) and (4.13) imply (4.4) and (4.7), respectively.

The boundary conditions are defined at the end points $s = 0$ and $s = 1$. They are listed as pairs in Table 4.1. Kinematic boundary conditions restrict displacement components or rotation and natural boundary conditions define tangential stresses, shear forces or bending moments. Below, we assume that the kinematic boundary conditions are homogeneous. This is performed only for the sake of simplicity. Problems with non-homogeneous boundary conditions can be analyzed quite similarly. Together with the regularity requirements, kinematic boundary conditions define the space of admissible displacements

$$V_0 := \{v \in V \mid v \text{ satisfies the kinematic boundary conditions}\},$$

where $V := H^1((0, 1)) \times H^2((0, 1))$.

Additionally, at the endpoints of the beam, stresses must satisfy the natural boundary conditions, namely

$$p_1 + cp_2 = N, \qquad p_2 = M, \quad \text{and} \quad p_2' = F. \tag{4.14}$$

Stresses satisfying these relations form the space of admissible stresses,

$$Q_0 := \big\{y \in H^1\big((0, 1)\big) \times H^2\big((0, 1)\big) \mid y \text{ satisfies (4.14)}\big\}. \tag{4.15}$$

The problem is called statically determined if for $p = 0$, the equations (4.11) imply $u = 0$. It is not difficult to see that the kernel of equations (4.11) consists of

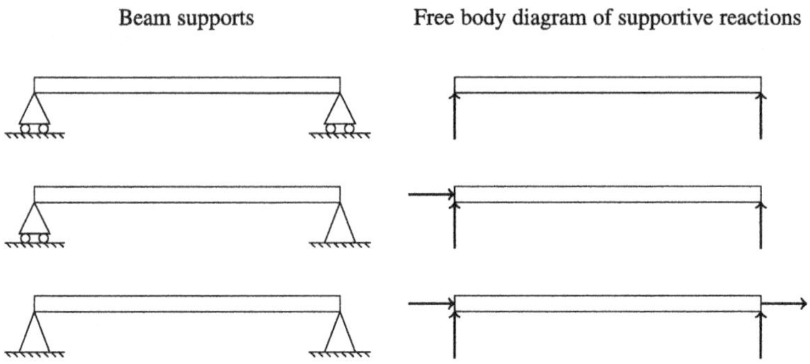

Fig. 4.4 Examples of different boundary condition types

rigid body motions (see [Cia78a]) that can be presented in the form

$$v = \begin{bmatrix} \cos(\theta) & \sin(\theta) \\ -\sin(\theta) & \cos(\theta) \end{bmatrix} \begin{bmatrix} d\psi_2 + b_1 \\ d\psi_1 + b_2 \end{bmatrix}, \tag{4.16}$$

where d, b_1 and b_2 are constants. Thus, the beam is statically determined (or overdetermined) if the kinematic boundary conditions forbid any rigid body motion.

It is natural to classify boundary conditions into three main groups:

- *Statically indetermined* cases, where there are not enough kinematic boundary conditions to restrict rigid body movements (see Fig. 4.4 top). These problems do not have unique solutions and cannot be analyzed within the framework of statical models.
- *Statically determined* cases, where there are three kinematic boundary conditions (which restrict rigid body movement, see Fig. 4.4 middle). In these cases, the constitutive equations (4.11) and the equilibrium equations (4.13) can be solved separately in a consecutive manner.
- *Statically determined* cases, where there are more than three kinematic boundary conditions (see Fig. 4.4 bottom). In these cases, equations (4.11) and (4.13) must be solved together as a single fourth order system. These kinds of boundary conditions allow the existence of non-zero stresses for an unloaded beam.

The energy of the arch is presented by the functional

$$\begin{aligned} J(u) &= \frac{1}{2}\int_0^1 \left\{ EA\left(u_1' - cu_2\right)^2 + EI\left(cu_1 + u_2'\right)'^2 \right\} \mathrm{d}s \\ &\quad - \int_0^1 f \cdot u \,\mathrm{d}s - Nu_1\Big|_0^1 + Fu_2\Big|_0^1 - Mu_2'\Big|_0^1. \end{aligned} \tag{4.17}$$

The solution $u \in V_0$ minimizes $J(u)$ over V_0 and satisfies the integral relation

$$a(u, w) = \ell(w), \quad \forall w \in V_0, \tag{4.18}$$

where

$$a(u, w) = \int_0^1 \begin{bmatrix} EA(u_1' - cu_2) \\ EI(cu_1 + u_2')' \end{bmatrix} \cdot \begin{bmatrix} w_1' - cw_2 \\ (cw_1 + w_2')' \end{bmatrix} \mathrm{d}s$$

and

$$\ell(w) := \int_0^1 f \cdot w \, \mathrm{d}s + Nw_1 \Big|_0^1 - Fw_2 \Big|_0^1 + Mw_2' \Big|_0^1 . \tag{4.19}$$

At each endpoint either homogeneous kinematic boundary condition is imposed or the corresponding load (N, F, or M) is defined. Hence, the functional ℓ in (4.19) is fully defined. The energy norm for the problem is $|||w|||^2 := \sqrt{a(w, w)}$.

For the existence of the solution of the Kirchhoff–Love arch problem we must show the ellipticity of $a : V_0 \times V_0 \to \mathbb{R}$ (see Sect. B.2), which is proved in [Cia78a] (Theorem 8.1.2, p. 433),

Theorem 4.2 *If the function c is continuously differentiable over the interval* I, *then the bilinear form*

$$a(u, v) = \int_{\mathrm{I}} \{ (u_1' - cu_2)(v_1' - cv_2) + (u_2' + cu_1)'(v_2 + cu_1)' \} \, \mathrm{d}s$$

is $H_0^1(\mathrm{I}) \times (H^2(\mathrm{I}) \cap H_0^1(\mathrm{I}))$-elliptic, and thus, it is a fortiori $H_0^1(\mathrm{I}) \times H_0^2(\mathrm{I})$-elliptic.

Theorem 4.2 states that for a statically determinate beam, there exists a positive constant C_{KL} such that

$$\int_0^1 (w_1^2 + w_2^2 + w_1'^2 + w_2'^2 + w_2''^2) \, \mathrm{d}s$$
$$\leq C_{\mathrm{KL}} \int_0^1 \left((w_1' - cw_2)^2 + (cw_1 + w_2')'^2 \right) \mathrm{d}s, \tag{4.20}$$

for all $w \in V_0$.

4.1.3.1 Estimates of Deviations for the Kirchhoff–Love Arch Model

Estimates of deviations from the exact solution of the Kirchhoff–Love arch model were firstly presented in [Mal09]. We follow the lines of this paper and show how to derive the error majorant with the help of transformations of the integral relation (4.18).

Theorem 4.3 *Let u be a solution of* (4.18) *and $v \in V_0$. Then,*

$$|||u - v|||^2 \leq \overline{\mathrm{M}}_{\mathrm{KL}}^2(v, y, \beta), \quad y \in Q_0, \beta > 0, \tag{4.21}$$

where

$$\overline{\mathsf{M}}^2_{\mathrm{KL}}(v, y, \beta)$$
$$:= \left(1+\frac{1}{\beta}\right)\frac{C_{\mathrm{KL}}}{\alpha}\int_0^1 \{(f_1 + (y_1' + cy_2'))^2 + (f_2 - (cy_1 + y_2''))^2\}\,\mathrm{d}s$$
$$+ (1+\beta)\int_0^1 \left\{\frac{1}{EA}\big(y_1 - EA(v_1' - cv_2)\big)^2 + \frac{1}{EI}\big(y_2 - EI(cv_1 + v_2')'\big)^2\right\}\mathrm{d}s, \tag{4.22}$$

C_{KL} *is defined by* (4.20), *and* $\alpha := \min\{EA, EI\}$.

Proof We begin with the integral identity that defines the generalized solution of the arch problem. We note that

$$\int_0^1 \begin{bmatrix} w_1' - cw_2 \\ (cw_1 + w_2')' \end{bmatrix} \cdot \begin{bmatrix} y_1 \\ y_2 \end{bmatrix} \mathrm{d}s$$
$$= \int_0^1 \begin{bmatrix} -y_1' - cy_2' \\ -cy_1 + y_2'' \end{bmatrix} \cdot \begin{bmatrix} w_1 \\ w_2 \end{bmatrix} \mathrm{d}s + w_1 y_1\Big|_0^1 + (cw_1 + w_2')y_2\Big|_0^1 - w_2 y_2'\Big|_0^1 \tag{4.23}$$

for any $w \in H^1((0,1)) \times H^2((0,1))$ and $y \in H^1((0,1)) \times H^2((0,1))$.

By (4.18) and (4.23), we obtain

$$\begin{aligned} a(u - v, w) = &\int_0^1 f \cdot w\,\mathrm{d}s + Nw_1\Big|_0^1 - Fw_2\Big|_0^1 + Mw_2'\Big|_0^1 \\ &- \int_0^1 \begin{bmatrix} EA(v_1' - cv_2) \\ EI(cv_1 + v_2')' \end{bmatrix} \cdot \begin{bmatrix} w_1' - cw_2 \\ (cw_1 + w_2')' \end{bmatrix} \mathrm{d}s \\ &+ \int_0^1 \begin{bmatrix} w_1' - cw_2 \\ (cw_1 + w_2')' \end{bmatrix} \cdot \begin{bmatrix} y_1 \\ y_2 \end{bmatrix} \mathrm{d}s \\ &- \int_0^1 \begin{bmatrix} -y_1' - cy_2' \\ cy_1 + y_2'' \end{bmatrix} \cdot \begin{bmatrix} w_1 \\ w_2 \end{bmatrix} \mathrm{d}s \\ &- w_1 y_1\Big|_0^1 - (cw_1 + w_2')y_2\Big|_0^1 + w_2 y_2'\Big|_0^1. \end{aligned} \tag{4.24}$$

We rewrite (4.24) in the form

$$a(u - v, w) = I_1 + I_2 + I_3, \tag{4.25}$$

where

$$I_1 = \int_0^1 \begin{bmatrix} f_1 + (y_1' + cy_2') \\ f_2 - (cy_1 + y_2'') \end{bmatrix} \cdot \begin{bmatrix} w_1 \\ w_2 \end{bmatrix} \mathrm{d}s,$$

$$I_2 = \int_0^1 \begin{bmatrix} y_1 - EA(v_1' - cv_2) \\ y_2 - EI(cv_1 + v_2')' \end{bmatrix} \cdot \begin{bmatrix} w_1' - cw_2 \\ (cw_1 + w_2')' \end{bmatrix} \mathrm{d}s,$$

and

$$I_3 = (N - y_1 - cy_2)w_1\Big|_0^1 + (-F + y_2')w_2\Big|_0^1 + (M - y_2)w_2'\Big|_0^1.$$

After imposing the boundary conditions, $w \in V_0$ and $y \in Q_0$, I_3 vanishes.

I_1 and I_2 we estimate from the above. By the Cauchy–Schwartz inequality, we have

$$I_1 \le \left(\int_0^1 (f_1 + (y_1' + cy_2'))^2 + (f_2 - (cy_1 + y_2''))^2 \,\mathrm{d}s\right)^{1/2} \times \left(\int_0^1 (w_1^2 + w_2^2)\,\mathrm{d}s\right)^{1/2}. \tag{4.26}$$

We can estimate the L^2-norm of w from the above by the full Sobolev norm and apply (4.20), then

$$I_1 \le \left(\int_0^1 (f_1 + (y_1' + cy_2'))^2 + (f_2 - (cy_1 + y_2''))^2 \,\mathrm{d}s\right)^{1/2} \times \frac{\sqrt{C_{\mathrm{KL}}}}{\sqrt{\alpha}} \left(\int_0^1 EA(w_1' - cw_2)^2 + EI(cw_1 + w_2')'^2 \,\mathrm{d}s\right)^{1/2}, \tag{4.27}$$

where $\alpha = \min\{EA, EI\}$. Now, we apply the Cauchy–Schwartz inequality again and find that

$$\begin{aligned} I_2 &= \int_0^1 \begin{bmatrix} \frac{1}{\sqrt{EA}}(y_1 - EA(v_1' - cv_2)) \\ \frac{1}{\sqrt{EI}}(y_2 - EI(cv_1 + v_2')') \end{bmatrix} \cdot \begin{bmatrix} \sqrt{EA}(w_1' - cw_2) \\ \sqrt{EI}(cw_1 + w_2')' \end{bmatrix} \mathrm{d}s \\ &\le \left(\int_0^1 \frac{1}{EA}(y_1 - EA(v_1' - cv_2))^2 \right. \\ &\quad \left. + \frac{1}{EI}(y_2 - EI(cv_1 + v_2')')^2 \,\mathrm{d}s\right)^{1/2} |||w|||. \end{aligned} \tag{4.28}$$

We apply (4.28) and (4.28) them to (4.25) and set $w = u - v$. Then, we arrive at

$$\begin{aligned} |||u - v||| &\le \frac{\sqrt{C_{\mathrm{KL}}}}{\sqrt{\alpha}} \left(\int_0^1 (f_1 + (y_1' + cy_2'))^2 + (f_2 - (cy_1 + y_2''))^2 \,\mathrm{d}s\right)^{1/2} \\ &\quad + \left(\int_0^1 \frac{1}{EA}(y_1 - EA(v_1' - cv_2))^2 \right. \\ &\quad \left. + \frac{1}{EI}(y_2 - EI(cv_1 + v_2')')^2 \,\mathrm{d}s\right)^{1/2}. \end{aligned} \tag{4.29}$$

Squaring both sides and (A.4) leads to (4.21). □

Two terms of the error majorant have a clear meaning. The first part

$$\overline{\mathsf{M}}_{\mathrm{KL}}^{\mathrm{equi}} := \int_0^1 \left(f_1 + (y_1' + cy_2')\right)^2 + \left(f_2 - (cy_1 + y_2'')\right)^2 \mathrm{d}s$$

is the error in the equilibrium condition (4.13) and the second part

$$\overline{\mathsf{M}}_{\mathrm{KL}}^{\mathrm{const}} := \int_0^1 \frac{1}{EA}\left(y_1 - EA(v_1' - cv_2)\right)^2 + \frac{1}{EI}\left(y_2 - EI(cv_1 + v_2')'\right)^2 \mathrm{d}s$$

is the violation of the constitutive relation (4.11). If we set $y := p$, then $\overline{\mathsf{M}}_{\mathrm{KL}}^{\mathrm{equi}} = 0$ and $\overline{\mathsf{M}}_{\mathrm{KL}}^{\mathrm{const}}$ coincides with the exact error.

Remark 4.2 If we define,

$$\begin{aligned} \Lambda u &:= \begin{bmatrix} u_1' - cu_2 \\ (cu_1 + u_2')' \end{bmatrix}, \qquad \Lambda^* p := \begin{bmatrix} -p_1' - cp_2' \\ -cp_1 + p_2'' \end{bmatrix}, \quad \text{and} \\ \mathcal{A}p &:= \begin{bmatrix} EAp_1 \\ EIp_2 \end{bmatrix}, \end{aligned} \tag{4.30}$$

then (4.11) and (4.13) can be written as

$$\mathcal{A}\Lambda u = p \tag{4.31}$$

and

$$\Lambda^* p = f, \tag{4.32}$$

respectively. Under the definitions (4.30) the majorant has the typical structure discussed in Sect. 4.4 for a generalized model problem

$$\begin{aligned} \overline{\mathsf{M}}_{\Lambda}^2(v, y, \beta) &:= \left(1 + \frac{1}{\beta}\right)\frac{C_{\mathrm{KL}}}{\alpha}\int_0^1 \left|f - \Lambda^* y\right|^2 \mathrm{d}s \\ &\quad + (1+\beta)\int_0^1 \mathcal{A}^{-1}(y - \mathcal{A}\Lambda v)\cdot(y - \mathcal{A}\Lambda v)\,\mathrm{d}s. \end{aligned} \tag{4.33}$$

Example 4.1 Consider a half-circular beam

$$\psi(t) = \begin{bmatrix} \cos(\pi t) \\ \sin(\pi t) \end{bmatrix}, \quad t \in [0, 1],$$

where the curvature is $c = 1$ and $EA = EI = 1$. Let both ends of the beam be clamped, i.e.,

$$u_1(0) = u_2(0) = u_2'(0) = u_1(1) = u_2(1) = u_2'(1) = 0.$$

The basis that satisfies the boundary conditions can be easily constructed using Fourier type basis functions. Let

$$V_0^N := \operatorname{span}\left\{ \begin{bmatrix} \sin(k\pi t) \\ 0 \end{bmatrix}, \begin{bmatrix} 1 - \cos(2k\pi t) \\ 0 \end{bmatrix}, \begin{bmatrix} 0 \\ 1 - \cos(2k\pi t) \end{bmatrix} \right\}_{k=1}^{N}. \tag{4.34}$$

We set $C_{\mathrm{KL}} = 1$ based on the approximated general eigenvalue problem: Find $v_k \in V_0^N$ such that

$$\int_0^1 A\Lambda v_k \cdot \Lambda w \, \mathrm{d}s = \lambda_k \int_0^1 v_k \cdot w \, \mathrm{d}s, \quad \forall w \in V_0^N.$$

We increased N and observed how the approximated value of the smallest eigenvalue develops.

Consider a polynomial solution

$$u(t) = \begin{bmatrix} t(t-1) \\ t^2(t-1)^2 \end{bmatrix}, \tag{4.35}$$

which satisfies the kinematic boundary conditions. We generate an approximate solution v by perturbing u. To measure the efficiency of the a posteriori error estimates, we introduce efficiency indexes,

$$I_{\mathrm{eff}}^{\oplus} := \frac{\overline{\mathrm{M}}_{\mathrm{KL}}}{|||u - v|||} \tag{4.36}$$

and

$$I_{\mathrm{eff}}^{\ominus} := \frac{\underline{\mathrm{M}}_{\mathrm{KL}}}{|||u - v|||}. \tag{4.37}$$

Since the estimates are guaranteed, we know that

$$I_{\mathrm{eff}}^{\ominus} \leq 1 \leq I_{\mathrm{eff}}^{\oplus}.$$

Let y in the majorant be defined using a Fourier basis, i.e.,

$$y \in Q_N := \operatorname{span}\left\{ \begin{bmatrix} \sin(k\pi t) \\ 0 \end{bmatrix}, \begin{bmatrix} \cos(k\pi t) \\ 0 \end{bmatrix}, \begin{bmatrix} 0 \\ \sin(k\pi t) \end{bmatrix}, \begin{bmatrix} 0 \\ \cos(k\pi t) \end{bmatrix} \right\}_{k=1}^{N}.$$

We minimize the majorant with respect to $y \in Q_N$ as in Algorithm 3.2. Regardless of the dimension N, the iteration converged within five to six steps. In Table 4.2, we can observe how the majorant improves as N increases. The efficiency index

Table 4.2 Efficiency index and parts of the majorant with different N

	N						
	4	6	8	9	10	11	12
$I^{\oplus}_{\text{eff}}$	1.0887	1.0728	1.0101	1.0024	1.0005	1.0001	1.0000
$\overline{\text{M}}^2_{\text{KL}}$	60.143	58.398	51.774	50.985	50.7949	50.752	50.742
$\overline{\text{M}}^{\text{equi}}_{\text{KL}}$	0.01320	0.00229	0.00485	0.00029	0.00002	$7.2\cdot 10^{-7}$	$3.3\cdot 10^{-8}$
$\overline{\text{M}}^{\text{const}}_{\text{KL}}$	58.374	57.658	50.776	50.742	50.740	50.740	50.740

Table 4.3 Efficiency index of the minorant with different values of N

	N				
	2	3	4	5	6
$I^{\ominus}_{\text{eff}}$	0.99985	0.99994	0.99997	0.99998	0.99999
$\underline{\text{M}}^2_{\text{KL}}(v, w^N)$	50.725	50.7341	50.737	50.7382	50.7388

Table 4.4 Efficiency index of the majorant with different N and C_{KL}

C_{KL}	N						
	4	6	8	9	10	11	12
1	1.0887	1.0728	1.0101	1.0024	1.0005	1.0001	1.0000
10	1.1219	1.0771	1.0296	1.0075	1.0017	1.0004	1.0001
100	1.2244	1.0839	1.0683	1.0224	1.0054	1.0012	1.0003
1000	1.5475	1.0966	1.0767	1.0577	1.0163	1.0037	1.0008

tends to one as the majorant approaches the exact deviation. Moreover, the equilibrium part of the majorant tends to zero as the constitutive part approaches the exact deviation error.

Next, we study the minorant. We solve the problem (4.18), using the Galerkin method with the Fourier type subspace V_0^N in (4.34). Then, we compute the energy of the obtained approximation w^N and estimate the error from below by comparing it to the energy of v as follows:

$$|||u - v|||^2 \geq 2\big(J(v) - J\big(w^N\big)\big) =: \underline{\text{M}}^2_{\text{KL}}\big(v, w^N\big). \tag{4.38}$$

The resulting lower bounds are presented in Table 4.3. The reason for the efficiency of the minorant is that the applied basis (4.34) can represent the exact solution (4.35) very well with a relatively small number of basis functions.

Remark 4.3 Since the term related to the equilibrium relation tends to zero, even a substantial overestimation of the constant C_{KL} does not seriously affect the efficiency of these estimates. In Table 4.4, we show the efficiency indexes obtained by different values for the constant C_{KL}.

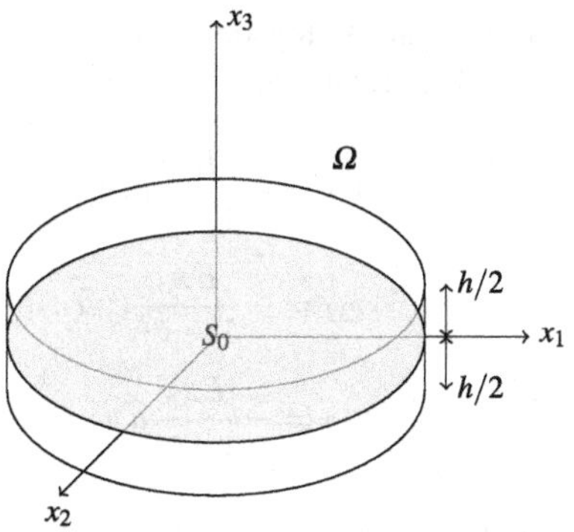

Fig. 4.5 Plate type elastic *body* Ω and middle *surface* S_0

4.1.4 The Kirchhoff–Love Plate

Plates are commonly used structures in many engineering applications. Here we consider two of the most common type plates, the Kirchhoff–Love plate model and the Reissner–Mindlin plate model. They can be viewed as two-dimensional counterparts of the Euler–Bernoulli beam model and the Timoshenko beam model, respectively.

We consider a bounded three-dimensional elastic body occupying the domain

$$\Omega := \left\{(x_1, x_2) \in \hat{\Omega}, x_3 \in \left(-\frac{h}{2}, \frac{h}{2}\right)\right\},$$

where $\hat{\Omega}$ is a bounded open domain in the x_1, x_2-plane with Lipschitz continuous boundary $\hat{\Gamma}$ and h is a positive constant (see Fig. 4.5), which is assumed to be small with respect to the size of $\hat{\Omega}$. We refer to the domain $\hat{\Omega}$ as a plate and denote elements of x_1, x_2-plane by $\widehat{x}$. $S_0 := \{x \in \Omega \mid x_3 = 0\}$ is the *middle surface* of the plate, which is deflected by volume and surface loads $f = (0, 0, \hat{f})$ and $F = (0, 0, \hat{F})$, respectively. The volume load $\hat{f} = \hat{f}(\hat{x})$ is distributed inside Ω, and the surface load $\hat{F} = \hat{F}(\hat{x})$ is distributed on the upper face $\Gamma_{\oplus} := \{x \in \Omega \mid x_3 = \frac{h}{2}\}$. In the classical theory of Kirchhoff–Love plates, the corresponding 3D model is replaced by a simplified 2D problem by means of the so-called Kirchhoff–Love hypothesizes. The first hypothesis states that the unit normal to the middle surface remains unstretched during the deformation of the plate. It means that the displacement vector is presented in the form

$$\begin{aligned}
u_1(x) &= -x_3\widehat{u}_{,1}(\widehat{x}),\\
u_2(x) &= -x_3\widehat{u}_{,2}(\widehat{x}),\\
u_3(x) &= \widehat{u}(\widehat{x}),
\end{aligned}$$

where $\widehat{u}$ is *a scalar-valued function* that represents deflections of S_0.

Another (static) hypothesis is that the components σ_{i3}, $i = 1, 2, 3$, are negligibly small compared to σ_{11}, σ_{12}, and σ_{22} so that they are set to zero. Thus, only the plane part of the stress tensor is considered. For the case of isotropic media, the Hooke's law leads at relations

$$\sigma_{11} = -\frac{Ex_3}{1-\nu^2}(\widehat{u}_{,11} + \nu\widehat{u}_{,22}) = -\frac{2\mu x_3}{1-\nu}(\widehat{u}_{,11} + \nu\widehat{u}_{,22}), \tag{4.39}$$

$$\sigma_{22} = -\frac{Ex_3}{1-\nu^2}(\nu\widehat{u}_{,11} + \widehat{u}_{,22}) = -\frac{2\mu x_3}{1-\nu}(\nu\widehat{u}_{,11} + \widehat{u}_{,22}), \tag{4.40}$$

$$\sigma_{12} = -\frac{Ex_3}{1+\nu}\widehat{u}_{,12} = -2\mu x_3\widehat{u}_{,12}. \tag{4.41}$$

In order to deduce the equation for $\widehat{u}$, we substitute relations (4.39)–(4.41) to the virtual energy relation

$$\int_\Omega \sigma : \varepsilon(w)\,\mathrm{d}x = \int_\Omega f \cdot w\,\mathrm{d}x + \int_{\Gamma_\oplus} F \cdot w\,\mathrm{d}x, \tag{4.42}$$

where the test functions are of the form

$$w = \big(-x_3\varphi_{,1}(\widehat{x}), -x_3\varphi_{,2}(\widehat{x}), \varphi(\widehat{x})\big),$$

where $\varphi \in H^2(\hat{\Omega}, \mathbb{R})$ is an arbitrary function vanishing on $\hat{\Gamma}$ together with its first derivatives. In view of the static hypothesis, the left-hand side of (4.79) contains only plane components and can be rewritten as follows:

$$\begin{aligned}\int_\Omega \sigma_{11}(u)\varepsilon_{11}(w)\,\mathrm{d}x &= \int_\Omega \frac{Ex_3^2}{1-\nu^2}(\widehat{u}_{,11} + \nu\widehat{u}_{,22})\varphi_{,11}\,\mathrm{d}x \\ &= \int_{\hat{\Omega}} D(\widehat{u}_{,11} + \nu\widehat{u}_{,22})\widehat{\varphi}_{,11} d\widehat{x},\end{aligned} \tag{4.43}$$

where $D := \frac{Eh^3}{12(1-\nu^2)} = \frac{\mu h^3}{6(1-\nu)}$. Analogously,

$$\int_\Omega \sigma_{22}(u)\varepsilon_{22}(w)\,\mathrm{d}x = \int_{\hat{\Omega}} D(\nu\widehat{u}_{,11} + \nu\widehat{u}_{,22})\varphi_{,22}\mathrm{d}\widehat{x}$$

and

$$\int_\Omega \sigma_{12}(u)\varepsilon_{12}(w)\,\mathrm{d}x = (1-\nu)\int_{\hat{\Omega}} D\widehat{u}_{,12}\varphi_{,12}\mathrm{d}\widehat{x}.$$

Hence, we arrive at the following problem: Find

$$\widehat{u} \in V_{00}(\hat{\Omega}) := \left\{\widehat{\eta} \in H^2(\hat{\Omega}) \mid \widehat{\eta} = \widehat{\eta}_{,n} = 0 \text{ on } \hat{\Gamma}\right\}$$

such that

$$\int_{\hat{\Omega}} D\big((\widehat{u}_{,11} + \nu\widehat{u}_{,22})\widehat{\varphi}_{,11} + (\nu\widehat{u}_{,11} + \nu\widehat{u}_{,22})\widehat{\varphi}_{,22} + 2(1-\nu)\widehat{u}_{,12}\widehat{\varphi}_{,12}\big)\mathrm{d}\widehat{x}$$
$$= \int_{\hat{\Omega}} \widehat{g}\widehat{\varphi}\mathrm{d}\widehat{x}, \quad \forall\widehat{\varphi} \in V_{00}(\hat{\Omega}), \tag{4.44}$$

where $\widehat{g}(\widehat{\mathbf{x}}) = h\widehat{f} + \widehat{F}$.

If $\widehat{u}$ is sufficiently regular, then (4.44) implies the classical plate equation (see, e.g., [DL72])

$$\widehat{u}_{,1111} + 2\widehat{u}_{,1122} + \widehat{u}_{,2222} = \frac{\widehat{g}}{D}, \tag{4.45}$$

a weak form of which is

$$\int_{\hat{\Omega}} D\widehat{\Delta}\widehat{u}\widehat{\Delta}\widehat{\varphi}\mathrm{d}\widehat{x} = \int_{\hat{\Omega}} \widehat{g}\widehat{\varphi}\mathrm{d}\widehat{x}, \quad \forall\widehat{\varphi} \in V_{00}(\hat{\Omega}). \tag{4.46}$$

This simplified 2D model is often used for numerical analysis of plate-type elastic bodies (see, e.g., [Bra07]). Concerning latest results on asymptotic convergence of KL type solutions, a priori error estimates, and a systematic bibliography, we refer to [BSS11].

The problem (4.46) belongs to the class of biharmonic problems, i.e.,

$$\begin{aligned} \operatorname{div}\operatorname{Div} B\nabla\nabla u &= f \quad \text{in } \Omega \subset \mathbb{R}^2 \\ u &= 0 \quad \text{on } \Gamma \\ \nabla u \cdot \nu &= 0 \quad \text{on } \Gamma, \end{aligned}$$

where B is a symmetric positive definite tensor, i.e.,

$$B_{jikm} = B_{ijkm} = B_{kmij}, \quad i, j, k, m \in \{1, 2\}$$

and

$$\alpha_1|\Upsilon|^2 \le B\Upsilon : \Upsilon \le \alpha_2|\Upsilon|^2, \quad \forall\Upsilon \in \mathbb{M}_s^{2\times 2}.$$

The generalized solution $u \in V_{00}$ satisfies the integral relation,

$$\int_\Omega B\nabla\nabla u : \nabla\nabla w \,\mathrm{d}x = \int_\Omega f w \,\mathrm{d}x, \quad \forall w \in V_{00}. \tag{4.47}$$

For this problem the majorant has been discussed in [NR01].

Theorem 4.4 *Let $u \in V_{00}$ be the solution of* (4.46) *and $v \in V_{00}$ be arbitrary. Then,*

$$\int_\Omega B\nabla\nabla(u - v) : \nabla\nabla(u - v)\,\mathrm{d}x \le \overline{\mathrm{M}}^2_{\nabla\nabla}(v, \kappa, y, \beta), \quad \forall\kappa \in Z_\nabla, y \in Y_{\mathrm{div}}, \beta > 0,$$

where

$$\overline{\mathsf{M}}^2_{\nabla\nabla}(v,\kappa,y,\beta) := (1+\beta)\|B\nabla\nabla v-\kappa\|^2_{B^{-1}}$$
$$+\left(\frac{1+\beta}{\beta}\right)\frac{C^2_{\nabla\Omega}}{\alpha_1}\left(\|\operatorname{Div}\kappa-y\|+C_{F\Omega}\|\operatorname{div}y-f\|\right)^2, \tag{4.48}$$
$$Z_{\mathrm{Div}} := \left\{\kappa\in L^2\left(\Omega,\mathbb{M}^{d\times d}\right) \mid \operatorname{Div}\kappa\in L^2\left(\Omega,\mathbb{R}^2\right)\right\},$$
$$Y_{\mathrm{div}} := \left\{y\in L^2\left(\Omega,\mathbb{R}^2\right) \mid \operatorname{div}y\in L^2(\Omega)\right\},$$

and $C_{\nabla\Omega}$ *is a constant in the inequality*

$$\|\nabla w\|\leq C_{\nabla\Omega}\|\nabla\nabla w\|, \quad \forall w\in V_{00}.$$

It is easy to see that the majorant contains the terms, which can be viewed as penalties for violations of the relations $\operatorname{div}y=f$, $y=\operatorname{Div}\kappa$, and $\kappa=B\nabla\nabla v$. Other variants of the majorant and a discussion of results related to biharmonic type problems can be found in [Fro04a, Fro04b, NR01, NR04, Rep08].

4.1.5 The Reissner–Mindlin Plate

The theory of Reissner–Mindlin plates is a generalization of the classical theory of Kirchhoff–Love thin plates. Again, we consider a plate type domain. For simplicity, in this section we omit the symbol ˆ in all formulas (so that Ω denotes a two-dimensional domain).

In the Reissner–Mindlin model, a plate is described in terms of the following two variables: a scalar-valued function $w(x)$ (which is the transverse displacement of the middle plane at point x) and a vector-valued function $\xi(x)$ (which is the rotation of fibers normal to the middle plane). In contrast to the Kirchhoff–Love model, the rotation of fibers is considered and presented by a special function. This yields better results in modeling of relatively "thick" plates.

The energy functional for this problem has the following form:

$$J(w,\xi) := \int_\Omega\left(\frac{1}{2}\mathbb{C}\varepsilon(\xi):\varepsilon(\xi)+\frac{1}{2\alpha}|\nabla w-\xi|^2-fw\right)\mathrm{d}x,$$

where $\mathbb{C}$ is a positive definite tensor (of the fourth-order), $\varepsilon(\xi)$ is the symmetric part of $\nabla\xi$, α is a positive parameter proportional to h^2 (h is the thickness) and the function f is related to the transverse loading of the plate.

The classical equations for the Reissner–Mindlin plate problem are as follows:

$$-\operatorname{div}p=f \quad \text{in } \Omega, \tag{4.49}$$
$$-\operatorname{Div}\mathbb{C}\varepsilon(\phi)-p=0 \quad \text{in } \Omega, \tag{4.50}$$

where $p := \frac{1}{\alpha}(\nabla u - \phi) \in L^2(\Omega, \mathbb{R}^2) =: Q$. The function p has an important physical meaning: it is the so-called shear stress vector (see, e.g. [AMZ02]). For simplicity, henceforth we consider the homogeneous Dirichlet boundary conditions, i.e.,

$$u = 0, \qquad \phi = 0 \quad \text{on } \Gamma. \tag{4.51}$$

Problems with mixed boundary conditions can be analyzed within the framework of the same scheme. We note that the existence of the minimizer $(u, \phi) \in S$ (where the set $S =: V_0 \times Y_0$ is the subset of $H^1(\Omega) \times H^1(\Omega, \mathbb{R}^2)$ containing functions satisfying the boundary conditions) follows from the known results of calculus of variations (see Sect. B.2). One can show that if $\alpha \to 0$, then the minimizers of this variational problem tend to the corresponding solution of the variational problem for the respective Kirchhoff–Love plate (see, e.g., [BF86]).

As before, one analysis is based on the weak formulation of the problem, which is as follows: Find $(u, \phi, p) \in V_0 \times Y_0 \times Q$ such that

$$\int_\Omega \mathbb{C}\varepsilon(u) : \varepsilon(\xi)\,dx - \int_\Omega p \cdot \xi\,dx = 0, \quad \forall \xi \in Y_0, \tag{4.52}$$

$$\int_\Omega p \cdot \nabla w\,dx - \int_\Omega f w\,dx = 0, \quad \forall \xi \in V_0, \tag{4.53}$$

$$\int_\Omega \left(p - \frac{1}{\alpha}(\nabla u - \phi)\right) \cdot q\,dx = 0, \quad \forall q \in Q. \tag{4.54}$$

It is not difficult to show that the solution of (4.52)–(4.54) coincides with the minimizer.

Numerical methods for the problem (4.52)–(4.54) are discussed in many publications (see, e.g., [AF89, AMZ02, BF86, BS98, LNS07, SS09] and references therein). In the next section, we derive computable majorants of deviations from the exact solution of the Reissner–Mindlin plate model, which can be used in order to measure errors encompassed in numerical approximations. Our exposition follows the lines of [RF04], where error majorants are obtained by the variational method (see, e.g., [NR04, Rep00b]), which is based on the dual energy formulation. Numerical tests can be found in [Fro04b, FNR06].

4.1.5.1 Error Majorants

Assume that $(v, \psi) \in S$ is an approximate solution. By $y := \frac{1}{\alpha}(\nabla v - \psi)$ we denote the corresponding shear stress vector and consider the deviations

$$e_v := v - u,$$

$$e_\psi := \psi - \phi,$$

$$e_y := y - p.$$

First, we define the *primal* problem,

$$J(u,\phi) = \inf_{(w,\xi)\in S} I(w,\xi) = \inf \mathcal{P}.$$

Then, we establish a generalized form of the Miklin's identity.

Proposition 4.1 *For any approximation* $(v,\psi)\in S$, *it holds*

$$|||e_\psi|||^2 + \alpha\|e_y\|^2 = 2\big(J(v,\psi) - J(u,\phi)\big), \tag{4.55}$$

where

$$\|e_y\|^2 = \int_\Omega |e_y|^2\,\mathrm{d}x \quad \textit{and} \quad |||e_\psi|||^2 := \int_\Omega \mathbb{C}\varepsilon(e_\psi) : \varepsilon(e_\psi)\,\mathrm{d}x.$$

Proof We have

$$\begin{aligned}
&2\big(J(v,\psi) - J(u,\phi)\big)\\
&= \int_\Omega \Big(\mathbb{C}\varepsilon(\psi) : \varepsilon(\psi) + \frac{1}{\alpha}|\nabla v - \psi|^2 - 2fv\Big)\,\mathrm{d}x\\
&\quad - \int_\Omega \Big(\mathbb{C}\varepsilon(\phi) : \varepsilon(\phi) + \frac{1}{\alpha}|\nabla u - \phi|^2 - 2fu\Big)\,\mathrm{d}x.
\end{aligned} \tag{4.56}$$

By (4.52),

$$\begin{aligned}
&\int_\Omega \mathbb{C}\varepsilon(\psi) : \varepsilon(\psi)\,\mathrm{d}x - \int_\Omega \mathbb{C}\varepsilon(\phi) : \varepsilon(\phi)\,\mathrm{d}x\\
&= \int_\Omega \mathbb{C}\varepsilon(\psi - \phi) : \varepsilon(\psi - \phi)\,\mathrm{d}x + 2\int_\Omega \mathbb{C}\varepsilon(\psi - \phi) : \varepsilon(\phi)\,\mathrm{d}x\\
&= |||e_\psi|||^2 + 2\int_\Omega \mathbb{C}\varepsilon(e_\psi) : \varepsilon(\phi)\,\mathrm{d}x\\
&= |||e_\psi|||^2 + 2\int_\Omega p : e_\psi\,\mathrm{d}x.
\end{aligned} \tag{4.57}$$

Now we recall definitions of y and p and find that

$$\begin{aligned}
\frac{1}{\alpha}\big(\|\nabla v - \psi\|^2 - \|\nabla u - \phi\|^2\big) &= \alpha\big(\|y\|^2 - \|p\|^2\big)\\
&= \alpha\Big(\|y - p\|^2 + 2\int_\Omega y\cdot(y - p)\,\mathrm{d}x\Big)\\
&= \alpha\|e_y\|^2 + 2\int_\Omega p\cdot(\alpha e_y)\,\mathrm{d}x.
\end{aligned} \tag{4.58}$$

Substituting (4.57) and (4.58) to (4.56) yields the relation

$$
\begin{aligned}
2\big(J(v,\psi) - J(u,\phi)\big) &= |||e_\psi|||^2 + 2\int_\Omega p : e_\psi \,\mathrm{d}x + \alpha\|e_y\|^2 \\
&\quad + 2\int_\Omega p\cdot(\alpha e_y)\,\mathrm{d}x - \int_\Omega f(v-u)\,\mathrm{d}x \\
&= |||e_\psi|||^2 + \alpha\|e_y\|^2 + 2\int_\Omega \big(p\cdot(e_\psi + \alpha e_y) - f(v-u)\big)\,\mathrm{d}x.
\end{aligned}
$$

Note that

$$
e_\psi + \alpha e_y = \psi - \phi + (\nabla v - \psi) - (\nabla u - \phi) = \nabla v - \nabla u = \nabla e_v \tag{4.59}
$$

and by (4.53) we obtain

$$
\int_\Omega \big(p\cdot(e_\psi + \alpha e_y) - f(v-u)\big)\,\mathrm{d}x = \int_\Omega (p\cdot\nabla e_v - f e_v)\,\mathrm{d}x = 0.
$$

Thus, we arrive at (4.55). □

Now, we discuss the so-called dual energy principle, which allows us to estimate the right-hand side of (4.55) from the above. Consider the Lagrangian (see Sect. B.3.1)

$$
L(w,\xi,q) := \int_\Omega \left(\frac{1}{2}\mathbb{C}\varepsilon(\xi) : \varepsilon(\xi) - f w + q\cdot(\nabla w - \xi) - \frac{1}{2}\alpha|q|^2\right)\mathrm{d}x. \tag{4.60}
$$

It generates the primal problem as follows:

$$
J(u,\psi) = \inf \mathcal{P} = \inf_{(w,\xi)\in S} J(w,\xi) = \inf_{(w,\xi)\in S}\sup_{q\in Q} L(w,\xi,q).
$$

Also, it generates the dual problem,

$$
I(p) = \sup \mathcal{P}^* = \sup_{q\in Q} I(q) = \sup_{q\in Q}\inf_{(w,\xi)\in S} L(w,\xi,q).
$$

From the theory of saddle points, it follows that

$$
J(u,\psi) = I(p). \tag{4.61}
$$

Then, Proposition 4.1, (4.61), and the definition of the dual problem yield the estimate

$$
|||e_\psi|||^2 + \alpha\|e_y\|^2 \le 2\big(J(v,\phi) - I(q)\big), \quad \forall q\in Q. \tag{4.62}
$$

The dual functional

$$
I(q) = \inf_{(w,\xi)\in S}\int_\Omega \left(\frac{1}{2}\mathbb{C}\varepsilon(\xi) : \varepsilon(\xi) - f w + q\cdot(\nabla w - \xi) - \frac{1}{2}\alpha|q|^2\right)\mathrm{d}x,
$$

is bounded from below only if

$$q \in Q_f := \left\{ q \in Q \,\middle|\, \int_\Omega q \cdot \nabla w \,\mathrm{d}x = \int_\Omega f w \,\mathrm{d}x, \forall w \in V_0 \right\}.$$

Thus, for $q \in Q_f$ we obtain

$$I(q) = \inf_{\xi \in Y_0} \int_\Omega \left(\frac{1}{2} \mathbb{C}\varepsilon(\xi) : \varepsilon(\xi) - q \cdot \xi - \frac{1}{2} \alpha |q|^2 \right) \mathrm{d}x.$$

In the right-hand side of this relation, we have a new variational problem (with respect to $\xi \in Y_0$, which contains H^1 vector-valued functions vanishing on Γ). It is a quadratic problem with the source term q. For a given q, we know that (cf. (B.39)–(B.41) with $u_0 = 0$)

$$\inf_{\xi \in Y_0} \int_\Omega \left(\frac{1}{2} \mathbb{C}\varepsilon(\xi) : \varepsilon(\xi) - q \cdot \xi \right) \mathrm{d}x = \sup_{\tau \in Z_q} \int_\Omega -\frac{1}{2} \mathbb{C}^{-1} \tau : \tau \,\mathrm{d}x,$$

where

$$Z_q := \left\{ \tau \in L^2(\Omega, \mathbb{M}_s^{2\times 2}) \,\middle|\, \int_\Omega \tau : \varepsilon(\xi) \,\mathrm{d}x = \int_\Omega q \cdot \xi \,\mathrm{d}x, \forall \xi \in Y_0 \right\}.$$

Therefore,

$$I(q) = \sup_{\tau \in Z_q} \tilde{I}(\tau, q) = \sup_{\tau \in Z_q} \int_\Omega \left(-\frac{1}{2} \mathbb{C}^{-1} \tau : \tau - \frac{1}{2} \alpha |q|^2 \right) \mathrm{d}x,$$

Since

$$I(q) \geq \tilde{I}(\tau, q), \quad \forall \tau \in Z_q, q \in Q_f,$$

we use (4.62) and find that

$$|||e_\psi|||^2 + \alpha \|e_y\|^2 \leq 2\big(J(v, \phi) - \tilde{I}(\tau, q)\big), \quad \forall \tau \in Z_q, q \in Q_f. \tag{4.63}$$

It is a matter of algebraic manipulation to show that (4.63) yields

$$|||e_\psi|||^2 + \alpha \|e_y\|^2 \leq \|\mathbb{C}\varepsilon(\psi) - \tau\|_{\mathbb{C}^{-1}}^2 + \alpha \|y - q\|^2, \quad \forall \tau \in Z_q, q \in Q_f, \tag{4.64}$$

where

$$\|\mathbb{C}\varepsilon(\psi) - \tau\|_{\mathbb{C}^{-1}}^2 := \int_\Omega \big(\mathbb{C}\varepsilon(\psi) - \tau\big) : \big(\varepsilon(\psi) - \mathbb{C}^{-1}\tau\big) \,\mathrm{d}x.$$

We see that if the right-hand side of (4.64) vanishes, then the relations (4.49) and (4.50) hold. The drawback of (4.63) and (4.64) is that these estimates are valid only for equilibrated functions $\tau \in Z_q$ and $q \in Q_f$, which are difficult to construct in practice. Next we modify the estimate (4.64) in such a way that it can be written in terms of auxiliary function in $\Sigma_{\rm div}$ and $Q_{\rm div}$, where

$$\Sigma_{\rm div} = \{\tau \in L^2(\Omega, \mathbb{M}_s^{2\times 2}) \mid {\rm Div}\,\tau \in Q\} \tag{4.65}$$

and

$$Q_{\rm div} = \{y \in L^2(\Omega, \mathbb{R}^2) \mid {\rm div}\, y \in L^2(\Omega)\}. \tag{4.66}$$

We introduce $\kappa \in \Sigma_{\rm div}$ and use the estimate

$$\|\mathbb{C}\varepsilon(\psi) - \tau\|_{\mathbb{C}^{-1}} \le \|\mathbb{C}\varepsilon(\psi) - \kappa\|_{\mathbb{C}^{-1}} + \|\tau - \kappa\|_{\mathbb{C}^{-1}}.$$

We square both sides, apply (A.4), and obtain the estimate

$$\|\mathbb{C}\varepsilon(\psi) - \tau\|^2_{\mathbb{C}^{-1}} \le (1+\beta_1)\|\mathbb{C}\varepsilon(\psi) - \kappa\|^2_{\mathbb{C}^{-1}} + \left(\frac{1+\beta_1}{\beta_1}\right)\|\tau - \kappa\|^2_{\mathbb{C}^{-1}}, \tag{4.67}$$

which is valid for any positive β_1. In order to estimate the last norm we need the following auxiliary result.

Proposition 4.2 *For any $\kappa \in Z_{\rm div}$,*

$$\inf_{\tau\in Z_q} \|\tau - \kappa\|_{\mathbb{C}^{-1}} \le C_{1\Omega}\|q + {\rm Div}\,\kappa\|, \tag{4.68}$$

where $C_{1\Omega}$ is a constant in the inequality

$$\|\xi\| \le C_{1\Omega}|||\xi|||, \quad \forall \xi \in Y_0.$$

Proof Let $\sigma := \tau - \kappa$. Since $\tau \in Z_q$ and $\kappa \in Z_{\rm div}$, we see that

$$\sigma \in Z_r := \left\{\sigma \in L^2(\Omega, \mathbb{M}_s^{2\times 2}) \,\Big|\, \int_\Omega \sigma : \varepsilon(\xi)\,dx = \int_\Omega r\cdot\xi\,dx, \forall \xi \in Y_0,\right\},$$

where $r := q + {\rm Div}\,\kappa$. We have

$$\begin{aligned}\inf_{\tau\in Z_q} \|\tau - \kappa\|^2_{\mathbb{C}^{-1}} &= \inf_{\sigma\in Z_r} \|\sigma\|^2_{\mathbb{C}^{-1}} \\ &= -\sup_{\sigma\in Z_r}\{-\|\sigma\|^2_{\mathbb{C}^{-1}}\} = -\inf_{\xi\in Y_0}\int_\Omega (\mathbb{C}\varepsilon(\xi):\varepsilon(\xi) - 2q\cdot\xi)\,dx.\end{aligned}$$

We estimate the right-hand side from the above to obtain the statement

$$\begin{aligned}-\inf_{\xi\in Y_0}\int_\Omega (\mathbb{C}\varepsilon(\xi):\varepsilon(\xi) - 2q\cdot\xi)\,dx &\le -\inf_{\xi\in Y_0}(|||\xi|||^2 - 2C_{1\Omega}\|r\|\,|||\xi|||) \\ &= -\inf_{a\ge 0}(a^2 - 2C_{1\Omega}\|r\|a) = C^2_{1\Omega}\|r\|^2. \quad \square\end{aligned}$$

In order to exclude $q \in Q_f$ from the estimate, we introduce $\gamma \in Q_{\text{div}}$ and use similar arguments, namely,

$$\|y - q\|^2 \leq (1 + \beta_2)\|y - \gamma\|^2 + \left(\frac{1 + \beta_2}{\beta_2}\right)\|q - \gamma\|^2 \tag{4.69}$$

and

$$\|q + \operatorname{Div}\kappa\|^2 \leq (1 + \beta_3)\|\operatorname{Div}\kappa + \gamma\|^2 + \left(\frac{1 + \beta_2}{\beta_2}\right)\|q - \gamma\|^2, \tag{4.70}$$

where β_2 and β_3 are arbitrary positive parameters.

Proposition 4.3 *For any* $\gamma \in Q_{\text{div}}$,

$$\inf_{q \in Q_f} \|q - \gamma\| \leq C_{F\Omega}\|\operatorname{div}\gamma + f\|. \tag{4.71}$$

Proof is similar to the previous case.

Applying (4.68)–(4.71) to (4.67) yields the desired majorant:

$$|||e_\psi|||^2 + \alpha\|e_y\|^2 \leq \overline{\mathsf{M}}^2_{\text{RM}}(\psi, y, \gamma, \kappa), \quad \forall \gamma \in Q_{\text{div}}, \kappa \in Z_{\text{div}}, \tag{4.72}$$

where

$$\begin{aligned} &\overline{\mathsf{M}}^2_{\text{RM}}(\psi, y, \gamma, \kappa; \beta_1, \beta_2, \beta_3) \\ &\quad := (1 + \beta_1)\left\|\mathbb{C}\varepsilon(\psi) - \kappa\right\|^2_{\mathbb{C}^{-1}} \\ &\qquad + (1 + \beta_2)\alpha\|y - \gamma\|^2 + \left(\frac{1 + \beta_1}{\beta_1}\right)(1 + \beta_3)C^2_{1\Omega}\|\gamma + \operatorname{Div}\kappa\|^2 \\ &\qquad + \left(\left(\frac{1 + \beta_1}{\beta_1}\right)\left(\frac{1 + \beta_3}{\beta_3}\right)C^2_{1\Omega} + (1 + \beta_2)\alpha\right)C^2_{F\Omega}\|\operatorname{div}\gamma + f\|^2. \end{aligned} \tag{4.73}$$

This estimate (4.73) consists of the violations of relations (4.52)–(4.53) and additional terms resulting from the fact that auxiliary functions are not "equilibrated", i.e., $\gamma \neq Q_f$ and $\kappa \neq Z_q$.

4.1.6 3D Linear Elasticity

Let the domain $\Omega \subset \mathbb{R}^3$ have the boundary Γ consisting of two disjoint parts Γ_D and Γ_N. We assume that Γ_D has the positive measure, e.i. $|\Gamma_D| > 0$. The classical formulation of the linear elasticity problem is to find a tensor-valued function σ (stress) and a vector-valued function u (displacement) that satisfy the system of equations (see, e.g., [TG51])

$$\sigma = \mathbb{L}\varepsilon \quad \text{in } \Omega, \tag{4.74}$$

$$\operatorname{Div}\sigma + f = 0 \quad \text{in } \Omega, \tag{4.75}$$

$$u = g \quad \text{on } \Gamma_D, \tag{4.76}$$

$$\sigma\nu = F \quad \text{on } \Gamma_N. \tag{4.77}$$

Here g defines the Dirichlet boundary condition, f is the volume force, ν is the unit normal vector outward to Γ, F is the surface load on Γ_N, and $\mathbb{L} = \{L_{ijkm}\}$ is the tensor of elasticity constants. We assume that it is positive definite and bounded, i.e.,

$$c_1|\varepsilon|^2 \leq \mathbb{L}\varepsilon : \varepsilon \leq c_2|\varepsilon|^2, \quad \forall \varepsilon \in \mathbb{M}_s^{3\times 3},$$

and satisfies the symmetry conditions

$$L_{jikm} = L_{ijkm} = L_{kmij}, \quad i, j, k, m \in \{1, 2, 3\}$$

(the Hooke's law), which presupposes a linear dependence between strains and stresses. The relation (4.75) means that internal stresses are in *equilibrium* with body forces.

A generalized solution of (4.74)–(4.77) is a function $u \in V_0 + g$, where

$$V_0 := \left\{ w \in H^1\left(\Omega, \mathbb{R}^d\right) \mid w = 0 \text{ on } \Gamma_D \right\} \tag{4.78}$$

that minimizes the energy functional

$$J(w) := \frac{1}{2}\mathcal{E}(w) - \langle \ell, w \rangle,$$

where

$$\langle \ell, w \rangle = \int_\Omega f \cdot w \,\mathrm{d}x + \int_{\Gamma_N} F \cdot w \,\mathrm{d}s,$$

and

$$\mathcal{E}(u) = \int_\Omega \sigma(u) : \varepsilon(u)\,\mathrm{d}x = \int_\Omega \mathbb{L}\varepsilon(u) : \varepsilon(u)\,\mathrm{d}x := |||u|||_{\mathbb{L}}^2$$

is the elastic energy of the deformation. The generalized solution satisfies the integral relation

$$\int_\Omega \mathbb{L}\varepsilon(u) : \varepsilon(w)\,\mathrm{d}x = \langle \ell, w \rangle, \quad \forall w \in V_0. \tag{4.79}$$

In the special case of an isotropic medium, the tensor $\mathbb{L}$ can be expressed with the help of only two material parameters. Usually, they are the Lame constants that lead to the form

$$\mathbb{L}\varepsilon(u) = \lambda \operatorname{div} u\mathbb{I} + 2\mu\varepsilon(u).$$

Another pair of constants is Young's modulus E and Poisson's ratio ν. They are related to the Lame constants as follows:

$$\lambda = \frac{E\nu}{(1+\nu)(1-2\nu)}, \qquad \mu = \frac{E}{2(1+\nu)}. \tag{4.80}$$

The tensor $\mathbb{L}$ can be expressed also, using the bulk modulus K,

$$\mathbb{L}\varepsilon(u) = K \operatorname{div} u \mathbb{I} + 2\mu\varepsilon^D(u), \tag{4.81}$$

where strain is decomposed to a volumetric and a deviatoric part

$$\varepsilon = \frac{1}{3}\operatorname{tr}\varepsilon\mathbb{I} + \varepsilon^D = \frac{1}{3}\operatorname{div}(u)\mathbb{I} + \varepsilon^D(u).$$

4.1.6.1 Estimates of Deviations from the Exact Displacement Field

The majorant for the problem (4.74)–(4.77) has the form (see [MR03, NR04, Rep01b, Rep08])

$$\overline{\mathsf{M}}^2_{\mathrm{EL}}(v,\tau;\beta) := (1+\beta)\int_\Omega \big(\varepsilon(v) - \mathbb{C}^{-1}\tau\big) : \big(\mathbb{C}\varepsilon(v) - \tau\big)\,\mathrm{d}x$$
$$+ \frac{1+\beta}{\beta} C\big(\|\operatorname{Div}\tau + f\|^2 + \|F - \tau n\|^2_{\Gamma_N}\big), \tag{4.82}$$

where $C > 0$ is a constant in the inequality

$$\|w\|^2_\Omega + \|w\|^2_{\Gamma_N} \le C|||w|||^2_{\mathbb{L}}, \quad \forall w \in V_0,$$

the existence of which follows from the Korn's inequality (see Sect. A.3.3). It is easy to see that the majorant has the same structure as other majorants we have considered earlier.

Three terms of the majorant penalize violations of the Hooke's law (4.74), equilibrium relations (4.75), and the Neumann boundary condition (4.77).

Theorem 4.5 *For any $v \in V_0$,*

$$|||u - v|||^2 \le \overline{\mathsf{M}}^2_{\mathrm{EL}}(v,\tau;\beta), \quad \forall\tau \in H_{\Gamma_N}(\mathrm{Div},\Omega), \beta > 0, \tag{4.83}$$

where

$$H_{\Gamma_N}(\mathrm{Div},\Omega) := \big\{\tau \in L^2(\Omega,\mathbb{M}^{d\times d}) \mid \operatorname{Div}\tau \in L^2(\Omega,\mathbb{R}^d), \tau n \in L^2(\Gamma_N,\mathbb{R}^d)\big\}$$

is the space of admissible stresses. The majorant vanishes if and only if $v = u$.

Theorem 4.5 was proved in [Rep01a], see also [NR04] and [Rep08]. For the case of isotropic media, the estimate follows from the result of [RX96] as a particular case.

We consider the special case of isotropic media and material parameter pair K and μ. For them, we have

$$\mathbb{L}^{-1}\tau = \frac{1}{9K}\operatorname{tr}\tau\mathbb{I} + \frac{1}{2\mu}\tau^D, \tag{4.84}$$

and

$$\varepsilon(v) - \mathbb{L}^{-1}\tau = \left(\frac{1}{3}\operatorname{div} v - \frac{1}{9K}\operatorname{tr}\right)\mathbb{I} + \left(\varepsilon(v)^D - 2\mu\tau^D\right),$$

$$\mathbb{L}\varepsilon(v) - \tau = \left(K\operatorname{div} v - \frac{1}{3}\operatorname{tr}\right)\mathbb{I} + \left(2\mu\varepsilon^D(v) - \tau^D\right).$$

Since for any deviatoric tensor τ^D, $\operatorname{tr}\tau^D = 0$ and $\tau^D : \mathbb{I} = 0$, we have

$$\left(\varepsilon(v) - \mathbb{L}^{-1}\tau\right) : \left(\mathbb{L}\varepsilon(v) - \tau\right) = K\left(\operatorname{div} v - \frac{1}{3K}\operatorname{tr}\tau\right)^2 + 2\mu\left|\varepsilon^D(v) - \frac{1}{2\mu}\tau^D\right|^2.$$

Hence, for the isotropic media the estimate (4.83) reads as follows:

$$\int_\Omega \left(K\operatorname{div}(u-v)^2 + 2\mu|\varepsilon^D(u-v)|^2\right)\mathrm{d}x$$
$$\leq \overline{\mathsf{M}}^2_{\mathrm{EL,iso}}(v,\tau,\beta), \quad \forall \tau \in H_{\Gamma_N}(\mathrm{Div},\Omega), \beta > 0,$$

where

$$\overline{\mathsf{M}}^2_{\mathrm{EL,iso}}(v,\tau;\beta) := (1+\beta)\int_\Omega \left(K\left(\operatorname{div} v - \frac{1}{3K}\operatorname{tr}\tau\right)^2 + 2\mu\left|\varepsilon(v)^D - \frac{1}{2\mu}\tau^D\right|^2\right)\mathrm{d}x$$
$$+ \frac{1+\beta}{\beta}C\left(\|\operatorname{Div}\tau + f\|^2 + \|F - \tau n\|^2_{\Gamma_N}\right). \tag{4.85}$$

In (4.85), the part of the majorant related to the Hooke's law (4.74) is decomposed into two terms, which measure possible violations of the Hooke's law for the volumetric and deviatoric.

Remark 4.4 Linear elasticity equations are often considered together with other physical equations (e.g., those that describe diffusion). We arrive at coupled models, for which error majorants should be obtained by combining estimates deduced for each sub-problem separately. Sometimes (as, e.g., for thermoelasticity) this method easily yields the estimate for the whole system. Sometimes, (as in the case of the Barenblatt–Biot poroelastic model) deducing the estimate requires some efforts (see [NRRV10]). In general, getting fully reliable, computable, and efficient bounds for coupled models is a complicated problem, which may be much more difficult than getting such bounds for each sub-problem separately (see Sect. 6.9).

4.1.7 The Plane Stress Model

Consider an isotropic material occupying the domain $\Omega := \hat{\Omega} \times (-\frac{h}{2}, \frac{h}{2})$, where $\hat{\Omega}$ belongs to the x_1x_2-plane and h is "much smaller" compared to $\hat{\Omega}$. As before, we denote plane coordinates by $\hat{x} = (x_1, x_2)$.

In the plane stress model, it is assumed that the stress tensor is planar, i.e.,

$$\sigma_{i3} = 0, \quad i = \{1, 2, 3\} \tag{4.86}$$

and we have $\hat{\sigma} \in \mathbb{M}^{2\times 2}$. Moreover, displacement and body forces are planar, i.e.,

$$u_3 = 0 \quad \text{and} \quad u := \hat{u} = (u_1, u_2).$$

Also, it is assumed that $f_3 = 0$ and $f := \hat{f} = (f_1, f_2)$. Strictly speaking, these assumptions violate the 3D elasticity, but they are often used because they make the model simpler. Due to (4.86), we have $\varepsilon_{13} = \varepsilon_{23} = 0$. The condition (4.81)

$$0 = \sigma_{33} = \left(K \operatorname{tr} \varepsilon \mathbb{I} + 2\mu \varepsilon^D(u)\right)_{33} = K \operatorname{tr} \varepsilon + 2\nu \frac{2\varepsilon_{33} - \varepsilon_{11} - \varepsilon_{12}}{3}$$

and $e_{33} = \operatorname{tr} \varepsilon - \operatorname{tr} \hat{\varepsilon}$ lead to

$$\operatorname{tr} \varepsilon = \frac{6\nu}{3K + 4\nu} \operatorname{tr} \hat{\varepsilon}.$$

Note that the strain tensor is not planar, but the e_{33}-component was eliminated from the problem by the Hooke's law (4.81). This leads to the Hooke's law for the plane stress model,

$$\hat{\sigma} = \hat{K} \operatorname{tr} \hat{\varepsilon} \hat{\mathbb{I}} + 2\nu \varepsilon^D,$$

where $\hat{K} := \frac{9K\nu}{3K+4\nu}$. The relations

$$-\operatorname{Div} \hat{\sigma} = \hat{f} \quad \text{in } \hat{\Omega}, \tag{4.87}$$

$$\hat{u} = \hat{u}_g \quad \text{on } \hat{\Gamma}_D, \tag{4.88}$$

$$\hat{\sigma}\hat{n} = \hat{F} \quad \text{on } \hat{\Gamma}_N, \tag{4.89}$$

(where, $|\Gamma_D| > 0$, and Γ_N denote two parts of the boundary of $\hat{\Omega}$) form the plane stress problem. By repeating the derivation of the majorant for 3D problem, we arrive at the estimate

$$\int_{\hat{\Omega}} \left(\hat{K}\operatorname{div}(\hat{u}-\hat{v})^2 + 2\mu|\hat{\varepsilon}^D(\hat{u}-\hat{v})|^2\right) \mathrm{d}x \leq \overline{\mathrm{M}}^2_{2\mathrm{D},\sigma}(v,\hat{\tau};\beta), \quad \forall\hat{\tau}\in\mathbb{M}^{2\times 2}, \beta>0,$$

where

$$\overline{\mathrm{M}}^2_{2\mathrm{D},\sigma}(v,\hat{\tau};\beta) := (1+\beta)\int_{\hat{\Omega}} \left(\hat{K}\left(\operatorname{div}\hat{v} - \frac{1}{2\hat{K}}\operatorname{tr}\hat{\tau}\right)^2 + 2\mu\left|\hat{\varepsilon}^D(v) - \frac{1}{2\mu}\hat{\tau}^D\right|^2\right)\mathrm{d}x$$
$$+ \frac{1+\beta}{\beta}\hat{C}\left(\|\operatorname{Div}\hat{\tau} + \hat{f}\|^2 + \|\hat{F} - \hat{\tau}\hat{n}\|^2_{\hat{\Gamma}_N}\right).$$

We see that the majorant for the plane stress model $\overline{\mathrm{M}}_{2\mathrm{D},\sigma}$ is a planar analog of $\overline{\mathrm{M}}_{\mathrm{EL}}$.

4.1.8 The Plane Strain Model

In the plane strain model, it is assumed that the strain tensor is planar, i.e.,

$$\varepsilon_{i3} = 0, \quad i = \{1,2,3\}, \tag{4.90}$$

and $\hat{\varepsilon}\in\mathbb{M}^{2\times 2}$. From (4.90), it follows that $\sigma_{13} = \sigma_{23} = 0$. The component σ_{33} is eliminated by the Hooke's law, and we obtain the planar Hooke's law for the model,

$$\hat{\sigma} = \hat{K}\operatorname{tr}\hat{\varepsilon}\mathbb{I} + 2\mu\varepsilon^D,$$

where $\hat{K} = \frac{E}{2(1+\nu)(1-2)\nu}$. Together with relations (4.87)–(4.89) they form the plane strain problem. The respective error majorant has the familiar form:

$$\overline{\mathrm{M}}^2_{2\mathrm{D},\varepsilon}(v,\hat{\tau},\beta) := (1+\beta)\int_{\hat{\Omega}} \left(\hat{K}\left(\operatorname{div}\hat{v} - \frac{1}{2\hat{K}}\operatorname{tr}\hat{\tau}\right)^2 + 2\mu\left|\hat{\varepsilon}^D(v) - \frac{1}{2\mu}\hat{\tau}^D\right|^2\right)\mathrm{d}x$$
$$+ \frac{1+\beta}{\beta}\hat{C}\left(\|\operatorname{Div}\hat{\tau} + \hat{f}\|^2 + \|\hat{F} - \hat{\tau}\hat{n}\|^2_{\hat{\Gamma}_N}\right) \tag{4.91}$$

and generates the estimate

$$\int_{\hat{\Omega}} \left(\hat{K}\operatorname{div}(\hat{u}-\hat{v})^2 + 2\mu|\hat{\varepsilon}^D(\hat{u}-\hat{v})|^2\right) \mathrm{d}x \leq \overline{\mathrm{M}}^2_{2\mathrm{D},\varepsilon}(v,\hat{\tau},\beta), \quad \forall\hat{\tau}\in\mathbb{M}^{2\times 2}, \beta>0.$$

Example 4.2 The following examples were computed by A. Muzalevski and are presented in [MR03]. The approximate solution is computed by a finite element method using linear nodal elements, and the majorant (4.91) is applied as an error

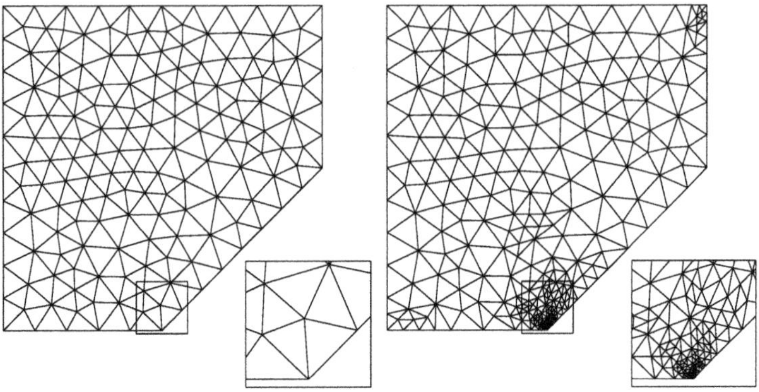

Fig. 4.6 Initial mesh (*left*) and refined mesh (*right*)

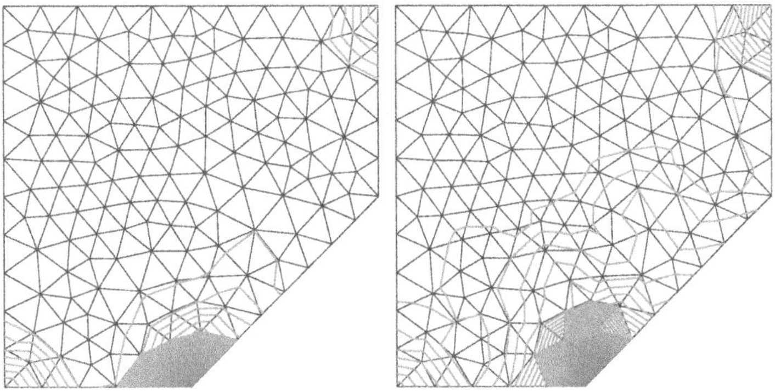

Fig. 4.7 True error distribution (*left*) and majorant indicator distribution (*right*)

indicator to guide a local mesh refinement. Then, a new approximation is computed, and the adaptive procedure continues.

First we consider a square domain with cuts. One fourth of the domain $\hat{\Omega}$ is depicted in Fig. 4.6, where the initial mesh and the one obtained by adaptive refinement procedure are presented. Moreover, in Fig. 4.7 we show the exact error distribution and the one obtained by the element-wise contributions of the majorant integral for the initial mesh. As expected, the error is concentrated around corners. The value of the error and the guaranteed upper bound (4.91) in each iteration stage are presented in Table 4.5. The efficiency index I_{eff} is defined in (2.3).

As an another example, we consider a square domain with a horizontal hexagonal slot. One fourth of the domain is depicted in Fig. 4.8 as well as the initial and adaptively refined mesh. The values of error estimate and the true error are presented in Table 4.6. A systematic discussion of the numerical tests can be found in [MR03].

Table 4.5 Error estimates for the triangular cut example

Nodes	$\|\|\|\varepsilon(u-v)\|\|\|$	$\overline{\mathrm{M}}_{2\mathrm{D},\varepsilon}$	I_{eff}
166	0.417	0.547	1.31
173	0.324	0.537	1.66
179	0.289	0.539	1.86
185	0.268	0.537	2.01
206	0.242	0.530	2.18
241	0.207	0.506	2.44
623	0.081	0.139	1.71
2413	0.056	0.114	2.05

4.2 The Stokes Problem

Classical formulation of the Stokes problem consists of finding a vector-valued function u (velocity) and a scalar-valued function p (pressure) that satisfy the relations

$$-\operatorname{Div}\sigma = f - \nabla p \quad \text{in } \Omega, \tag{4.92}$$

$$\sigma = \nu\varepsilon(u) \quad \text{in } \Omega, \tag{4.93}$$

$$\operatorname{div} u = 0 \quad \text{in } \Omega, \tag{4.94}$$

$$u = u_0 \quad \text{on } \Gamma, \tag{4.95}$$

where u_0 is a given function such that $\operatorname{div} u_0 = 0$, ν is a positive constant (viscosity), $\varepsilon(u)$ is the tensor of small strains (see (A.33)), and, as before, Ω is a bounded set in $\mathbb{R}^d$ with Lipschitz boundary Γ.

To define the corresponding generalized solution of the Stokes problem we need to introduce extra notation. Henceforth, $\overset{\circ}{\mathrm{S}}$ denotes the closure of smooth solenoidal functions compactly supported in Ω with respect to the norm $\|\nabla w\|$.

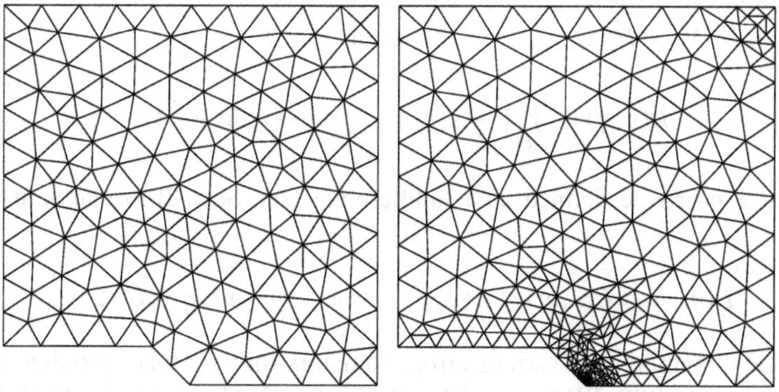

Fig. 4.8 Initial (*left*) and refined (*right*) meshes

Table 4.6 Error estimates for the hexagonal cut example

Nodes	$\|\|\|\varepsilon(u-v)\|\|\|$	$\overline{\mathrm{M}}_{2\mathrm{D},\varepsilon}$	I_{eff}
174	0.443	0.690	1.56
177	0.370	0.690	1.86
181	0.337	0.687	2.03
185	0.316	0.687	2.17
198	0.291	0.680	2.33
234	0.271	0.627	2.31
652	0.095	0.181	1.91
2523	0.068	0.151	2.19

With $\dot{J}^\infty(\Omega)$ we denote smooth solenoidal functions with compact supports in Ω. The space $\overset{\circ}{\mathrm{S}}$ is the closure of $\dot{J}^\infty(\Omega)$ with respect to the norm $\|\nabla v\|$ (in other words $\overset{\circ}{\mathrm{S}}$ contains solenoidal fields from $H^1(\Omega,\mathbb{R}^d)$ vanishing on the boundary). The set $\overset{\circ}{\mathrm{S}}+u_0$ contains functions of the form $w+u_0$. Let $V := H^1(\Omega,\mathbb{R}^d)$, $\Sigma := L^2(\Omega,\mathbb{M}^{d\times d})$ and V_0 be a subspace of V that contains functions with zero traces on Γ. Also we use the Hilbert space $\Sigma_{\mathrm{div}}(\Omega)$ (cf. (4.65)) and the space $\overset{\circ}{L}{}^2(\Omega)$, which contains square summable functions with zero mean.

Henceforth, we assume that $f \in L^2(\Omega,\mathbb{R}^d)$ and $u_0 \in H^1(\Omega,\mathbb{R}^d)$. Generalized solution of (4.92)–(4.95) is a function $u \in \overset{\circ}{\mathrm{S}}+u_0$ that meets the relation

$$\int_\Omega \nu\varepsilon(u):\varepsilon(w)\,\mathrm{d}x = \int_\Omega f\cdot w\,\mathrm{d}x, \quad \forall w \in \overset{\circ}{\mathrm{S}}. \tag{4.96}$$

It is well-known (e.g., see [Lad70, Tem79]) that u exists and is unique. It can be viewed as the minimizer of the variational problem

$$\inf_{v\in\overset{\circ}{\mathrm{S}}+u_0} J(v), \quad J(v) = \int_\Omega \left(\frac{\nu}{2}|\varepsilon(v)|^2 - f\cdot v\right)\mathrm{d}x. \tag{4.97}$$

In addition, the Stokes problem can be presented in a *minimax form.* Let the Lagrangian $L : V_0 \times \overset{\circ}{L}{}^2(\Omega) \to \mathbb{R}$ be defined as follows:

$$L(v,q) = \int_\Omega \left(\frac{\nu}{2}|\varepsilon(v)|^2 - f\cdot v - q\,\mathrm{div}\,v\right)\mathrm{d}x.$$

Now, u and p can be defined as a saddle-point (see Sect. B.3) that satisfies the relations

$$L(u,q) \le L(u,p) \le L(v,p), \quad \forall v \in V_0+u_0, q \in \overset{\circ}{L}{}^2(\Omega). \tag{4.98}$$

Numerical methods and various error estimators for the Stokes problem are discussed in, e.g., [AO00, BW91, CKP11, DA05, Fei93, GR86, PS11, PS08, Ran00, Ver96, Ver89] and in many other publications.

4.2.1 Divergence-Free Approximations

In this section, we show a simple way of deriving guaranteed bounds of deviations from the exact solution for any solenoidal approximation $v \in \overset{\circ}{\mathrm{S}} + u_0$, which follows the lines of [Rep02b, Rep04, Rep08]. First, we note that

$$\int_\Omega \nu\varepsilon(u-v) : \varepsilon(w)\,\mathrm{d}x = \int_\Omega \big(f \cdot w - \nu\varepsilon(v) : \varepsilon(w)\big)\,\mathrm{d}x, \quad \forall w \in \overset{\circ}{\mathrm{S}}. \tag{4.99}$$

Let $\tau \in H(\Omega, \mathrm{Div})$. Then, we represent (4.99) as follows:

$$\int_\Omega \nu\varepsilon(u-v) : \varepsilon(w)\,\mathrm{d}x = \int_\Omega (f + \mathrm{Div}\,\tau - \nabla q) \cdot w + \big(\tau - \nu\varepsilon(v)\big) : \varepsilon(w))\,\mathrm{d}x,$$

where $q \in \overset{\circ}{L}{}^2(\Omega)$ is a scalar-valued function (note that since w is a solenoidal field the product of this function and ∇q is zero). We set $w = u - v$ and arrive at the estimate

$$\nu\big\|\varepsilon(u-v)\big\| \leq \big\|\nu\varepsilon(v) - \tau\big\| + C_{F\Omega}\|\,\mathrm{Div}\,\tau + f - \nabla q\,\|. \tag{4.100}$$

We can square both parts and rewrite this estimate in the form

$$\nu^2\big\|\varepsilon(u-v)\big\|^2 \leq (1+\beta)\big\|\nu\varepsilon(v) - \tau\big\|^2 + \frac{(1+\beta)C_{F\Omega}^2}{\beta}\|\,\mathrm{Div}\,\tau + f - \nabla q\,\|^2. \tag{4.101}$$

As the estimates presented earlier in this book, it involves certain free parameters that should be chosen in such a way that the right-hand side of (4.101) is minimal. Moreover,

$$\nu^2\big\|\varepsilon(u-v)\big\|^2 = \inf_{\substack{\tau\in H(\Omega,\mathrm{Div})\\ q\in\widetilde{H}(\Omega),\beta>0}} \left\{(1+\beta)\big\|\nu\varepsilon(v) - \tau\big\|^2 + \frac{(1+\beta)}{\beta}C_{F\Omega}^2\|\,\mathrm{div}\,\tau + f - \nabla q\,\|^2\right\}.$$

In this estimate, it is required that $q \in \widetilde{H}(\Omega) := \overset{\circ}{L}{}^2(\Omega) \cap H^1(\Omega)$. If q has a weaker regularity, then we use a modification of (4.100), which is obtained by introducing a new tensor-valued function instead of τ (which is $\eta := \tau + q\mathbb{I}$). We have

$$\nu\big\|\varepsilon(u-v)\big\| \leq \big\|\nu\varepsilon(v) - \eta - q\mathbb{I}\big\| + C_{F\Omega}\|\,\mathrm{Div}\,\tau + f\,\|. \tag{4.102}$$

Estimates (4.100), (4.101) or (4.102) are applicable only if v belongs to the set of solenoidal functions. Typically, approximate solutions computed by numerical methods do not exactly satisfy this condition. For this reason, it is desirable to have a projection method, by which we can post-process an approximate solution to the

Fig. 4.9 Hsiesh–Clough–Tocher macroelement (*left*) and decomposed subelements (*right*)

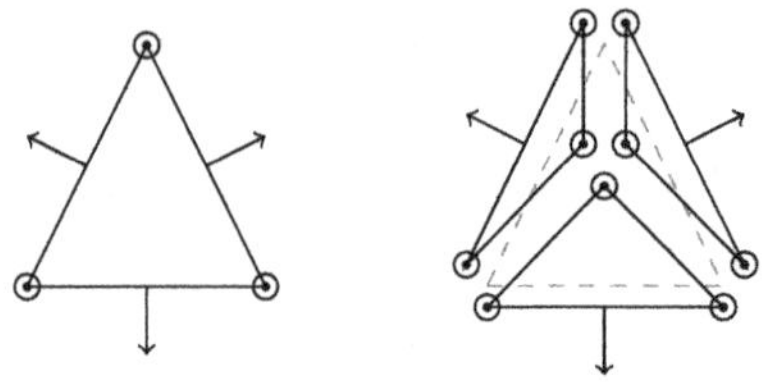

set of divergence-free fields. In general, this may be a rather difficult task, but for 2D problems such a projection can be constructed fairly easily by means of the stream function. Let $W \in H^2(\Omega)$ and

$$v = \left(\frac{\partial W}{\partial x_2}, -\frac{\partial W}{\partial x_1} \right). \tag{4.103}$$

It is easy to see that $\operatorname{div} v = 0$. If $\widehat{v}$ is not a solenoidal vector-valued function, then we can post-process it by a certain mapping to the set of functions of the form (4.103). For this purpose, the function W is constructed with the help of C^1-elements. One way is to use the Hsieh–Clough–Tocher elements, which are macroelements constructed by dividing a triangle into three sub-triangles T_i (see, e.g., [BH81]). On each of them, the function is presented by a 3rd order polynomial. These elements and the corresponding degrees of freedom are depicted in Fig. 4.9. Let $\mathring{S}_h$ denote the space of functions satisfying (4.103), where W is constructed by Hsieh–Clough–Tocher approximations. Obviously, $\mathring{S}_h \subset \mathring{S}$. Assume that $u_h \notin \mathring{S}$ is a finite element approximation computed by some numerical procedure. Define $v_h \in \mathring{S}_h$ such that

$$\|\nabla(u_h - v_h)\| = \min_{w_h \in \mathring{S}_h} \|\nabla(u_h - w_h)\|. \tag{4.104}$$

The computational cost of this projection procedure can be reduced by taking the values of the stream function directly from the values of u_h by (4.103). For example, if u_h is constructed by means of Taylor–Hood elements, then the degrees of freedom associated to the derivatives of W are defined as follows:

$$\frac{\partial W}{\partial x_1} = -u_{h2} \quad \text{and} \quad \frac{\partial W}{\partial x_2} = u_{h1},$$

at each node and

$$\frac{\partial W}{\partial n} = (-u_{h2}, u_{h1}) \cdot n$$

on each exterior edge. The remaining degrees of freedom are defined by local minimization in accordance with (4.104). If u_h is computed by means of Crouzeix–Raviart elements, then a similar procedure can be applied if the previously solution is smoothed (averaged).

Example 4.3 Analysis of numerical efficiency of the estimates discussed in Sect. 4.2.1 was performed in [Gor07]. Below we present one test example taken from there (see also [GNR06]). In it, the majorant (4.101) was applied to finite element approximations of the Stokes problem in the domain

$$\Omega := \big((-1,1)\times(-1,1)\big)\setminus\big([0,1]\times[-1,0]\big).$$

It is assumed that $f = 0$ and the boundary conditions coincide with the boundary traces generated by the exact solution

$$\vec{u}(r,\theta) = r^{\alpha}\big((1+\alpha)(\sin\theta, -\cos\theta)w + (\cos\theta, \sin\theta)\big)\frac{\partial w}{\partial\theta},$$

$$p(t,\theta) = -\frac{r^{\alpha-1}}{1-\alpha}\left((1+\alpha)^2\frac{\partial w}{\partial\theta} + \frac{\partial^3 w}{\partial\theta^3}\right),$$

$$w(\theta) = \frac{\sin((1+\alpha)\theta)\cos(\alpha\gamma)}{1+\alpha} - \cos\big((1+\alpha)\theta\big) - \frac{\sin((1-\alpha)\theta)\cos(\alpha\gamma)}{1-\alpha} + \cos\big((1-\alpha)\theta\big),$$

where $\alpha = 0.54448$ and $\gamma = \frac{3\pi}{2}$.

Approximations were computed with the help of Taylor–Hood elements and Uzawa iteration algorithm. The velocity field u_h obtained by this method is close to a solenoidal field, but does not exactly satisfy the condition $\operatorname{div} u_h = 0$. Therefore, it is necessary to apply the projection method described above and find a close solenoidal approximation. It is substituted in the majorant (4.101), which is then minimized with respect to the auxiliary functions τ and q. The auxiliary functions belong to finite dimensional subspaces constructed with the help of piecewise quadratic functions. For τ, the initial guess is computed by simple (patch-wise) averaging of $\nu\nabla u_h$, and the pressure p_h is computed by the Uzawa algorithm. This test problem was solved by an adaptive method, in which the integrand of the majorant (4.101) was used as an error indicator. In this simple example, marking was performed on the basis of the mean value principle: we refine (split) the elements, for which the indicated error is greater than one half of the maximum error.

The exact error, value of (4.101), and the respective efficiency index are presented in Table 4.7. The first column of the table shows the amount of Uzawa iterations used for finding the corresponding approximate solution, the second column containing the amount of mesh nodes. The error and the upper bound computed by the majorant are shown in the next two columns. By the efficiency indexes presented in the last column, we see that for all meshes the quality of error estimation was quite good.

4.2.2 Approximations with Nonzero Divergence

In many cases, it is more convenient to operate with numerical solutions which satisfy the divergence-free condition only approximately. Error majorants can be derived for such approximations as well, but they have an additional term and involve a new global constant. The motivation of this is as follows.

Table 4.7 Error estimates

# iterations	# nodes	$\|\|\|u - u_h\|\|\|$	Majorant	I_{eff}
5	472	0.94	1.288	1.37
9	2174	0.041	0.057	1.41
12	4303	0.026	0.03926	1.51
14	5734	0.013	0.0166	1.28
19	7893	0.0096	0.01373	1.43
26	12552	0.008	0.0095	1.19

An estimate of the distance between a function $\widehat{v} \in V_0$ and the space $\overset{\circ}{\mathrm{S}}$ follows from Lemma A.1. Set $f = \operatorname{div}\widehat{v}$. Lemma A.1 guarantees the existence of a function $u_f \in V_0$ such that

$$\operatorname{div}(\widehat{v} - u_f) = 0, \quad \text{and} \quad \|\nabla u_f\| \leq \kappa_\Omega \|\operatorname{div}\widehat{v}\|.$$

In other words, there exists a solenoidal field $w_0 := \widehat{v} - u_f \in \overset{\circ}{\mathrm{S}}$ such that

$$\left\|\nabla(\widehat{v} - w_0)\right\| \leq \kappa_\Omega \|\operatorname{div}\widehat{v}\|. \tag{4.105}$$

This fact can be presented in another form. Let $\widehat{v} \in V_0$, then

$$\inf_{v \in \overset{\circ}{\mathrm{S}}} \left\|\nabla(\widehat{v} - v)\right\| \leq \kappa_\Omega \|\operatorname{div}\widehat{v}\|. \tag{4.106}$$

Thus, the distance between $\widehat{v} \in V_0$ and $\overset{\circ}{\mathrm{S}}$ (in the strong norm generated by gradients) is controlled by the constant κ_Ω. Obviously, the same estimate holds if $\widehat{v} \in V_0 + u_0$ is projected on $\overset{\circ}{\mathrm{S}} + u_0$.

Now, we can easily deduce error majorant for a function $\widehat{v} \in V_0 + u_0$, which does not belong to $\overset{\circ}{\mathrm{S}} + u_0$. Using the triangle inequality twice, we find that

$$\nu\left\|\varepsilon(u - \widehat{v})\right\| \leq \left\|\nu\varepsilon(\widehat{v}) - \tau\right\| + C_\Omega \|\operatorname{div}\tau + f - \nabla q\| + 2\nu\kappa_\Omega \|\operatorname{div}\widehat{v}\|, \tag{4.107}$$

where the additional term that can be thought of as a penalty for possible violation of the divergence-free condition.

A similar estimate can be derived for the pressure field (see [Rep02b]). It has the form

$$\frac{1}{2\kappa_\Omega}\|p - q\| \leq \nu\kappa_\Omega \|\operatorname{div}\widehat{v}\| + \left\|\nu\varepsilon(\widehat{v}) - \tau\right\| + C_{F\Omega} \|\operatorname{Div}\tau + f - \nabla q\|. \tag{4.108}$$

It is easy to see that the right-hand side of (4.108) contains the same terms as the right-hand side of (4.107) and vanishes if and only if $\widehat{v} = u$, $\tau = \sigma$ and $p = q$. However, the penalty factors more strongly dependent on the value of κ_Ω.

The presence of the LBB-constant makes the estimates (4.107) and (4.108) less attractive because this constant may be difficult to estimate. On the other hand, if a method of estimation would be developed, then rather simple nonsolenoidal approximations of the velocity field could be analyzed by the estimate.

Remark 4.5 Estimates (4.107) and (4.108) have been extended to generalized forms of the Stokes problem in [RS07, RS08]. Further development of the method in the context of finite element approximations is presented in [HSV12]. A posteriori estimates for the Stokes problem in velocity-vorticity formulation are derived in [MR11b] and for the evolutionary Stokes problem in [NR10b].

4.2.3 Stokes Problem in Rotating System

As an example of a more complicated model, we consider the Stoke's problem in rotating frame:

$$
\begin{aligned}
-\nu \Delta u + \frac{1}{R_b} B \times u &= f - \nabla p \quad && \text{in } \Omega \subset \mathbb{R}^2, \\
\operatorname{div} u &= 0 && \text{in } \Omega, \\
u &= u_g && \text{on } \Gamma.
\end{aligned}
\tag{4.109}
$$

The term $\frac{1}{R_b} B \times u$ is related to the Coriolis force (which must be taken into account if, e.g., effects caused by the Earth rotation are significant). Here $B = be_z$ is the adimensional rotation speed and R_b is the so-called Rossby number. The term related to the centripetal force is included to the pressure term and is not considered in rotating fluid computations (see, e.g., [Var62]). Similar observations as for the Stokes problem are valid. We assume $f \in L^2(\Omega, \mathbb{R}^d)$ and $\operatorname{div} u_g = 0$. The generalized solution $u \in \overset{\circ}{\mathrm{S}} + u_g$ satisfies the integral relation

$$
\int_\Omega \left(\nu \nabla u : \nabla w + \frac{1}{R_b} (B \times u) \cdot w \right) \mathrm{d}x = \int_\Omega f \cdot w \, \mathrm{d}x, \quad \forall w \in \overset{\circ}{\mathrm{S}}.
$$

Error majorants for this problem were derived in [GMNR07]. Let u be the exact solution of (4.110) and v be a function in $\overset{\circ}{\mathrm{S}} + u_g$. Then,

$$
\left\| \nu \nabla (u - v) \right\| \leq \| \tau - \nu \nabla v \| + C_{F\Omega} \left\| f + \operatorname{div} \tau - \frac{1}{R_b} B \times v - \nabla q \right\|, \tag{4.110}
$$

where $\tau \in \Sigma_{\mathrm{div}}(\Omega)$ and $q \in H^1(\Omega) \cap \overset{\circ}{L}{}^2(\Omega)$. If $\xi \in \Sigma_{\mathrm{div}}(\Omega)$ and $q \in \overset{\circ}{L}{}^2(\Omega)$, then we have a somewhat different estimate

$$
\left\| \nu \nabla (u - v) \right\| \leq \| \xi - \nu \nabla v + q \mathbb{I} \| + C_{F\Omega} \left\| f + \operatorname{div} \xi - \frac{1}{R_b} B \times v \right\|. \tag{4.111}
$$

Error majorants applicable for non-solenoidal approximations and estimates for the pressure can be derived along the same lines as for the Stokes problem (see

[GMNR07]). The following numerical tests in Examples 4.4 and 4.5 were computed by E. Gorshkova (see [Gor07]).

Example 4.4 We consider viscous flow in a rotating container that consists of three connected cylinders with different radii, i.e., $\Omega = \Omega_1 \cup \Omega_2 \cup \Omega_3$, where

$$\Omega_1 := \left\{ (r, \theta, z) \in \mathbb{R}^3 \;\middle|\; 0 \le r \le R_{\text{top}}, 0 \le \theta < 2\pi, \frac{h}{2} \le z \le \frac{3h}{2} \right\},$$

$$\Omega_2 := \left\{ (r, \theta, z) \in \mathbb{R}^3 \;\middle|\; 0 \le r \le R_{\text{middle}}, 0 \le \theta < 2\pi, -\frac{h}{2} \le z \le \frac{h}{2} \right\},$$

$$\Omega_3 := \left\{ (r, \theta, z) \in \mathbb{R}^3 \;\middle|\; 0 \le r \le R_{\text{bottom}}, 0 \le \theta < 2\pi, -\frac{3h}{2} \le z \le -\frac{h}{2} \right\}.$$

The container rotates around the vertical e_z-axis. The income and outcome boundary conditions are defined as follows: $u_r = 0$ and $u_\theta = \frac{1}{R_b} br$ at the "top" and "bottom" sides, $u_z = \frac{R_{\text{top}}^2 - r^2}{R_{\text{top}}^3}$ at the top side, and $u_z = \frac{R_{\text{bottom}}^2 - r^2}{R_{\text{bottom}}^3}$ at the bottom side.

The problem by the finite element method using quadratic approximations for v and linear approximations for q. The velocity field is post-processed by the same method as for the Stokes problem. The resulting solenoidal field is considered as an approximation, the error of which is measured by the estimate (4.110). The majorant (4.110) is minimized numerically with respect to τ and q, which are constructed by means of quadratic elements (the initial values are taken from the approximate solution). This procedure results in an error bound and an error indicator that describes the distribution of local error. The marking is performed in accordance with the "bulk strategy" (see Algorithm 2.2 with $\theta_{\text{bulk}} = 0.6$). A series of adapted meshes is depicted in Fig. 4.10. In this example, we set $R_{\text{top}} = R_{\text{bottom}} = 0.6$, $R_{\text{middle}} = 1.0$, and $h = R_b = b = 1.0$. The exact solution for the test problem is not known. Therefore, we use a "referenced" solution u_{ref}, which is specially computed on a very fine mesh. The results obtained by the majorant (4.110) are presented in Table 4.8. Additionally, we study the accuracy of the indicator with respect to the marker used ("bulk" criterium). We apply Definition 2.3 and compute the value of $\mathcal{M}(\mathbb{E}(u_h), \mathbb{M})$, i.e., the percentage of the elements where the majorant generates wrong marking. For this test, we set $b = 100$ and $R_{\text{top}} = R_{\text{bottom}} = 0.8$. The results are collected to Table 4.9.

Example 4.5 Fig. 4.11 demonstrates results of numerical tests for rotating cylinder in which the efficiency of error indicators generated by the majorant was verified. Figure 4.11a is related to the value of rotation parameter 3. True relative error for the geometry and mesh presented in the figure is around 3.2 % (it was computed by comparing it with the reference solution), and the majorant gives 3.5 %, so that $I_{\text{eff}} \cong 1.09$. Comparison of marked elements shows 97.2 % coincidence for the maximum criterion (where the elements containing error larger than one half of

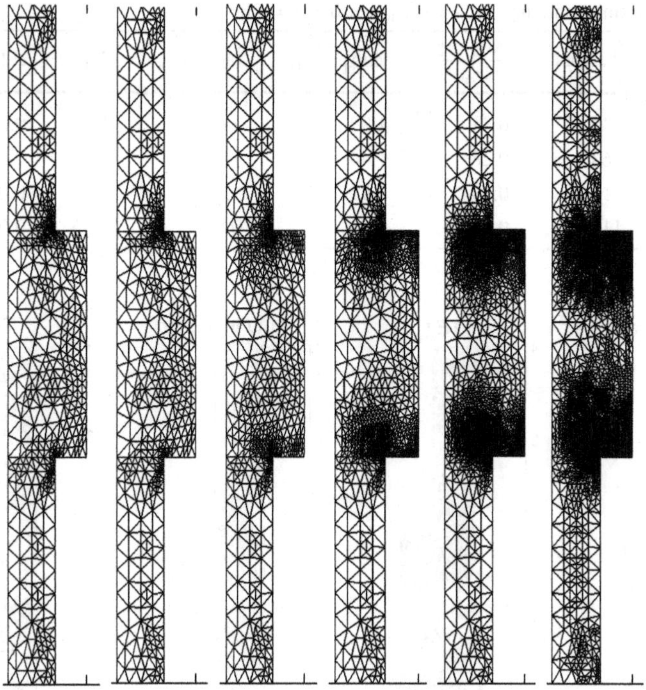

Fig. 4.10 Adaptation of meshes in the rotating container problem

Table 4.8 Rotating container ($R_{top} = R_{bottom} = 0.6$, $b = 1$)

# iterations	# nodes	$\|\|\|u_h - u_{ref}\|\|\|$	Majorant	I_{eff}
1	312	0.0087	0.0129	1.48
2	472	0.0066	0.0086	1.30
3	643	0.0059	0.0081	1.37
4	692	0.0051	0.0061	1.20
5	786	0.0033	0.0050	1.51
6	981	0.0023	0.0045	1.96
7	1240	0.0016	0.0027	1.69
8	2054	0.0011	0.0018	1.64

the largest occurring error are selected) and bulk criterion (Algorithm 2.3, where $\theta = 0.6$). Figure 4.11 presents similar information for a problem with a faster rotation (the rotation parameter is 100). In this case, the relative error is 1.8 %, the majorant is 2.18 %, and $I_{eff} \cong 1.21$. Maximum and bulk criteria show 94.2 % and 90.9 % coincidence, respectively.

Table 4.9 Rotating container $R_{\text{top}} = R_{\text{bottom}} = 0.8$, $b = 100$

# iterations	# nodes	$\frac{\|\|\|u^h_{\text{ref}}-v\|\|\|}{\|\|\|u^h_{\text{ref}}\|\|\|}$	$\|\|\|u^h_{\text{ref}} - v\|\|\|$	$\overline{\mathrm{M}}_{\text{Stokes,rot}}$	I_{eff}	$\mathcal{M}(E(u_h), \mathbb{M})$
1	425	0.38	1.93	2.63	1.35	0.03
2	637	0.28	1.44	1.87	1.29	0.05
3	956	0.19	0.997	1.18	1.19	0.04
4	1147	0.126	0.63	0.86	1.36	0.08

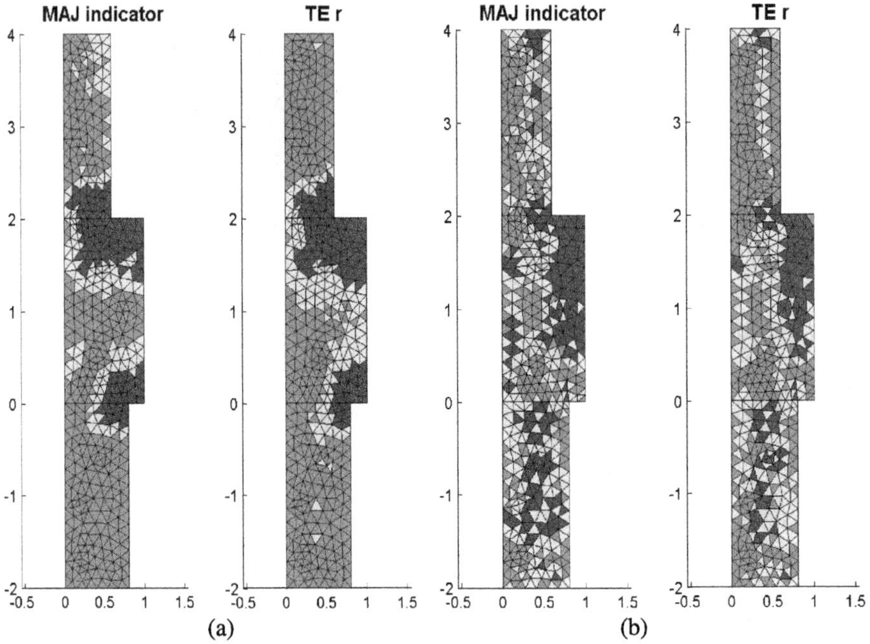

Fig. 4.11 Marking of elements in the rotating container example, where **a** the rotation parameter is 3 and **b** the rotation parameter is 100

4.3 A Simple Maxwell Type Problem

In classical settings, the Maxwell problem is defined by the functions E, D (which denote the electric field and induction, respectively) and the functions H and B (magnetic field and induction, respectively), which satisfy the system

$$\frac{\partial D}{\partial t} - \operatorname{curl} H = -J,$$

$$\frac{\partial B}{\partial t} + \operatorname{curl} E = 0$$

for all $(t, x) \in (0, T) \times \Omega$. Here J denotes the current and, as before, Ω is a bounded domain in $\mathbb{R}^d$ with Lipschitz boundary Γ. Moreover, in this section we assume that Ω is simply connected. Physical properties of the electromagnetic media are described by the constitutive relations

$$D = \varepsilon E \quad \text{and} \quad B = \mu H,$$

where $\varepsilon(x) > 0$ is the dielectric permittivity and $\mu(x) > 0$ is the magnetic permeability (both μ and ε are assumed to be positive constants or positive bounded functions). Then, the Maxwell's equations are presented in terms of E and H as follows:

$$\varepsilon \frac{\partial E}{\partial t} - \operatorname{curl} H = -J,$$

$$\mu \frac{\partial H}{\partial t} + \operatorname{curl} E = 0.$$

These equations must be accompanied by initial conditions and suitable boundary conditions. Here we assume that E satisfies the so-called perfect electric conductor boundary condition

$$E \times n = 0 \quad \text{on } \Gamma,$$

where n denotes the unit outward normal of Γ. Usually the time derivatives are replaced by incremental relations. Using the backward-Euler scheme we have

$$\frac{\varepsilon}{\Delta t}\left(E^n - E^{n-1}\right) - \operatorname{curl} H^n = -J,$$

$$\frac{\mu}{\Delta t}\left(H^n - H^{n-1}\right) + \operatorname{curl} E^n = 0, \quad n = 1, \ldots, N, N = \frac{T}{\Delta t},$$

where Δt is the time step. By eliminating H^n and transferring E^{n-1} and H^{n-1} to the right-hand side, we have

$$\operatorname{curl}\left(\mu^{-1} \operatorname{curl} E^n\right) + \frac{\varepsilon}{(\Delta t)^2} E^n = \frac{1}{\Delta t}\left(-J + \frac{\varepsilon}{\Delta t} E^{n-1} + \operatorname{curl} H^{n-1}\right).$$

We denote the right-hand side by $f \in L^2(\Omega, \mathbb{R}^d)$, set $\kappa = \varepsilon(\Delta t)^{-2}$ and arrive at the model problem

$$\operatorname{curl}\left(\mu^{-1} \operatorname{curl} E\right) + \kappa E = f \quad \text{in } \Omega, \tag{4.112}$$

$$E \times n = 0 \quad \text{on } \Gamma, \tag{4.113}$$

in which the superscript n is omitted.

Below, we study (4.112)–(4.113) in the 2D case, so that the double curl is understood as $\underline{\operatorname{curl}}\,\operatorname{curl}$, where

$$\operatorname{curl} w := \partial_1 w_2 - \partial_2 w_1, \qquad \underline{\operatorname{curl}} \varphi := \begin{pmatrix} \partial_2 \varphi \\ -\partial_1 \varphi \end{pmatrix}.$$

Let $V(\Omega)$ denote the space $H(\mathrm{curl};\Omega)$ (see Sect. A.2) and

$$V_0(\Omega) := \{w \in V \mid w \times n = 0 \text{ on } \Gamma\}.$$

The generalized solution $E \in V_0(\Omega)$ of (4.112)–(4.113) is defined by the integral relation

$$\int_\Omega \left(\mu^{-1} \operatorname{curl} E \operatorname{curl} w + \kappa E \cdot w\right) \mathrm{d}x = \int_\Omega f \cdot w \, \mathrm{d}x, \quad \forall w \in V_0(\Omega). \tag{4.114}$$

Henceforth, we assume that f is a solenoidal function and $0 < \mu_\ominus \le \mu(x) \le \mu_\oplus$.

4.3.1 Estimates of Deviations from Exact Solutions

Let $\widetilde{E} \in V_0$ be an approximation to the exact solution E. Our goal is to obtain computable and guaranteed bounds of the difference between E and any function $\widetilde{E} \in V_0$ measured in terms of the weighted (energy) norm

$$|[w]|_{(\gamma,\delta)} := \left(\int_\Omega \left(\gamma |\operatorname{curl} w|^2 + \delta |w|^2\right) \mathrm{d}x\right)^{1/2}.$$

We begin with auxiliary results, which are further used in the derivation of the upper bound. By the Helmholtz decomposition, $E = E_0 + \nabla\psi$, where E_0 is a solenoidal field and $\psi \in \overset{\circ}{H}{}^1(\Omega)$. Since $\operatorname{curl}\nabla\psi = 0$, we rewrite (4.114) as follows:

$$\int_\Omega \left(\mu^{-1} \operatorname{curl} E_0 \operatorname{curl} w + \kappa (E_0 + \nabla\psi) \cdot w\right) \mathrm{d}x = \int_\Omega f \cdot w \, \mathrm{d}x.$$

Next, we make the same decomposition for the trial function and set $w = w_0 + \nabla\phi$. Since

$$\int_\Omega f \cdot \nabla\phi \, \mathrm{d}x = \int_\Omega E_0 \cdot \nabla\phi \, \mathrm{d}x = \int_\Omega w_0 \cdot \nabla\psi \, \mathrm{d}x = 0,$$

we observe that

$$\int_\Omega \left(\mu^{-1} \operatorname{curl} E_0 \operatorname{curl} w_0 + \kappa E_0 \cdot w_0 + \kappa \nabla\psi \cdot \nabla\phi\right) \mathrm{d}x = \int_\Omega f \cdot w_0 \, \mathrm{d}x.$$

By setting $w_0 = 0$ and $\phi = \psi$, we find that $\|\nabla\psi\| = 0$. Hence, E is a solenoidal function.

Note that ϕ satisfies the relation

$$\int_\Omega \nabla\phi \cdot \nabla\xi \, \mathrm{d}x = \int_\Omega w \cdot \nabla\xi \, \mathrm{d}x = -\int_\Omega (\operatorname{div} w)\xi \, \mathrm{d}x, \quad \forall \xi \in \overset{\circ}{H}{}^1(\Omega),$$

which implies the estimate

$$\|\nabla\phi\| \le C_{F\Omega} \|\operatorname{div} w\|, \tag{4.115}$$

where $C_{F\Omega}$ is the constant in the Friedrichs inequality for the domain Ω (cf. (A.27)). For solenoidal fields we also have the estimate (see, e.g., [Sar82])

$$\|w_0\| \leq C_\Omega \| \operatorname{curl} w_0 \| = C_\Omega \| \operatorname{curl} w \|. \tag{4.116}$$

Constants in these inequalities appear in advanced a posteriori estimates for the problem (4.112)–(4.113), which we discuss later (cf. Proposition 4.4). However, first we mention simpler estimates derived in [Rep08] (similar estimates have been reported [Han08]). If $\kappa > 0$ and $\widetilde{E}$ is a conforming approximation of E, then a simple majorant of the error has the form

$$\begin{aligned}|[E - \widetilde{E}]|^2_{(\mu^{-1},\kappa)} \leq \overline{\mathsf{M}}^2_{\text{curl}}(v, y) \\ := \left\|\kappa^{-1/2}(f - \kappa\widetilde{E} - \operatorname{curl} y)\right\|^2 \\ + \left\|\mu^{1/2}\left(y - \mu^{-1} \operatorname{curl} \widetilde{E}\right)\right\|^2.\end{aligned} \tag{4.117}$$

It is easy to see that

$$\inf_{\substack{\widetilde{E}\in V_0,\\ y\in H(\Omega,\text{curl})}} \overline{\mathsf{M}}_{\text{curl}}(\widetilde{E}, y) = 0$$

and the exact upper bound is attained if and only if

$$\operatorname{curl} y + \kappa\widetilde{E} = f \quad \text{a.e. in } \Omega, \tag{4.118}$$

$$y = \mu^{-1} \operatorname{curl} \widetilde{E} \quad \text{a.e. in } \Omega. \tag{4.119}$$

Since $\widetilde{E} \times n = 0$ on Γ, the relations (4.118) and (4.119) mean that $\widetilde{E}$ coincides with the exact solution E and $y = \mu^{-1} \operatorname{curl} E$.

For any $y \in V_0$, the quantity $\overline{\mathsf{M}}_{\text{curl}}(\widetilde{E}, y)$ gives an upper bound of the error. It is clear that the function y should be selected in such a way that the majorant would be minimal. Since

$$\begin{aligned}\inf_{y\in V_0} \overline{\mathsf{M}}^2_{\text{curl}}(\widetilde{E}, y) &\leq \overline{\mathsf{M}}^2_{\text{curl}}\left(v, \mu^{-1} \operatorname{curl} E\right) \\ &= \left\|\kappa^{-1/2}\left(f - \kappa\widetilde{E} - \operatorname{curl} \mu^{-1} \operatorname{curl} E\right)\right\|^2 + \left\|\mu^{-1/2} \operatorname{curl}(E - \widetilde{E})\right\|^2 \\ &= \left\|\kappa^{-1/2}(E - \widetilde{E})\right\|^2 + \left\|\mu^{-1/2} \operatorname{curl}(E - \widetilde{E})\right\|^2 = |[E - \widetilde{E}]|^2_{(\mu^{-1},\kappa)},\end{aligned}$$

we see that computable quantities generated by the majorant can approximate $|[E - \widetilde{E}]|_{(\mu^{-1},\kappa)}$ with any desired accuracy.

An advanced form of the error majorant is presented by the estimate (4.120). The exposition below follows along the lines of [NR10a], where similar estimates are considered for a 3D problem.

Proposition 4.4 *Let $\widetilde{E} \in V_0 \cap H(\text{div}; \Omega)$ be an approximation of E. Then,*

$$|[E - \widetilde{E}]|^2_{\gamma,\delta} \leq \overline{\mathsf{M}}^2_{\text{curl}}(\lambda, \alpha_1, \alpha_2, \widetilde{E}, y), \quad y \in H^1(\Omega), \tag{4.120}$$

where

$$\overline{\mathsf{M}}^2_{\mathrm{curl}}(\lambda, \alpha_1, \alpha_2, \widetilde{E}, y) := R_1(\lambda, \widetilde{E}, y) + \frac{\alpha_1}{4} R_2^2(\lambda, \widetilde{E}, y) + \frac{\alpha_2}{4} R_3^2(\lambda, \widetilde{E}, y),$$

α_1 *and* α_2 *are arbitrary numbers in* $[1, +\infty)$,

$$\gamma = \left(1 - \frac{1}{\alpha_1}\right)\mu^{-1}, \qquad \delta = \left(1 - \frac{1}{\alpha_2}\right)\kappa,$$

$$\lambda \in I_{[0,1]} := \left\{\lambda \in L^\infty(\Omega) \mid \lambda(x) \in [0,1] \textit{ for a.e. } x \in \Omega\right\},$$

κ *is a positive constant, and the quantities* R_i, $i = 1, 2, 3$ *are defined by* (4.126)–(4.128).

Proof Subtracting $\int_\Omega (\mu^{-1} \operatorname{curl} \widetilde{E} \operatorname{curl} w + \kappa \widetilde{E} \cdot w)\,\mathrm{d}x$ from (4.114) leads at

$$\int_\Omega \left(\mu^{-1} \operatorname{curl}(E - \widetilde{E}) \operatorname{curl} w + \kappa (E - \widetilde{E}) \cdot w\right) \mathrm{d}x$$
$$= \int_\Omega \left(f \cdot w - \mu^{-1} \operatorname{curl} \widetilde{E} \operatorname{curl} w - \kappa \widetilde{E} \cdot w\right) \mathrm{d}x. \tag{4.121}$$

In view of the integration by parts formula (for the operator curl in 2D)

$$\int_\Omega y \operatorname{curl} w \,\mathrm{d}x = \int_\Omega \underline{\operatorname{curl}} y \cdot w \,\mathrm{d}x + \int_\Gamma y (w \times n)\,\mathrm{d}s,$$

which is valid for any $y \in H^1(\Omega)$ and $w \in H(\operatorname{curl}; \Omega)$, we find that

$$\int_\Omega (\underline{\operatorname{curl}} y \cdot w - y \operatorname{curl} w)\,\mathrm{d}x = 0, \quad \forall w \in V_0(\Omega). \tag{4.122}$$

Therefore, by (4.122) and (4.121) we obtain

$$\int_\Omega \left(\mu^{-1} \operatorname{curl}(E - \widetilde{E}) \operatorname{curl} w + \kappa (E - \widetilde{E}) \cdot w\right) \mathrm{d}x$$
$$= \int_\Omega r(\widetilde{E}, y) \cdot w \,\mathrm{d}x + \int_\Omega d(\widetilde{E}, y) \operatorname{curl} w \,\mathrm{d}x, \tag{4.123}$$

where

$$r(\widetilde{E}, y) := f - \underline{\operatorname{curl}} y - \kappa \widetilde{E},$$
$$d(\widetilde{E}, y) := y - \mu^{-1} \operatorname{curl} \widetilde{E}.$$

With the help of the weight function λ we decompose the integral identity (4.123) as follows:

$$\int_\Omega \big(\mu^{-1}\operatorname{curl}(E-\widetilde{E})\operatorname{curl} w + \kappa(E-\widetilde{E})\cdot w\big)\,\mathrm{d}x$$
$$= \int_\Omega \lambda r(\widetilde{E},y)\cdot w\,\mathrm{d}x + \int_\Omega (1-\lambda) r(\widetilde{E},y)\cdot w\,\mathrm{d}x$$
$$+ \int_\Omega d(\widetilde{E},y)\operatorname{curl} w\,\mathrm{d}x, \tag{4.124}$$

where $\lambda \in I_{[0,1]}$. Let $w = E - \widetilde{E}$. Since

$$\int_\Omega \lambda r(\widetilde{E},y)\cdot(E-\widetilde{E})\,\mathrm{d}x \le \|\lambda\kappa^{-1/2} r(\widetilde{E},y)\|\,\|\kappa^{1/2}(E-\widetilde{E})\|$$

and by inequalities (4.115) and (4.116)

$$\int_\Omega (1-\lambda) r(\widetilde{E},y)\cdot(E-\widetilde{E})\,\mathrm{d}x$$
$$\le \|(1-\lambda) r(\widetilde{E},y)\|\big(C_{F\Omega}\|\operatorname{div}\widetilde{E}\| + C_\Omega \mu_\oplus^{1/2}\|\mu^{-1/2}\operatorname{curl}(E-\widetilde{E})\|\big).$$

Now, (4.124) implies the estimate

$$\int_\Omega \big(\mu^{-1}|\operatorname{curl}(E-\widetilde{E})|^2 + \kappa|E-\widetilde{E}|^2\big)\,\mathrm{d}x$$
$$\le R_1 + R_2\|\mu^{-1/2}\operatorname{curl}(E-\widetilde{E})\| + R_3\|\kappa^{1/2}(E-\widetilde{E})\|, \tag{4.125}$$

where

$$R_1(\lambda,\widetilde{E},y) = C_{F\Omega}\|(1-\lambda) r(\widetilde{E},y)\|\,\|\operatorname{div}\widetilde{E}\|, \tag{4.126}$$
$$R_2(\lambda,\widetilde{E},y) = C_\Omega \mu_\oplus^{1/2}\|(1-\lambda) r(\widetilde{E},y)\| + \|\mu^{1/2} d(\widetilde{E},y)\|, \tag{4.127}$$
$$R_3(\lambda,\widetilde{E},y) = \|\lambda\kappa^{-1/2} r(\widetilde{E},y)\|. \tag{4.128}$$

By applying Young's inequality to the right-hand side of (4.125), we obtain

$$\int_\Omega \left(1-\frac{1}{\alpha_1}\right)\mu^{-1}|\operatorname{curl}(E-\widetilde{E})|^2\,\mathrm{d}x + \int_\Omega \left(1-\frac{1}{\alpha_2}\right)\kappa|E-\widetilde{E}|^2\,\mathrm{d}x$$
$$\le R_1 + \frac{\alpha_1}{4}R_2^2 + \frac{\alpha_2}{4}R_3^2,$$

which implies (4.120). □

Corollary 4.1 *If* $\alpha_1 = \alpha_2 = 2$ *then* (4.120) *has the following form*:

$$|[E-\widetilde{E}]|^2_{(\mu^{-1},\kappa)} \le \overline{\mathrm{M}}^2_{\mathrm{curl},\lambda}, \tag{4.129}$$

where

$$\overline{\mathsf{M}}^2_{\mathrm{curl},\lambda} := \overline{\mathsf{M}}^2_{\mathrm{curl}}(\lambda, \widetilde{E}, y) = 2R_1(\lambda, \widetilde{E}, y) + R_2^2(\lambda, \widetilde{E}, y) + R_3^2(\lambda, \widetilde{E}, y),$$

and this estimate is sharp.

Proof It holds that

$$\inf_{\substack{\lambda \in I_{[0,1]},\\ y \in H^1(\Omega)}} \overline{\mathsf{M}}^2_{\mathrm{curl},\lambda}(\lambda, \widetilde{E}, y) \le \inf_{y \in H^1(\Omega)} \overline{\mathsf{M}}^2_{\mathrm{curl},1}(\widetilde{E}, y) \le \overline{\mathsf{M}}^2_{\mathrm{curl},1}(\widetilde{E}, p),$$

where $p = \mu^{-1} \operatorname{curl} E$ and $\overline{\mathsf{M}}_{\mathrm{curl},1} := \overline{\mathsf{M}}^2_{\mathrm{curl}}(1, \widetilde{E}, y)$, i.e.,

$$\overline{\mathsf{M}}^2_{\mathrm{curl},1}(\widetilde{E}, p) = \left\| \mu^{-1/2} \operatorname{curl}(E - \widetilde{E}) \right\|^2 + \left\| \kappa^{1/2}(E - \widetilde{E}) \right\|^2 = \left|[E - \widetilde{E}]\right|^2_{(\mu^{-1},\kappa)},$$

so that the estimate is sharp. □

Remark 4.6 By setting $\lambda = 1$ and $\lambda = 0$ we arrive at two particular forms of the error majorant, namely,

$$\overline{\mathsf{M}}^2_{\mathrm{curl},1}(\widetilde{E}, y) = \left\| \kappa^{-1/2} r(\widetilde{E}, y) \right\|^2 + \left\| \mu^{1/2} d(\widetilde{E}, y) \right\|^2 \quad \text{cf. (4.117)}, \tag{4.130}$$

and

$$\begin{aligned} \overline{\mathsf{M}}^2_{\mathrm{curl},0}(\widetilde{E}, y) &= 2C_{F\Omega} \left\| r(\widetilde{E}, y) \right\| \| \operatorname{div} \widetilde{E} \| \\ &\quad + \left(C_\Omega \mu_\oplus^{1/2} \left\| r(\widetilde{E}, y) \right\| + \left\| \mu^{1/2} d(\widetilde{E}, y) \right\| \right)^2. \end{aligned} \tag{4.131}$$

It should be noted that $\overline{\mathsf{M}}_{\mathrm{curl},0}$ is robust with respect to small values of κ. However, this form of the error majorant may lead to a considerable overestimation if κ is large. In contrary to that, $\overline{\mathsf{M}}_{\mathrm{curl},1}$ is sensitive with respect to small κ, but it is well adapted to large values of this parameter. The combined majorant $\overline{\mathsf{M}}_{\mathrm{curl},\lambda}$ is applicable to both cases. This property is due to the presence of the function λ, which allows us to compensate small values of κ.

Now, our goal is to derive a lower bound of the error. We apply the same method that we have used several times before.

Proposition 4.5 *Assume that $\kappa > 0$ and $\widetilde{E} \in V_0$ is an approximation of E. For any $w \in V_0$ the following estimate holds:*

$$\left|[E - \widetilde{E}]\right|^2_{(\mu^{-1},\kappa)} \ge \underline{\mathsf{M}}^2_{\mathrm{curl}}(\widetilde{E}, w), \tag{4.132}$$

where

$$\underline{\mathsf{M}}^2_{\text{curl}}(\widetilde{E}, w) := \int_\Omega \big(2f \cdot w - \mu^{-1}|\operatorname{curl} w|^2 \\ - \kappa|w|^2 - 2\mu^{-1} \operatorname{curl} \widetilde{E} \operatorname{curl} w - 2\kappa \widetilde{E} \cdot w\big)\, \mathrm{d}x.$$

Moreover, the lower bound is sharp, i.e.,

$$|[E - \widetilde{E}]|^2_{(\mu^{-1},\kappa)} = \inf_{w \in V_0} \underline{\mathsf{M}}^2_{\text{curl}}(\widetilde{E}, w) = \underline{\mathsf{M}}^2_{\text{curl}}(\widetilde{E}, E - \widetilde{E}).$$

Proof Note that

$$\sup_{w \in V_0} \int_\Omega \Big(\mu^{-1} \operatorname{curl}(E - \widetilde{E}) \operatorname{curl} w + \kappa w \cdot (E - \widetilde{E}) \\ - \frac{1}{2}\big(\mu^{-1} \operatorname{curl} w \operatorname{curl} w + \kappa w \cdot w\big)\Big)\, \mathrm{d}x$$

$$\leq \sup_{\substack{\tau \in H^1(\Omega) \\ w \in L^2(\Omega, \mathbb{R}^2)}} \int_\Omega \Big(\mu^{-1} \operatorname{curl}(E - \widetilde{E})\tau - \frac{1}{2}\mu^{-1}\tau^2 \\ + \kappa w \cdot (E - \widetilde{E}) - \frac{1}{2}\kappa w \cdot w\Big)\, \mathrm{d}x$$

$$= \frac{1}{2}|[E - \widetilde{E}]|^2_{(\mu^{-1},\kappa)}.$$

On the other hand,

$$\sup_{w \in V_0} \int_\Omega \Big(\mu^{-1} \operatorname{curl}(E - \widetilde{E}) \operatorname{curl} w + \kappa w \cdot (E - \widetilde{E}) \\ - \frac{1}{2}\big(\mu^{-1} \operatorname{curl} w \operatorname{curl} w + \kappa w \cdot w\big)\Big)\, \mathrm{d}x$$

$$\geq \int_\Omega \Big(\mu^{-1} \operatorname{curl}(E - \widetilde{E}) \operatorname{curl}(E - \widetilde{E}) + \kappa(E - \widetilde{E}) \cdot (E - \widetilde{E}) \\ - \frac{1}{2}\big(\mu^{-1}\big|\operatorname{curl}(E - \widetilde{E})\big|^2 + \kappa|E - \widetilde{E}|^2\big)\Big)\, \mathrm{d}x$$

$$= \frac{1}{2}|[E - \widetilde{E}]|^2_{(\mu^{-1},\kappa)}.$$

Thus, we conclude that

$$\frac{1}{2}|[E - \widetilde{E}]|^2_{(\mu^{-1},\kappa)} = \sup_{w \in V_0} \int_\Omega \Big(\mu^{-1} \operatorname{curl}(E - \widetilde{E}) \operatorname{curl} w \\ + \kappa w \cdot (E - \widetilde{E}) - \frac{1}{2}\big(\mu^{-1} \operatorname{curl} w \operatorname{curl} w + \kappa w \cdot w\big)\Big)\, \mathrm{d}x.$$

Now (4.114) yields (4.132).

Finally, we note that the sharpest bound is presented by the quantity

$$\underline{\mathsf{M}}^2_{\text{curl}}(\widetilde{E}) = \sup_{w \in V_0} \underline{\mathsf{M}}^2_{\text{curl}}(\widetilde{E}, w).$$

By setting $w = E - \widetilde{E}$, we obtain

$$\underline{\mathsf{M}}^2_{\text{curl}}(\widetilde{E}, E - \widetilde{E}) = \left|[E - \widetilde{E}]\right|^2_{(\mu^{-1},\kappa)},$$

so that the lower bound is sharp. □

4.3.2 Numerical Examples

Estimates derived in the previous section have been verified in a series of numerical tests which are discussed in this section. Approximations for the model problem were calculated by means of the lowest order Nédélec's elements of the first type (e.g., see [Mon03, Néd80]).

In the derivation of the upper bound we have used the Helmholz decomposition for the numerical approximation of the exact solution. In view of this fact, we assume that numerical approximations belong not only to $H(\text{curl})$ but also to $H(\text{div})$. The lowest order Nédélec's elements do not preserve the continuity of the normal component across element edges, so the divergence of approximate solutions is not square summable. To overcome this difficulty we chose to force the normal continuity by post-processing numerical solutions. An alternative way is to use standard Courant type elements and generate solution in $H^1(\Omega) \times H^1(\Omega)$. It is worth noting that these difficulties do not arise in the derivation of $\overline{\mathsf{M}}_{\text{curl},1}$, because it can be derived separately without using Helmholz decomposition (see [Han08, NR10a, Rep07]). Also the lower bound does not require the square summability of the divergence of the numerical approximation.

The free function y can be obtained by a global minimization of the majorant, which requires solving a finite dimensional problem with respect to y. Increasing the order of elements or using a more refined mesh than the mesh on which the approximate solution was computed results in better values of the majorant. As before, we measure the efficiency of the majorant in terms of the efficiency index (cf. (2.3)).

$$I_{\text{eff}} = \frac{\overline{\mathsf{M}}_{\text{curl},\lambda}}{|[E - \widetilde{E}]|_{(\mu^{-1},\kappa)}}.$$

For the *first test example* we take

$$\Omega = (0,1)^2, \qquad \mu \equiv 1, \qquad \kappa > 0, \qquad f = \left(\pi^2 + \kappa\right)\begin{pmatrix}\sin(\pi x_2)\\ \sin(\pi x_1)\end{pmatrix}. \tag{4.133}$$

Table 4.10 Problem (4.133): Efficiency indexes for different values of κ

κ	Linear y			Quadratic y		
	$\overline{\mathrm{M}}_{\mathrm{curl},1}$	$\overline{\mathrm{M}}_{\mathrm{curl},0}$	$\overline{\mathrm{M}}_{\mathrm{curl},\lambda}$	$\overline{\mathrm{M}}_{\mathrm{curl},1}$	$\overline{\mathrm{M}}_{\mathrm{curl},0}$	$\overline{\mathrm{M}}_{\mathrm{curl},\lambda}$
10^{-3}	103.79	1.98	1.98	6.35	1.07	1.07
10^{-1}	10.42	1.98	1.98	1.18	1.07	1.06
10^{0}	3.42	1.98	1.91	1.02	1.08	1.02
10^{1}	1.42	1.96	1.42	1.00	1.18	1.00
10^{3}	1.00	7.14	1.00	1.00	7.05	1.00

For this problem, we know the exact solution $u = (\sin(\pi x_2), \sin(\pi x_1))$. Table 4.10 shows the efficiency of the error majorants $\overline{\mathrm{M}}_{\mathrm{curl},1}$, $\overline{\mathrm{M}}_{\mathrm{curl},0}$, and $\overline{\mathrm{M}}_{\mathrm{curl},\lambda}$ for different κ. In this series of tests, for each κ the approximate solution was calculated on a mesh with 82 elements, and post-processed so that the divergence of the approximate solution becomes square summable. In the left part of Table 4.10, the results correspond to the case in which y is computed by minimizing majorants over the space of piecewise affine continuous functions generated by the same mesh that has been used to compute the approximate solution. The right part exposes results obtained by applying piece-wise quadratic approximations for y on a refined mesh.

It is not surprising that the efficiency indexes in the quadratic case are smaller, because in the tests the number of degrees of freedom for a quadratic y is approximately four times larger than for the linear y. Another observation, which follows from Table 4.10 consists of that the majorants $\overline{\mathrm{M}}_{\mathrm{curl},1}$ and $\overline{\mathrm{M}}_{\mathrm{curl},0}$ may essentially overestimate the error, while $\overline{\mathrm{M}}_{\mathrm{curl},\lambda}$ keeps small values of the efficiency index for all κ. The behavior of the majorants with respect to κ is also depicted in Fig. 4.12. The left picture corresponds to linear approximations of y and the right one shows

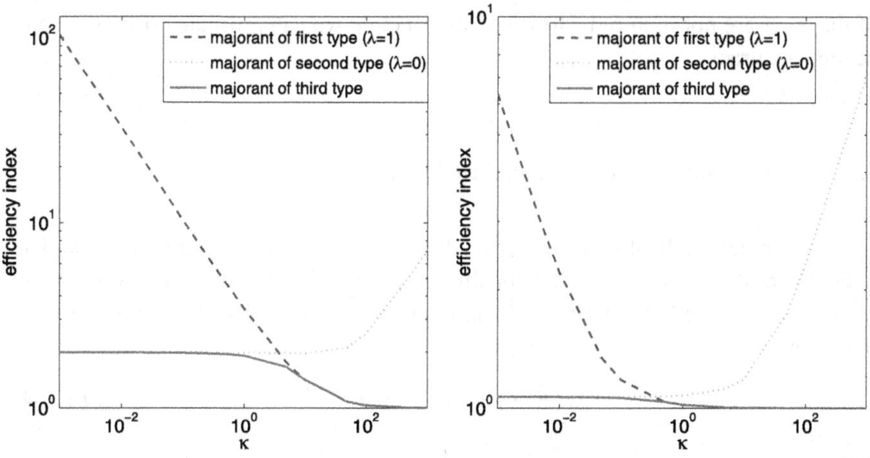

Fig. 4.12 Problem (4.133): Efficiency indexes of the majorants $\overline{\mathrm{M}}_{\mathrm{curl},1}$, $\overline{\mathrm{M}}_{\mathrm{curl},0}$ and $\overline{\mathrm{M}}_{\mathrm{curl},\lambda}$ for different κ

Table 4.11 Problem (4.133) with $\kappa = 10^{-3}$: The efficiency of $\overline{\mathsf{M}}_{\mathrm{curl},1}$ and $\underline{\mathsf{M}}_{\mathrm{curl}}$

# elem	$\|[E - \widetilde{E}]\|^2$	$\underline{\mathsf{M}}^2_{\mathrm{curl}}$	Linear y		Quadratic y	
			$\overline{\mathsf{M}}^2_{\mathrm{curl},1}$	I_{eff}	$\overline{\mathsf{M}}^2_{\mathrm{curl},1}$	I_{eff}
82	0.11908	–	1897.90	126.25	7.04419	7.69
328	0.11908	0.08914	486.837	63.94	0.55972	2.17
1312	0.11908	0.11158	123.000	32.14	0.14689	1.11
5248	0.11908	0.11721	30.9403	16.12	0.12083	1.01

results computed with the help of quadratic approximations of y. From these results we also see that $\overline{\mathsf{M}}_{\mathrm{curl},1}$ significantly benefits from using quadratic elements to approximate y.

Even though $\overline{\mathsf{M}}_{\mathrm{curl},1}$ seriously overestimates the error with small values of κ, the theory claims that it is sharp. In principle, with $\overline{\mathsf{M}}_{\mathrm{curl},1}$ one should be able to get as low efficiency index values as with $\overline{\mathsf{M}}_{\mathrm{curl},\lambda}$. To verify this, we took the case $\kappa = 10^{-3}$ and calculated the numerical approximation in a mesh with 82 elements. For this test we did not post-process the numerical approximation, because this majorant does not require that approximate solution belongs to $H(\mathrm{div})$. To test the sharpness of this majorant, we calculated the free parameter y on subsequently refined meshes. The results presented in Table 4.11 are in good agreement with the theory. The efficiency of $\overline{\mathsf{M}}_{\mathrm{curl},1}$ using linear y is low, but using quadratic y for y we clearly see that the upper bound converges to the exact error. Also, calculations related to $\underline{\mathsf{M}}_{\mathrm{curl}}$ show that it is rather efficient.

From these results we can conclude that it is possible to achieve high accuracy of two-sided bounds of error. This is only a matter of computational resources that we are ready to invest in error analysis. Certainly, in many cases very sharp bounds of the error are not required. However, in principle we can estimate errors with any desired accuracy.

4.3.2.1 Error Indicators Generated by $\overline{\mathsf{M}}_{\mathrm{curl},1}$

Getting efficient indicators of the error distribution is another important task. The majorant $\overline{\mathsf{M}}_{\mathrm{curl},1}$ is most suitable for this purpose because it does not contain any constants. Using the two terms in $\overline{\mathsf{M}}_{\mathrm{curl},1}$ separately, we define the following error indicators

$$\mathbb{E}_r(\widetilde{E}, y) = \left\| \kappa^{-1/2} r(\widetilde{E}, y) \right\| \tag{4.134}$$

and

$$\mathbb{E}_d(\widetilde{E}, y) = \left\| \mu^{1/2} d(\widetilde{E}, y) \right\|. \tag{4.135}$$

If $y \approx \mu^{-1} \operatorname{curl} E$, then the indicator (4.134) should give a good error distribution for the weighed L^2-norm of the error

$$\left\|\kappa^{1/2}(E - \widetilde{E})\right\|.$$

Respectively, the indicator (4.135) should give a good error distribution for the weighed $H(\text{curl})$-seminorm of the error

$$\left\|\mu^{-1/2} \operatorname{curl}(E - \widetilde{E})\right\|.$$

Our numerical tests applied two different techniques to compute y in the indicators (4.134) and (4.135): (a) y_{glo} denotes the function obtained by global minimization of the majorant $\overline{\mathrm{M}}_{\text{curl},1}$, where y_{glo} was calculated with linear elements in the same mesh on which the approximate solution was calculated; (b) y_{avg} denotes the function obtained by a simple averaging procedure, where for each node we calculate the approximate solution's curl values on the surrounding elements and weight them by the sizes of respective elements. Then, we average the values to obtain a value for the node.

For the second test example we take

$$\begin{aligned} &\Omega = (0,1)^2 \setminus \left(\left[\frac{1}{2}, 1\right] \times \left[0, \frac{1}{2}\right]\right), \qquad \mu \equiv 1, \\ &\kappa = 1, \qquad f = \begin{pmatrix} 1 \\ 0 \end{pmatrix}. \end{aligned} \tag{4.136}$$

For this problem we do not know the exact solution: instead, we introduce a reference solution, which was calculated in a mesh with 286114 elements.

Figures 4.13 and 4.14 present typical results for the indicators (4.134) and (4.135). In them, we denote, by $I_r(y)$ and $I_d(y)$, the markings indicated by $I_r(y) := \mathbb{E}_r(\widetilde{E}, y)$ and $I_d(y) := \mathbb{E}_d(\widetilde{E}, y)$, respectively. Moreover, we have marked with black color all those elements for which the indicated error is greater than the indicated average error. The first row is related to the indicator (4.134) and the second one to (4.135). The first picture of each row depicts marking of elements based upon the true distribution of the error. The second picture shows the marking based on the error indication computed with the help of y_{glo}, and the third picture shows the marking where the error indication is based on y_{avg}. We observe that the latter method may provide a suitable indication in some cases and rather coarse results in others.

We end up this section with literature comments. Error indicators studied in the literature devoted to Maxwell type equations are usually based on residual approach. In particular, residual type estimates were studied in [BHHW00, Mon98, Mon03], and an equilibrated residual approach was presented in [BS08]. A posteriori estimates for nonconforming approximations for $H(\text{curl})$-elliptic partial differential equations were considered in [HPS07]. A Zienkiewicz–Zhu type error estimate was analyzed in [Nic05].

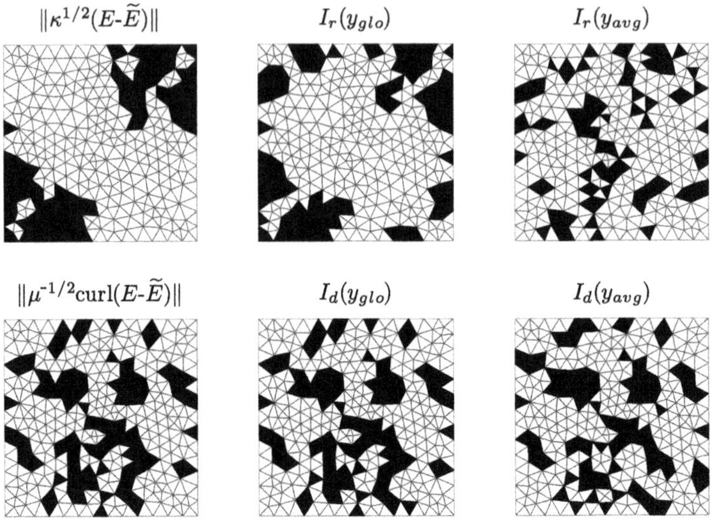

Fig. 4.13 Problem (4.133): Performance of error indicators

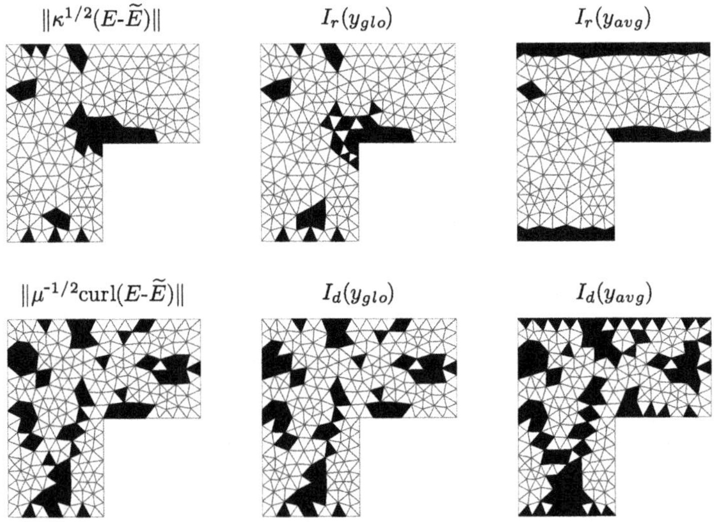

Fig. 4.14 Problem (4.136): Performance of error indicators

Functional type estimates for the problem (4.112)–(4.113) were derived in [Han08, Rep07, Rep08] and numerically studied in [AMM$^+$09] (the material of this section and numerical results are based on the latter paper). [Han08] presented a majorant for the case of complex κ, $\mathcal{R}(\kappa) \geq 0$. However, sharpness of this upper bound was not proved. In [NR10a], a sharp lower bound for $\kappa > 0$ and two new upper bounds were presented. The first upper bound is valid for $\kappa \geq 0$, and it is insensitive with respect to small values of κ. However, this estimate is sensitive with respect to large values of κ, and the sharpness of this estimate was not proved. The second upper bound is derived in a more sophisticated way and provides a more general upper bound. Also, it behaves well with respect to small and large values of κ.

4.4 Generalizations

The reader may have noticed that almost all the above-discussed estimates contain similar terms, which penalize possible violations of the fundamental physical relations generating the problem. However, there are certain differences and the majorants $\overline{\mathsf{M}}_2$, $\overline{\mathsf{M}}_{\mathrm{div}\,A\nabla}$, $\overline{\mathsf{M}}_{\Delta}$, $\overline{\mathsf{M}}_{\mathrm{EL}}$, $\overline{\mathsf{M}}_{2D,\sigma}$, $\overline{\mathsf{M}}_{2D,\varepsilon}$, $\overline{\mathsf{M}}_{\nabla\nabla}$, $\overline{\mathsf{M}}_{\mathrm{RM}}$, $\overline{\mathsf{M}}_{\mathrm{KL}}$, and $\overline{\mathsf{M}}_{\mathrm{curl},0}$ form a group of their own, and the majorants $\overline{\mathsf{M}}_{1D}$, $\overline{\mathsf{M}}_1$, and $\overline{\mathsf{M}}_{\mathrm{curl},1}$ form a somewhat different group. Below, we discuss a general framework which encompasses all the above-mentioned linear problems (a general framework of a posteriori error estimation applicable for a wide set on nonlinear problems generated by convex energy functionals is presented in [NR04, Rep08, Rep00b]).

Let $\mathcal{V}$ and U be two Hilbert spaces with the inner products $(\cdot,\cdot)_{\mathcal{V}}$ and $(\cdot,\cdot)_U$ respectively. These products generate the norms $\|\cdot\|_{\mathcal{V}}$ and $\|\cdot\|_U$. We introduce a positive definite self-adjoint linear operator $\mathcal{A}: U \to U$ and a semi-positive definite self-adjoint operator $\mathcal{B}: \mathcal{V} \to \mathcal{V}$.

In addition, we introduce a bounded linear operator $\Lambda: V \to U$, where $V \subset \mathcal{V}$ is a Hilbert space generated by the inner product $(w,v)_V := (w,v)_{\mathcal{V}} + (\Lambda w, \Lambda v)_U$. Henceforth V_0 denotes a convex, closed and non-empty subspace of V such that $V_0 \subset V \subset \mathcal{V} \subset V_0^*$.

Typically, V is a Sobolev space associated with the differential operator Λ and V_0 contains the functions which satisfy homogeneous Dirichlet boundary conditions on part of the boundary.

Using these definitions, we can represent almost all energy functionals associated with the problems discussed in Chaps. 3 and 4 in the following common form:

$$J(w) := \frac{1}{2}(\mathcal{A}\Lambda w, \Lambda w)_U + \frac{1}{2}(\mathcal{B}w, w)_{\mathcal{V}} - (f, w)_{\mathcal{V}}, \tag{4.137}$$

where $f \in \mathcal{V}$. We assume that

$$(\mathcal{A}y, y)_U \geq c_1 \|y\|_U^2, \quad \forall y \in U, \tag{4.138}$$

and

$$\|w\|_{\mathcal{V}} \leq C_F \|\Lambda w\|_U, \quad \forall w \in V_0. \tag{4.139}$$

The adjoint operator $\Lambda^*: U \to V_0^*$ is defined by the relation

$$\langle \Lambda^* y, w \rangle = (y, \Lambda w)_U, \quad \forall y \in U, w \in V_0, \tag{4.140}$$

where $\langle \cdot, \cdot \rangle$ denotes the pairing of V_0, and its conjugate V_0^* and $\langle \Lambda^* y, w \rangle$ is the value of the functional $\Lambda^* y \in V_0^*$ at $w \in V_0$.

For our analysis, it is convenient to introduce the form $a : V_0 \times V_0 \to \mathbb{R}$,

$$a(u, w) := (\mathcal{A}\Lambda u, \Lambda w)_U + (\mathcal{B}u, w)_{\mathcal{V}}, \tag{4.141}$$

which is symmetric and bilinear. Under the assumptions made above, the form a is V-elliptic (cf. (B.4) and (B.5)). The form a defines the norm

$$|||w||| := \sqrt{a(w, w)}, \tag{4.142}$$

that is, the energy norm. Since $\mathcal{A}$ is self-adjoint and positive definite, we can define additional equivalent norms in U

$$\|y\|^2_{\mathcal{A}} := (\mathcal{A}y, y)_U \quad \text{and} \quad \|y\|^2_{\mathcal{A}^{-1}} := \left(\mathcal{A}^{-1}y, y\right)_U.$$

Using the form a, we rewrite the energy in the form

$$J(w) := \frac{1}{2}a(w, w) - (f, w)_{\mathcal{V}}. \tag{4.143}$$

The (generalized) solution u is the minimizer of the variational problem

$$J(u) = \min_{w \in V_0} J(w), \tag{4.144}$$

which (see Theorem B.5) exists and is unique. Moreover, it satisfies the relation

$$a(u, w) = (f, w)_{\mathcal{V}}, \quad \forall w \in V_0. \tag{4.145}$$

We introduce the function

$$p := \mathcal{A}\Lambda u.$$

It can be considered as a generalization of the notion "flux", which is typical in analysis of diffusion type problems. Assume that $\mathcal{B} = 0$. Then, (4.145) reads as follows:

$$(p, \Lambda w)_U = (f, w)_{\mathcal{V}}, \quad \forall w \in V_0.$$

This means that $p \in Q$, where

$$Q := \left\{y \in U \mid \Lambda^* y \in \mathcal{V}\right\}.$$

This allows us to rewrite the problem as the system

$$\Lambda^* p = f, \tag{4.146}$$

$$p = \mathcal{A}\Lambda u. \tag{4.147}$$

The relations (4.146) and (4.147) are the generalized form of the equilibrium relation and the constitutive (duality) relation, respectively. Usually, these key relations have a clear physical motivation. If $\mathcal{B} \neq 0$, then it enters the first equation and the system has the form

$$\Lambda^* p + \mathcal{B}u = f, \tag{4.148}$$

$$p = \mathcal{A}\Lambda u. \tag{4.149}$$

These relations motivate the main terms entering error majorants, which we consider below.

4.4.1 Error Majorant

The goal of this section is to generalize the method that has been used for the reaction diffusion problem (see Sect. 3.2.1) and deduce several generalized forms of the error majorant (which encompass $\overline{\mathsf{M}}_{\mathrm{div}\,A\nabla}$ defined in (3.38), $\overline{\mathsf{M}}_{\mathrm{EL}}$, and $\overline{\mathsf{M}}_{\mathrm{KL}}$).

First, we consider the case $\mathcal{B} = 0$. The solution is defined by the integral identity

$$(\mathcal{A}\Lambda u, \Lambda w)_U = (f, w)_{\mathcal{V}}, \quad \forall w \in V_0, \tag{4.150}$$

and the respective error majorant is presented by the following theorem.

Theorem 4.6 *Let u be the solution of the problem* (4.150) *and $v \in V_0$. Then,*

$$|||u - v|||^2 \leq \overline{\mathsf{M}}^2_\Lambda(v, y, \beta), \quad \forall y \in Q \text{ and } \beta > 0, \tag{4.151}$$

where

$$\overline{\mathsf{M}}^2_\Lambda(v, y, \beta) := (1+\beta)\|\mathcal{A}\Lambda v - y\|^2_{\mathcal{A}^{-1}} + \left(1 + \frac{1}{\beta}\right)\frac{C_F^2}{c_1}\|f - \Lambda^* y\|^2_{\mathcal{V}}, \tag{4.152}$$

where C_F and c_1 are defined by (4.139) *and* (4.138) *respectively, and y is an arbitrary function in Q. Moreover, this bound is sharp, i.e.,*

$$|||u - v|||^2 = \inf_{\substack{y \in Q,\\ \beta > 0}} \overline{\mathsf{M}}^2_\Lambda(v, y, \beta). \tag{4.153}$$

Proof By subtracting $(\mathcal{A}\Lambda v, \Lambda w)_U$ from (4.150) and applying (4.140), we find that

$$\begin{aligned}\left(\mathcal{A}\Lambda(u - v), \Lambda w\right)_U &= (f, w)_{\mathcal{V}} - (\mathcal{A}\Lambda v, \Lambda w)_U - \left(\Lambda^* y, w\right)_{\mathcal{V}} + (y, \Lambda w)_U \\ &= (y - \mathcal{A}\Lambda v, \Lambda w)_U + \left(-\Lambda^* y + f, w\right)_{\mathcal{V}},\end{aligned}$$

where $y \in Q$. We estimate the first term from the above by (A.8) and the second term by the Cauchy–Schwartz inequality. Then, the estimates (4.138) and (4.139) imply

$$\left(\mathcal{A}\Lambda(u-v),\Lambda w\right)_U \le \|\mathcal{A}\Lambda v - y\|_{\mathcal{A}^{-1}}\|\Lambda w\|_{\mathcal{A}} + \left\|-\Lambda^* y + f\right\|_{\mathcal{V}}\|w\|_{\mathcal{V}}$$
$$\le \|\mathcal{A}\Lambda v - y\|_{\mathcal{A}^{-1}} |||w||| + \frac{C_F}{\sqrt{c_1}}\left\|-\Lambda^* y + f\right\|_{\mathcal{V}} |||w|||.$$

We set $w := u - v$ and obtain

$$|||u-v||| \le \|\mathcal{A}\Lambda v - y\|_{\mathcal{A}^{-1}} + \frac{C_F}{\sqrt{c_1}}\left\|f - \Lambda^* y\right\|_{\mathcal{V}}.$$

Squaring both sides of the estimate and applying (A.6) we arrive at (4.151). If $y := \mathcal{A}\Lambda u$, then the second term vanishes. Therefore, (4.153) holds if we tend β to zero. □

If we compare the structure of (4.151) with (4.146) and (4.147), we see that two parts of $\overline{\mathsf{M}}_\Lambda$ are measures of errors in the basic relations (4.146) and (4.147).

Now, we consider the case $\mathcal{B} \neq 0$. Moreover, we assume that $\mathcal{B}$ is positive definite and define the norms

$$\|w\|^2_{\mathcal{B}} := (\mathcal{B}w, w)_{\mathcal{V}} \quad \text{and} \quad \|w\|^2_{\mathcal{B}^{-1}} := \left(\mathcal{B}^{-1}w, w\right)_{\mathcal{V}}.$$

In order to derive an analog of the majorant $\overline{\mathsf{M}}_{\widehat{\alpha}}$ (cf. (3.30)), we apply the same method and introduce a weight function $\mu : \mathcal{V} \to \mathcal{V}$, which attains values from 0 to 1. The class of such functions is denoted by Υ.

Theorem 4.7 *For any* $v \in V_0$,

$$|||u-v|||^2 \le \overline{\mathsf{M}}^2_{\Upsilon}(v, y, \beta, \mu), \quad \forall y \in Q, \beta > 0, \mu \in \Upsilon, \tag{4.154}$$

where

$$\overline{\mathsf{M}}^2_{\Upsilon}(v, y, \beta, \mu) := (1+\beta)\|\mathcal{A}\Lambda v - y\|^2_{\mathcal{A}^{-1}} + \left(1 + \frac{1}{\beta}\right)\frac{C_F^2}{c_1}\left\|\mu\mathcal{R}(v, y)\right\|^2_{\mathcal{V}}$$
$$+ \left\|(1-\mu)\mathcal{R}(v, y)\right\|_{\mathcal{B}^{-1}}, \tag{4.155}$$

and $\mathcal{R}(v, y) := f - \Lambda^* y - \mathcal{B}v$. *The estimate is sharp, i.e.,*

$$|||u-v|||^2 = \inf_{\substack{y \in Q,\\ \beta > 0}} \overline{\mathsf{M}}^2_{\Upsilon}(v, y, \beta, 0).$$

Proof We apply (4.145) and (4.140), and obtain

$$\begin{aligned} a(u-v,w) &= (f-\mathcal{B}v,w)_{\mathcal{V}} - (\mathcal{A}\Lambda v,\Lambda w)_U - \left(\Lambda^* y,w\right)_{\mathcal{V}} + (y,\Lambda w)_U \\ &= (y-\mathcal{A}\Lambda v,\Lambda w)_U + \left(\mathcal{R}(v,y),w\right)_{\mathcal{V}}. \end{aligned}$$

By means of the function μ we rewrite this identity as follows:

$$a(u-v,w) = (y-\mathcal{A}\Lambda v,\Lambda w)_U + \left(\mu\mathcal{R}(v,y),w\right)_{\mathcal{V}} + \left((1-\mu)\mathcal{R}(v,y),w\right)_{\mathcal{V}}.$$

We apply (A.8), the Cauchy–Schwartz inequality, (4.138), and (4.139) and obtain

$$\begin{aligned} a(u-v,w) &\le \|\mathcal{A}\Lambda v - y\|_{\mathcal{A}^{-1}}\|\Lambda w\|_{\mathcal{A}} + \left\|\mu\mathcal{R}(v,y)\right\|_{\mathcal{V}}\frac{C_F}{\sqrt{c_1}}\|\Lambda w\|_{\mathcal{A}} \\ &\quad + \left\|(1-\mu)\mathcal{R}(v,y)\right\|_{\mathcal{B}^{-1}}\|w\|_{\mathcal{B}} \\ &= \left(\|\mathcal{A}\Lambda v - y\|_{\mathcal{A}^{-1}} + \frac{C_F}{\sqrt{c_1}}\|\mu r\|_{\mathcal{V}}\right)\|\Lambda w\|_{\mathcal{A}} \\ &\quad + \left\|(1-\mu)\mathcal{R}(v,y)\right\|_{\mathcal{B}^{-1}}\|w\|_{\mathcal{B}} \\ &\le \left(\left(\|\mathcal{A}\Lambda v - y\|_{\mathcal{A}^{-1}} + \frac{C_F}{\sqrt{c_1}}\|\mu r\|_{\mathcal{V}}\right)^2\right. \\ &\quad \left. + \left\|(1-\mu)\mathcal{R}(v,y)\right\|^2_{\mathcal{B}^{-1}}\right)^{1/2} |||w|||. \end{aligned} \tag{4.156}$$

Setting $w := u - v$, using (A.6), and squaring both sides results in the estimate

$$\begin{aligned} |||u-v|||^2 &\le \left(\|\mathcal{A}\Lambda v - y\|_{\mathcal{A}^{-1}} + \frac{C_F}{\sqrt{c_1}}\left\|\mu\mathcal{R}(v,y)\right\|_{\mathcal{V}}\right)^2 \\ &\quad + \left\|(1-\mu)\mathcal{R}(v,y)\right\|^2_{\mathcal{B}^{-1}} \\ &\le (1+\beta)\|\mathcal{A}\Lambda v - y\|^2_{\mathcal{A}^{-1}} + \left(1+\frac{1}{\beta}\right)\frac{C_F^2}{c_1}\left\|\mu\mathcal{R}(v,y)\right\|^2_{\mathcal{V}} \\ &\quad + \left\|(1-\mu)\mathcal{R}(v,y)\right\|^2_{\mathcal{B}^{-1}}. \end{aligned}$$

It remains to show that the estimate is sharp. Let us set $y = \mathcal{A}\Lambda u$, then

$$\|\mathcal{A}\lambda v - y\|_{\mathcal{A}^{-1}} = \left\|\Lambda(v-u)\right\|_{\mathcal{A}} \quad \text{and} \quad \mathcal{R}(v,y) = \mathcal{B}(u-v).$$

If $\mu = 0$, then the right-hand side of (4.154) coincides with the left-hand one. □

Remark 4.7 The function μ is at our disposal: if we select $\mu = 0$ or $\mu = 1$, we obtain the following estimates of a special type:

$$|||u-v|||^2 \le \|\mathcal{A}\Lambda v - y\|^2_{\mathcal{A}^{-1}} + \left\|(1-\mu)\mathcal{R}(v,y)\right\|^2_{\mathcal{B}^{-1}} \tag{4.157}$$

and

$$|||u - v|||^2 \leq (1+\beta)\|\mathcal{A}\Lambda v - y\|^2_{\mathcal{A}^{-1}} + \left(1 + \frac{1}{\beta}\right)\frac{C_F^2}{c_1}\left\|\mathcal{R}(v, y)\right\|^2_{\mathcal{V}}, \tag{4.158}$$

which are analogs of $\overline{\mathsf{M}}_1$ (3.28) and $\overline{\mathsf{M}}_2$ (3.29), respectively. The bound (4.157) has no "gap", i.e., for $y := p = \mathcal{A}\Lambda u$ it coincides with the exact error, whereas (4.158) does not have this property. The benefits of the bound (4.158) are the same as those of $\overline{\mathsf{M}}_2$ (see comments in Sect. 3.2.1).

4.4.2 Error Minorant

The methods of deriving error minorants, which we have used in previous sections can also be written in terms of our abstract setting.

Theorem 4.8 *Let u be the solution of* (4.145) *and $v \in V_0$. Then,*

$$|||u - v|||^2 \geq \underline{\mathsf{M}}^2_\Lambda(v, w), \quad \forall w \in V_0, \tag{4.159}$$

where

$$\underline{\mathsf{M}}^2_\Lambda(v, w) := -a(w, w + 2v) + 2(f, w)_{\mathcal{V}}. \tag{4.160}$$

The minorant is sharp, i.e.,

$$|||u - v|||^2 = \sup_{w \in V_0} \underline{\mathsf{M}}^2_\Lambda(v, w). \tag{4.161}$$

Proof In view of (B.72), we have

$$|||u - v|||^2 = 2\big(J(v) - J(u)\big).$$

Since (4.144), $J(u) \leq J(w + v)$, for all $w \in V_0$, we have

$$|||u - v|||^2 \geq 2\big(J(v) - J(w + v)\big) = -a(w, w + 2v) + 2(f, w)_{\mathcal{V}}.$$

It is easy to see that $\underline{\mathsf{M}}^2_\Lambda(v, u - v) = |||u - v|||$ and (4.161) holds. □

Remark 4.8 In computations, it may be easier to use a different form of the minorant

$$|||u - v|||^2 = 2\big(J(v) - J(u)\big) \geq 2\big(J(v) - J(w)\big) =: \underline{\mathsf{M}}^2_{\Lambda,J}(v, w), \quad \forall w \in V_0.$$

Let v be a Galerkin approximation computed on a subspace $V_0^h \subset V_0$. Then, the maximizer $\tilde{w} \in \tilde{V}_0^h$ of $\underline{\mathsf{M}}^2_{\Lambda,J}(v, w)$ can be computed from the problem

$$a(\tilde{w}, w) = (f, w)_{\mathcal{V}}, \quad \forall w \in \tilde{V}_0^h$$

using the same solver which was used to compute the approximation v. However, in order to obtain a positive lower bound, the subspace $\tilde{V}_0^h$ must contain more trial functions than V_0^h.

Chapter 5
Errors Generated by Uncertain Data

Abstract In this chapter, we study effects caused by incompletely known data. In practice, the data are never known exactly, therefore the results generated by a mathematical model also have a limited accuracy. Then, the whole subject of error analysis should be treated in a different manner, and accuracy of numerical solutions should be considered within a framework of a more complicated scheme, which includes such notions as maximal and minimal distances to the solution set and its radius.

5.1 Mathematical Models with Incompletely Known Data

Incompletely known data are constantly present in mathematical modeling. Typically, a mathematical model describing a physical phenomenon contains some parameters defined by measurements or other means. For example, material parameters often belong to this class. The measurements are performed with a limited accuracy, and it is important to be able to estimate how much this inaccuracy affects the results. This inaccuracy does not depend on the choice of a particular numerical scheme, but presents a fundamental property of the problem itself.

The main approach usually used for controlling uncertainty in a model is the so-called probabilistic approach, which leads to stochastic PDEs. These indeterminate data occurring in a PDE are considered random variables with the known probability density. The aim is then to find or approximate the mean value, variance, and other probabilistic quantities related to the solution. An overview of the theory and related numerical methods (dating back to [Bab61]) can be found in [Sch97]. The most popular numerical method is the Monte Carlo method [Eli83]. The idea of the method is to generate samples of input data within an uncertainty range and to compute respective solutions. Then, the probabilistic features are studied via a statistical analysis.

Probability distributions are not the only way of modeling the uncertainty. In the evidence theory (also known as the Dempster–Shafer theory) [Dem67, Sha76], the requirements for the probability measure are relaxed and the obtained probability assignments are applied instead.

In the theory of fuzzy sets [Zad65] (and evolved possibility theory [Zad78]), the uncertainty is introduced via a membership function. In the classical theory, there

O. Mali et al., *Accuracy Verification Methods*,

Computational Methods in Applied Sciences 32, DOI 10.1007/978-94-007-7581-7_5,

are only two options: an element is a member of a set or it is not. The membership function determines the degree of truth of the statement that an element belongs to a set. The idea is to analyze how the fuzziness of data is inherited by a solution. An introduction and examples can be found in [Ber99].

The application of these theories to physical models is discussed in [HCB04], where the question is studied in the framework of the worst-case scenario method.

An overview of reliability engineering, which is related to system analysis and risk analysis, is given in [Zio09], where uncertainty is classified either as aleatory or as epistemic. The first class is related to the uncertainty due to the inherent variability of the system and the second one is associated with a lack of knowledge.

Sensitivity analysis is another concept related to incompletely known data. It qualitatively indicates the influence of a particular input parameter on the exact solution or another quantity of interest. Typically, for this purpose derivatives (gradients) of the solution norm or other quantities with respect to input parameters are investigated. This analysis can be carried out either for the original PDE or for an approximated finite dimensional model. The corresponding theory is exposed in, e.g., [Hay79, Lit00, Rou97]. Also, the sensitivity analysis can be conducted numerically by Monte Carlo type simulations, where the scattering of the results may indicate the level of correlation between input and output data (see, e.g., [KH99]). The sensitivity of the solution with respect to the geometrical factors is studied in [HM03, HN96], where the information on the sensitivity is used to solve the optimization problem.

5.2 The Accuracy Limit

Below, we suggest a mathematical framework for studying the effects of incompletely known data, which differs from the above-mentioned approaches. It is clear that, owing to limited knowledge of the data, a solution of PDE (or another quantity of interest) can be known only with a certain limited accuracy. This *accuracy limit* has a profound impact for simulation practice:

> If the accuracy of an approximate solution is within the accuracy limit dictated by indeterminacy (data uncertainty), then efforts to improve this solution (e.g., by enriching the set of basis functions) are not consistent.

Henceforth, the set of admissible data is denoted by $\mathcal{D}$. It is described by setting a certain "mean" data $D_\circ$ and admissible perturbations around it. Here and later on, the subscript $\circ$ means that the corresponding objects (data or solutions) are related to $D_\circ$. We assume that all admissible data in $\mathcal{D}$ are such that the corresponding problems are well defined, i.e., any such problem possesses a unique solution. Moreover, we define a solution mapping (see Fig. 5.1)

$$\mathcal{S} : \mathcal{D} \to \mathcal{S}(\mathcal{D}) \subset V,$$

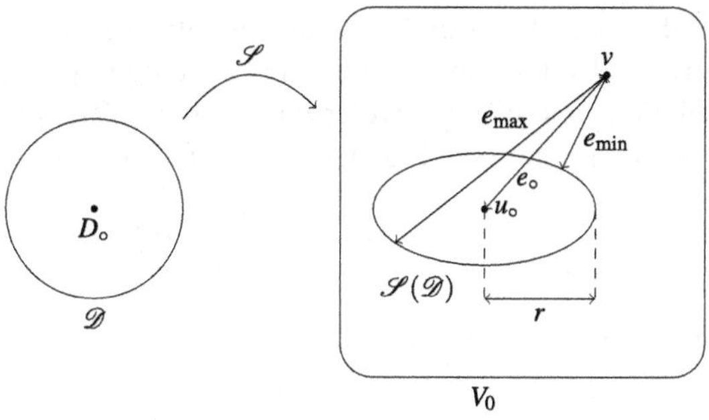

Fig. 5.1 Illustration of definitions

where $\mathcal{S}(\mathcal{D})$ is a set of possible solutions, which belongs to a suitable (energy) space V common for all data.

Analysis of problems with incompletely known data motivates investigation of several topics related to size and structure of the set $\mathcal{S}(\mathcal{D})$. One topic concerns the size of $\mathcal{S}(\mathcal{D})$. To measure it, we choose a "mean" solution $u_\circ := \mathcal{S}(D_\circ)$ as an "anchor" and define the *radius of the solution set* (see Fig. 5.1) as follows:

$$r := \sup_{u \in \mathcal{S}(\mathcal{D})} |||u_\circ - u|||_\circ. \tag{5.1}$$

In practice, it is better to use its normalized counterpart

$$\bar{r} := \sup_{u \in \mathcal{S}(\mathcal{D})} \frac{|||u_\circ - u|||_\circ}{|||u_\circ|||_\circ}, \tag{5.2}$$

where $|||\cdot|||_\circ$ denotes the energy norm generated by the "mean problem" (i.e., by the boundary-value problem with the data $D_\circ$). Other definitions of r can be defined in terms of dual solution or some suitable functionals.

The radius of the solution set $\mathcal{S}(\mathcal{D})$ is a fundamental quantity, which is not related to a particular approximation. It shows the accuracy limit dictated by data uncertainty.

Another topic is related to *sensitivity* of solutions (or some quantities generated by solutions) with respect to variations in data. Here, the goal is to study the variability of certain quantities of interest with respect to the data. It is of special interest to see how the variability depends on the value of $D_\circ$ and how various factors affect its magnitude.

Denote the distance between $v \in V$ and the mean solution u_0 by $e_\circ := |||u_\circ - v|||_\circ$. Let

$$\bar{e}_\circ := \frac{|||u_\circ - v|||_\circ}{|||u_\circ|||_\circ}.$$

In the case of incompletely known data, this quantity cannot fully characterize the error. Indeed, the exact solution is not uniquely determined (we have many different solutions and can not prefer one of them). Therefore, it is necessary to define two other quantities (see Fig. 5.1): the worst case (maximal) error

$$e_{\max} := \sup_{u \in \mathcal{S}(\mathcal{D})} |||u - v|||_\circ \tag{5.3}$$

and the best case (minimal) error

$$e_{\min} := \inf_{u \in \mathcal{S}(\mathcal{D})} |||u - v|||_\circ . \tag{5.4}$$

Evaluation of $e_{\max}$ and $e_{\min}$ is another important topic arising in the situation where $\mathcal{S}(\mathcal{D})$ contains more than one element.

If the distance from v to the set $\mathcal{S}(\mathcal{D})$ is much larger than $\mathrm{diam}(\mathcal{S}(\mathcal{D}))$ (in this case $e_\circ \approx e_{\max} \approx e_{\min}$), then the distance from v to any function in the solution set mainly represents the approximation error. If this is not the case, i.e., $e_{\max} \gg e_{\min}$, then any further efforts to improve the approximative solution v are useless, because v is already close to the solution set and therefore the accuracy limit has been reached.

Example 5.1 In order to illustrate the definitions, we consider the following elementary example generated by a system of linear equations.

$$Ax = b,$$

where the matrix A is not completely known but belongs to a set

$$A \in \mathcal{D} := \left\{ A \in \mathbb{M}^{2\times 2} \mid A = A_\circ + \delta E,\, E \in \mathbb{M}^{2\times 2},\, |E| \le 1, \delta > 0 \right\}.$$

The respective solution set is

$$\mathcal{S}(\mathcal{D}) := \left\{ x \in \mathbb{R}^2 \mid Ax = b,\, A \in \mathcal{D} \right\}.$$

We use the Monte-Carlo algorithm to compute elements of the solution set, i.e., we randomly generate a large number of symmetric matrices such that $|E| < 1$ and compute the respective solutions. Two “mean” matrices

$$A_\circ^{(1)} = \begin{bmatrix} 1 & 0 \\ 0 & 1 \end{bmatrix} \quad \text{and} \quad A_\circ^{(2)} = \frac{1}{3}\begin{bmatrix} 2 & 1 \\ 1 & 2 \end{bmatrix}$$

generate the sets $\mathcal{D}_1$ and $\mathcal{D}_2$, respectively. Figure 5.2 depicts the corresponding sets $\mathcal{S}(\mathcal{D}_1)$ and $\mathcal{S}(\mathcal{D}_2)$, where $\delta = 0.05$ and $b = [1, 1]^T$. The computations also produce a lower bound for the radius of the solution set, i.e.,

$$r_i := \left\{ \max_{x \in \mathcal{S}(\mathcal{D}_i)} \frac{|x_\circ - x|}{|x_\circ|} \right\},$$

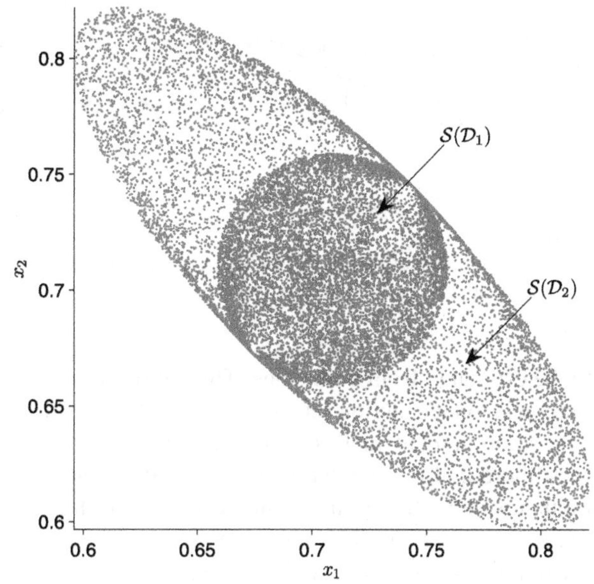

Fig. 5.2 Representations of solution sets $\mathcal{S}(\mathcal{D}_1)$ (*darker*) and $\mathcal{S}(\mathcal{D}_2)$ (*lighter*)

where $x_\circ$ satisfies $A_\circ x_\circ = b$. For our example, they are $r_1 = 0.053$ and $r_2 = 0.152$, respectively. It is easy to see that in the second case the perturbation of coefficients by five percent generates approximately three times larger difference between the most distant elements of $\mathcal{S}(\mathcal{D})$.

5.3 Estimates of the Worst and Best Case Scenario Errors

At first glance, the computation of worst and best case errors for problems generated by PDEs seems to be an unfeasible task, since elements of the set $\mathcal{S}(\mathcal{D})$ are generally unknown. The remedies are the functional a posteriori error estimates of deviations from exact solutions, which are guaranteed and *explicitly* depend on the problem data.

Estimates of deviations from exact solutions considered in Chaps. 3 and 4 explicitly depend on the problem data, which makes it possible to estimate quantities including an extremum over the solution set. Thus, taking the supremum over the solution set (which is generally unknown) can be replaced by taking the supremum over the set of admissible data (which are known).

The main motivation to investigate worst and best case errors is that the ratio

$$\upsilon := \frac{e_{\max} - e_{\min}}{e_{\min}} \tag{5.5}$$

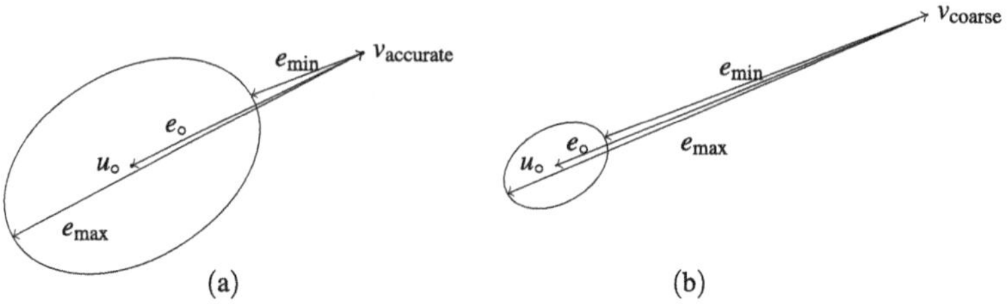

Fig. 5.3 Illustration of two cases: **a** Error generated by data uncertainty dominates ($\upsilon \approx 2$ and $\upsilon_{\circ} \approx 0.5$); **b** approximation error dominates ($\upsilon \approx 0$ and $\upsilon_{\circ} \approx 0$)

can be estimated from both sides. This ratio indicates how close an approximative solution is to the solution set (see Fig. 5.3). If this ratio is close to one, then the size of the solution set is small compared to the approximation error and it makes sense to invest more in the computation. However, if the ratio is large, then the approximate solution is very close to the solution, set and further adaptation of meshes or other means to improve the approximation makes no sense.

Often it is computationally more feasible to study the ratio

$$\upsilon_{\circ} := \frac{e_{max} - e_{\circ}}{e_{\circ}}. \tag{5.6}$$

Since e_{max} encompasses the approximation error, and the error due to incomplete knowledge of the data and $e_{\circ}$ is the approximation error, their ratio indicates relative magnitudes of these error sources. The information obtained from υ and $\upsilon_{\circ}$ is illustrated in Fig. 5.3. Certainly, there is no unified criterion showing when it is necessary to stop computations, and it is somewhat up to the analyst to decide whether, e.g., $\upsilon \approx 2$ or $\upsilon \approx 1$ is a proper condition to cease all efforts to acquire an improved approximate solution.

The benefits of using majorants and minorants of the functional type are transparently seen in Theorem 5.1, where the worst and best case errors are estimated from both sides. Here, we write the majorant (minorant) as $\overline{\mathsf{M}}(v, y, \beta; D)$ ($\underline{\mathsf{M}}(v, w; D)$) to highlight the fact that the majorant (minorant) depends explicitly on the problem data D. We recall that the majorant and minorant are guaranteed, i.e., for any $v \in V$,

$$\underline{\mathsf{M}}^2(v, w; D) \leq |||u - v|||_D^2 \leq \overline{\mathsf{M}}^2(v, y, \beta; D), \quad \forall w \in V_0, y \in Q, \tag{5.7}$$

where u is the solution of the boundary-value problem generated by the data D and w and y are auxiliary functions in respective function spaces.

Theorem 5.1 *Let $v \in V$ and u be the exact solution of a boundary value problem generated by D. Then, e_{max} (5.3) and e_{min} (5.4) can be controlled as follows:*

$$\underline{K} \sup_{w} \sup_{D \in \mathcal{D}} \underline{\mathsf{M}}^2(v, w; D) \leq e_{max}^2 \leq \overline{K} \inf_{y,\beta} \sup_{D \in \mathcal{D}} \overline{\mathsf{M}}^2(v, y, \beta; D)$$

and

$$\underline{K} \sup_{w} \inf_{D \in \mathcal{D}} \underline{\mathsf{M}}^2(v, w; D) \le e_{\min}^2 \le \overline{K} \inf_{y,\beta} \inf_{D \in \mathcal{D}} \overline{\mathsf{M}}^2(v, y, \beta; D).$$

Here the constants $\underline{K}$ and $\overline{K}$ satisfy the inequalities

$$\underline{K} |||w|||_D^2 \le |||w|||_\circ^2 \le \overline{K} |||w|||_D^2, \quad \forall w \in V_0, D \in \mathcal{D}, \tag{5.8}$$

where $|||\cdot|||_D$ denotes the energy norm generated by the problem data D, $|||\cdot|||_\circ$ is the norm generated by $D_\circ$, and V_0 is a subspace of V used in the minorant.

Proof We apply the right-hand side of (5.7), (A.6), and (5.8) to estimate the worst case (maximal) error from the above,

$$e_{\max}^2 = \sup_{u \in \mathcal{S}(\mathcal{D})} |||u - v|||_\circ^2 \le \overline{K} \sup_{D \in \mathcal{D}} \inf_{y,\beta} \overline{\mathsf{M}}^2(v, y, \beta; D) \le \overline{K} \inf_{y,\beta} \sup_{D \in \mathcal{D}} \overline{\mathsf{M}}^2(v, y, \beta; D).$$

Alternatively, we can apply the left-hand side of (5.7) and (5.8),

$$e_{\max}^2 = \sup_{u \in \mathcal{S}(\mathcal{D})} |||u - v|||_\circ^2 \ge \underline{K} \sup_{D \in \mathcal{D}} \sup_{w} \underline{\mathsf{M}}^2(v, w; D) = \underline{K} \sup_{w} \sup_{D \in \mathcal{D}} \underline{\mathsf{M}}^2(v, w; D).$$

The estimation of the best case (minimal) error is similar:

$$e_{\min}^2 = \inf_{u \in \mathcal{S}(\mathcal{D})} |||u - v|||_\circ^2 \le \overline{K} \inf_{D \in \mathcal{D}} \inf_{y,\beta} \overline{\mathsf{M}}^2(v, y, \beta; D) = \overline{K} \inf_{y,\beta} \inf_{D \in \mathcal{D}} \overline{\mathsf{M}}^2(v, y, \beta; D)$$

and

$$e_{\min}^2 = \inf_{u \in \mathcal{S}(\mathcal{D})} |||u - v|||_\circ^2 \ge \underline{K} \inf_{D \in \mathcal{D}} \sup_{w} \underline{\mathsf{M}}^2(v, w; D) \ge \underline{K} \sup_{w} \inf_{D \in \mathcal{D}} \underline{\mathsf{M}}^2(v, w; D). \qquad \square$$

Example 5.2 In order to demonstrate practical applications of Theorem 5.1, we consider the equation $\operatorname{div}(a\nabla u) + f = 0$ with homogeneous Dirichlet boundary conditions in a bounded domain $\Omega \subset \mathbb{R}^2$. Assume that the function a is defined with an uncertainty and the information we possess is that

$$a \in \mathcal{D}_a := \left\{ a \in L_\infty(\Omega) \mid 0 < a_\ominus \le a(x) \le a_\oplus, \forall x \in \Omega \right\}.$$

The energy norm for the problem is defined by the relation

$$|||w|||_a^2 := \int_\Omega a |\nabla w|^2 \, dx,$$

and the constants $\underline{K} = \frac{a_\ominus + a_\oplus}{2a_\oplus}$ and $\overline{K} = \frac{a_\ominus + a_\oplus}{2a_\ominus}$ serve the inequality

$$\underline{K} |||w|||_a^2 \le |||w|||_{a_\circ}^2 \le \overline{K} |||w|||_a^2, \quad \forall w \in V_0, a \in \mathcal{D}_a,$$

where $a_\circ := \frac{1}{2}(a_\ominus + a_\oplus)$. We recall that for this problem

$$\overline{\mathsf{M}}^2_{\operatorname{div} a\nabla}(v, y, \beta) := \frac{1+\beta}{\beta} \int_\Omega (a\nabla v - y) \cdot \left(v - \frac{1}{a} y \right) \mathrm{d}x \\ + (1+\beta) \frac{C_F}{a_\ominus} \int_\Omega (\operatorname{div} y + f)^2 \, \mathrm{d}x$$

and

$$\underline{\mathsf{M}}^2_{\operatorname{div} a\nabla}(v, w) := -\int_\Omega a\nabla(w + 2v) \cdot \nabla w \, \mathrm{d}x + 2\int_\Omega f w \, \mathrm{d}x.$$

Our goal is to compute the supremum and infimum of the majorant and minorant over the set of uncertain data, i.e., the function a.

The upper bound of the maximal error is presented by the quantity

$$\sup_{a \in \mathcal{D}_a} \overline{\mathsf{M}}^2_{\operatorname{div} a\nabla}(v, y, \beta) = \frac{1+\beta}{\beta} \sup_{a \in \mathcal{D}_a} \int_\Omega \left(a|\nabla v|^2 + \frac{|y|^2}{a} - 2\nabla v \cdot y \right) \mathrm{d}x \\ + (1+\beta) \frac{C_F}{a_\ominus} \int_\Omega (\operatorname{div} y + f)^2 \, \mathrm{d}x.$$

Since $a|\nabla v|^2 + \frac{|y|^2}{a}$ is convex with respect to a, taking the supremum yields

$$\sup_{a \in \mathcal{D}_a} \overline{\mathsf{M}}^2_{\operatorname{div} a\nabla}(v, y, \beta) = \frac{1+\beta}{\beta} \int_\Omega g_{\max}(\nabla v, y) \, \mathrm{d}x + (1+\beta) \frac{C_F}{a_\ominus} \int_\Omega (\operatorname{div} y + f)^2 \, \mathrm{d}x,$$

where

$$g_{\max}(\nabla v, y) := \max\left\{ a_\oplus |\nabla v|^2 + \frac{|y|^2}{a_\oplus} - 2\nabla v \cdot y, \, a_\ominus |\nabla v|^2 + \frac{|y|^2}{a_\ominus} - 2\nabla v \cdot y \right\}.$$

In order to compute the lower bound, we find

$$\inf_{a \in \mathcal{D}_a} \overline{\mathsf{M}}^2_{\operatorname{div} a\nabla}(v, y, \beta) = \frac{1+\beta}{\beta} \inf_{a \in \mathcal{D}_a} \int_\Omega \left(a|\nabla v|^2 + \frac{|y|^2}{a} - 2\nabla v \cdot y \right) \mathrm{d}x \\ + (1+\beta) \frac{C_F}{a_\ominus} \int_\Omega (\operatorname{div} y + f)^2 \, \mathrm{d}x \\ = \frac{1+\beta}{\beta} \int_\Omega g_{\min}(\nabla v, y) \, \mathrm{d}x + (1+\beta) \frac{C_F}{a_\ominus} \int_\Omega (\operatorname{div} y + f)^2 \, \mathrm{d}x,$$

where

$$g_{\min}(\nabla v, y) := \begin{cases} a_\oplus |\nabla v|^2 + \frac{|y|^2}{a_\oplus} - 2\nabla v \cdot y & \text{if } a_\oplus < \frac{|y|}{|\nabla v|}, \\ 2(|\nabla v||y| - \nabla v \cdot y) & \text{if } a_\ominus \le \frac{|y|}{|\nabla v|} \le a_\oplus, \\ a_\ominus |\nabla v|^2 + \frac{|y|^2}{a_\ominus} - 2\nabla v \cdot y & \text{if } \frac{|y|}{|\nabla v|} < a_\ominus. \end{cases}$$

Similarly we compute the respective quantities associated with the minorant. They are

$$\sup_{a \in \mathcal{D}_a} \underline{\mathsf{M}}^2_{\operatorname{div} a \nabla}(v, w) = \sup_{a \in \mathcal{D}_a} -\int_\Omega \big(a(\nabla w + 2\nabla v) \cdot \nabla w + 2fw\big)\, dx$$
$$= \int_\Omega \big(z_{\max}(v, w) + 2fw\big)\, dx,$$

where

$$z_{\max}(v, w) := \begin{cases} a_\oplus(\nabla w + 2\nabla v) \cdot \nabla w & \text{if } (\nabla w + 2\nabla v) \cdot \nabla w > 0, \\ a_\ominus(\nabla w + 2\nabla v) \cdot \nabla w & \text{if } (\nabla w + 2\nabla v) \cdot \nabla w < 0, \end{cases}$$

and

$$\inf_{a \in \mathcal{D}_a} \underline{\mathsf{M}}^2_{\operatorname{div} a \nabla}(v, w) = \inf_{a \in \mathcal{D}_a} -\int_\Omega \big(a(\nabla w + 2\nabla v) \cdot \nabla w + 2fw\big)\, dx$$
$$= \int_\Omega \big(z_{\min}(v, w) + 2fw\big)\, dx,$$

where

$$z_{\min}(v, w) := \begin{cases} a_\oplus(\nabla w + 2\nabla v) \cdot \nabla w & \text{if } (\nabla w + 2\nabla v) \cdot \nabla w < 0, \\ a_\ominus(\nabla w + 2\nabla v) \cdot \nabla w & \text{if } (\nabla w + 2\nabla v) \cdot \nabla w > 0. \end{cases}$$

In the tests below, $u^h_\circ$ (approximation of $u_\circ$ associated to the problem generated by $a_\circ$) is computed on different meshes with the help of the lowest order (linear) Courant elements. Then, we estimate the approximation error, the worst case error, and the best case error from both sides in order to study the relation between $u^h_\circ$ and the solution set. For the functions w and y, we apply the quadratic Courant elements.

The computation of the upper bound of the "mean error" $e_\circ$ is performed by globally minimizing the majorant with respect to y (see Sect. 3.3.1). We denote the respective upper bound by $e^\oplus_\circ$. For the lower bound, we compute the minorant by comparing the energies of $u^h_\circ$ and u^h_{ref}, where u^h_{ref} is the reference solution computed on a very fine mesh with second order elements, i.e.,

$$|\!|\!| u_\circ - u^h_\circ |\!|\!|^2_\circ \geq 2\big(J(u^h_\circ) - J(u^h_{\text{ref}})\big) =: e^\ominus_\circ .$$

Table 5.1 shows the upper and lower bounds of $e_{\max}$ and $e_{\min}$, i.e., the quantities $e^\oplus_{\max}$, $e^\ominus_{\max}$, $e^\oplus_{\min}$, and $e^\ominus_{\min}$. In these tests problems, $\Omega := (0, 1) \times (0, 1)$, $f = 1$, $a_\ominus = 0.98$ and $a_\oplus = 1.02$, and finite element approximations were computed on uniform meshes.

In Fig. 5.4, we depict three zones formed by two-sided bounds for $e_{\max}$, $e_{\min}$, and $e_\circ$, respectively. One can observe how these different errors start to deviate from each other as the approximate solution approaches the solution set. Moreover, in

Table 5.1 Error quantities

# nodes	$e_\circ^\ominus$	$e_\circ^\oplus$	$\bar{e}_\circ^\ominus$	$\bar{e}_\circ^\oplus$	$e_{\max}^\ominus$	$e_{\max}^\oplus$	$e_{\min}^\ominus$	$e_{\min}^\oplus$	$\upsilon_\ominus$	$\upsilon_\oplus$	$\upsilon_\ominus^\circ$	$\upsilon_\oplus^\circ$
25	0.080	0.080	0.42	0.42	0.081	0.083	0.077	0.078	0.03	0.08	0.01	0.04
81	0.041	0.041	0.22	0.221	0.043	0.044	0.039	0.040	0.08	0.14	0.04	0.07
169	0.028	0.028	0.148	0.149	0.0298	0.031	0.0255	0.0261	0.14	0.20	0.07	0.10
625	0.0138	0.0141	0.0735	0.0748	0.0161	0.0167	0.0118	0.0122	0.32	0.41	0.16	0.22
1369	0.0089	0.0094	0.0475	0.0500	0.0116	0.0121	0.0073	0.0077	0.50	0.64	0.24	0.37
3481	0.0050	0.0058	0.0267	0.0311	0.00818	0.00865	0.00400	0.00440	0.86	1.16	0.40	0.73

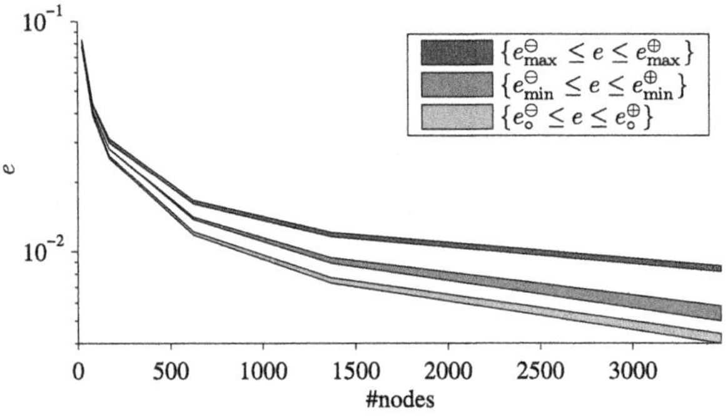

Fig. 5.4 Bounds for $e_{\max}$, $e_{\min}$, and $e_\circ$, respectively

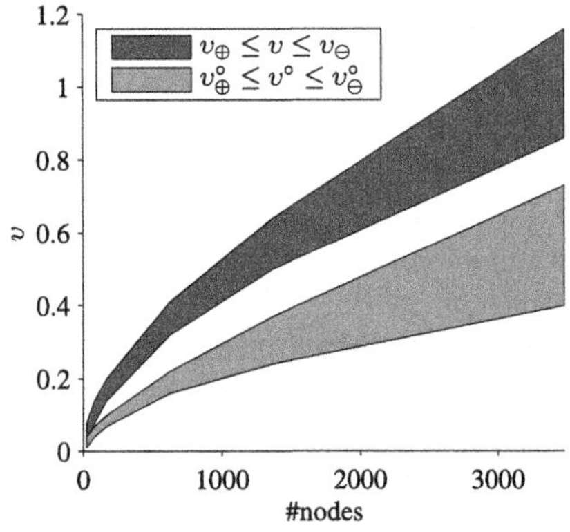

Fig. 5.5 Development of υ and $\upsilon_\circ$ as the amount of nodes is increased

Fig. 5.5, we depict the bounds of υ and $\upsilon_\circ$ (cf. (5.5) and (5.6)). It is easy to see that υ is close to zero if the amount of nodes used to compute the approximate solution is small. The impact of indeterminacy becomes more significant as the amount of nodes increases, i.e., accuracy of the approximate solution improves. For this particular problem, the bounds of υ and $\upsilon_\circ$ indicate that computations on meshes having more than 3481 nodes are disputable, since the generated approximate solution is already close to the set of possible solutions. In other words, the situation arising for # nodes = 3481 and finer meshes is close to the case (a) in Fig. 5.3.

5.4 Two-Sided Bounds of the Radius of the Solution Set

The quantity (5.2) shows the accuracy limit associated with the incompletely known data. In this section, we consider a way of estimating $\bar{r}$. We follow along the lines of [MR11a, MR10, MR08] and use the notation from Sect. 4.4.

Consider the problem (4.145): Find $u \in V_0$ such that

$$(\mathcal{A}\Lambda u, \Lambda w)_U = (f, w)_\mathcal{V}, \quad \forall w \in V_0.$$

Assume that the operator $\mathcal{A}$ is not completely known but we know that it belongs to a set of admissible operators

$$\mathcal{D} := \{\mathcal{A} \in \mathcal{L}(U, U) \mid \mathcal{A} = \mathcal{A}_\circ + \delta\Psi, \|\Psi\|_\mathcal{L} \leq 1, \delta \geq 0\}, \tag{5.9}$$

where $A_\circ$ is the "mean" operator, $\delta \geq 0$ is the indeterminacy magnitude and

$$\|\Psi\|_\mathcal{L} := \sup_{\substack{y \in U \\ y \neq 0}} \frac{\|\Psi y\|_U}{\|y\|_U}.$$

We assume that the non-perturbed operator is elliptic and bounded,

$$\underline{c}\|y\|_U^2 \leq (\mathcal{A}_\circ y, y)_U \leq \overline{c}\|y\|_U^2, \quad \forall y \in U. \tag{5.10}$$

Then, all the problems generated by $\mathcal{A} \in \mathcal{D}$ are elliptic provided that

$$\theta := \frac{\delta}{\underline{c}} < 1. \tag{5.11}$$

Proposition 5.1 *Let the set $\mathcal{D}$ be defined as in* (5.9), *where $\mathcal{A}_\circ$ satisfies* (5.10). *Then,*

$$\underline{K}|||w|||_\mathcal{A}^2 \leq |||w|||_\circ^2 \leq \overline{K}|||w|||_\mathcal{A}^2, \quad \forall w \in V, \mathcal{A} \in \mathcal{D},$$

where

$$\underline{K} := \max\left\{\frac{1}{\operatorname{cond}(\mathcal{A}_\circ) + \theta}, \frac{1 - 2\theta}{1 - \theta}\right\}, \quad \operatorname{cond}(\mathcal{A}_\circ) = \frac{\overline{c}}{\underline{c}}, \quad \textit{and} \quad \overline{K} := \frac{1}{1 - \theta}.$$

Proof We note that

$$(\underline{c} - \delta)\|y\|_U^2 \leq (\mathcal{A}y, y)_U \leq (\overline{c} + \delta)\|y\|_U^2, \quad \forall y \in U, \mathcal{A} \in \mathcal{D}.$$

By the definition (5.9),

$$|||w|||_\circ^2 = |||w|||_\mathcal{A}^2 - \delta(\Psi\Lambda w, \Lambda w)_U.$$

Therefore,

$$
\begin{aligned}
|||w|||_\circ^2 &\leq |||w|||_{\mathcal{A}}^2 + \delta \|\Psi \Lambda w\|_U \|\Lambda w\|_U \\
&\leq |||w|||_{\mathcal{A}}^2 + \delta |||w|||^2 \leq \left(1 + \frac{\delta}{\underline{c} - \delta}\right) |||w|||_{\mathcal{A}}^2.
\end{aligned} \tag{5.12}
$$

For the lower bound, we have

$$
\begin{aligned}
|||w|||_\circ^2 &\geq |||w|||_{\mathcal{A}}^2 - \delta \|\Psi \Lambda w\|_U \|\Lambda w\|_U \\
&\geq |||w|||_{\mathcal{A}}^2 - \delta |||w|||^2 \geq \left(1 - \frac{\delta}{\underline{c} - \delta}\right) |||w|||_{\mathcal{A}}^2.
\end{aligned} \tag{5.13}
$$

An alternative lower bound is

$$
|||w|||_\circ^2 \geq \underline{c} |||w|||^2 \geq \frac{\underline{c}}{\overline{c} + \delta} |||w|||_{\mathcal{A}}^2.
$$

Clearly, the maximum of lower bounds is also a lower bound. The definition of θ (5.11) leads to the statement. □

The normalized radius can be estimated from both sides, using only the data of the mean operator and the perturbation magnitude. Alternatively, if the non-perturbed solution $u_\circ$ is at our disposal, more accurate bounds can be constructed.

Theorem 5.2 *The radius of the solution set for the problem* (4.145), *where* $\mathcal{A} \in \mathcal{D}$ *and* $u_\circ = \mathcal{S}(\mathcal{A}_\circ)$, *can be estimated as follows*:

$$
\bar{r}_\ominus \leq \bar{R}_\ominus(u_\circ) \leq \bar{r} \leq \bar{R}_\oplus(u_\circ) \leq \bar{r}_\oplus,
$$

where

$$
\bar{r}_\ominus := \sqrt{\underline{K}} \frac{\operatorname{cond}^{-1}(\mathcal{A}_\circ)\theta}{\sqrt{1 - \operatorname{cond}^{-1}(\mathcal{A}_\circ)\theta}}, \qquad \bar{r}_\oplus := \sqrt{\overline{K}} \frac{\theta}{\sqrt{1-\theta}},
$$

$$
\bar{R}_\ominus(u_\circ) := \sqrt{\underline{K}} \frac{\delta(|||u_\circ|||^2 / |||u_\circ|||_\circ^2)}{\sqrt{1 - \delta(|||u_\circ|||^2 / |||u_\circ|||_\circ^2)}}, \qquad \bar{R}_\oplus(u_\circ) := \sqrt{\overline{K}} \frac{\delta}{\sqrt{\underline{c} - \delta}} \frac{|||u_\circ|||}{|||u_\circ|||_\circ},
$$

and $\underline{K}$ *and* $\overline{K}$ *are constants from Proposition* 5.1.

Proof Let us consider lower bounds first. We use the left-hand side of Proposition 5.1 and Theorem 4.8 to estimate $\bar{r}$ from below as follows:

$$
\begin{aligned}
r^2 &\geq \underline{K} \sup_{u \in \mathcal{S}(\mathcal{D})} |||u - u_\circ|||_{\mathcal{A}}^2 \\
&= \underline{K} \sup_{\mathcal{A} \in \mathcal{D}} \sup_{w \in V_0} \underline{\mathsf{M}}_{\Lambda}^2(u_\circ, w) \\
&= \underline{K} \sup_{w \in V_0} \sup_{\mathcal{A} \in \mathcal{D}} \underline{\mathsf{M}}_{\Lambda}^2(u_\circ, w).
\end{aligned} \tag{5.14}
$$

The minorant can be written as

$$\begin{aligned}\underline{\mathsf{M}}^2_\Lambda(u_\circ, w) &= -\big((\mathcal{A}_\circ + \delta\Psi)\Lambda w, \Lambda w\big)_U - 2\big((\mathcal{A}_\circ + \delta\Psi)\Lambda u_\circ, \Lambda w\big)_U + 2(f, w)_\mathcal{V} \\ &= -|||w|||^2_\circ - \delta\big(\Psi\Lambda(w + 2u_\circ), \Lambda w\big)_U \\ &\quad - 2\big((\mathcal{A}_\circ\Lambda u_\circ, \Lambda w)_U - (f, w)_\mathcal{V}\big),\end{aligned} \tag{5.15}$$

where the last term vanishes, because $u_\circ$ is the solution of the non-perturbed problem. We estimate $\sup_{w\in V_0}\sup_{\mathcal{A}\in\mathcal{D}}\underline{\mathsf{M}}^2_\Lambda(u_\circ, w)$ from below and set $w := \alpha u_\circ$ $(\alpha > 0)$. Then,

$$\underline{\mathsf{M}}^2_\Lambda(u_\circ, \alpha u_\circ) = -\alpha^2|||u_\circ|||^2_\circ - \delta(\alpha+2)\alpha(\Psi\Lambda u_\circ, \Lambda u_\circ)_U$$

and

$$\begin{aligned}\sup_{\|\Psi\|_\mathcal{L}\le 1}\underline{\mathsf{M}}^2_\Lambda(u_\circ, \alpha u_\circ) &= -\alpha^2|||u_\circ|||^2_\circ + \delta(\alpha+2)\alpha|||u_\circ|||^2 \\ &= \alpha\big(2\delta|||u_\circ|||^2 + \alpha\big(\delta|||u_\circ|||^2 - |||u_\circ|||^2_\circ\big)\big).\end{aligned} \tag{5.16}$$

The expression on the right-hand side attains the maximum if

$$a = \tilde{\alpha} := \frac{\delta|||u_\circ|||^2}{|||u_\circ|||^2_\circ - \delta|||u_\circ|||^2}.$$

Hence,

$$\sup_{\|\Psi\|_\mathcal{L}\le 1}\underline{\mathsf{M}}^2_\Lambda(u_\circ, \tilde{\alpha}u_\circ) = \frac{\delta^2|||u_\circ|||^4}{|||u_\circ|||^2_\circ - \delta|||u_\circ|||^2}. \tag{5.17}$$

Since

$$\sup_{w\in V_0}\sup_{\mathcal{A}\in\mathcal{D}}\underline{\mathsf{M}}^2_\Lambda(u_\circ, w) \ge \sup_{\|\Psi\|_\mathcal{L}\le 1}\underline{\mathsf{M}}^2_\Lambda(u_\circ, \tilde{\alpha}u_\circ),$$

we substitute this expression in (5.14), divide by $|||u_\circ|||^2_{\mathcal{A}_\circ}$, and take the square root. As a result, we find the lower bound $\bar{R}_\ominus(u_\circ)$. To obtain the lower bound $\bar{r}_\ominus$, we note that

$$\frac{\delta^2|||u_\circ|||^4}{|||u_\circ|||^2_\circ - \delta|||u_\circ|||^2} \ge \frac{(\delta/\overline{c})^2}{1 - (\delta/\overline{c})}|||u_\circ|||^2_\circ,$$

divide the expression by $|||u_\circ|||^2_\circ$, and take the square root.

Similarly, we estimate r from the above by the right-hand side of Proposition 5.1, Theorem 4.6, and (A.6):

$$\begin{aligned}r^2 &\le \overline{K}\sup_{u\in\mathcal{S}(\mathcal{D})}|||u - u_\circ|||^2_\mathcal{A} \\ &= \overline{K}\sup_{\mathcal{A}\in\mathcal{D}}\inf_{\substack{y\in Q\\ \beta>0}}\overline{\mathsf{M}}^2_\Lambda(u_\circ, y, \beta) \le \overline{K}\inf_{\substack{y\in Q\\ \beta>0}}\sup_{\mathcal{A}\in\mathcal{D}}\overline{\mathsf{M}}^2_\Lambda(u_\circ, y, \beta).\end{aligned} \tag{5.18}$$

To estimate the infimum from the above we set $y := \mathcal{A}_\circ \Lambda u_\circ$. Then,

$$\overline{\mathsf{M}}^2_\Lambda(u_\circ, \mathcal{A}_\circ \Lambda u_\circ, \beta) = (1+\beta)\big(\Lambda u_\circ - \mathcal{A}^{-1}\mathcal{A}_\circ \Lambda u_\circ, \mathcal{A}\Lambda u_\circ - \mathcal{A}_\circ \Lambda u_\circ\big)_U + \left(1+\frac{1}{\beta}\right)\frac{C_F^2}{\underline{c}-\delta}\|f - \Lambda^* \mathcal{A}_\circ \Lambda u_\circ\|^2_{\mathcal{V}}, \tag{5.19}$$

where the last term vanishes, since the exact flux satisfies the equilibrium condition and we can take β arbitrarily close to zero. Thus,

$$\begin{aligned} &\overline{\mathsf{M}}^2_\Lambda(u_\circ, \mathcal{A}_\circ \Lambda u_\circ, 0) \\ &= \big(\mathcal{A}^{-1}((\mathcal{A}_\circ + \delta\Psi)\Lambda u_\circ - \mathcal{A}_\circ \Lambda u_\circ), (\mathcal{A}_\circ + \delta\Psi)\Lambda u_\circ - \mathcal{A}_\circ \Lambda u_\circ\big)_U \\ &\le \frac{\delta^2}{\underline{c}-\delta}(\Psi \Lambda u_\circ, \Psi \Lambda u_\circ)_U \le \frac{\delta^2}{\underline{c}-\delta}|||u_\circ|||^2. \end{aligned} \tag{5.20}$$

Substituting this estimate in (5.18), dividing by $|||u_\circ|||^2_\circ$, and taking the square root yield $\bar{R}_\oplus(u_\circ)$. To obtain $\bar{r}_\oplus$, we use the estimate

$$\frac{\delta^2}{\underline{c}-\delta}|||u_\circ|||^2 \le \frac{\delta^2}{\underline{c}(\underline{c}-\delta)}|||u_\circ|||^2_\circ = \frac{\theta^2}{1-\theta}|||u_\circ|||^2_\circ$$

and again divide by $|||u_\circ|||^2_\circ$. □

The bounds $\bar{r}_\ominus$ and $\bar{r}_\oplus$ in Theorem 5.2 are valid for a wide range of problems and depend only on the perturbation magnitude and spectral range of the non-perturbed operator. Thus, if $\underline{c} \ll \overline{c}$, they may be very coarse. If the non-perturbed solution $u_\circ$ is at our disposal, then we can apply the bounds $\bar{R}_\ominus(u_\circ)$ and $\bar{R}_\oplus(u_\circ)$ to obtain sharper estimates, which are more related to a particular problem.

Example 5.3 Consider the problem

$$-\operatorname{div}(A\nabla u) = f \quad \text{in } \Omega, \qquad u = 0 \quad \text{on } \Gamma, \tag{5.21}$$

where

$$A \in \mathcal{D}_A := \big\{A \in L_\infty\big(\Omega, \mathbb{M}^{2\times 2}\big) \mid A = A_\circ + \delta\Psi\big\}, \tag{5.22}$$

$\|\Psi\|_{L_\infty(\Omega,\mathbb{M}^{2\times2})} \le 1$, $\delta \ge 0$ satisfies (5.11), and $\Omega := (0,1)\times(0,1)$.

Our aim is to compare the estimates of the radius provided by Theorem 5.2. Consider the case where the solution related to the non-perturbed matrix is

$$u_\circ = \sin(10\pi x)y(y-1)$$

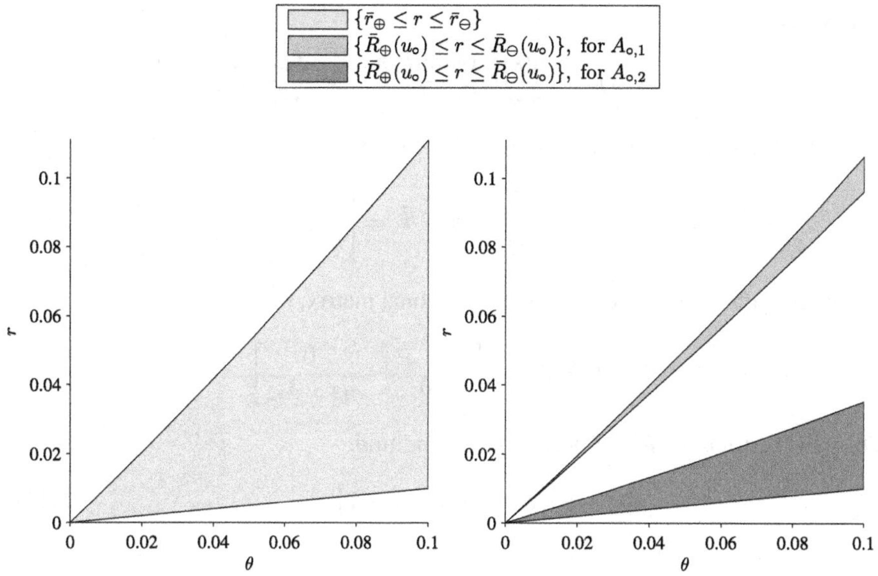

Fig. 5.6 Bounds for $\bar{r}$ as a function of θ in test problems generated by $A_{\circ 1}$ and $A_{\circ 2}$

and the "mean matrices" are

$$A_{\circ 1} = \begin{bmatrix} 1 & 0 \\ 0 & 10 \end{bmatrix} \quad \text{and} \quad A_{\circ 2} = \begin{bmatrix} 10 & 0 \\ 0 & 1 \end{bmatrix},$$

so that $\operatorname{cond}(A_{\circ 1}) = \operatorname{cond}(A_{\circ 2}) = 10$. The bounds $\bar{r}_\ominus$ and $\bar{r}_\oplus$ do depend only on the condition number of $A_\circ$, and thus they are identical for both matrices. The intervals generated by the bounds for the radius are depicted in Fig. 5.6.

It is easy to see that the bounds $\bar{R}_\ominus(u_\circ)$ and $\bar{R}_\oplus(u_\circ)$ indicate very different radii of the solution sets related to $A_{\circ 1}$ and $A_{\circ 2}$ respectively.

Example 5.4 Consider another example, which has an analytical solution. Our goal is to compute the radius of the solution set for a certain type of perturbation and compare it with the upper bound from Theorem 5.2.

Again, we use the problem (5.21) and assume that f coincides with the eigenfunction, i.e.,

$$f(x, y) := \sin(k_1 \pi x) \sin(k_2 \pi y), \quad k_1, k_2 \in \mathbb{N}. \tag{5.23}$$

Let

$$A_\circ := \begin{bmatrix} a_1 & 0 \\ 0 & a_2 \end{bmatrix}, \quad a_1, a_2 > 0.$$

In this case,

$$u_\circ(x,y) = \frac{\sin(k_1\pi x)\sin(k_2\pi y)}{\pi^2(a_1k_1^2 + a_2k_2^2)} = \frac{f(x,y)}{\pi^2\bar{a}\cdot\bar{k}},$$

where for the sake of convenience we use the notation

$$\bar{a} := \begin{bmatrix} a_1 \\ a_2 \end{bmatrix} \quad \text{and} \quad \bar{k} := \begin{bmatrix} k_1^2 \\ k_2^2 \end{bmatrix}.$$

Let the perturbations be generated by a diagonal matrix, i.e.,

$$\tilde{A} = A_\circ + \delta\Psi = \begin{bmatrix} a_1 + \delta\varepsilon_1 & 0 \\ 0 & a_2 + \delta\varepsilon_2 \end{bmatrix}.$$

Then, the restriction $|\Psi| \leq 1$ leads to the condition

$$\bar{\varepsilon} \in \mathcal{E} := \left\{\varepsilon_1^2 + \varepsilon_2^2 \leq 1\right\}, \tag{5.24}$$

where

$$\bar{\varepsilon} := \begin{bmatrix} \varepsilon_1 \\ \varepsilon_2 \end{bmatrix}.$$

The respective solution is given by the relation

$$\tilde{u}(x,y) = \frac{f(x,y)}{\pi^2(\bar{a} + \delta\bar{\varepsilon})\cdot\bar{k}}.$$

Now, we can observe how perturbations of the matrix affect the solution. Since

$$\frac{|||u_\circ - \tilde{u}|||}{|||u_\circ|||} = \frac{|\delta\bar{\varepsilon}\cdot\bar{k}|}{|\bar{a}\cdot\bar{k} + \delta\bar{\varepsilon}\cdot\bar{k}|} \tag{5.25}$$

we find that

$$\max_{\bar{\varepsilon}\in\mathcal{E}} \frac{|\delta\bar{\varepsilon}\cdot\bar{k}|}{|\bar{a}\cdot\bar{k} + \delta\bar{\varepsilon}\cdot\bar{k}|} = \frac{\delta}{\bar{a}\cdot\bar{k}/|\bar{k}| - \delta}. \tag{5.26}$$

The maximal value in (5.26) is attained if $\bar{\varepsilon} = -\frac{\bar{k}}{|\bar{k}|}$. Without a loss of generality we assume that $a_1 \leq a_2$. In this case, the normalized perturbation is $\theta = \frac{\delta}{a_1}$. Thus, for this model problem, the radius of the solution set is given by the relation

$$\hat{r}_{\text{mod}} := \max_{\bar{\varepsilon}\in\mathcal{E}} \frac{|||u_\circ - \tilde{u}|||}{|||u_\circ|||} = \frac{\theta}{\bar{a}\cdot\bar{k}/(a_1|\bar{k}|) - \theta}. \tag{5.27}$$

In Fig. 5.7, we plot r_{mod} for different values of $\bar{k}$ and compare them with the upper bound established in Theorem 5.2. It is easy to see that the radius approaches the upper bound derived in Theorem 5.2, as the value of k_1 increases. This observation shows that there exist such elliptic problems for which the relation between

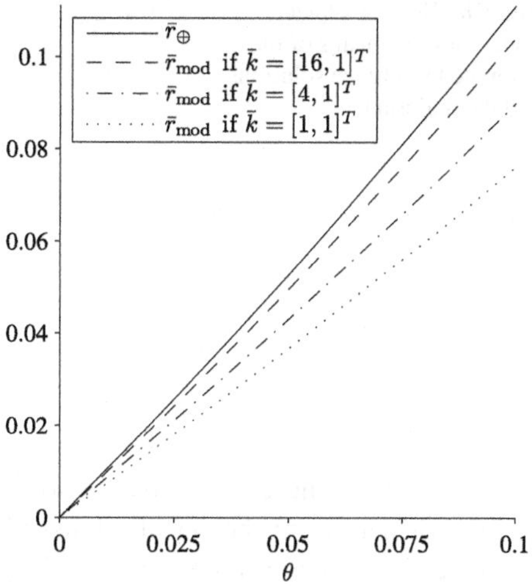

Fig. 5.7 Comparison of $\bar{r}_\oplus$ from Theorem 5.2 and $\hat{r}_{\rm mod}$ (5.27) for different $\bar{k}$

the radius of the solution set and the magnitude of indeterminacy coincide with the theoretical upper bound in Theorem 5.2. Indeed, if k_1 tends to infinity, then the ratio

$$\frac{(1/a_1)\bar{a}\cdot\bar{k}}{|\bar{k}|} = \frac{k_1^2 + \mathrm{cond}(A_\circ)k_2^2}{\sqrt{k_1^4 + k_2^4}}$$

tends to one and the right-hand side of (5.27) tends to the upper bound established in Theorem 5.2.

5.5 Computable Estimates of the Radius of the Solution Set

The a priori type error bounds $\bar{r}_\ominus$ and $\bar{r}_\oplus$ in Theorem 5.2 may be coarse if the condition number of $\mathcal{A}_\circ$ is large. This difficulty can be overcome by using $\bar{R}_\ominus$ and $\bar{R}_\oplus$, which requires certain computational efforts. These efforts are motivated if we wish to define the accuracy limit generated by indeterminate data and investigate sensitivity of solutions with respect to data variations. Below we discuss possible numerical strategies focused on this problem.

5.5.1 Using the Majorant

Assume that for any $D \in S(\mathcal{D})$ we have an error majorant $\overline{\mathsf{M}}(v, y; D)$. First, we compute an approximation $u_\circ^h$ of the non-perturbed solution $u_\circ$ associated with the mean data $D_\circ$ and apply the majorant to estimate from the above the respective approximation error

$$e_\circ := |||u_\circ - u_\circ^h|||_\circ \leq e_\circ^\oplus,$$

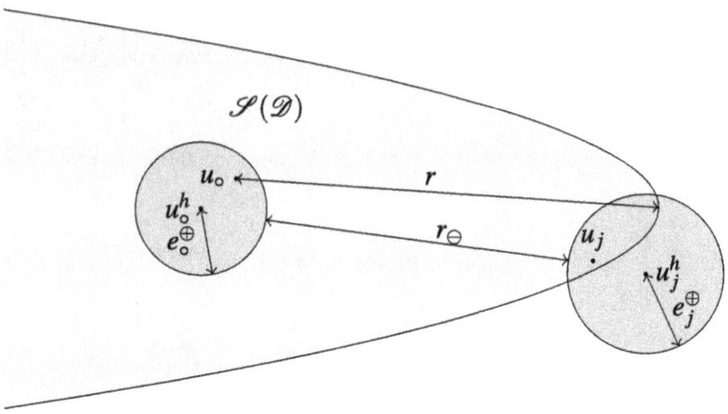

Fig. 5.8 Getting a lower bound of r by means of the computed reference solutions and the majorant

where $|||\cdot|||_\circ$ is the energy norm associated with $D_\circ$. Then, we select a certain amount of data from the set of admissible data, i.e., $D_j \in \mathcal{D}$, $j = 1, \ldots, N$, and compute the respective approximate solutions u_j^h supplied with the respective guaranteed upper bounds error

$$|||u_j - u_j^h|||_{D_j} \leq e_j^\oplus,$$

where $u_j = \mathcal{S}(D_j)$ and where $|||\cdot|||_{D_j}$ is the energy norm associated with D_j. Let $\kappa_{\circ j}$ be the coefficient in the relation establishing equivalence of norms $|||\cdot|||_\circ$ and $|||\cdot|||_{D_j}$. Then, $\kappa_{\circ j}|||u_j - u_j^h|||_\circ \leq e_j^\oplus$. Since the computed upper bounds for the approximation error are guaranteed, we know that the solution set $\mathcal{S}(\mathcal{D})$ intersects with every ball $B(u_j^h, e_j^\oplus)$ (Fig. 5.8). Thus,

$$r \geq \max_{j=1,2,\ldots,N} |||u_\circ - u_j|||_\circ \geq \max_{j=1,2,\ldots,N} \left\{ |||u_\circ^h - u_j^h|||_\circ - \frac{e_j^\oplus}{\kappa_{\circ j}} \right\} - e_\circ^\oplus.$$

The corresponding numerical procedure is formalized in Algorithm 5.1 where the selection method for D_j is not specified. This method depends on the problem and on the accuracy we wish to have. Certainly, the procedure becomes computationally expensive if N is large. On the other hand, the method is very general and can be applied to any problem (including non-linear problems), provided that the corresponding error majorant of the functional type has been established.

5.5.2 Using a Reference Solution

Usually, the exact solution of the non-perturbed problem is not at our disposal. However, we can use a numerical solution $u_\circ^h$. The upper bound of the error arising due to this change is evaluated by the majorant

$$|||u_\circ - u_\circ^h|||_\circ \leq \overline{\mathsf{M}}(u_\circ^h, y, D_\circ).$$

Algorithm 5.1 Numerically computed lower bound for the radius using majorant

Input: N {Number of computed solutions}
Compute approximate solution $u_\circ^h$ related to data $D_\circ$.
Compute upper bound for the approximation error, $|||u - u_\circ^h|||_\circ \le e_\circ^\oplus$.
$r_\ominus^h = 0$ {initial value for the lower bound}
for $j = 1$ **to** N **do**
  Select $D_j \in \mathcal{D}$. {select admissible perturbation of the data}
  Compute approximate solution u_j^h related to data D_j.
  $e_j^\oplus$ {compute approximation error by the majorant}
  $r_j^\ominus := |||u_\circ^h - u_j^h|||_\circ - \frac{e_j^\oplus}{\kappa_{\circ j}} - e_\circ^\oplus$
  if $r_j^\ominus \ge r_\ominus^h$ **then**
    $r_\ominus^h = r_j^\ominus$
  end if
end for
Output: $r_\ominus^h$ {Lower estimate for the radius}

Then, the ratio $\frac{|||u_\circ|||}{|||u_\circ|||_\circ}$ can be estimated from both sides as follows:

$$\frac{|||u_\circ|||}{|||u_\circ|||_\circ} \ge \frac{|||u_\circ^h||| - |||u_\circ^h - u_\circ|||}{|||u_\circ^h|||_\circ + |||u_\circ^h - u_\circ|||_\circ} \ge \frac{|||u_\circ^h||| - (1/\sqrt{\underline{c}})\overline{\mathsf{M}}}{|||u_\circ^h|||_\circ + \overline{\mathsf{M}}}$$

and

$$\frac{|||u_\circ|||}{|||u_\circ|||_\circ} \le \frac{|||u_\circ^h||| + |||u_\circ^h - u_\circ|||}{|||u_\circ^h|||_\circ - |||u_\circ^h - u_\circ|||_\circ} \le \frac{|||u_\circ^h||| + (1/\sqrt{\underline{c}})\overline{\mathsf{M}}}{|||u_\circ^h|||_\circ - \overline{\mathsf{M}}}.$$

This observation applied to $\bar{R}_\ominus(u_\circ)$ and $\bar{R}_\oplus(u_\circ)$ from Theorem 5.2 motivates Algorithm 5.2, which produces two-sided bounds of the radius.

If the non-perturbed solution can be approximated with high accuracy and an efficient majorant can be computed, then Algorithm 5.2 finds bounds which are as efficient as $\bar{R}_\ominus(u_\circ)$ and $\bar{R}_\oplus(u_\circ)$ from Theorem 5.2. Whether this is possible depends on the particular problem, computational methods and computer resources. We emphasize that in this method (unlike the previous one) only a single reference solution $u_\circ^h$ and the respective error bounds must be computed.

5.5.3 An Advanced Lower Bound

In the proof of Theorem 5.2, we have selected the free function w in a special form. An advanced form of the lower bound can be obtained if we estimate maximal

Algorithm 5.2 Estimates of the radius obtained by the reference solution

Input: δ {Perturbation magnitude}, $\underline{K}$, $\overline{K}$ {Constants in Proposition 5.1}
Compute the reference solution $u_\circ^h$ related to the non-perturbed problem.
Compute a majorant $\overline{\mathsf{M}}$, e.g., by Algorithm 3.2.
Compute the numbers

$$Z := \frac{|||u_\circ^h||| - (1/\sqrt{\overline{c}})\overline{\mathsf{M}}}{|||u_\circ^h|||_\circ + \overline{\mathsf{M}}} \quad \text{and} \quad G := \frac{|||u_\circ^h||| + (1/\sqrt{\overline{c}})\overline{\mathsf{M}}}{|||u_\circ^h|||_\circ - \overline{\mathsf{M}}}$$

and obtain the estimates

$$R_\ominus^h(u_\circ^h) := \sqrt{\underline{K}}\frac{\delta Z}{\sqrt{1-\delta Z}} \quad \text{and} \quad R_\oplus^h(u_\circ^h) := \sqrt{\overline{K}}\frac{\delta G}{\sqrt{\underline{c}-\delta}}.$$

Output: $R_\ominus^h(u_\circ^h)$, $R_\oplus^h(u_\circ^h)$ {Estimates for the radius}

value the minorant attained on the whole set of admissible functions. This method is discussed below with the paradigm of the problem (5.21)–(5.22). As before, we assume that the mean matrix $A_\circ$ is positive definite and bounded, i.e.,

$$\underline{c}\|\xi\|^2 \leq A_\circ\xi \cdot \xi \leq \overline{c}\|\xi\|^2, \quad \forall \xi \in \mathbb{R}^2, \tag{5.28}$$

$0 < \delta < \underline{c}$, and $|\Psi| \leq 1$. Proposition 5.1 holds for the norm $|||w|||_A^2 := \int_\Omega A\nabla w \cdot \nabla w \, dx$, and we have

$$\underline{K}|||w|||_A^2 \leq |||w|||_\circ^2 \leq \overline{K}|||w|||_A^2, \quad \forall w \in V_0 := H_0^1(\Omega),\, A \in \mathcal{D}_A. \tag{5.29}$$

An advanced form of the lower bound is presented by (5.30).

Theorem 5.3 *Let $u_\circ$ be the solution of* (5.21)–(5.22) *generated by $A_\circ \in \mathcal{D}$. Then,*

$$r^2 \geq \underline{K} \sup_{w \in V_0} \mathcal{R}_\ominus(u_\circ, w), \tag{5.30}$$

where w is an arbitrary function in V_0, $\underline{K}$ is from (5.28) *and*

$$\mathcal{R}_\ominus(u_\circ, w) := -|||w|||_\circ^2 + \delta \int_\Omega |\nabla w + 2\nabla u_\circ||\nabla w| \, dx. \tag{5.31}$$

Proof We have

$$r^2 = \sup_{u \in \mathcal{S}(\mathcal{D})} |||u_\circ - u|||_\circ^2 \geq \underline{K} \sup_{u \in \mathcal{S}} |||u_\circ - u|||_A^2.$$

On the other hand,

$$\sup_{u \in \mathcal{S}(\mathcal{D})} |||u_\circ - u|||_A^2 = \sup_{A \in \mathcal{D}} \left\{ \sup_{w \in V_0(\Omega)} \underline{\mathsf{M}}^2_{\operatorname{div} A\nabla}(u_\circ, w) \right\}$$
$$= \sup_{w \in V_0} \left\{ \sup_{A \in \mathcal{D}} \underline{\mathsf{M}}_{\operatorname{div} A\nabla}(u_\circ, w) \right\},$$

and we conclude that

$$r^2 \geq \underline{K} \sup_{w \in V_0(\Omega)} \left\{ \sup_{A \in \mathcal{D}} \underline{\mathsf{M}}^2_{\operatorname{div} A\nabla}(u_\circ, w) \right\}, \tag{5.32}$$

where

$$\underline{\mathsf{M}}^2_{\operatorname{div} A\nabla}(u_\circ, w) = -\int_\Omega (A_\circ + \delta\Psi)(\nabla w + 2\nabla u_\circ) \cdot \nabla w \, \mathrm{d}x + 2\int_\Omega f w \, \mathrm{d}x.$$

Since $u_\circ$ is the exact solution of the non-perturbed problem, we have

$$\underline{\mathsf{M}}^2_{\operatorname{div} A\nabla}(u_\circ, w) = -\int_\Omega A_\circ \nabla w \cdot \nabla w \, \mathrm{d}x - \delta \int_\Omega \Psi(\nabla w + 2\nabla u_\circ) \cdot \nabla w \, \mathrm{d}x, \tag{5.33}$$

and

$$\sup_{A \in \mathcal{D}} \underline{\mathsf{M}}^2_{\operatorname{div} A\nabla}(u_\circ, w) = -|||w|||_\circ^2 + \sup_{|\Psi| \leq 1} \left\{ -\delta \int_\Omega \Psi(\nabla w + 2\nabla u_\circ) \cdot \nabla w \, \mathrm{d}x \right\}. \tag{5.34}$$

Set $\Psi = -\frac{\nabla w \otimes (\nabla w + 2\nabla u_\circ)}{|\nabla w \otimes (\nabla w + 2\nabla u_\circ)|}$, where $\otimes$ stands for the dyad product (see (A.1)). Then,

$$\big(\nabla w \otimes (\nabla w + 2\nabla u_\circ)\big)(\nabla w + 2\nabla u_\circ) = \nabla w |\nabla w + 2\nabla u_\circ|^2,$$
$$\big|\nabla w \otimes (\nabla w + 2\nabla u_\circ)\big|^2 = |\nabla w + 2\nabla u_\circ|^2 |\nabla w|^2,$$

and we find that

$$-\delta\Psi(\nabla w + 2\nabla u_\circ) \cdot \nabla w = \delta \frac{|\nabla w|^2 |\nabla w + 2\nabla u_\circ|^2}{|(\nabla w + 2\nabla u_\circ) \otimes \nabla w|} = \big|(\nabla w + 2\nabla u_\circ) \otimes \nabla w\big|.$$

Now (5.34) implies (5.30). □

Theorem 5.3 suggests a way of computing a lower bound by maximizing $\mathcal{R}_\ominus(u_\circ, w)$ with respect to $w \in V_h \subset H_0^1(\Omega)$, where the subspace V_h is at our disposal. Let $V_h := \operatorname{span}(\psi_i)$, $i = 1, \ldots, M$, where the ψ_i satisfy Dirichlet boundary conditions. Then, we compute the lower estimate of the radius by solving the following (non-linear) maximization problem:

$$r^2 \geq \underline{K} \max_{\alpha_i \in \mathbb{R}} \mathcal{R}_\ominus \left(u_\circ, \sum_{i=1}^N \alpha_i \psi_i \right)$$

Algorithm 5.3 Numerically computed lower bound for the radius

---

Input: θ {normalized perturbation}, $\underline{K}$ {constant from Proposition 5.1}
Compute a reference solution $u_\circ^h$ for the "mean problem".
Compute the majorant $\overline{\mathsf{M}}$ by Algorithm 3.2.
Solve the optimization problem

$$\mathcal{G}(\hat{w}) = \max_{w \in V_h \subset H_0^1(\Omega)} \left(\mathcal{R}_\ominus\left(u_\circ^h, w\right) - 2\overline{\mathsf{M}}\theta \|\nabla w\|\right).$$

Output: $\sqrt{\underline{K}\mathcal{G}(\hat{w})}$ {lower bound for the radius}

---

with respect to the coefficients α_i. If $u_\circ$ is not at our disposal, then we can introduce a reference solution $u_\circ^h$ and use a modified estimate.

Corollary 5.1 *Under the assumptions of Theorem* 5.3,

$$r^2 \geq \underline{K}\left(\mathcal{R}_\ominus\left(u_\circ^h, w\right) - 2\left\|\left|u_\circ - u_\circ^h\right|\right\|_\circ \theta \|\nabla w\|\right), \quad \forall w \in V_0,$$

where θ *is defined in* (5.11) *and*

$$\left\|\left|u_\circ - u_\circ^h\right|\right\|_\circ \leq \overline{\mathsf{M}}_{\operatorname{div} A\nabla}\left(u_\circ^h, y, \beta\right), \quad \forall y \in H(\Omega, \operatorname{div}), \beta > 0.$$

Proof We estimate the lower bound $\mathcal{R}_\ominus(u_\circ, w)$ in Theorem 5.3 from below as follows:

$$\begin{aligned}\mathcal{R}_\ominus(u_\circ, w) &= -\|\|w\|\|_\circ^2 + \delta \int_\Omega \left|\nabla w + 2\nabla u_\circ^h + 2\nabla\left(u_\circ - u_\circ^h\right)\right| |\nabla w| \, dx \\ &\geq \mathcal{R}_\ominus\left(u_\circ^h, w\right) - 2\delta \int_\Omega \left|\nabla\left(u_\circ - u_\circ^h\right)\right| |\nabla w| \, dx. \end{aligned} \tag{5.35}$$

Note that

$$\int_\Omega \left|\nabla\left(u_\circ - u_\circ^h\right)\right| |\nabla w| \, dx \leq \frac{1}{\underline{c}_1} \left\|\left|\nabla\left(u_\circ - u_\circ^h\right)\right|\right\|_\circ \|\nabla w\|.$$

Substituting these estimates to (5.35) and applying the Cauchy–Schwartz inequality yield the statement. □

In Corollary 5.1, we have the lower bound of the solution set in fully computable quantities (see Algorithm 5.3). The main benefit of Algorithm 5.3 (compared to Algorithm 5.1) is that it operates only with the non-perturbed problem. However, it requires solving a non-linear optimization problem. The efficiency of the estimate provided by Algorithm 5.3 depends on the accuracy of the computed reference solution, efficiency of the respective error majorant, and structure of V_h.

5.6 Multiple Sources of Indeterminacy

So far we have considered only the cases where the coefficients of the elliptic problem are not completely known. However, also the data associated to the right-hand and/or the boundary conditions may not be completely known. In this section, we generalize our analysis to these cases.

5.6.1 Incompletely Known Right-Hand Side

We generalize the estimates $\bar{R}_\ominus(u_\circ)$ and $\bar{R}_\oplus(u_\circ)$ of Theorem 5.2 derived for the Recall problem (4.145). Assume that uncertainty comes out of two sources, so that $\mathcal{A} \in \mathcal{D}_\mathcal{A}$ (see (5.9)) and $f \in \mathcal{D}_f$, where

$$\mathcal{D}_f := \{f \in \mathcal{V} \mid f = f_\circ + \xi\}, \quad \|\xi\|_\mathcal{V} \le \varepsilon.$$

Theorem 5.4 *The radius of the solution set for the problem* (4.145), *where* $\mathcal{A} \in \mathcal{D}_\mathcal{A}$, *satisfies the following estimate*:

$$R_{\ominus,RHS}(u_\circ) \le r \le R_{\oplus,RHS}(u_\circ),$$

where

$$R_{\ominus,RHS}(u_\circ) := \sqrt{\underline{K}}\frac{\delta |||u_\circ|||^2 + \varepsilon\|u_\circ\|_\mathcal{V}}{\sqrt{|||u_\circ|||_\circ^2 - \delta|||u_\circ|||^2}},$$

$$R_{\oplus,RHS}(u_\circ) := \sqrt{\overline{K}}\frac{1}{\sqrt{\underline{c} - \delta}}\big(\delta\|\Lambda u_\circ\|_U + C_F\varepsilon\big),$$

where the constants $\underline{K}$ *and* $\overline{K}$ *are defined in Proposition* 5.1, $\underline{c}$ *and* $\overline{c}$ *are from the inequality* (5.10) *and* C_F *is from* (4.139).

Proof The proof is similar to that of Theorem 5.2. We apply Proposition 5.1 and Theorem 4.8 as follows:

$$\begin{aligned} r^2 &\ge \underline{K} \sup_{u \in \mathcal{S}(\mathcal{D})} |||u - u_\circ|||_\mathcal{A}^2 \\ &= \underline{K} \sup_{\substack{\mathcal{A}\in\mathcal{D}_\mathcal{A} \\ f\in\mathcal{D}_f}} \sup_{w \in V_0} \underline{\mathsf{M}}_\Lambda^2(u_\circ, w) = \underline{K} \sup_{w \in V_0} \sup_{\substack{\mathcal{A}\in\mathcal{D}_\mathcal{A} \\ f\in\mathcal{D}_f}} \underline{\mathsf{M}}_\Lambda^2(u_\circ, w). \end{aligned} \tag{5.36}$$

Here

$$\begin{aligned} \underline{\mathsf{M}}_\Lambda^2(u_\circ, w) = &-\big((\mathcal{A}_\circ + \delta\Psi)\Lambda w, \Lambda w\big)_U \\ &- 2\big((\mathcal{A}_\circ + \delta\Psi)\Lambda u_\circ, \Lambda w\big)_U + 2(f_\circ + \xi, w)_\mathcal{V} \end{aligned}$$

$$= -|||w|||_\circ^2 - \delta\big(\Psi\Lambda(w+2u_\circ), \Lambda w\big)_U + (\xi, w)_\mathcal{V}$$
$$- 2\big((\mathcal{A}_\circ \Lambda u_\circ, \Lambda w)_U - (f_\circ, w)_\mathcal{V}\big),$$

where the last term vanishes. We estimate $\sup_{w\in V_0} \sup_{(A,l)\in D} \underline{\mathsf{M}}^2_\Lambda(u_\circ, w)$ from below by setting $w := \alpha u_\circ$ $(\alpha > 0)$. Then,

$$\underline{\mathsf{M}}^2_\Lambda(u_\circ, \alpha u_\circ) = -\alpha^2 |||u_\circ|||^2_{A_\circ} - \delta(\alpha+2)\alpha(\Psi\Lambda u_\circ, \Lambda u_\circ)_U + 2\alpha(\xi, u_\circ)_\mathcal{V}.$$

Taking the supremum over possible perturbations leads to

$$\sup_{\|\Psi\|_\mathcal{L}\le 1, \|\xi\|_\mathcal{V}\le\varepsilon} \underline{\mathsf{M}}^2_\Lambda(u_\circ, \alpha u_\circ)$$
$$= -\alpha^2 |||u_\circ|||_\circ^2 + \delta(\alpha+2)\alpha |||u_\circ|||^2 + 2\alpha\varepsilon\|u_\circ\|_\mathcal{V}$$
$$= \alpha\big(2(\delta|||u_\circ|||^2 + \varepsilon\|u_\circ\|_\mathcal{V}) + \alpha(\delta|||u_\circ|||^2 - |||u_\circ|||_\circ^2)\big). \tag{5.37}$$

The expression attains the maximum if

$$\alpha = \tilde{\alpha} := \frac{\delta|||u_\circ|||^2 + \varepsilon\|u_\circ\|_\mathcal{V}}{|||u_\circ|||_\circ^2 - \delta|||u_\circ|||^2}.$$

Substituting this value yields

$$\sup_{\|\Psi\|_\mathcal{L}\le 1, \|\xi\|_\mathcal{V}\le\varepsilon} \underline{\mathsf{M}}^2_\Lambda(u_\circ, \tilde{\alpha} u_\circ) = \frac{(\delta|||u_\circ|||^2 + \varepsilon\|u_\circ\|_\mathcal{V})^2}{|||u_\circ|||_\circ^2 - \delta|||u_\circ|||^2},$$

and (5.36) yields a lower bound.

By Proposition 5.1 and Theorem 4.6 we find that

$$r^2 \le \overline{K} \sup_{u\in\mathcal{S}(\mathcal{D})} |||u - u_\circ|||^2_\mathcal{A}$$
$$= \overline{K} \sup_{\substack{\mathcal{A}\in\mathcal{D}_\mathcal{A} \\ f\in\mathcal{D}_f}} \inf_{\substack{y\in Q \\ \beta>0}} \overline{\mathsf{M}}^2_\Lambda(u_\circ, y, \beta) = \overline{K} \inf_{\substack{y\in Q \\ \beta>0}} \sup_{\substack{\mathcal{A}\in\mathcal{D}_\mathcal{A} \\ f\in\mathcal{D}_f}} \overline{\mathsf{M}}^2_\Lambda(u_\circ, y, \beta). \tag{5.38}$$

Here

$$\overline{\mathsf{M}}^2_\Lambda(u_\circ, y, \beta) := (1+\beta)\big(\mathcal{A}\Lambda u_\circ - y, \Lambda u_\circ - \mathcal{A}^{-1}y\big)_U$$
$$+ \left(1 + \frac{1}{\beta}\right)\frac{C_F^2}{\underline{c} - \delta}\|f - \Lambda^* y\|^2_\mathcal{V}$$
$$= (1+\beta)\big(\mathcal{A}^{-1}(\mathcal{A}\Lambda u_\circ - y), \mathcal{A}\Lambda u_\circ - y\big)_U$$
$$+ \left(1 + \frac{1}{\beta}\right)\frac{C_F^2}{\underline{c} - \delta}\|f - \Lambda^* y\|^2_\mathcal{V}. \tag{5.39}$$

We substitute $\mathcal{A} = \mathcal{A}_\circ + \delta\Psi$, $f = f_\circ + \xi$, and $y = \mathcal{A}_\circ \Lambda u_\circ$. Then,

$$\begin{aligned}
\overline{\mathsf{M}}^2_\Lambda(u_\circ, \mathcal{A}_\circ\Lambda u, \beta) &= (1+\beta)\big(\mathcal{A}^{-1}(\mathcal{A}\Lambda u_\circ - \mathcal{A}_\circ\Lambda u_\circ), \mathcal{A}\Lambda u_\circ - \mathcal{A}_\circ\Lambda u_\circ\big)_U \\
&\quad + \left(1+\frac{1}{\beta}\right)\frac{C_F^2}{\underline{c}-\delta}\| f_\circ - \Lambda^*\mathcal{A}_\circ\Lambda u_\circ + \xi\|^2_{\mathcal{V}} \\
&\le \frac{\delta^2}{\underline{c}-\delta}\|(\mathcal{A}_\circ + \delta\Psi)\Lambda u_\circ - \mathcal{A}_\circ\Lambda u_\circ\|^2_U + \left(1+\frac{1}{\beta}\right)\frac{C_F^2}{\underline{c}-\delta}\|\xi\|^2_{\mathcal{V}} \\
&= \frac{1}{\underline{c}-\delta}\left((1+\beta)\delta^2\|\Psi\Lambda u_\circ\|^2_U + \left(1+\frac{1}{\beta}\right)C_F^2\|\xi\|^2_{\mathcal{V}}\right)
\end{aligned}$$

and

$$\sup_{\|\Psi\|_{\mathcal{L}}\le 1, \|\xi\|_{\mathcal{V}}\le\varepsilon} \overline{\mathsf{M}}^2_\Lambda(u_\circ, \mathcal{A}_\circ\Lambda u, \beta) \le \frac{1}{\underline{c}-\delta}\left((1+\beta)\delta^2\|\Lambda u_\circ\|^2_U + \left(1+\frac{1}{\beta}\right)C_F^2\varepsilon^2\right).$$

Set $\beta := \frac{C_F\varepsilon}{\delta\|\Lambda u_\circ\|}$. We arrive at the inequality

$$\sup_{\|\Psi\|_{\mathcal{L}}\le 1, \|\xi\|_{\mathcal{V}}\le\varepsilon} \overline{\mathsf{M}}^2_\Lambda(u_\circ, \mathcal{A}_\circ\Lambda u, \beta) \le \frac{1}{\underline{c}-\delta}\big(\delta\|\Lambda u_\circ\|_U + C_F\varepsilon\big)^2,$$

which together with (5.38) leads to the result. □

Remark 5.1 The comparison of $R_{\ominus,RHS}(u_\circ)$ and $R_{\oplus,RHS}(u_\circ)$ in Theorem 5.4 with $\bar{R}_\ominus(u_\circ)$ and $\bar{R}_\oplus(u_\circ)$ in Theorem 5.2 reveals that the right-hand side indeterminacy produces additional linearly growing terms to the respective estimates.

5.6.2 The Reaction Diffusion Problem

As an addition to the previous discussion, we consider a problem which has several sources of incompletely known data (see [MR10]). The reaction diffusion problem with mixed Dirichlét–Robin boundary conditions is defined by the system

$$-\operatorname{div}(A\nabla u) + \rho u = f \quad \text{in } \Omega, \tag{5.40}$$

$$u = 0 \quad \text{on } \Gamma_D, \tag{5.41}$$

$$n \cdot A\nabla u = F \quad \text{on } \Gamma_N, \tag{5.42}$$

$$\alpha u + n \cdot A\nabla u = G \quad \text{on } \Gamma_R. \tag{5.43}$$

The generalized solution is the function $u \in V_0$ satisfying the integral identity,

$$\int_\Omega A\nabla u \cdot \nabla w \, \mathrm{dx} + \int_{\Gamma_R} \alpha u w \, \mathrm{ds} = \int_\Omega f w \, \mathrm{dx} - \int_{\Gamma_N} F w \, \mathrm{ds} - \int_{\Gamma_R} G w \, \mathrm{ds}, \quad \forall w \in V_0,$$

where

$$V_0 := \{w \in H_0^1(\Omega) \mid w = 0 \text{ on } \Gamma_D\}. \tag{5.44}$$

Here, $\Omega \subset \mathbb{R}^d$ has the boundary $\Gamma_D \cup \Gamma_N \cup \Gamma_R$, $A \in L_\infty(\Omega, \mathbb{M}^{d\times d})$ (symmetric and positive definite), $\rho \in L_\infty(\Omega, \mathbb{R}_+)$, and $\alpha \in L_\infty(\Gamma_R, \mathbb{R}_+)$ are related to the properties of the medium.

The system has several functions dependent on the material properties, namely A, ρ, and α. We assume that

$$\mathcal{D}_A := \{A \in L_\infty(\Omega, \mathbb{M}^{d\times d}) \mid A = A_\circ + \delta_1 \Psi\}, \tag{5.45}$$

$$\mathcal{D}_\rho := \{\rho \in L_\infty(\Omega, \mathbb{R}) \mid \rho = \rho_\circ + \delta_2 \psi_\rho\}, \tag{5.46}$$

$$\mathcal{D}_\alpha := \{\alpha \in L_\infty(\Gamma_3, \mathbb{R}) \mid \alpha = \alpha_\circ + \delta_3 \psi_\alpha\}, \tag{5.47}$$

where $\|\Psi\|_{L_\infty(\Omega,\mathbb{M}^{d\times d})} \le 1$, $\|\psi_\rho\|_{L_\infty(\Omega)} \le 1$, and $\|\psi_\alpha\|_{L_\infty(\Gamma_3)}$. Thus, in the case considered, the set of indeterminate data is

$$\mathcal{D} := \mathcal{D}_A \times \mathcal{D}_\rho \times \mathcal{D}_\alpha.$$

Let

$$\begin{aligned} \underline{c}_1 |\xi|^2 \le A_\circ \xi \cdot \xi \le \overline{c}_1 |\xi|^2, &\quad \forall \xi \in \mathbb{R}^d, \text{ on } \Omega, \\ \underline{c}_2 \le \rho_\circ \le \overline{c}_2 &\quad \text{on } \Omega, \\ \underline{c}_3 \le \alpha_\circ \le \overline{c}_3 &\quad \text{on } \Gamma_R, \end{aligned}$$

where $\underline{c}_i > 0$. Then, the "mean" problem has a unique solution $u_\circ$. The condition

$$0 \le \delta_i < \underline{c}_i, \quad i = 1, 2, 3,$$

guarantees that the perturbed problem remains elliptic and possesses a unique solution u. We define normalized perturbations and the corresponding "condition numbers":

$$\theta_i := \frac{\delta_i}{\underline{c}_i} \quad \text{and} \quad \text{cond}_i := \frac{\overline{c}_i}{\underline{c}_i}, \quad i = 1, 2, 3.$$

First, we establish a technical result (analogous to Proposition 5.1).

Proposition 5.2 *Let A, ρ, and α be defined by* (5.45), (5.46), *and* (5.47), *respectively. Then,*

$$\underline{\mathrm{C}} |||w|||^2_{(A,\rho,\alpha)} \le |||w|||^2_\circ \le \overline{\mathrm{C}} |||w|||^2_{(A,\rho,\alpha)}, \quad \forall w \in V_0,$$

where

$$|||w|||^2_{(A,\rho,\alpha)} := \int_\Omega (A\nabla w \cdot \nabla w + \rho w^2)\, dx + \int_{\Gamma_R} \alpha w^2 \, ds,$$

$$|||w|||_\circ := |||w|||_{(A_\circ,\rho_\circ,\alpha_\circ)},$$

$$\underline{\mathrm{C}} := \max\left\{\min_{i\in\{1,2,3\}} \frac{1}{\mathrm{cond}_i + \theta_i}, \min_{i\in\{1,2,3\}} \frac{1-2\theta_i}{1-\theta_i}\right\}, \quad \textit{and}$$

$$\overline{\mathrm{C}} := \max_{i\in\{1,2,3\}} \frac{1}{1-\theta_i}.$$

Proof We note that

$$(\underline{c}_1 - \delta_1)|||w|||^2 \le |||w|||_A^2 \le (\overline{c}_1 + \delta_1)|||w|||^2, \quad \forall w \in V_0, A \in \mathcal{D}_A.$$

$$(\underline{c}_2 - \delta_2)|||w|||^2 \le |||w|||_\rho^2 \le (\overline{c}_2 + \delta_2)|||w|||^2, \quad \forall w \in V_0, \rho \in \mathcal{D}_\rho.$$

$$(\underline{c}_3 - \delta_3)|||y|||^2 \le |||w|||_\alpha^2 \le (\overline{c}_3 + \delta_3)|||y|||^2, \quad \forall w \in V_0, \alpha \in \mathcal{D}_\alpha.$$

By the definitions (5.45), (5.46), and (5.47), we have

$$|||w|||_\circ^2 = |||w|||_{(A,\rho,\alpha)}^2 - \int_\Omega (\delta_1 \Psi \nabla w \cdot \nabla w + \delta_2 \psi_\rho w^2)\,\mathrm{d}x - \delta_3 \int_{\Gamma_R} \psi_\alpha w^2\,\mathrm{d}s.$$

This implies

$$\begin{aligned}
|||w|||_\circ^2 &\le |||w|||_{(A,\rho,\alpha)}^2 + \delta_1\|\nabla w\| + \delta_2\|w\|^2 + \delta_3\|w\|_{\Gamma_R}\\
&\le |||w|||_{(A,\rho,\alpha)}^2 + \frac{\delta_1}{\underline{c}_1 - \delta_1}\|\nabla w\|_A^2 + \frac{\delta_2}{\underline{c}_2 - \delta_2}\|\sqrt{\rho} w\|^2 + \frac{\delta_3}{\underline{c}_3 - \delta_3}\|\sqrt{\alpha} w\|_{\Gamma_R}^2\\
&\le \max_{i\in\{1,2,3\}}\left(1 + \frac{\delta_i}{\underline{c}_i - \delta_i}\right)|||w|||_{(A,\rho,\alpha)}^2.
\end{aligned}$$

We derive a similar estimate for the lower bound:

$$\begin{aligned}
|||w|||_\circ^2 &\ge |||w|||_{(A,\rho,\alpha)}^2 - \delta_1\|\nabla w\| - \delta_2\|w\|^2 - \delta_3\|w\|_{\Gamma_R}\\
&\ge |||w|||_{(A,\rho,\alpha)}^2 + \frac{\delta_1}{\underline{c}_1 - \delta_1}\|\nabla w\|_A^2 + \frac{\delta_2}{\underline{c}_2 - \delta_2}\|\sqrt{\rho} w\|^2 + \frac{\delta_3}{\underline{c}_3 - \delta_3}\|\sqrt{\alpha} w\|_{\Gamma_R}^2\\
&\ge \min_{i\in\{1,2,3\}}\left(1 - \frac{\delta_i}{\underline{c}_i - \delta_i}\right)|||w|||_{(A,\rho,\alpha)}^2.
\end{aligned}$$

Another lower bound is as follows:

$$\begin{aligned}
|||w|||_\circ^2 &\ge \underline{c}_1\|\nabla w\|^2 + \underline{c}_2\|w\|^2 + \underline{c}_3\|w\|_{\Gamma_R}^2\\
&\ge \frac{\underline{c}_1}{\overline{c}_1 + \delta_1}\|\nabla w\|_A^2 + \frac{\underline{c}_2}{\overline{c}_2 + \delta_2}\|\sqrt{\rho} w\|^2 + \frac{\underline{c}_3}{\overline{c}_3 + \delta_3}\|\sqrt{\alpha} w\|_{\Gamma_R}^2\\
&\ge \min_{i\in\{1,2,3\}} \frac{\underline{c}_i}{\overline{c}_i + \delta_i}|||w|||_{(A,\rho,\alpha)}^2.
\end{aligned}$$

Clearly, the maximum of lower bounds is also a lower bound. The definitions of θ_i lead to the statement. □

Theorem 5.5 *Let $u_\circ$ be the solution of the non-perturbed problem* (5.40)–(5.43) *and A, ρ, and α be defined by* (5.45), (5.46), *and* (5.47), *respectively. Then,*

$$\bar{r}_{\ominus,RD} \leq \bar{R}_{\ominus,RD}(u_\circ) \leq \bar{r} \leq \bar{R}_{\oplus,RD}(u_\circ) \leq \bar{r}_{\oplus,RD}, \tag{5.48}$$

where

$$\bar{r}_{\ominus,RD} := \sqrt{\underline{\mathrm{C}}} \min_{i\in\{1,2,3\}} \frac{\mathrm{cond}_i^{-1}\theta_i}{\sqrt{1-\mathrm{cond}_i^{-1}\theta_i}}, \tag{5.49}$$

$$\bar{r}_{\oplus,RD} := \sqrt{\overline{\mathrm{C}}} \max_{i\in\{1,2,3\}} \frac{\theta_i}{\sqrt{1-\theta_i}}, \tag{5.50}$$

$$\bar{R}_{\ominus,RD}(u_\circ) := \sqrt{\underline{\mathrm{C}}} \frac{|||u_\circ|||_\delta^2/|||u_\circ|||_\circ^2}{\sqrt{1-|||u_\circ|||_\delta^2/|||u_\circ|||_\circ^2}}, \tag{5.51}$$

$$\bar{R}_{\oplus,RD}(u_\circ) := \sqrt{\overline{\mathrm{C}}} \frac{\sqrt{\Upsilon_\delta(u_\circ)}}{|||u_\circ|||_\circ}, \tag{5.52}$$

$\underline{\mathrm{C}}$ *and* $\overline{\mathrm{C}}$ *are from Proposition* 5.2, $|||u_\circ|||_\delta^2 := \delta_1\|\nabla u_\circ\|_\Omega^2 + \delta_2\|u_\circ\|_\Omega^2 + \delta_3\|u_\circ\|_{\Gamma_R}^2$, *and*

$$\Upsilon_\delta(u_\circ) := \frac{\delta_1^2}{\underline{c}_1(\underline{c}_1-\delta_1)}\int_\Omega A\nabla u_\circ\cdot\nabla u_\circ\,\mathrm{d}x + \frac{\delta_2^2}{\underline{c}_2(\underline{c}_2-\delta_2)}\|\sqrt{\rho_\circ}u_\circ\|_\Omega^2$$
$$+ \frac{\delta_3^2}{\underline{c}_3(\underline{c}_3-\delta_3)}\|\sqrt{\alpha_\circ}u_\circ\|_{\Gamma_R}^2.$$

Proof Consider the lower bound first. We have

$$r^2 = \sup_{u\in\mathcal{S}} |||u_\circ - u|||_0^2 \geq \underline{\mathrm{C}} \sup_{u\in\mathcal{S}(\mathcal{D})} |||u_\circ - u|||_{(A,\rho,\alpha)}^2.$$

Since (see [MR10])

$$\sup_{u\in\mathcal{S}(\mathcal{D})} |||u_\circ - u|||_{(A,\rho,\alpha)}^2 = \sup_{(A,\rho,\alpha)\in\mathcal{D}}\Big\{\sup_{w\in V_0}\underline{\mathrm{M}}_{A,\rho,\alpha}^2(u_\circ,w)\Big\}$$
$$= \sup_{w\in V_0}\Big\{\sup_{(A,\rho,\alpha)\in\mathcal{D}}\underline{\mathrm{M}}_{A,\rho,\alpha}^2(u_\circ,w)\Big\},$$

we conclude that

$$r^2 \geq \underline{\mathrm{C}} \sup_{w\in V_0}\Big\{\sup_{(A,\rho,\alpha)\in\mathcal{D}}\underline{\mathrm{M}}_{A,\rho,\alpha}^2(u_\circ,w)\Big\}. \tag{5.53}$$

Now our goal is to estimate the right-hand side of (5.53) from below. For this purpose, we employ the structure of the minorant, which is as follows:

$$\begin{aligned}\underline{\mathsf{M}}^2_{A,\rho,\alpha}(u_\circ, w) = &-\int_\Omega (A_\circ + \delta_1 \Psi)(\nabla w + 2\nabla u_\circ)\cdot \nabla w \,\mathrm{d}x \\ &- \int_\Omega (\rho_\circ + \delta_2 \psi_\rho)(w + 2u_\circ) w \,\mathrm{d}x \\ &- \int_{\Gamma_R} (\alpha_\circ + \delta_3 \psi_\alpha)(w + 2u_\circ) w \,\mathrm{d}s + 2l(w), \end{aligned} \tag{5.54}$$

where $l(w) := \int_\Omega f w \,\mathrm{d}x - \int_{\Gamma_N} F w \,\mathrm{d}s - \int_{\Gamma_R} G w \,\mathrm{d}s$. Note that

$$\int_\Omega (A_\circ \nabla u_\circ \cdot \nabla w \,\mathrm{d}x + \rho_\circ u_\circ w)\,\mathrm{d}x + \int_{\Gamma_R} \alpha_\circ u_\circ w \,\mathrm{d}s = l(w).$$

Hence,

$$\begin{aligned}\underline{\mathsf{M}}^2_{A,\rho,\alpha}(u_\circ, w) = &-\int_\Omega A_\circ \nabla w \cdot \nabla w \,\mathrm{d}x - \int_\Omega \rho_\circ w^2 \,\mathrm{d}x - \int_{\Gamma_R} \alpha_\circ w^2 \,\mathrm{d}s \\ &- \delta_1 \int_\Omega \Psi(\nabla w + 2\nabla u_\circ)\cdot \nabla w \,\mathrm{d}x - \delta_2 \int_\Omega \psi_\rho (w + 2u_\circ) w \,\mathrm{d}x \\ &- \delta_3 \int_{\Gamma_R} \psi_\alpha (w + 2u_\circ) w \,\mathrm{d}s. \end{aligned} \tag{5.55}$$

We substitute $w := \lambda u_\circ$ $(\lambda > 0)$ and obtain

$$\begin{aligned}\underline{\mathsf{M}}^2_{A,\rho,\alpha}(u_\circ, \lambda u_\circ) = -\lambda^2 |||u_\circ|||^2_\circ - \lambda(\lambda+2)\Big(&\delta_1 \int_\Omega \Psi \nabla u_\circ \cdot \nabla u_\circ \,\mathrm{d}x \\ &+ \delta_2 \int_\Omega \psi_\rho u_\circ^2 \,\mathrm{d}x + \delta_3 \int_{\Gamma_R} \psi_\alpha u_\circ^2 \,\mathrm{d}s\Big). \end{aligned} \tag{5.56}$$

Since,

$$\sup_{|\Psi|\le 1} -\int_\Omega \Psi \nabla u_\circ \cdot \nabla u_\circ \,\mathrm{d}x = \int_\Omega |\nabla u_\circ|^2 \,\mathrm{d}x = \|\nabla u_\circ\|^2_\Omega$$

$$\sup_{|\psi_\rho|\le 1} -\int_\Omega \psi_\rho u_\circ^2 \,\mathrm{d}x = \|u_\circ\|^2_\Omega$$

$$\sup_{|\psi_\rho|\le 1} -\int_{\Gamma_R} \psi_\alpha u_\circ^2 \,\mathrm{d}x = \|u_\circ\|^2_{\Gamma_R},$$

we have

$$\sup_{(A,\rho,\alpha)\in\mathcal{D}} \underline{\mathsf{M}}^2_{A,\rho,\alpha}(u_\circ, \lambda u_\circ) = -\lambda^2 |||u_\circ|||^2_\circ + \lambda(\lambda+2)\,|||u_\circ|||^2_\delta.$$

The expression attains the maximum at

$$\tilde{\lambda} = \frac{|||u_\circ|||_\delta^2}{|||u_\circ|||_\circ^2 - |||u_\circ|||_\delta^2}.$$

Substituting this yields

$$\sup_{(A,\rho,\alpha)\in\mathcal{D}} \underline{\mathsf{M}}^2_{A,\rho,\alpha}(u_\circ, \tilde{\lambda}u_\circ) = \frac{|||u_\circ|||_\delta^4}{|||u_\circ|||_\circ^2 - |||u_\circ|||_\delta^2}. \tag{5.57}$$

We apply (5.57) to estimate (5.53) from below, divide by $|||u_\circ|||_\circ^2$, and take the square root to obtain (5.51).

Note that

$$\begin{aligned}
|||u_\circ|||_\circ^2 &= \int_\Omega \left(A_\circ \nabla u_\circ \cdot \nabla u_\circ + \rho_\circ u_\circ^2\right) \mathrm{d}x + \int_{\Gamma_R} \alpha_\circ u_\circ \,\mathrm{d}s \\
&\geq \int_\Omega \left(\underline{c}_1 \nabla u_\circ \cdot \nabla u_\circ + \underline{c}_2 u_\circ^2\right) \mathrm{d}x + \int_{\Gamma_R} \underline{c}_3 u_\circ \,\mathrm{d}s \\
&> \delta_1 \|\nabla u_\circ\|_\Omega^2 + \delta_2 \|u_\circ\|_\Omega^2 + \delta_3 \|u_\circ\|_{\Gamma_R}^2 \\
&= |||u_\circ|||_\delta^2,
\end{aligned}$$

so that the respective lower bound is positive. Moreover,

$$\begin{aligned}
|||u_\circ|||_\delta^2 &\geq \frac{\delta_1}{\overline{c}_1} \int_\Omega A_\circ \nabla u_\circ \cdot \nabla u_\circ \,\mathrm{d}x + \frac{\delta_2}{\overline{c}_2} \int_\Omega \rho_\circ u_\circ^2 \,\mathrm{d}x + \frac{\delta_3}{\overline{c}_3} \int_{\Gamma_R} \alpha_\circ u_\circ^2 \,\mathrm{d}s \\
&\geq \min_{i\in\{1,2,3\}} \frac{\delta_i}{\overline{c}_i} |||u_\circ|||_\circ^2.
\end{aligned} \tag{5.58}$$

By applying (5.58) to the estimate (5.57), we arrive at (5.49).

Now we deduce an upper bound. We have

$$\begin{aligned}
\sup_{u\in\mathcal{S}(\mathcal{D})} |||u_\circ - \tilde{u}|||_{(A,\rho,\alpha)}^2 &= \sup_{(A,\rho,\alpha)\in\mathcal{D}} \left\{ \inf_{y,\mu_i,\gamma_j} \overline{\mathsf{M}}^2_{A,\rho,\alpha}(u_\circ, y, \gamma, \mu_1, \mu_2) \right\} \\
&\leq \inf_{y,\mu_i,\gamma_j} \left\{ \sup_{(A,\rho,\alpha)\in\mathcal{D}} \overline{\mathsf{M}}^2_{A,\rho,\alpha}(u_\circ, y, \gamma, \mu_1, \mu_2) \right\},
\end{aligned}$$

where (see [MR10])

$$\begin{aligned}
\overline{\mathsf{M}}^2_{A,\rho,\alpha}(u_\circ, y, \gamma, \mu_1, \mu_2) = \kappa \Bigg(\gamma_1 D(\nabla v, y) &+ \gamma_2 \left\| \frac{\sqrt{\kappa^2\gamma_2^2\rho + 1}}{\kappa\gamma_2\rho + 1} r_1(v, y) \right\|_\Omega^2 \\
&+ \gamma_3 \left\| \frac{\sqrt{\kappa^2\gamma_3^2\alpha + 1}}{\kappa\gamma_3\alpha + 1} r_2(v, y) \right\|_{\Gamma_3}^2 + \gamma_4 \|F - y\cdot\nu\|_{\Gamma_2}^2 \Bigg).
\end{aligned}$$

By Proposition 5.2, we obtain

$$r^2 \le \overline{\mathrm{C}} \inf_{y,\mu_i,\gamma_j} \Big\{ \sup_{(A,\rho,\alpha)\in\mathcal{D}} \overline{\mathrm{M}}^2_{A,\rho,\alpha}(u_\circ, y, \gamma, \mu_1, \mu_2) \Big\}. \tag{5.59}$$

We need to estimate explicitly the term in the brackets. For this purpose we estimate from the above the last two terms of the majorant and represent them in the form

$$\begin{aligned}
&\overline{\mathrm{M}}^2_{A,\rho,\alpha}(u_\circ, y, \gamma, \mu_1, \mu_2) \\
&\le \kappa \Bigg(\gamma_1 D(\nabla v, y) + \bigg\| \sqrt{\gamma_2 \kappa (1-\mu_1)^2 + \frac{\mu_1^2}{\kappa(\underline{c}_2 - \delta_2)}}\, r_1(v,y) \bigg\|^2_\Omega \\
&\quad + \bigg\| \sqrt{\gamma_3 \kappa (1-\mu_2) + \frac{\mu_2}{\kappa(\underline{c}_3 - \delta_3)}}\, r_2(v,y) \bigg\|^2_{\Gamma_R} + \gamma_4 \|F - y\cdot\nu\|^2_{\Gamma_N} \Bigg).
\end{aligned} \tag{5.60}$$

Now we find the upper bounds with respect to $A \in \mathcal{D}_A$, $\rho \in \mathcal{D}_\rho$, and $\alpha \in \mathcal{D}_\alpha$ separately.

First, we consider the term D generated by A and A^{-1}:

$$\begin{aligned}
&\sup_{A\in\mathcal{D}_A} D(\nabla u_\circ, y) \\
&= \sup_{|\Psi|<1} \int_\Omega (A_\circ + \delta_1\Psi)^{-1} \big|(A_\circ + \delta\Psi)\nabla u_\circ - y\big|^2 \,dx \\
&\le \frac{1}{\underline{c}_1 - \delta_1} \\
&\quad \times \sup_{|\Psi|<1} \Big\{ \|A_\circ \nabla u_\circ - y\|^2 + 2\delta_1 \int_\Omega \Psi \nabla u_\circ \cdot (A_\circ \nabla u_\circ - y)\,dx + \delta_1^2 \|\Psi \nabla u_\circ\|^2 \Big\} \\
&\le \frac{1}{\underline{c}_1 - \delta_1} \Big(\|A_\circ \nabla u_\circ - y\|^2_\Omega + 2\delta_1 \int_\Omega |\nabla u_\circ| |A_\circ \nabla u_\circ - y| \,dx + \delta_1^2 \|\nabla u_\circ\|^2_\Omega \Big).
\end{aligned} \tag{5.61}$$

For the term related to the error in the equilibrium equation, we have

$$\begin{aligned}
&\sup_{\rho\in\mathcal{D}_\rho} \big\| r_1^\rho(u_\circ, y) \big\|^2_\Omega \\
&= \sup_{|\psi_2|<1} \int_\Omega \big(f - (\rho_\circ + \delta_2\psi_2)u_\circ + \operatorname{div} y\big)^2 \,dx \\
&= \sup_{|\psi_2|<1} \int_\Omega \big(\operatorname{div} y - \operatorname{div}(A_\circ \nabla u_\circ) - \delta_2 \psi_2 u_\circ\big)^2 \,dx \\
&\le \big\| \operatorname{div}(y - A_\circ \nabla u_\circ) \big\|^2_\Omega + 2\delta_2 \int_\Omega \big|\operatorname{div}(y - A_\circ \nabla u_\circ)\big| |u_\circ| \,dx + \delta_2 \|u_\circ\|^2.
\end{aligned}$$

(5.62)

Similarly, for the term related to the error in the Robin boundary condition we have

$$\sup_{\alpha \in \mathcal{D}_\alpha} \left\| r_2^\alpha(u_\circ, y) \right\|_{\Gamma_R}^2 \leq \left\| \frac{\partial(y - A_\circ \nabla u_\circ)}{\partial \nu} \right\|_{\Gamma_R}^2 + 2\delta_3 \int_{\Gamma_R} \left| \frac{\partial(y - A_\circ \nabla u_\circ)}{\partial \nu} \right| |u_\circ| \, ds + \delta_3^2 \|u_\circ\|_{\Gamma_R}^2. \quad (5.63)$$

It is clear that for $y = y_0 := A_\circ \nabla u_\circ$, the estimates (5.61)–(5.63) attain minimal values. In addition, we set in (5.60) $\mu_1 = \mu_2 = 1$ and derive that

$$\overline{\mathsf{M}}^2_{A,\rho,\alpha}(u_\circ, A_\circ \nabla u_\circ, \gamma, 1, 1) \leq \kappa \left(\frac{\delta_1^2 \gamma_1}{\underline{c}_1 - \delta_1} \|\nabla u_\circ\|_\Omega^2 + \frac{\delta_2^2}{\underline{c}_2 - \delta_2} \|u_\circ\|_\Omega^2 + \frac{\delta_3^2}{\underline{c}_3 - \delta_3} \|u_\circ\|_{\Gamma_R}^2 \right). \quad (5.64)$$

Now we tend γ_2, γ_3, and γ_4 (which are contained in κ) to infinity. Then, (5.64) and (5.60) imply (5.52). An upper bound for the normalized radius follows from the relation

$$\begin{aligned}\overline{\mathsf{M}}^2_{A,\rho,\alpha}&(u_\circ, A_\circ \nabla u_\circ, \gamma, 1, 1) \\ &\leq \frac{\delta_1^2}{\underline{c}_1(\underline{c}_1 - \delta_1)} \int_\Omega A_\circ \nabla u_\circ \cdot \nabla u_\circ \, dx \\ &\quad + \frac{\delta_2^2}{\underline{c}_2(\underline{c}_2 - \delta_2)} \|\sqrt{\rho_\circ} u_\circ\|_\Omega^2 + \frac{\delta_3^2}{\underline{c}_3(\underline{c}_3 - \delta_3)} \|\sqrt{\alpha_\circ} u_\circ\|_{\Gamma_R}^2 \\ &\leq \max_{i \in \{1,2,3\}} \frac{\delta_i^2}{\underline{c}_i(\underline{c}_i - \delta_i)} |||u_\circ|||^2,\end{aligned}$$

which leads to (5.50). □

The main difference between Theorem 5.5 and Theorem 5.2, where only a single source of indeterminacy is considered, is that the bounds $\bar{R}_{\ominus,RD}(u_\circ)$ and $\bar{R}_{\oplus,RD}(u_\circ)$ depend on the indeterminacy magnitude of A, ρ and α weighted by $\|\nabla u_\circ\|_{A_\circ}$, $\|\sqrt{\rho_\circ} u_\circ\|_\Omega$, and $\|\sqrt{\alpha_\circ} u_\circ\|_{\Gamma_R}$, respectively. The bounds $\bar{r}_{\ominus,RD}$ and $\bar{r}_{\oplus,RD}$, which do not employ any knowledge of the non-perturbed solution, are likely to be less accurate. In these estimates, the most uncertain parameter dominates. Also the other results exposed earlier for the incompletely known operator $\mathcal{A}$ can be extended to the reaction diffusion problem.

5.7 Error Indication and Indeterminate Data

Error indicators used in numerical analysis of partial differential equations usually assume that data of the problem are known exactly. In this case, a good error indicator can suggest efficient reconstructions of meshes, which lead to accurate numerical

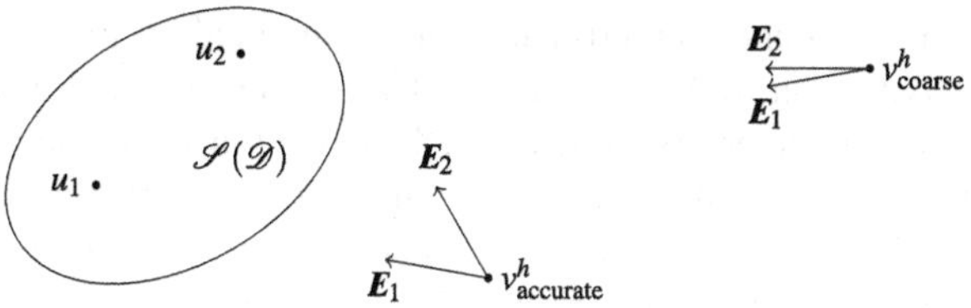

Fig. 5.9 Error indications $\boldsymbol{E}_1$ and $\boldsymbol{E}_2$ oriented toward two different solutions u_1 and u_2 in the solution set $\mathcal{S}(\mathcal{D})$

solutions. In this section, we discuss how this process may be affected by incompletely known data. Certainly this discussion is based upon rather simple examples. However, to the best of our knowledge, such type studies are quite new, and our goal is to show some principal difficulties arising if error indicators are applied to problems with uncertain data. It is clear that similar difficulties arise in many other problems. Computational results of this section were performed by I. Anjam and, in general, the exposition follows along the lines of [AMNR12].

We begin with observations motivated by Fig. 5.9 where we depict two different "error indication directions", $\boldsymbol{E}_1$ and $\boldsymbol{E}_2$. These directions are computed by means of the indicator $\boldsymbol{E}$ with the data D_1 and D_2, which lead to two different exact solutions u_1 and u_2, respectively. If our approximate solution v^h is far from $\mathcal{S}(\mathcal{D})$, then the directions are close (in other words, if we have a coarse approximation, then good error indicators are robust with respect to small variations of data). However, this may be not true for accurate approximations. This fact does not depend on the quality of an error indicator and takes place even for the best one based on comparison of approximations and exact solutions. In practice, the arrows depicted in Fig. 5.9 mean certain reconstructions of meshes. It is easy to see that if the approximate solution lies in the vicinity of $\mathcal{S}(\mathcal{D})$, then error indicators provide very different results if the data vary within admissible bounds. Therefore, the process of sensible mesh adaptation has a limit beyond which further refinements become unreliable. Below we demonstrate this fact on several simple examples. For this purpose, we use the model problem (5.21)–(5.22) again. Our goal is to study how incomplete knowledge of the diffusion coefficients impacts the reliability of error indication.

5.7.1 Numerical Experiments

In our numerical experiments, we again consider the stationary diffusion equation $\operatorname{div} A\nabla u + f = 0$ with small disturbances of the diffusion matrix $A = A_\circ + \delta B$, where the magnitude of variations δ satisfies $A_\circ\xi \cdot \xi \geq \underline{c} > \delta$ for all $|\xi| = 1$. For each element $T \in \mathcal{T}_h$, the matrix B (which defines disturbances) is symmetric and its coefficients may attain one of the three values: $\{-1, 0, 1\}$. A perturbation generated in this way is clearly an extreme one. It suits our purposes, since we are trying to find perturbations generating the worst case situation, which may occur with different diffusion matrices A that belong to the set $\mathcal{D}$.

We note that since the number of matrices contained in $\mathcal{D}$ is much larger than those representable in such a form, the sensitivity of error indicators with respect to data uncertainty is even higher than detected in our experiments.

Let $\mathbb{E}$ denote an error indicator computed on the set of elements $\mathcal{T}_h$ for an approximation u_h (see Chap. 2). The output of $\mathbb{E}$ is a vector $\{\mathbb{E}(u_h)\}$ that contains approximate errors value for each element in T. In computational practice, error indicators are used together with a marker $\mathbb{M}$. In this series of numerical experiments, we confine ourselves to the marker $\mathbb{M}$, which marks a certain predefined amount of elements with highest errors (denoted by N_{ref}).

Our analysis of effects caused by data uncertainty is based on the following strategy. We select a mesh $\mathcal{T}_h$ and a certain number of matrices $A_j = A_\circ + \delta B_j$ for some given δ. For each set of data associated with the exact solution $u_j = \mathcal{S}(A_j)$, we compute the corresponding approximations u_{jh} on the mesh $\mathcal{T}_h$. Then, for each u_{jh}, we calculate the error indicator $\mathbb{E}_j = \mathbb{E}(A_j, u_{jh})$ and the corresponding markings $\mathbb{M}(\mathbb{E}_j)$.

The difference of two markings is natural to evaluate by means of the boolean measure analogous to that we used in (2.4). We define the quantity

$$\operatorname{diff}(\mathbb{M}, \mathbb{E}_i, \mathbb{E}_j) := 1 - \frac{[\![\mathbb{M}(\mathbb{E}_i) \equiv \mathbb{M}(\mathbb{E}_j)]\!]}{N} \in [0, 1], \tag{5.65}$$

where $\equiv$ is the logical operator defined in Table 2.1.

The quantity

$$\Theta := \max_{i,j}\left\{\operatorname{diff}(\mathbb{M}, \mathbb{E}_i, \mathbb{E}_j)\right\} \tag{5.66}$$

shows the maximal difference produced by an error indicator with different diffusion matrices from the set $\mathcal{D}$. We have tested the following commonly used error indicators.

We test the error indicators based on the majorant (see (3.103)) generated by the following reconstructions of the flux: y_{G} obtained by nodal gradient-averaging (see Sect. 2.2.2.2), y_{RT}^j obtained by edge-wise gradient-averaging and j equilibration cycles (see Sect. 2.2.2.4), and y_{glo} obtained by global minimization of the majorant (see (3.102)). Additionally, we introduce "globally averaged" y_{Gglo}, which is calculated by globally minimizing $\|y_{\mathrm{Gglo}} - A\nabla u_h\|^2_{A^{-1}}$ (see, e.g., [CB02, BC02]) using Raviart–Thomas elements. We denote the corresponding indicated element-wise error distributions by $\mathbb{E}_{\overline{\mathrm{M}}}(u_h, y_{\mathrm{G}})$, $\mathbb{E}_{\overline{\mathrm{M}}}(u_h, y_{\mathrm{RT}}^j)$, $\mathbb{E}_{\overline{\mathrm{M}}}(u_h, y_{\mathrm{glo}})$, and $\mathbb{E}_{\overline{\mathrm{M}}}(u_h, y_{\mathrm{Gglo}})$, respectively. Moreover, we recall two residual type error indicators $\mathbb{E}(\eta_{RF})$ (see (3.108)) and $\mathbb{E}(\eta_{RJ})$ (see (3.109)).

5.7.2 Results and Conclusions

The approximate solutions of the model problem were computed by using standard Courant type finite element approximations. Indicators $\mathbb{E}_{\overline{\mathrm{M}}}(u_h, y_{\mathrm{glo}})$ and $\mathbb{E}_{\overline{\mathrm{M}}}(u_h, y_{\mathrm{Gglo}})$ were computed with the help of linear Raviart–Thomas finite elements. All the problems were solved on same regular meshes, and the arising systems of linear simultaneous equations were exactly solved by direct methods. In view of this fact, approximate solutions possess the Galerkin orthogonality property, and, therefore, the residual error indicator $\mathbb{E}(\eta_{RF})$ can be used. For the edge-averaging indicator $\mathbb{E}_{\overline{\mathrm{M}}}(u_h, y^j_{\mathrm{RT}})$, we set $j = 5$ (the number of times the quasi-equilibration cycle P_{RM} is applied).

N_{elem} denotes the overall amount of the elements. The marker $\mathbb{M}$ used selects 30 % of elements to be refined, i.e., $N_{\mathrm{ref}} = 0.3N_{\mathrm{elem}}$. Note that the maximal value of Θ for this marker is 0.6. Even if markings generated by two different indicators select completely different elements, for 40 % of all elements the marked value coincides (it is zero).

We studied how the magnitude of variations δ affects error indicators and discuss typical results with the example of a simple problem where $\Omega = (0,1)^2$, $A_\circ = I$, and $f = 2(x_1(1-x_1) + x_2(1-x_2))$. The exact solution of this "mean" problem is $u_\circ = x_1(1-x_1)x_2(1-x_2)$.

The results are shown in Table 5.2 and Fig. 5.10. They show the performance of indicators on six different meshes. It is worth outlining that the actual sensitivity of error indicators with respect to the data uncertainty is even higher than in these results, because we do not consider all problems with admissible data.

Table 5.2 shows how the values of Θ (associated with the indicators) depend on the number of elements N_{elem} and the parameter δ. It is easy to see that sufficiently small values of Θ (which correspond to relatively stable performance of an error indicator) are obtained only for small δ (such as 0.005 or 0.01) and a rather moderate number of elements. If the values of δ are not very small (e.g., 0.04), then all the indicators may generate quite different markings. We recall that $\Theta = 0.6$ if the indicators computed for different elements of the solution set $\mathcal{D}$ may generate completely opposite markings. Obviously, this situation arises if the corresponding approximate solution lies within (or is very close) the set $\mathcal{S}(\mathcal{D})$.

The curves in Fig. 5.10 represent these results graphically. We see that for $\delta > 0.01$ all indicators lose the reliability.

We observe that if the indeterminacy is significant compared with the approximation error, uncertainties in the matrix entries may seriously corrupt the process of error indication. This phenomenon does not depend on a particular error indicator.

Finally, we note that in this simple test problem the effect of indicator deterioration is easy to discover even for relatively coarse meshes. However, by our experience, similar effects eventually arises in all problems if more and more refined meshes are used. In other words, indeterminacy of data limits efficiency (and applicability) of error indicators.

Table 5.2 The values of Θ

	(a) $\mathbb{E}_{\overline{M}}(v, y_G)$, patch-wise averaging						(b) $\mathbb{E}_{\overline{M}}(v, y^0_{RT})$ edge averaging					
N_{elem}	δ						δ					
	0.005	0.01	0.02	0.03	0.04	0.05	0.005	0.01	0.02	0.03	0.04	0.05
800	0.09	0.16	0.31	0.40	0.48	0.51	0.09	0.16	0.30	0.40	0.45	0.50
3200	0.18	0.31	0.47	0.53	0.52	0.58	0.16	0.30	0.46	0.52	0.53	0.52
12800	0.32	0.48	0.52	0.59	0.60	0.60	0.30	0.46	0.53	0.56	0.59	0.59
51200	0.48	0.52	0.60	0.60	0.60	0.60	0.46	0.53	0.59	0.60	0.60	0.60
115200	0.53	0.59	0.60	0.60	0.60	0.60	0.52	0.57	0.59	0.60	0.60	0.60

	(c) $\mathbb{E}(\eta_{RF})$ residual, full						(d) $\mathbb{E}(\eta_{RJ})$ residual, jumps					
N_{elem}	δ						δ					
	0.005	0.01	0.02	0.03	0.04	0.05	0.005	0.01	0.02	0.03	0.04	0.05
800	0.16	0.24	0.39	0.48	0.54	0.56	0.09	0.15	0.30	0.40	0.44	0.52
3200	0.25	0.38	0.53	0.57	0.57	0.57	0.16	0.30	0.46	0.53	0.54	0.53
12800	0.38	0.53	0.57	0.58	0.59	0.59	0.30	0.45	0.54	0.53	0.59	0.59
51200	0.53	0.57	0.59	0.60	0.60	0.60	0.45	0.54	0.59	0.60	0.60	0.60
115200	0.56	0.58	0.60	0.60	0.60	0.60	0.53	0.53	0.60	0.60	0.60	0.60

	(e) $\mathbb{E}_{\overline{M}}(v, y_{Gglo})$, global averaging						(f) $\mathbb{E}_{\overline{M}}(v, y_{glo})$, majorant min					
N_{elem}	δ						δ					
	0.005	0.01	0.02	0.03	0.04	0.05	0.005	0.01	0.02	0.03	0.04	0.05
800	0.08	0.15	0.30	0.39	0.46	0.50	0.08	0.15	0.30	0.39	0.45	0.50
3200	0.17	0.30	0.46	0.53	0.54	0.52	0.16	0.30	0.46	0.53	0.53	0.52
12800	0.30	0.46	0.54	0.57	0.60	0.60	0.30	0.46	0.53	0.57	0.60	0.60
51200	0.46	0.53	0.59	0.60	0.60	0.60	0.46	0.53	0.60	0.60	0.60	0.60
115200	0.52	0.57	0.60	0.60	0.60	0.60	0.52	0.57	0.60	0.60	0.60	0.60

5.8 Linear Elasticity with Incompletely Known Poisson Ratio

We consider the isotropic linear elasticity problem discussed in Sect. 4.1.6. Our main goal is to analyze sensitivity of the energy with respect to the Poisson ratio ν and to show that for some classes of linear elasticity problems the overall energy (and the corresponding exact solutions) are extremely sensitive to small variations of ν. In the worst case, we may be faced with a phenomenon which can be called a "blow-up" of the indeterminacy error caused by incomplete information on the true value of ν. Certainly if some other parameters are also defined with uncertainties, then this effect may arise even in a more significant form.

Henceforth, we assume that the uncertainty of material parameters is generated by one factor: the uncertainty of Poisson's ratio ν. In practice, values of the Young's

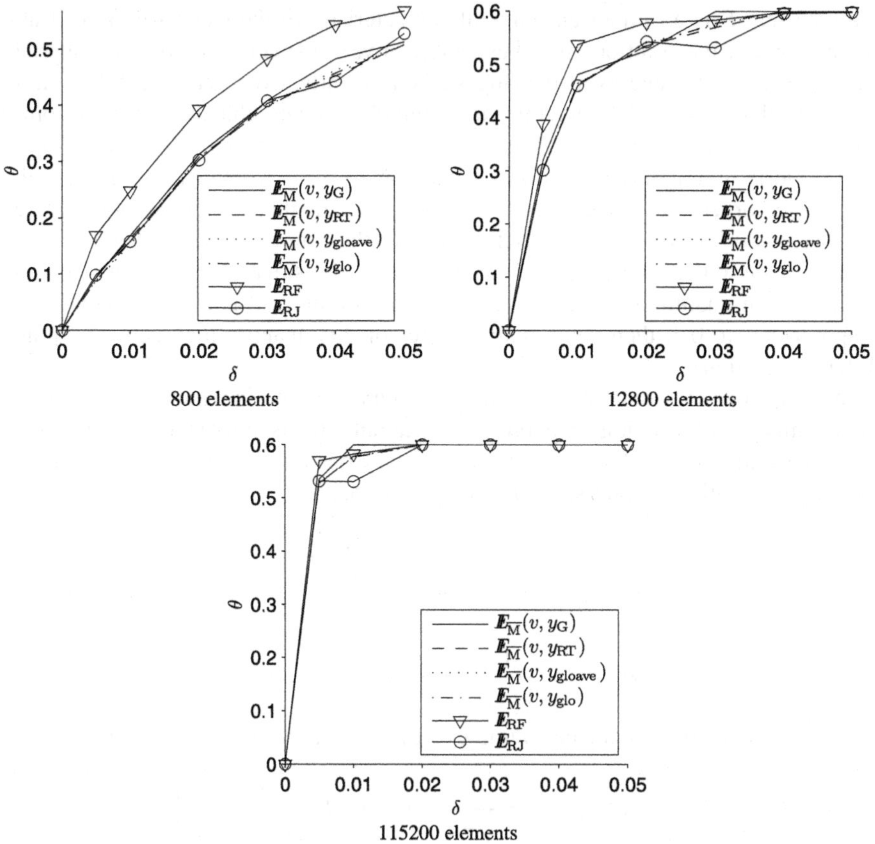

Fig. 5.10 Values of Θ for different δ for three meshes

modulus E are often known only within some interval, but this constant occurs in the equation as a multiplier. In view of this fact, the corresponding effects are easy to evaluate (they are proportional to the indeterminacy range). In this study, we neglect these effects. Moreover, we assume that solutions are normalized with respect to E, which effectively means that E is replaced by one.

By the superscript ν we denote the quantities and functions associated with Poisson's ratio ν (e.g., C^ν). Similarly, we denote the energy, strain, displacement or stress related to the exact solution (e.g., u^ν denotes the exact solution of the linear elasticity problem, and ε^ν stands for $\varepsilon(u^\nu)$). We estimate differences of quantities related to the exact solutions obtained for the ratios ν and $\nu + \delta$. For this purpose, it is convenient to use incremental type quantities, e.g., for the energy $\mathcal{E}$ we use the quantity

$$\Delta_\delta^\nu \mathcal{E} := \frac{\mathcal{E}^{\nu+\delta} - \mathcal{E}^\nu}{\delta}. \tag{5.67}$$

We note that $\mathcal{E}$ is an important integral characteristic of the exact solution. If all components of the solution are robust (insensitive) to small variations of material parameters, then the energy also changes insignificantly. However, if $\Delta_\delta^\nu \mathcal{E}$ becomes large for relatively small δ, then this fact definitely indicates that at least some components of the exact solution are highly sensitive with respect to small changes of ν.

It is worth noting that difficulties arising in analysis of problems associated with almost incompressible media as well as some special methods invented to overcome them are well-known (see, e.g., [KRW07] and the references cited therein). Our goal is to show that in the case of uncertain data the situation is even worse and difficulties in quantitative analysis arise before any method is applied, namely, if ν approaches to 0.5, then very small changes of this quantity may generate quite different solutions.

We begin our consideration with the elementary example, in which effects caused by incomplete information on parameters generate effects similar to those we discuss later with the paradigm of linear elasticity. Consider the following minimization problem with constraints: Find $x^0 \in K$ such that

$$Q^\gamma\left(x^0\right) = \min_{x \in K \subset \mathbb{R}^2} Q^\gamma(x),$$

where

$$Q^\gamma(x) := \frac{1}{1-\gamma}(x_1 - x_2)^2 + x_1^2 + x_2^2.$$

The function Q^γ represents the overall energy. It can be written in the form,

$$Q^\gamma(x) = y \cdot x, \quad y = \frac{1}{1-\gamma} Bx + x, \; B = \begin{bmatrix} 1 & -1 \\ -1 & 1 \end{bmatrix},$$

where x and y can be regarded as simple analogs of strain and stress, respectively. Let K be an affine set

$$K := \left\{x \in \mathbb{R}^2 \mid x_2 = a x_1 + b,\, a, b \in \mathbb{R}\right\}.$$

Assume that only the parameter γ is defined with an uncertainty, but the coefficients a and b are exactly known. It is not difficult to find that

$$x^0 = \frac{b}{(\gamma-2)a^2 + 2a + \gamma - 2} \begin{bmatrix} -1 - a(\gamma-2) \\ a + \gamma - 2 \end{bmatrix}$$

solves the problem and the corresponding values of Q^γ and y are

$$Q^\gamma\left(x^0\right) = \frac{b^2(\gamma-3)}{(\gamma-2)a^2 + 2a + \gamma - 2}$$

and

$$y\left(x^0\right) = \frac{b(\gamma-3)}{(\gamma-2)a^2 + 2a + \gamma - 2} \begin{bmatrix} -a \\ 1 \end{bmatrix}.$$

Sensitivity of these quantities with respect to γ is measured by the derivatives

$$\frac{\mathrm{d}x^0}{\mathrm{d}\gamma} = \frac{b(a-1)(a+1)}{(2a-2a^2+\gamma(a^2+1)-2)^2}\begin{bmatrix}-1\\-a\end{bmatrix}, \tag{5.68}$$

$$\frac{\mathrm{d}Q^\gamma(x^0)}{\mathrm{d}\gamma} = \frac{b^2(a+1)^2}{(2a-2a^2+\gamma(a^2+1)-2)^2}, \tag{5.69}$$

$$\frac{\mathrm{d}y(x^0)}{\mathrm{d}\gamma} = \frac{b(a+1)^2}{(2a-2a^2+\gamma(a^2+1)-2)^2}\begin{bmatrix}-a\\1\end{bmatrix}. \tag{5.70}$$

We note that Q^γ can be written in the form

$$Q^\gamma(x) = \frac{1}{1-\gamma}|Ex|^2 + |x|^2,$$

where

$$E = \frac{1}{\sqrt{2}}\begin{bmatrix}1 & -1\\1 & -1\end{bmatrix}.$$

The behavior of derivatives (5.68)–(5.70) is drastically different in two cases. If the kernel of the matrix E, i.e., $\mathrm{Ker}(E) = \{x \in \mathbb{R}^2 \mid x_1 = x_2\}$, and K do not have a common point (i.e., if $a = 1$), then the values, derivatives, and logarithmic derivatives of $Q^\gamma(x^0)$ and $y(x^0)$ tend to infinity as γ tends to one. In the other case, where $\mathrm{Ker}(E)$ and K do have a common point, the values of x^0, $y(x^0)$, $Q^\gamma(x^0)$ and the respective derivatives are bounded when γ tends to one.

This study indicates that for this elementary model problem, different constraints (defined by the set K) have a crucial effect on the reliability of a quantitative analysis. We see that if γ contains a very small uncertainty, e.g., $\gamma \in [\gamma_\circ - \varepsilon, \gamma_\circ + \varepsilon]$, where ε is very small, but $\gamma_\circ$ is close to one, then nevertheless the errors caused by this uncertainty may be so high that quantitative analysis of the problem is hardly possible.

The analogy between this algebraic example and a similar problem in the linear elasticity theory is presented in Table 5.3. The function $Q : \mathbb{R}^2 \to \mathbb{R}$ is an analog of the energy functional $\mathcal{E}$ in linear elasticity theory. The term $\frac{1}{\gamma}x^T Ex$ behaves like the divergence term $\int_\Omega \lambda|\operatorname{div}(u)|^2\,\mathrm{d}x$ of the energy functional. The coefficient tends to infinity as the parameter γ tends to 1 in a similar way as λ. Indeed, the coefficient λ tends to infinity as ν tends to $\frac{1}{2}$ (analogously the first part of Q^γ tends to infinity as γ tends to one). The kernel of $\operatorname{div}(u)$ is nontrivial: it contains solenoidal fields. The second term $|x|^2$ is positive definite. Its analog in linear elasticity is $\int_\Omega \mu|\varepsilon(u)|^2\,\mathrm{d}x$.

As we will see in the next section, the two cases $K \cap \mathrm{Ker}(E) \neq \emptyset$ and $K \cap \mathrm{Ker}(E) = \emptyset$ are analogs of two different types of boundary conditions: the first one admits the existence of divergence-free solutions and the second one does not.

Table 5.3 Analogy of algebraic example and isotropic linear elasticity

Example	Linear elasticity	Physical description
γ	ν	Poisson ratio
x	$\varepsilon(u)$	Strains
y	σ	Stresses
K	V_0	Set of admissible solutions
Q	$\mathcal{E}$	Energy functional
$\frac{1}{1-\gamma}x^T Ex$	$\int_\Omega \lambda \operatorname{div}(u)^2 \,\mathrm{d}x$	First part of the energy functional
$\operatorname{Ker}(E)$	$\operatorname{Ker}(\operatorname{div})$	Kernel of the "blow-up" term
$x^T x$	$\int_\Omega \mu\lvert\varepsilon(u)\rvert^2 \,\mathrm{d}x$	Second part of the energy functional

5.8.1 Sensitivity of the Energy Functional

Henceforth, we assume that

$$0 \le \nu < \frac{1}{2} \quad \text{and} \quad \delta \le \frac{1}{2}\Big(\frac{1}{2} - \nu\Big). \tag{5.71}$$

This condition guarantees that the problems with different Poisson's coefficients are uniformly elliptic. Also we assume that $\ell = 0$ (this assumption is made only for simplicity).

First, we establish a relation that serves as a basis for our subsequent analysis. We have

$$\begin{aligned}
|||\varepsilon^{\nu+\delta} - \varepsilon^{\nu}|||^2_{C^{\nu+\delta}} &= \int_\Omega C^{\nu+\delta}\varepsilon^{\nu} : \varepsilon^{\nu}\,\mathrm{d}x - \int_\Omega C^{\nu+\delta}\varepsilon^{\nu+\delta} : \varepsilon^{\nu+\delta}\,\mathrm{d}x \\
&= \delta \int_\Omega (\Delta^{\nu}_{\delta} C)\varepsilon^{\nu} : \varepsilon^{\nu}\,\mathrm{d}x - \delta\big(\Delta^{\nu}_{\delta}\mathcal{E}\big).
\end{aligned}$$

Hence

$$\Delta^{\nu}_{\delta}\mathcal{E} = \int_\Omega (\Delta^{\nu}_{\delta} C)\varepsilon^{\nu} : \varepsilon^{\nu}\,\mathrm{d}x - \frac{1}{\delta}|||\varepsilon^{\nu+\delta} - \varepsilon^{\nu}|||^2_{C^{\nu+\delta}}. \tag{5.72}$$

For isotropic media we have

$$\int_\Omega (\Delta^{\nu}_{\delta} C)\varepsilon^{\nu} : \varepsilon^{\nu}\,\mathrm{d}x = \int_\Omega \Big((\Delta^{\nu}_{\delta}\lambda)\big|\operatorname{div} u^{\nu}\big|^2 + 2(\Delta^{\nu}_{\delta}\mu)\big|\varepsilon^{\nu}\big|^2\Big)\,\mathrm{d}x.$$

In view of (5.71), we have

$$\begin{aligned}
\kappa_{\ominus} := \frac{4(1+2\nu^2)}{(5+2\nu)(1+\nu)(1-2\nu)^2} &\le \Delta^{\nu}_{\delta}\lambda \le \frac{2(1+2\nu^2+2\nu\delta)}{(1+\nu)^2(1-2\nu)^2} \\
&\le \frac{2+\nu+2\nu^2}{(1+\nu)^2(1-2\nu)^2} := \kappa_{\oplus}
\end{aligned}$$

and

$$m_\ominus := -\frac{1}{(1+\nu)^2} \leq 2\Delta_\delta^\nu \mu \leq -\frac{4}{(5+2\nu)(1+\nu)} := m_\oplus.$$

Now we find that

$$\int_\Omega (\Delta_\delta^\nu C)\varepsilon^\nu : \varepsilon^\nu \, dx \geq \int_\Omega \left(\kappa_\ominus |\operatorname{div} u^\nu|^2 - \frac{1}{(1+\nu)^2} |\varepsilon^\nu|^2 \right) dx. \qquad (5.73)$$

In order to estimate the second term on the right-hand side of (5.72), we use the majorant for the linear elasticity problem,

$$\overline{\mathsf{M}}_{\mathrm{EL}}(v,\tau) = \left(\int_\Omega \left(C^{\nu+\delta}\varepsilon(u^\nu) - \tau\right) : \left(\varepsilon(u^\nu) - \left(C^{\nu+\delta}\right)^{-1}\tau\right) dx \right)^{1/2}$$
$$+ c_K \| \operatorname{Div} \tau + f \|,$$

where c_K is the constant related to Korn's inequality and τ is an auxiliary stress function that is at our disposal. Here, we consider u^ν as an approximation v for the problem defined by $u^{\nu+\delta}$. For $\tau := C^\nu \varepsilon(u^\nu)$, the equilibrium term vanishes and the estimate reads as follows:

$$|||\varepsilon\left(u^{\nu+\delta} - u^\nu\right)|||^2_{C^{\nu+\delta}} \leq \overline{\mathsf{M}}^{\nu+\delta}_{\mathrm{EL}}(u^\nu,\tau), \qquad (5.74)$$

where

$$\overline{\mathsf{M}}^{\nu+\delta}_{\mathrm{EL}}(u^\nu,\tau) := \int_\Omega \left(C^{\nu+\delta}\varepsilon(u^\nu) - \tau\right) : \left(\varepsilon(u^\nu) - \left(C^{\nu+\delta}\right)^{-1}\tau\right) dx.$$

Lemma 5.1

$$|||\varepsilon^{\nu+\delta} - \varepsilon^\nu|||^2_{C^{\nu+\delta}} \leq \delta^2 \left(c_1^\nu \|\operatorname{div} u^\nu\|^2 + c_2^\nu \|\varepsilon(u^\nu)\|^2\right),$$

where

$$c_1^\nu := \frac{\lambda^\nu}{\nu} n \kappa_\oplus \quad \textit{and} \quad c_2^\nu := \frac{1}{4}(5+2\nu) m_\ominus^2.$$

Proof It is easy to see that

$$C^{\nu+\delta}\varepsilon(u^\nu) - C^\nu\varepsilon(u^\nu) = \left(\lambda^{\nu+\delta} - \lambda^\nu\right) \operatorname{div} u^\nu \mathbb{I} + 2\left(\mu^{\nu+\delta} - \mu^\nu\right)\varepsilon(u^\nu). \qquad (5.75)$$

Since

$$C^{-1}\tau = \frac{1}{2\mu}\left(\tau - \frac{\lambda}{3\lambda + 2\mu} \operatorname{tr}(\tau)\mathbb{I}\right),$$

we have

$$
\begin{aligned}
&\left(C^{\nu+\delta}\right)^{-1} C^{\nu} \varepsilon\left(u^{\nu}\right) \\
&= \frac{1}{2\mu^{\nu+\delta}} \left(\lambda^{\nu} \operatorname{div} u^{\nu} \mathbb{I} + 2\mu^{\nu} \varepsilon\left(u^{\nu}\right) - \frac{\lambda^{\nu+\delta}}{3\lambda^{\nu+\delta} + 2\mu^{\nu+\delta}} (3\lambda^{\nu} + 2\mu^{\nu}) \operatorname{div} u^{\nu} \mathbb{I} \right) \\
&= \left(\frac{\lambda^{\nu}}{2\mu^{\nu+\delta}} - \frac{\lambda^{\nu+\delta}(3\lambda^{\nu} + 2\mu^{\nu})}{2\mu^{\nu+\delta}(3\lambda^{\nu+\delta} + 2\mu^{\nu+\delta})} \right) \operatorname{div} u^{\nu} \mathbb{I} + \frac{\mu^{\nu}}{\mu^{\nu+\delta}} \varepsilon\left(u^{\nu}\right).
\end{aligned}
$$

We note that

$$
3\lambda^{\nu} + 2\mu^{\nu} = \frac{1}{1-2\nu}, \qquad \frac{\lambda^{\nu+\delta}}{2\mu^{\nu+\delta}} = \frac{\nu+\delta}{1-2\nu-2\delta},
$$

$$
\frac{\lambda^{\nu}}{2\mu^{\nu+\delta}} - \frac{\lambda^{\nu+\delta}(3\lambda^{\nu} + 2\mu^{\nu})}{2\mu^{\nu+\delta}(3\lambda^{\nu+\delta} + 2\mu^{\nu+\delta})} = -\frac{\delta}{(1+\nu)(1-2\nu)} =: -\frac{\delta}{\nu}\lambda^{\nu},
$$

and

$$
\varepsilon\left(u^{\nu}\right) - \left(C^{\nu+\delta}\right)^{-1} C^{\nu} \varepsilon\left(u^{\nu}\right) = \delta \frac{\lambda^{\nu}}{\nu} \operatorname{div} u^{\nu} \mathbb{I} + \frac{\mu^{\nu+\delta} - \mu^{\delta}}{\mu^{\nu+\delta}} \varepsilon\left(u^{\nu}\right). \tag{5.76}
$$

By (5.75) and (5.76), we obtain

$$
\begin{aligned}
&\overline{\mathsf{M}}_{\mathrm{EL}}^{\nu+\delta}(u^{\nu}, \tau) \\
&= n\delta \frac{\lambda^{\nu}}{\nu} \left(\lambda^{\nu+\delta} - \lambda^{\nu}\right) \left\| \operatorname{div} u^{\nu} \right\|^2 + \frac{(\lambda^{\nu+\delta} - \lambda^{\nu})(\mu^{\nu+\delta} - \mu^{\nu})}{\mu^{\nu+\delta}} \left\| \operatorname{div} u^{\nu} \right\|^2 \\
&\quad + 2\delta \frac{\lambda^{\nu}}{\nu} \left(\mu^{\nu+\delta} - \mu^{\nu}\right) \left\| \operatorname{div} u^{\nu} \right\|^2 + 2 \frac{(\mu^{\nu+\delta} - \mu^{\nu})^2}{\mu^{\nu+\delta}} \left\| \varepsilon\left(u^{\nu}\right) \right\|^2.
\end{aligned}
$$

Since the second and third terms are negative, we find that

$$
\begin{aligned}
\overline{\mathsf{M}}_{\mathrm{EL}}^{\nu+\delta}(u^{\nu}, \tau) &\le n\delta \frac{\lambda^{\nu}}{\nu} \left(\lambda^{\nu+\delta} - \lambda^{\nu}\right) \left\| \operatorname{div} u^{\nu} \right\|^2 + 2 \frac{(\mu^{\nu+\delta} - \mu^{\nu})^2}{\mu^{\nu+\delta}} \left\| \varepsilon\left(u^{\nu}\right) \right\|^2 \\
&\le \delta^2 \left(\frac{\lambda^{\nu}}{\nu} n \kappa_{\oplus} \left\| \operatorname{div} u^{\nu} \right\|^2 + \frac{m_{\ominus}^2}{2\mu^{\nu+\delta}} \left\| \varepsilon\left(u^{\nu}\right) \right\|^2 \right),
\end{aligned}
$$

which, together with (5.74), leads to the statement. □

Theorem 5.6 *Let $u_g \in V$ be a function with minimal divergence norm, i.e.,*

$$
\| \operatorname{div} u_g \| = \min_{u \in V} \| \operatorname{div} u \|. \tag{5.77}
$$

Then, for $0 \le \delta \le \min\{\frac{1}{2c_1^{\nu}}\kappa_{\ominus}, \bar{\delta}\}$, the following estimate is valid:

$$
\Delta_{\delta}^{\nu} \mathcal{E} \ge \frac{1}{2} \kappa_{\ominus} \| \operatorname{div} u_g \|^2 - c_3^{\nu} \left\| \varepsilon(u_g) \right\|^2, \tag{5.78}
$$

where

$$c_3^\nu := \frac{1}{(1+\nu)^2} + \frac{m_\ominus^2(5+2\nu)}{4}\frac{\nu\kappa_\ominus}{2\lambda^\nu n\kappa_\oplus}$$
$$= \frac{1}{(1+\nu)^2}\left(1 + \frac{(1+2\nu^2)(1-2\nu)}{2n(2+\nu+2\nu^2)}\right). \tag{5.79}$$

Proof By applying the estimate (5.73) and Lemma 5.1 to (5.72), we have

$$\Delta_\delta^\nu \mathcal{E} = \int_\Omega (\Delta_\delta^\nu C)\varepsilon^\nu : \varepsilon^\nu \,\mathrm{d}x - \frac{1}{\delta}\left|\!\left|\!\left|\varepsilon^{\nu+\delta} - \varepsilon^\nu\right|\!\right|\!\right|^2_{C^{\nu+\delta}}$$
$$\geq \left(\kappa_\ominus - \delta c_1^\nu\right)\left\|\operatorname{div} u^\nu\right\|^2 - \left(\frac{1}{(1+\nu)^2} + c_2^\nu\delta\right)\left\|\varepsilon\left(u^\nu\right)\right\|^2.$$

Let $\delta \leq \frac{\kappa_\ominus}{2c_1^\nu}$. Then, we obtain

$$\Delta_\delta^\nu \mathcal{E} \geq \frac{1}{2}\kappa_\ominus\left\|\operatorname{div} u^\nu\right\|^2 - \left(\frac{1}{(1+\nu)^2} + \frac{m_\ominus^2(5+2\nu)}{4}\frac{\nu\kappa_\ominus}{2\lambda^\nu n\kappa_\oplus}\right)\left\|\varepsilon\left(u^\nu\right)\right\|^2$$
$$= \frac{1}{2}\kappa_\ominus\left\|\operatorname{div} u^\nu\right\|^2 - c_3^\nu\left\|\varepsilon\left(u^\nu\right)\right\|^2. \tag{5.80}$$

Since u_g satisfies (5.77) and u^ν minimizes the energy functional $\mathcal{E}^\nu$, we find that

$$\int_\Omega \left(\lambda^\nu\left|\operatorname{div}(u_g)\right|^2 + 2\mu^\nu\left|\varepsilon\left(u^\nu\right)\right|^2\right)\mathrm{d}x$$
$$\leq \mathcal{E}^\nu\left(u^\nu\right) \leq \mathcal{E}^\nu(u_g) = \int_\Omega \left(\lambda^\nu\left|\operatorname{div}(u_g)\right|^2 + 2\mu^\nu\left|\varepsilon(u_g)\right|^2\right)\mathrm{d}x \tag{5.81}$$

and, therefore,

$$\left\|\varepsilon\left(u^\nu\right)\right\| \leq \left\|\varepsilon(u_g)\right\|. \tag{5.82}$$

By applying (5.77) and (5.82), we estimate the right-hand side of (5.80) from below and arrive at the statement. □

Corollary 5.2 *Since the right-hand side does not depend on* δ, *we can pass to the limit as* $\delta \to 0$ *and obtain*

$$\frac{\partial \mathcal{E}^\nu}{\partial \nu} \geq \frac{1}{2}\kappa_\ominus\|\operatorname{div} u_g\|^2 - c_3^\nu\left\|\varepsilon(u_g)\right\|^2. \tag{5.83}$$

Theorem 5.6 shows *the sensitivity of the internal energy associated with the exact solution* u^ν *with respect to small variations of Poisson's ratio* ν. If it is large, then ν must be known with a high accuracy; otherwise a quantitative analysis of the problem is not motivated. For this reason the estimate (5.83) deserves special discussion. First of all, we note that the right-hand side of (5.83) is easy to compute.

The asymptotic properties of the estimate depend crucially on the first term. A distinctive feature of the set V (of admissible displacements) is whether the boundary conditions are "compatible" (in the sense that there exists a divergence-free function u_g that satisfies these conditions) or not.

If $\operatorname{div} u_g = 0$, then (5.81) shows that

$$\mathcal{E}^\nu(u^\nu) \le \mathcal{E}^\nu(u_g) = \frac{1}{1+\nu}\|\varepsilon(u_g)\|^2,$$

which means that for all $\nu \in [0, \frac{1}{2})$ such quantities as $\mathcal{E}^\nu$ and $\|\varepsilon(u^\nu)\|$ are uniformly bounded. Moreover,

$$\int_\Omega |\operatorname{div}(u^\nu)|^2 \, dx \le \int_\Omega \left(2\frac{\mu^\nu}{\lambda^\nu}|\varepsilon(u_g)|^2\right) dx \to 0 \quad \text{as } \nu \to \frac{1}{2}.$$

In this case, small variations of Poisson's ratio do not lead to large changes in the solution.

Let us consider another case. Assume that the boundary conditions are non-compatible, i.e.,

$$\|\operatorname{div} u_g\| > 0. \tag{5.84}$$

Then, even the normalized energy increment blows up.

Corollary 5.3 *Under the assumptions of Theorem* 5.6,

$$\frac{\Delta_\delta^\nu \mathcal{E}^\nu}{\mathcal{E}^\nu} \ge \mathbb{C}(\nu, u_g) \quad \textit{and} \quad \frac{1}{\mathcal{E}^\nu}\frac{\partial \mathcal{E}^\nu}{\partial \nu} \ge \mathbb{C}(\nu, u_g),$$

where

$$\mathbb{C}(\nu, u_g) := (1+\nu)\frac{c_4^\nu \|\operatorname{div} u_g\|^2 - c_3^\nu(1-2\nu)\|\varepsilon(u_g)\|^2}{\nu\|\operatorname{div} u_g\|^2 + (1-2\nu)\|\varepsilon(u_g)\|^2}$$

and

$$c_4^\nu := \frac{2(1+2\nu^2)}{(5+2\nu)(1+\nu)(1-2\nu)}. \tag{5.85}$$

Proof By (5.78) and (5.81) we see that the normalized energy increment is subject to the relation

$$\begin{aligned}\frac{\Delta_\delta^\nu \mathcal{E}^\nu}{\mathcal{E}^\nu} &\ge \frac{(1/2)\kappa_\ominus\|\operatorname{div} u_g\|^2 - c_3^\nu\|\varepsilon(u_g)\|^2}{\lambda^\nu\|\operatorname{div}(u_g)\|^2 + 2\mu^\nu\|\varepsilon(u_g)\|^2} \\ &= (1+\nu)\frac{c_4^\nu\|\operatorname{div} u_g\|^2 - c_3^\nu(1-2\nu)\|\varepsilon(u_g)\|^2}{\nu\|\operatorname{div} u_g\|^2 + (1-2\nu)\|\varepsilon(u_g)\|^2}.\end{aligned} \tag{5.86}$$

Since the right-hand side of (5.86) does not depend on δ, it also follows that the logarithmic derivative of the energy is bounded by the same constant. □

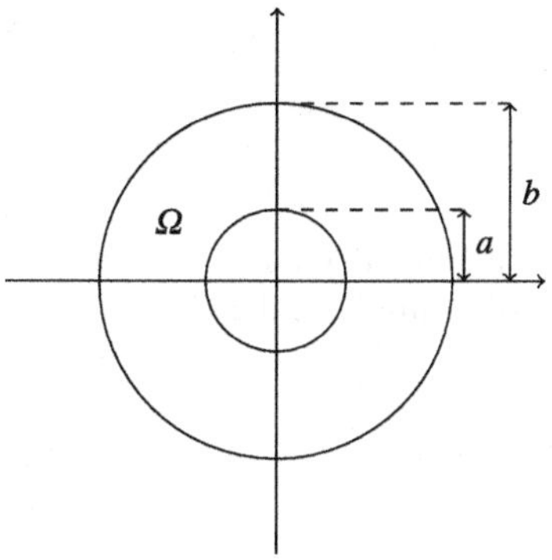

Fig. 5.11 Domain of the axisymmetric problem

$\mathbb{C}(\nu, u_g)$ is the sensitivity constant, which depends only on the geometry, ν, and the "minimal divergence function" u_g. If $\|\operatorname{div} u_g\| > 0$, then $\mathbb{C}(\nu, u_g)$ blows up as ν tends to $1/2$. Thus, the value of the energy increment normalized by the value of $\mathcal{E}^\nu$ (and the logarithmic derivative) also becomes highly sensitive to small changes in the Poisson's ratio.

Remark 5.2 It is worth noting that the estimate (5.86) and the constant $\mathbb{C}(\nu, u_g)$ can be used to estimate the effects caused by variations of ν around a given value. Indeed,

$$\frac{\mathcal{E}^{\nu+\delta} - \mathcal{E}^\nu}{\mathcal{E}^\nu} \geq \delta\mathbb{C}(\nu, u_g). \tag{5.87}$$

From (5.87) it follows that if γ denotes the upper limit of acceptable uncertainty in terms of the energy, then the value of the Poisson's ratio must be known with an accuracy of $\gamma\mathbb{C}^{-1}(\nu, u_g)$. Obviously, in the case of a blow-up situation, this condition may be impossible to satisfy in practice.

5.8.2 Example: Axisymmetric Model

In this section, we study exact solutions of an axisymmetric problem and use them to demonstrate effects caused by an incomplete knowledge of Poisson's ratio. The geometry of the problem is presented in Fig. 5.11. Let (r, θ) be polar coordinates; then

$$\Omega := \{a < r < b, 0 < \theta < 2\pi\}.$$

In polar coordinates,

$$\varepsilon_r = \frac{du_r}{dr}, \qquad \varepsilon_\theta = \frac{u_r}{r}, \qquad \varepsilon_{r\theta} = \frac{r}{2}\frac{d}{dr}\left(\frac{u_\theta}{r}\right), \tag{5.88}$$

and the constitutive relations read as follows:

$$\sigma_r = 2\mu\varepsilon_r + \lambda(\varepsilon_r + \varepsilon_\theta), \tag{5.89}$$

$$\sigma_\theta = 2\mu\varepsilon_\theta + \lambda(\varepsilon_r + \varepsilon_\theta), \tag{5.90}$$

$$\sigma_{r\theta} = 2\mu\varepsilon_{r\theta}. \tag{5.91}$$

We assume that the volume and surface loads are zero. In this case, the equations of equilibrium have the form

$$\frac{d\sigma_r}{dr} + \frac{\sigma_r - \sigma_\theta}{r} = 0, \tag{5.92}$$

$$\frac{d\sigma_{r\theta}}{dr} + 2\frac{\sigma_{r\theta}}{r} = 0, \tag{5.93}$$

and

$$\mathcal{E} := \int_\Omega \sigma : \varepsilon \, \mathrm{d}x = \int_\Omega \left(\lambda \, \mathrm{tr}^2(\varepsilon) + 2\mu|\varepsilon|^2\right) \mathrm{d}x.$$

First note that

$$\sigma_r - \sigma_\theta = 2\mu(\varepsilon_r - \varepsilon_\theta) = 2\mu\left(\frac{du_r}{dr} - \frac{u_r}{r}\right) = 2\mu r \frac{d}{dr}\left(\frac{u_r}{r}\right).$$

Substituting this relation into (5.92), we obtain

$$\sigma_r + 2\mu\frac{u_r}{r} = \gamma_1, \tag{5.94}$$

where γ_1 is constant. In view of (5.89), we arrive at the differential equation

$$\frac{d}{dr}(ru_r) = r\frac{\gamma_1}{\lambda + 2\mu},$$

which implies that

$$u_r = \gamma_1 r + \frac{\gamma_2}{r}. \tag{5.95}$$

From (5.93) we find that $\sigma_{r\theta} = \frac{\gamma_3}{r^2}$ and $u_\theta = \gamma_3 + \frac{\gamma_4}{r}$. Next,

$$\sigma_r = 2(\mu + \lambda)\gamma_1 - 2\mu\frac{\gamma_2}{r^2} \quad \text{and} \quad \sigma_\theta = 2(\mu + \lambda)\gamma_1 + 2\mu\frac{\gamma_2}{r^2}.$$

The energy is

$$\mathcal{E} = 4\pi\left((\lambda + \mu)(b^2 - a^2)\gamma_1^2 + \mu\left(\frac{1}{a^2} - \frac{1}{b^2}\right)\gamma_2^2\right).$$

5.8.2.1 Compatibility of the Set *V*

As we have discussed earlier, the boundary conditions distinguish whether the set of admissible functions *V* has (is "compatible") or does not have (is "non-compatible") any divergence-free members. The divergence-free functions for the model problem can be directly computed. They all have the form

$$w(r) := \frac{c}{r},$$

where c is constant.

For the Dirichlet–Neumann boundary conditions,

$$u_r(a) = g_a \quad \text{and} \quad \sigma_r(b) = F_b,$$

the divergence-free minimizer

$$u_g^{DN} = \frac{g_a a}{r}$$

always belongs to the set V. The exact solution of the problem is

$$u^{DN} = \gamma_1^{DN} r + \frac{\gamma_2^{DN}}{r},$$

where

$$\gamma_1^{DN} = \frac{ab^2(2g_a(\lambda+\mu) - bF_b)}{2(b^2(\lambda+\mu)+a^2\mu)} \quad \text{and} \quad \gamma_2^{DN} = \frac{F_b b^2 + 2g_a a\mu}{2(b^2(\lambda+\mu)+a^2\mu)}.$$

For the Dirichlet–Dirichlet conditions,

$$u_r(a) = g_a \quad \text{and} \quad u_r(b) = g_b,$$

the divergence minimizer u_g coincides with the solution of the problem and is

$$u^{DD} = u_g^{DD} = \gamma_1^{DD} r + \frac{\gamma_2^{DD}}{r},$$

where

$$\gamma_1^{DD} = \frac{g_b b - g_a a}{b^2 - a^2} \quad \text{and} \quad \gamma_2^{DD} = \frac{ab(g_a b - g_b a)}{b^2 - a^2}.$$

The function u_g is divergence-free only if $g_a a = g_b b$. This condition defines whether the Dirichlet–Dirichlet conditions are "compatible" or not.

Next, we observe the blow-up that occurs with "non-compatible" boundary conditions. We study the behavior of the energy quotient at the incompressibility limit. For our model problem, we can compute the derivative of energy with respect to the Poisson's ratio,

$$\begin{aligned}\frac{\partial \mathcal{E}^\nu}{\partial \nu} &= 4\pi \left\{ (b^2 - a^2)\left(\frac{\partial \lambda^\nu}{\partial \nu} + \frac{\mu^\nu}{\partial \nu}\right)\gamma_1^2 + \left(\frac{1}{a^2} - \frac{1}{b^2}\right)\frac{\partial \mu^\nu}{\partial \nu}\gamma_2^2 \right\} \\ &= \frac{2\pi(b^2 - a^2)}{(1+\nu)^2}\left\{ \frac{1+4\nu}{(1-2\nu)^2}\gamma_1^2 - \frac{1}{a^2 b^2}\gamma_2^2 \right\}.\end{aligned}$$

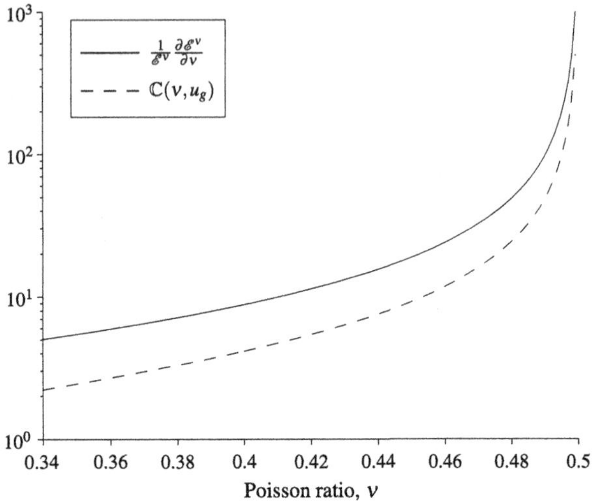

Fig. 5.12 Exact values of $\frac{1}{\mathcal{E}}\frac{\partial \mathcal{E}^\nu}{\partial \nu}$ and the lower estimate of Corollary 5.3 for the model problem with pure Dirichlet conditions and parameter values $a = 0.2$, $b = 1.0$, $g_a = 0.01$, and $g_b = -0.03$

Moreover, for the logarithmic derivative we have

$$\frac{1}{\mathcal{E}^\nu}\frac{\partial \mathcal{E}^\nu}{\partial \nu} = \frac{1}{(1+\nu)(1-2\nu)} \frac{(1+4\nu)a^2b^2\gamma_1^2 - (1-2\nu)^2\gamma_2^2}{a^2b^2\gamma_1^2 + (1-2\nu)\gamma_2^2}.$$

It is easy to see that $|\frac{\partial \mathcal{E}^\nu}{\partial \nu}|$ and $|\frac{1}{\mathcal{E}^\nu}\frac{\partial \mathcal{E}^\nu}{\partial \nu}|$ tend to ∞ if $\gamma_1^2 > 0$, i.e., the boundary conditions are non-compatible. If the boundary conditions are compatible, i.e., $g_a a = g_b b$ and $\gamma_1 = 0$, then

$$\left|\frac{1}{\mathcal{E}^\nu}\frac{\partial \mathcal{E}^\nu}{\partial \nu}\right| = \frac{1}{1+\nu}.$$

Moreover, since u_g is known for the applied boundary conditions, we can compute the sensitivity constant in Corollary 5.3,

$$\mathbb{C}(\nu, u_g) = (1+\nu)\frac{a^2b^2(2c_4^\nu - (1-2\nu)c_3^\nu)\gamma_1 - c_3^\nu(1-2\nu)\gamma_2}{a^2b^2\gamma_1^2 + (1-2\nu)\gamma_2},$$

where c_3^ν and c_4^ν are defined by (5.79) and (5.85). For pure Dirichlet conditions, the blow-up of the exact logarithmic derivative and the sensitivity constant can be observed from Fig. 5.12. For Dirichlet–Neumann conditions, a similar plot provides an exact quotient, which is almost zero, while the lower bound crudely underestimates it.

5.8.2.2 Numerically Constructed Solution Set

Here we demonstrate how the sensitivity of the solution depends on the compatibility of the boundary conditions. We do not restrict this study only to demonstrate

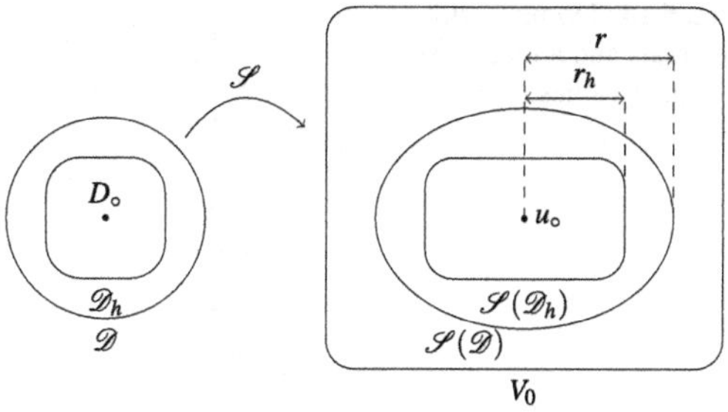

Fig. 5.13 Illustration of the numerical experiment

the blow-up phenomenon. In particular, it is interesting to observe how far from the incompressibility limit the sensitivity of the solution renders any quantitative results too inaccurate for engineering purposes.

In reality, ν is an unknown function defined on (a, b) with values in $[\nu_\circ - \delta, \nu_\circ + \delta] \subset (0, \frac{1}{2})$. Thus, in our case the set of admissible parameters $\mathcal{D}$ (cf. Sect. 5.2) has the following simple form:

$$\nu \in \mathcal{D} := \left\{\nu : (a, b) \to [\nu_\circ - \delta, \nu_\circ + \delta]\right\}.$$

Our interest is to study the set of solutions $\mathcal{S}(\mathcal{D})$ associated with all members of $\mathcal{D}$. Obviously it is impossible to derive analytical solutions for arbitrary $\nu \in \mathcal{D}$. In order to obtain a reasonable representation of $\mathcal{S}(\mathcal{D})$, we consider the set of piecewise constant functions $\mathcal{D}_h \subset \mathcal{D}$. For $\nu \in \mathcal{D}_h$, we can compute exact solutions and obtain $\mathcal{S}(\mathcal{D}_h)$. The sets are depicted in Fig. 5.13. A procedure for computing $\mathcal{S}(\mathcal{D}_h)$ is described below.

Consider the case where the material parameters on the entire domain (a, b) are piecewise constants. Let the interval (a, b) be divided into N non-intersecting subintervals $I_k := (r_k, r_{k+1})$, where r_k $(k = 1, \ldots, N+1)$ are grid points. We assume that the Poisson's ratio ν_k is constant on each interval I_k. Moreover, we allow only M different constants from the interval $[\nu_\circ - \delta, \nu_\circ + \delta]$. Consequently, Lamé's parameters λ_k and μ_k are piecewise constants as well.

Without body forces, on each interval I_k the displacement and stresses have the form

$$u_r = \gamma_1^k r + \frac{\gamma_2^k}{r}, \qquad \sigma_r = 2(\lambda_k + \mu_k)\gamma_1^k - 2\mu_k \frac{\gamma_2^k}{r^2}.$$

The solution must satisfy continuity conditions at the junctions of each interval. Thus, for every r_k $(k = 2, \ldots, N+1)$,

$$\gamma_1^{k-1} r_k + \frac{\gamma_2^{k-1}}{r_k} = \gamma_1^k r_k + \frac{\gamma_2^k}{r_k}, \tag{5.96}$$

$$2(\lambda_{k-1} + \mu_{k-1})\gamma_1^{k-1} - 2\mu_{k-1}\frac{\gamma_2^{k-1}}{r_k^2} = 2(\lambda_k + \mu_k)\gamma_1^k - 2\mu_k \frac{\gamma_2^k}{r_k^2}. \tag{5.97}$$

The boundary conditions and continuity conditions (5.96) and (5.97) together form a set of linear equations, from which the coefficients γ_1^k and γ_2^k can be found.

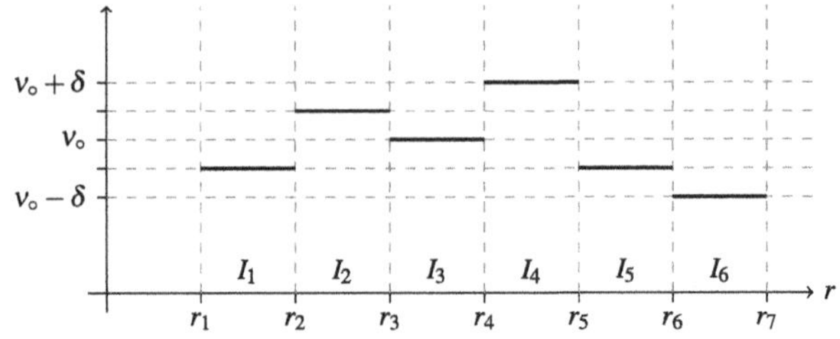

Fig. 5.14 Domain with intervals and a piecewise constant values of ν

The experiments are performed as follows: we compute a solution associated with all combinations generated by a piecewise continuous $\nu \in \mathcal{D}_h$. A particular member of $\mathcal{D}_h$ is depicted in Fig. 5.14. Each of these solutions (displacement and stresses denoted by subscript pert), are then compared with the non-perturbed solution. We examine the relative perturbations of various solution-dependent quantities defined as follows:

$$e^u_{L_2} := \max_{u_{\text{pert}} \in \mathcal{S}(\mathcal{D}_h)} \frac{\|u_{r0} - u_{\text{pert}}\|_{L_2}}{\|u_{r0}\|_{L_2}}, \qquad e^u_{L_\infty} := \max_{u_{\text{pert}} \in \mathcal{S}(\mathcal{D}_h)} \frac{\|u_{r0} - u_{\text{pert}}\|_{L_\infty}}{\|u_{r0}\|_{L_\infty}},$$

$$e^\sigma_{L_2} := \max_{u_{\text{pert}} \in \mathcal{S}(\mathcal{D}_h)} \frac{\|\sigma_{r0} - \sigma_{\text{pert}}\|_{L_2}}{\|\sigma_{r0}\|_{L_2}}, \qquad e^\sigma_{L_\infty} := \max_{u_{\text{pert}} \in \mathcal{S}(\mathcal{D}_h)} \frac{\|\sigma_{r0} - \sigma_{\text{pert}}\|_{L_\infty}}{\|\sigma_{r0}\|_{L_\infty}}.$$

and similarly for stress. Moreover, we compute relative perturbations in the energy:

$$e^J := \max_{u_{\text{pert}} \in \mathcal{S}(\mathcal{D}_h)} \frac{|J(u_{\text{pert}}) - J(u_\circ)|}{J(u_\circ)}.$$

These quantities are computed analytically on each interval. In the experiments discussed below, we set $(a,b) := (0.4, 1.0)$, Young's modulus was set to one, the Dirichlet conditions were $g_a = 0.01$ and $g_b = 0.02$, and the Neumann condition was $F_r = 0.2$. The perturbation parameters were $N = 10$ and $M = 2$. We compare the behavior of predefined quantities of interest in the case of non-compatible (Dirichlet) and compatible (Neumann) boundary conditions.

Figure 5.15 depicts results of experiments for $\nu_\circ = 0.33$ and $\nu_\circ = 0.45$. Indeterminacy is restricted to variations up to 5 %, i.e., $\frac{\delta}{\nu_\circ} \leq 0.05$. For each perturbation level, we compute the approximated solution set and compute the quantities of interest. In the first case ($\nu_\circ = 0.33$), the difference between the results obtained for non-compatible and compatible boundary conditions is not very drastic. The dependence of the radius of the solution set measured by the L_2-norm of displacements or stresses is almost linear. However, the rate of growth is somewhat bigger in the non-compatible case.

In the second case ($\nu_\circ = 0.45$), the non-compatible and compatible boundary conditions generate quite different results. In this case, only displacements measured in the L_2-norm are almost unaffected by the increase of $\nu_\circ$. All other quantities contain large uncertainty measured in dozens of percents as ν is perturbed. However,

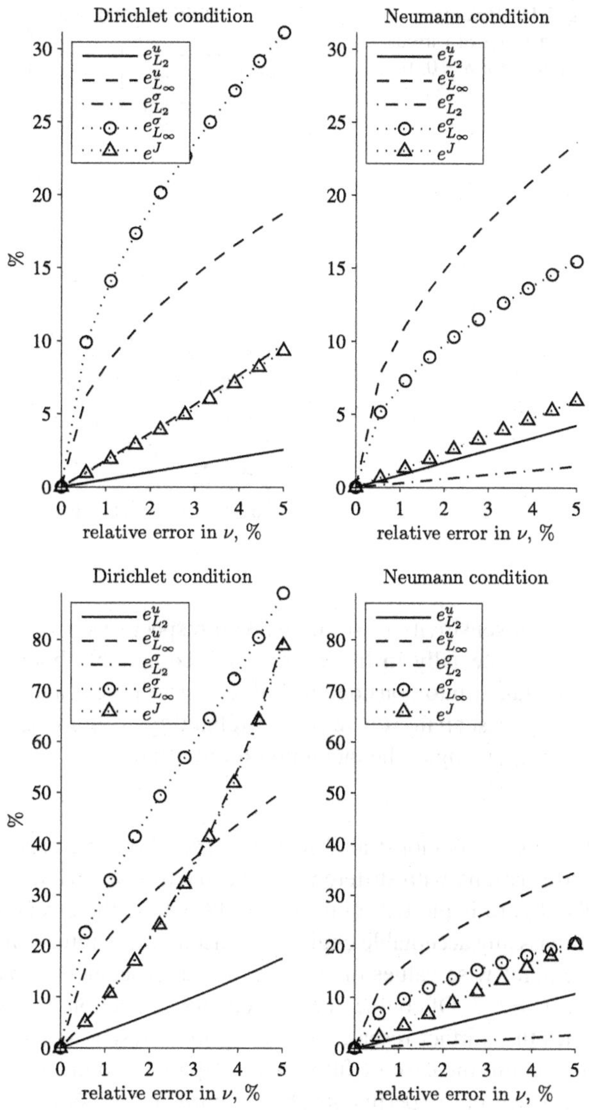

Fig. 5.15 Relative perturbations of various quantities with respect to the relative perturbation in Poisson's ratio for $\nu_\circ = 0.33$ (*top*) and $\nu_\circ = 0.45$ (*bottom*)

in the case of Neumann boundary conditions these quantities behave as in the case $\nu_\circ = 0.33$.

We note that in every test example the accuracy required for computing pointwise values (L_∞-norm) is considerably higher than the accuracy required to compute the integral quantities. From these plots, one can depict the accuracy required for the Poisson's ratio with respect to the desired accuracy and relevant quantity.

These and other experiments lead us to the following conclusion:

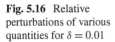

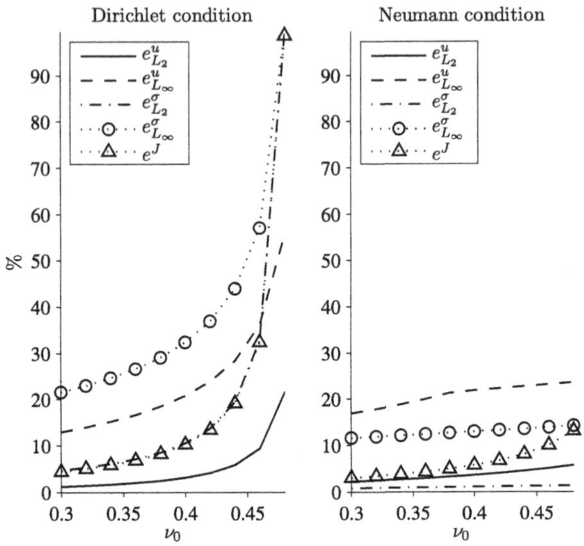

Fig. 5.16 Relative perturbations of various quantities for $\delta = 0.01$

High sensitivity of solutions with respect to small variations of Poisson's ratio (especially in the case of non-compatible boundary conditions) should be taken into consideration in engineering computations. This phenomenon may essentially decrease the reliability of numerical solutions even before approaching to the incompressibility limit.

This fact is demonstrated in Fig. 5.16, where we present the behavior of relative perturbations with different $\nu_\circ$ and fixed $\delta = 0.01$ for both boundary conditions. The figure is plotted up to $\nu_\circ = 0.48$, where the errors in all quantities of interest surpass any acceptable engineering accuracy. Another interesting observation is that the point-wise values of stresses (and displacements) may be very inaccurate even for $\nu_\circ = 0.3$. Of course, the studied model example is rather elementary, but it indicates that further research including more test examples is required. Certainly, one may count the case of fully non-compatible boundary conditions as a special one. But since our experiments show significant effect due to the incompletely known Poisson ratio way before incompressibility limit is reached, it is relevant to take into account these effects also if the boundary conditions are "almost non-compatible". For example, it is natural to await strong sensitivity of solutions with respect to ν if the Neumann (or Robin) conditions are set on a small part of the boundary and non-homogeneous Dirichlet conditions are set on the remainder part of the boundary.

Chapter 6
Overview of Other Results and Open Problems

Abstract This chapter presents an overview of results related to error control methods, which were not considered in previous chapters. In the first part, we discuss possible extensions of the theory exposed in Chaps. 3 and 4 to nonconforming approximations and certain classes of nonlinear problems. Also, we shortly discuss some results related to explicit evaluation of modeling errors. The remaining part of the chapter is devoted to a posteriori estimates of errors in iteration methods. Certainly, the overview is not complete. A posteriori error estimation methods are far from having been fully explored and this subject contains many unsolved problems and open questions, some of which we formulate in the last section.

6.1 Error Estimates for Approximations Violating Conformity

Nonconforming approximations (see Appendix B) are widely used in modern numerical analysis. This fact is motivated by the following reasons.

- If a problem has a complicated geometry, then it may be difficult to create meshes (and the corresponding sets of trial functions) that exactly satisfy the prescribed boundary conditions.
- In conforming methods, the meshes must satisfy additional conditions (e.g., the so-called "hanging nodes" are forbidden). This fact may seriously hamper mesh generation procedures.
- Very often nonconforming methods offer more efficient approximations of exact solutions.

The reader will find many results related to a posteriori error control of nonconforming approximations in Sects. B.4.6 and B.4.7. Here, we discuss how numerical results computed by nonconforming approximations can be verified within the framework of the above-discussed a posteriori estimates if a suitable post-processing procedure P (projecting approximations to the energy space) is known. If we use a posteriori estimates of the functional type (which do not exploit special properties of approximations), then the corresponding error can be evaluated by the same method. Indeed, let $\widehat{v} \in \widehat{V}$ be a nonconforming approximation of the solution $u \in V_0 + u_0$ to $\Delta u + f = 0$ (we recall that V_0 contains functions vanishing on Γ_N and u_0 defines

O. Mali et al., *Accuracy Verification Methods*,
Computational Methods in Applied Sciences 32, DOI 10.1007/978-94-007-7581-7_6,

the boundary condition on Γ). For any $v \in V_0 + u_0$, we have the triangle inequality

$$\|[\nabla(u - \widehat{v})]\| \leq \|[\nabla(u - v)]\| + \|[\nabla(\widehat{v} - v)]\|, \quad \forall v \in V_0 + u_0. \tag{6.1}$$

Here $\|[\cdot]\|$ is the "broken norm" defined on subsets of Ω where the conformity condition holds (e.g., $\|[\nabla w]\|^2 = \sum_i \|\nabla w\|^2_{\Omega_i}$, where Ω_i are non-overlapping subsets of Ω). Since $u - v \in V_0$, the broken norm of $u - v$ coincides with the usual norm $\|\nabla(u - v)\|$, for which we can apply the error majorant $\overline{\mathsf{M}}$. Then,

$$\|[\nabla(u - \widehat{v})]\| \leq \inf_{\substack{v \in V_0 + u_0 \\ y \in H(\Omega, \mathrm{div})}} \left\{\overline{\mathsf{M}}(v, y) + \|[\nabla(\widehat{v} - v)]\|\right\}. \tag{6.2}$$

If $v = u$ and $y = \nabla u$ then $\overline{\mathsf{M}}(v, y) = 0$. Thus, the estimate has no gap. In practice, we set $v = \widetilde{v} := \mathsf{P}_{V_0 + u_0}(\widehat{v})$, and replace (6.2) by

$$\|[\nabla(u - \widehat{v})]\| \leq \overline{\mathsf{M}}(\widetilde{v}, y) + \|[\nabla(\widehat{v} - v)]\|, \tag{6.3}$$

where $y \in H(\Omega, \mathrm{div})$ should be defined by reconstruction of the numerical flux (or by minimization of the right-hand side of (6.2)).

Lower bounds of the approximation errors are derived analogously. By the evident inequality

$$\|[\nabla(u - \widehat{v})]\| \geq \|[\nabla(u - v)]\| - \|[\nabla(\widehat{v} - v)]\|, \quad \forall v \in V_0 + u_0, \tag{6.4}$$

we find that

$$\|[\nabla(u - \widehat{v})]\| \geq \sup_{\substack{v \in V_0 + u_0 \\ w \in V_0}} \left\{\underline{\mathsf{M}}(v, w) - \|[\nabla(\widehat{v} - v)]\|\right\} \tag{6.5}$$

and, therefore,

$$\|[\nabla(u - \widehat{v})]\| \geq \underline{\mathsf{M}}(\widetilde{v}, w) - \|[\nabla(\widehat{v} - v)]\|. \tag{6.6}$$

In [RSS03], this way was used in order to obtain error estimates for approximations violating the main (Dirichlét) boundary conditions. A consequent discussion and numerical examples for DG approximations can be found in [LRT09, TR09]. Error estimates for the finite volume method were derived and numerically tested in [CDNR09]. Another method based on Helmgholtz decomposition is presented in [Rep08, RT11].

6.2 Linear Elliptic Equations

At present, guaranteed and computable bounds are known for a much wider set of problems than those considered in Chaps. 3 and 4. Below, we shortly overview some of these results. Boundary-value problems related to high order differential operators are interesting from the mathematical viewpoint and arise in numerous applications. Numerical analysis of these problems is often more complicated than of problems generated by the harmonic operator and other differential operators of the second order. For this class of problems, residual-based error indicators were derived in several publications (see, e.g., [CDP97] and the literature cited in this paper). Functional type error majorants have been first derived by the variational technique in [NR01] for the following problem: Find a function $u = u(x)$ such that

$$\nabla \cdot \nabla \cdot (B\nabla\nabla u) = f \quad \text{in } \Omega,$$
$$u = \frac{\partial u}{\partial \nu} = 0 \quad \text{on } \partial\Omega,$$

where $B \in \mathcal{L}(\mathrm{M}_s^2, \mathrm{M}_s^2)$ is a positive operator having certain symmetry properties. By a non-variational technique, the same result was obtained in [Rep08]. The biharmonic equation $\Delta\Delta u = f$ is a special case of this problem (estimates for this problem have been studied in [Fro04b]). The corresponding error estimates are considered in Sect. 4.1.4 (see (4.48)). In [BMR08], error majorants have been derived for a class of variational inequalities associated with differential operators of the 4th order.

Elliptic problems in exterior domains form another practically important class of problems. Here, guaranteed and computable bounds of deviations from the exact solutions can be derived by transformations of the corresponding integral identity. However, on this way some specific features of problems in exterior domains must be taken into account. In this case, the estimates are derived in terms of weighted Sobolev spaces (see [PR09]).

6.3 Time-Dependent Problems

Consider the classical initial-boundary value problem for the heat equation: find $u(x,t)$ such that

$$u_t - \Delta u = f \quad (x,t) \in Q_T, \tag{6.7}$$

$$u(x,0) = \phi, \quad x \in \Omega, \tag{6.8}$$

$$u(x,t) = 0, \quad (x,t) \in S_T, \tag{6.9}$$

where $Q_T := \Omega \times (0,T)$ is the space-time cylinder and $S_T := \Gamma \times [0,T]$. This problem is one of the simplest evolutionary problems which is often used to model various diffusion type processes (e.g., heat transfer). A posteriori error estimates for this and other parabolic type problems were obtained by many authors within the framework of the residual method (e.g., see [SV08, SW10, Ver98, Ver03, Ver05]) or dual-weighted residual method (see [FNP09, Ran02, Ran00, RV10] and the references therein).

The method is based on transformations of integral identities which was applied to the problem (6.7)–(6.9) in [Rep02a], where two forms of the error majorant were derived. The simplest one has the following form:

$$\begin{aligned}|||w|||^2_{(1,2-\delta)} \leq \int_{\Omega} \left|v(x,0)-\phi(x)\right|^2 \mathrm{d}x \\ +\frac{1}{\delta}\int_{Q_T}\Big((1+\beta)|y-\nabla v|^2 \\ +C^2_{F\Omega}\Big(1+\frac{1}{\beta}\Big)|f-v_t+\operatorname{div} y|^2\Big)\,\mathrm{d}x\,\mathrm{d}t,\end{aligned} \tag{6.10}$$

where

$$|||w|||^2_{(\kappa,\varepsilon)} := \kappa\left\|w(\cdot,T)\right\|^2_{\Omega} + \varepsilon\|\nabla w\|^2_{Q_T}, \qquad \mu,\varepsilon > 0.$$

Here $y \in Y_{\text{div}}(Q_T) = \{y \in L^2(Q_T,\mathbb{R}^d), \operatorname{div} y \in L^2(\Omega) \text{ for a.e., } t \in [0,T]\}$ and β is a positive bounded function of t. A modification of this estimate valid for a wider class of problems was tested in [GR05]. In [NR10b], this technique was applied to the evolutionary Stokes problem. Error majorants for the wave equation were derived in [Rep09a] (the method relies on analysis of the corresponding integral identity and Gronwall inequalities). By the same method, error majorants were derived for a class of hyperbolic Maxwell equations (see [PRR11]).

6.4 Optimal Control and Inverse Problems

Estimates of deviations from exact solutions of partial differential equations open new ways of error estimation for some classes of optimal control problems.

Let $\psi \in L^\infty(\Omega)$, $\sigma^d \in L^2(\Omega,\mathbb{R}^d)$, $f \in L^2(\Omega)$ be given functions, and

$$K := \left\{v \in L^2(\Omega) \mid v \leq \psi \text{ a.e. in } \Omega\right\}.$$

The goal is to find a control function $u \in K$ and a state function η_u defined by the boundary-value problem

$$-\Delta\eta_v = v + f \quad \text{in } \Omega, \tag{6.11}$$

$$v = 0 \quad \text{on } \Gamma \tag{6.12}$$

such that the cost functional

$$J_1(\eta,v) := \frac{1}{2}\left\|\nabla\eta - \sigma^d\right\|^2 + \frac{a}{2}\left\|v-u^d\right\|^2, \quad a>0,$$

attains its minimal value $J(\eta_u,u)$.

Another version of the problem is generated by the functional

$$J_2(\eta,v) := \frac{1}{2}\left\|\eta_v - \eta^d\right\|^2 + \frac{a}{2}\left\|v-u^d\right\|^2,$$

where $\eta^d \in L^2(\Omega)$.

Adaptive methods and a posteriori error indicators were intensively studied in the last decade mainly in the context of residual type and goal-oriented approaches (see, e.g., [BKR00, BR01, GH08, HH08, HHIK08, HK10, VW08] and numerous references cited in these publications). In [GHR07, Rep08], it was shown that with the help of error majorants these optimal control problems can be transformed into new forms, in which the differential equations are taken into account in a weak sense (as penalties). In this way, fully computable, guaranteed, and two-sided estimates for the cost functional were deduced. Also, in these publications the convergence of the respective minimizing sequence to the optimal solution was proved and two-sided error bounds in terms of the combined (state-control) norm

$$|[u-v]|^2 := \frac{1}{2}\|\nabla(\eta_u - \eta_v)\|^2 + \frac{a}{2}\|u-v\|^2$$

were obtained.

Finally, we note that in [RR12] it was shown that minimization of error majorants with special conditions imposed on the variables lead to solutions of inverse problems associated with the corresponding differential equation.

6.5 Nonlinear Boundary Value Problems

A unified theory of a posteriori error estimation exists for the class of variational problems

$$\inf_{v\in V} J(v, \Lambda v), \quad J(v) := G(\Lambda v) + F(v),\ \forall v \in V, \tag{6.13}$$

where G and F are convex continuous functionals, V is a reflexive Banach space and Λ is a linear continuous operator which maps V to another reflexive Banach space Y. In the special case

$$G(y) = \frac{1}{2}(\mathcal{A}y, y), \qquad F(v) = \langle \ell, v\rangle,$$

the problem (6.13) leads to a quadratic type functional. We assume that the operator Λ is coercive on V and J is coercive on V, i.e.,

$$\|\Lambda w\|_Y \geq c\|w\|_V, \quad \forall v \in V, \tag{6.14}$$

$$J(w, \Lambda w) \to +\infty \quad \text{as } \|w\|_V \to +\infty. \tag{6.15}$$

In this case, the problem (6.13) has a solution u.

The operator $\Lambda^* : V^* \to Y^*$ satisfying the relation

$$\langle\!\langle y^*, \Lambda w\rangle\!\rangle = \langle \Lambda^* y, w\rangle, \quad \forall w \in V \tag{6.16}$$

is conjugate to Λ. Here, Y^* is the space topologically dual to Y with the respective duality pairing $\langle\!\langle y^*, y\rangle\!\rangle$ and the pairing of $v \in V$ and $v^* \in V^*$ is denoted by $\langle v^*, v\rangle$.

Let G^* denote the Fenchel conjugate of G (cf. Appendix A), which is defined by the relation

$$G^*(y^*) = \sup_{y \in Y}(\langle\langle y^*, y\rangle\rangle - G(y)).$$

If the functionals G and G^* are uniformly convex with the forcing functionals Φ_δ and $\Phi^*_{\delta^*}$, respectively (see Sect. A.4), then the following general estimate holds (see [NR04, Rep97a, Rep99a, Rep00b, RX97]):

$$\Phi_\delta\left(\frac{\Lambda(v-u)}{2}\right) + \Phi^*_{\delta^*}\left(\frac{y^* - p^*}{2}\right) \leq \frac{1}{2}\left(D_G(\Lambda v, y^*) + D_F(v, \Lambda^* y^*)\right), \tag{6.17}$$

where

$$D_F(v, \Lambda^* y^*) := F(v) + F^*(-\Lambda^* y^*) + \langle \Lambda^* y^*, v\rangle \tag{6.18}$$

and

$$D_G(\Lambda v, y^*) := G(\Lambda v) + G^*(y^*) - \langle\langle y^*, y\rangle\rangle \tag{6.19}$$

are the so called *compound* functionals D_F and D_G. These functionals are nonnegative. Moreover, the relation

$$F(v) + F^*(-\Lambda^* y^*) + \langle \Lambda^* y^*, v\rangle = 0 \tag{6.20}$$

is equivalent to

$$-\Lambda^* y^* \in \partial F(v) \tag{6.21}$$

and

$$G(\Lambda v) + G^*(y^*) - \langle\langle y^*, \Lambda v\rangle\rangle = 0 \tag{6.22}$$

is equivalent to

$$y^* \in \partial G(\Lambda v). \tag{6.23}$$

We recall that (6.21) and (6.23) are the *duality relations*, which hold if and only if v and y^* coincide with u and p^*, respectively. Therefore, the right-hand side of (6.17) vanishes if and only if v and y^* coincide with the corresponding exact solutions. A systematic exposition of this approach to a posteriori error estimation based upon convex duality method can be found in [NR04, Rep08].

6.5.1 Variational Inequalities

An overview of basic facts related to variational inequalities is contained in Appendix B. We follow the notation used there and introduce a bilinear form $a : V \times V \to \mathbb{R}$, and a convex continuous functional $j : V \to \mathbb{R}$ defined on a Banach space V.

Consider the following problem (variational inequality): Find $u \in K$ such that

$$a(u, w-u) + j(w) - j(u) \geq \langle \ell, w-u \rangle$$

for any $w \in K$, where K is a convex closed subset of V and $\ell \in V^*$. Let

$$a(u, w) = \int_\Omega A\nabla u \cdot \nabla w \, dx, \qquad j = 0, \qquad \langle \ell, w \rangle := \int_\Omega f w \, dx, \quad f \in L^2(\Omega),$$

and

$$K = K_{\phi\psi} := \left\{ v \in V_0 := \overset{\circ}{H}{}^1(\Omega) \mid \phi(x) \leq v(x) \leq \psi(x) \text{ a.e. in } \Omega \right\},$$

where ϕ, $\psi \in H^2(\Omega)$ are two given functions. In this case, we arrive at the classical obstacle problem, which is a typical representative of one class of variational inequalities (cf. Sect. B.3.3). This problem has a variational setting: Find $u \in V_0 + u_0$ such that

$$J(u) = \inf_{w \in V_0 + u_0} J(w), \quad J(w) = \frac{1}{2} a(w, w) - \int_\Omega f w \, dx.$$

The generalized solution of the problem satisfies the variational inequality

$$\int_\Omega A\nabla u \cdot \nabla(w-u) \, dx \geq \int_\Omega f(w-u) \, dx, \quad \forall w \in K_{\phi\psi}, \tag{6.24}$$

and generates the following three sets:

$$\begin{aligned}
\Omega^u_\oplus &:= \left\{ x \in \Omega \mid u(x) = \psi(x) \right\} \quad \text{(upper coincidence set)}, \\
\Omega^u_\ominus &:= \left\{ x \in \Omega \mid u(x) = \phi(x) \right\} \quad \text{(lower coincidence set)}, \\
\Omega^u_0 &:= \left\{ x \in \Omega \mid \phi(x) < u(x) < \psi(x) \right\}.
\end{aligned}$$

Here Ω^u_0 is an open set, where a solution satisfies the differential equation. The configuration of Ω^u_0 is not known a priori, so that this problem contains unknown free boundaries. Moreover, u has a limited regularity $u \in H^2(\Omega)$ even for smooth data (in the best case $u \in W^{2,\infty}$) and unknown free boundaries.

A priori estimates of approximation errors generated by low order finite element approximations of obstacle problems were derived in [Fal74]. Various residual type estimates and error indicators can be found in [Bra05, CN00, HK94, Kor96], and estimates based on a version of the hypercircle estimate and equilibration are discussed in [BHS08]. For a systematic overview of numerical methods and error estimates for contact type problems, we refer the reader to [Woh11].

Guaranteed error majorants can be derived by the two methods, which we have earlier discussed in the context of linear problems, namely, by a variational method based on specially constructed perturbed problems and by nonvariational method based on direct analysis of (6.24).

The variational method uses a generalized version of the Mikhlin's estimate

$$|||v-u||| \le J(v) - \inf \mathcal{P} \le J(v) - J^*(\tau^*), \tag{6.25}$$

where $J^* : Y^* \to \mathbb{R}$ is the functional of the corresponding dual problem. It is worth noting that unlike the linear case, (6.25) holds as an inequality. The main difficulty in the derivation of a posteriori estimates using (6.25) is that J^* does not have an explicit form. To overcome it, we consider the so-called *perturbed functional*

$$J_\lambda(v) := J(v) - \int_\Omega \lambda \cdot (\mathbf{v} - \Phi)\, dx,$$

where

$$\begin{aligned} &\Phi = (\phi, -\psi), \\ &\mathbf{v} = (v, -v), \\ &\lambda \in L_\oplus := \big\{(\lambda_1, \lambda_2) \parallel \lambda_i \in L_2(\Omega), \lambda_i(x) \ge 0 \text{ a.e. in } \Omega\big\}. \end{aligned}$$

It is easy to see that

$$\sup_{\lambda \in L_\oplus} J_\lambda(v) = J(v) - \inf_{\lambda \in L_\oplus} \int_\Omega \lambda \cdot (\mathbf{v} - \Phi)\, dx = \begin{cases} J(v) & \text{if } v \in K_{\phi,\psi}, \\ +\infty & \text{if } v \notin K_{\phi,\psi}. \end{cases}$$

Thus, we arrive at the following perturbed *Problem* $\mathcal{P}_\lambda$: Find $u_\lambda \in V_0$ such that

$$J_\lambda(u_\lambda) = \inf_{v \in V_0} J_\lambda(v) := \inf \mathcal{P}_\lambda. \tag{6.26}$$

Since

$$\inf_{v \in V_0} J_\lambda(v) \le \inf_{v \in K_{\phi,\psi}} J_\lambda(v) = \inf_{v \in K_{\phi,\psi}} J(v) = \inf \mathcal{P},$$

we see that $\inf \mathcal{P}_\lambda \le \inf \mathcal{P}$ for any $\lambda \in L_\oplus$ and, therefore,

$$\frac{1}{2} |||v-u|||^2 \le J(v) - \inf \mathcal{P}_\lambda. \tag{6.27}$$

To estimate the right-hand side of (6.27), we use the problem dual to $\mathcal{P}_\lambda$. For a given λ, the problem $\mathcal{P}_\lambda$ is a quadratic problem so that the corresponding dual functional is easy to construct. As a result, we can deduce an upper bound, which is valid for any $\lambda \in L_\oplus$. A special choice of λ leads to the following estimate (see the details in [BR00, NR04, Rep00a]):

$$\begin{aligned} \big|\big|\big|\nabla(u-v)\big|\big|\big| \le & \int_\Omega \big(A\nabla v \cdot \nabla v + A^{-1} y \cdot y - 2y \cdot \nabla v\big)\, dx \\ & + C_{F\Omega} \big\|[f + \operatorname{div} y]_v\big\|, \end{aligned} \tag{6.28}$$

where $C_{F\Omega}$ is the constant in (A.27) for functions vanishing on Γ (or an upper bound of this constant) and

$$[f + \operatorname{div} y]_v := \begin{cases} (f + \operatorname{div} y)_\ominus & \text{for a.e. } x \in \Omega^v_\oplus, \\ f + \operatorname{div} y & \text{for a.e. } x \in \Omega^v_0, \\ (f + \operatorname{div} y)_\oplus & \text{for a.e. } x \in \Omega^v_\ominus. \end{cases}$$

We outline that this estimate does not require an information on the structure of $\Omega^u_\oplus$ and $\Omega^u_\ominus$ and instead operates with the sets $\Omega^v_\oplus$ and $\Omega^v_\ominus$ (which are known).

It is easy to see that the right-hand side of (6.28) vanishes if and only if v coincides with u. Indeed, assume that it is equal to zero. Then,

$$\begin{aligned} y - A\nabla v &= 0, \\ (f + \operatorname{div} y)_\ominus &= 0 \quad \text{for a.e. } x \in \Omega^v_\oplus, \\ f + \operatorname{div} y &= 0 \quad \text{for a.e. } x \in \Omega^v_0, \\ (f + \operatorname{div} y)_\oplus &= 0 \quad \text{for a.e. } x \in \Omega^v_\ominus \end{aligned}$$

and

$$\begin{aligned} \int_\Omega A\nabla v \cdot \nabla(v - w)\, dx &= \int_\Omega y \cdot \nabla(v - w)\, dx \\ &= \int_\Omega (\operatorname{div} y + f)(w - v)\, dx + \int_\Omega f(v - w)\, dx \\ &= \int_{\Omega_\oplus} (f + \operatorname{div} y)_\oplus (w - v)\, dx \\ &\quad + \int_{\Omega_\ominus} (f + \operatorname{div} y)_\ominus (w - v)\, dx + \int_\Omega f(v - w)\, dx \\ &\leq \int_\Omega f(v - w)\, dx, \quad \forall w \in K_{\phi\psi}. \end{aligned}$$

In other words, the majorant vanishes if and only if $v = u$ and y is the exact flux $A\nabla u$.

The estimate (6.28) can also be deduced directly from (6.24) and represented in the form, where $C_{F\Omega}$ is replaced by a collection of Poincare constants associated with subdomains (cf. Sect. 3.2.6) (see [Rep08, Rep09b]).

In [RV08], the reader will find estimates of this type for variational inequalities generated by nonlinear boundary conditions and in [BMR08, FR10, FR06, NR04, Rep09b] for variational inequalities generated by other nonlinear problems.

6.5.2 Elastoplasticity

Various variational statements motivated by plasticity and nonlinear elasticity are generated by the energy functional

$$J(v) = \int_{\Omega} g\big(\varepsilon(v)\big)\,\mathrm{d}x + \int_{\Omega} f v\,\mathrm{d}x + \int_{\Gamma_N} F v\,\mathrm{d}s, \tag{6.29}$$

where $g = \sigma(\varepsilon(v)) : \varepsilon(v)$ is the internal energy function and f and F are the volume and surface forces, respectively. Consider the class of models, which are presented by the relations

$$\operatorname{div}\sigma + f = 0, \qquad \text{in } \Omega, \tag{6.30}$$

$$\sigma = \Phi'\big(\varepsilon(v)\big), \qquad \text{in } \Omega, \tag{6.31}$$

$$u = u_0 \qquad \text{on } \Gamma_D, \tag{6.32}$$

$$\sigma n = F \qquad \text{on } \Gamma_N, \tag{6.33}$$

where n is the outward unit normal vector to the boundary $\partial\Omega$, ε is the tensor of small strains and the media is described by a convex potential Φ. If $\Phi = \mathbb{L}\varepsilon : \varepsilon$, then the system is a linear elastic problem. Other cases are related to various nonlinear models. In particular, models in the deformation plasticity theory of isotropic solid bodies are generated by the constitutive relation

$$\sigma = K_0 \operatorname{tr}(\varepsilon)\mathbb{I} + \gamma\big(\big|\varepsilon^D\big(u(x)\big)\big|\big)\varepsilon^D\big(u(x)\big), \tag{6.34}$$

where

$$\gamma(t) = \begin{cases} 2\mu & \text{if } t \le t_0 = k_*/\sqrt{2}\mu, \\ (2\mu - \delta)t_0 t^{-1} + \delta & \text{if } t > t_0, \end{cases}$$

$k_* > 0$ is the shear stress constant, K_0 and μ are positive (elasticity) constants, and $\delta > 0$ is the hardening parameter. Let $f \in L^2(\Omega, \mathbb{R}^d)$, $F \in L^2(\Gamma_2\Omega, \mathbb{R}^d)$, $u_0 \in H^1(\Omega, \mathbb{R}^d)$, and $V_0 := \{v \in H^1(\Omega, \mathbb{R}^d) \mid v = 0 \text{ on } \Gamma_D\}$. The variational problem associated with (6.34) is to find a displacement vector $u \in V_0 + u_0$ such that

$$J(u) = \inf_{v \in V_0 + u_0} J(v), \tag{6.35}$$

where

$$g(\varepsilon) := K_0 \operatorname{tr}(\varepsilon)^2 + \delta\big|\varepsilon^D\big|^2 + \left(1 - \frac{\delta}{2\mu}\right)\phi\big(\big|\varepsilon^D\big|\big),$$

and

$$\phi(t) = \begin{cases} \mu t^2 & \text{if } |t| \le t_0 = k_*/\sqrt{2}\mu, \\ k_*(\sqrt{2}t - k_*/2\mu) & \text{if } |t| > t_0. \end{cases}$$

It is easy to see that the functional (6.35) is a particular case of the functional $J(v) = G(\Lambda v) + F(v)$ and G is a uniformly convex functional. Thus, the corresponding errors are subject to (6.17). The reader can find a more detailed analysis of a posteriori error estimates for nonlinear variational problems related to deformation theory of hardening elasto-plastic materials in [RX96] and [RV09]. For numerical methods related to the problem, see, e.g., [BMR12, NW03, ST87] and references therein.

6.5.3 Problems with Power Growth Energy Functionals

Power growth functionals are also representable in the form $J(v) = G(\Lambda v) + F(v)$. They generate an important class of nonlinear problems. Consider the simplest case, in which $G(y) = \frac{1}{\alpha}\int_\Omega |y|^\alpha\,dx$ and $F(v) = \int_\Omega f v\,dx$. Then, the problem is to minimize the functional

$$I_\alpha(v) := \int_\Omega \left(\frac{1}{\alpha}|\nabla v|^\alpha + f v\right) dx, \quad \alpha > 1,$$

over the space $V = \{v \in H^\alpha(\Omega) \mid v = 0 \text{ on } \partial\Omega\}$. The derivation of error estimates is based upon Clarkson's inequalities, which guarantee uniform convexity of $G(y)$. For $\alpha \geq 2$, this fact follows from the first Clarkson's inequality (see [Sob50])

$$\int_\Omega \left|\frac{y_1 + y_2}{2}\right|^\alpha dx + \int_\Omega \left|\frac{y_1 - y_2}{2}\right|^\alpha dx \leq \frac{1}{2}\int_\Omega \left(|y_1|^\alpha + |y_2|^\alpha\right) dx, \tag{6.36}$$

which is valid for all $y_1, y_2 \in Y$ (thus, in this case, it is convenient to write estimates in terms of the gradient of the primal variable). In this case, we can use (6.17) with $\Phi(z) = \frac{1}{\alpha}\|z\|^\alpha_{\alpha,\Omega}$ (for any δ). Since F is a linear functional, the term $D_F(v, \Lambda^* y^*)$ is finite and it is equal to zero if and only if

$$y^* \in Q^*_f := \left\{y^* \in L^\alpha(\Omega), \operatorname{div} y^* + f = 0\right\}.$$

Then, we obtain the estimate

$$\frac{1}{\alpha 2^\alpha}\int_\Omega \left|\nabla(v - u)\right|^\alpha dx \leq \frac{1}{2}\left(J_\alpha(v) - I^*_\alpha(q^*)\right), \quad \forall q^* \in Q^*_f, \tag{6.37}$$

For $1 < \alpha \leq 2$, the functional J_α is also uniformly convex. This fact follows from the second Clarkson's inequality (see [Sob50])

$$\left(\int_\Omega \left(\frac{y_1 + y_2}{2}\right)^\alpha dx\right)^{1/(\alpha-1)} + \left(\int_\Omega \left(\frac{y_1 - y_2}{2}\right)^\alpha dx\right)^{1/(\alpha-1)}$$
$$\leq \left(\frac{1}{2}\int_\Omega \left(|y_1|^\alpha + |y_2|^\alpha\right) dx\right)^{1/(\alpha-1)}. \tag{6.38}$$

However, in this case, the functional Φ_δ depends on the radius δ of a ball $\mathfrak{B}(0_Y, \delta)$ that contains y_1 and y_2 (see [MM80]), so that (6.17) holds with the forcing functional $\Phi_\delta(z) = \delta^{(\alpha-2/\alpha-1)}\kappa\|z\|_{\alpha,\Omega}^{\alpha/(\alpha-1)}$ (where $\kappa = \frac{1}{\kappa_0+1}$ and κ_0 is the integer part of $\frac{1}{\alpha-1}$) provided that ∇v lies in the corresponding ball. For the case $1 < \alpha \leq 2$, an unconditional estimate of the deviation from the exact solution can be obtained for the dual variable (see [BR07]).

6.6 Modeling Errors

We have already discussed origins of modeling errors in Sect. 1.2. Errors caused by an incomplete knowledge of data (see Chap. 5) can be viewed as a special and important class of modeling errors [MR10, MR08].

Modeling errors may be very essential and, therefore, in reliable modeling we must take them into account. In particular, modeling errors establish a limit beyond which decreasing of approximation errors is senseless.

In contrast to approximation errors (which were the focus of numerous studies in the last 40–50 years), the amount of publications related to modeling errors is not that large. We do not aim to present an overview of the results and mention just several publications, which give an insight into recent advances and provide a link to other papers. In [BBRR07, HN96, JSV10, MRV09, OBN$^+$05, OPHK01, RV12, RWW10, SO97], the authors present a combined study of the accuracy as a result of several factors (numerical approximations, integrations, randomness in material coefficients and loads). Also, we refer to [BE03], where the authors use the concept of dual-weighted residuals in the context of modeling errors.

Estimates of deviations from exact solutions to boundary value problems (which follow from the methods of a posteriori error estimation considered in Chaps. 3 and 4 of this book) suggest a way of evaluating modeling errors explicitly. The underlying idea is rather transparent: since an estimate of the deviation from the exact solution is applicable to any admissible function, we can apply them to exact solutions of simplified models. The first result was obtained in this way in [Rep01a], where an upper bound of the difference between the exact solution of a three-dimensional elasticity problem (in a plate type domain) and the exact solution of a simplified (plane stress) model has been obtained. Also, we refer to [RSS04], where dimension reduction errors of elliptic problems in thin domains have been estimated explicitly. In [RS10a, RS10b], such estimates have been derived for the Kirchhoff–Love plate model. Estimates of modeling errors generated by linearization of nonlinear models in the theory of viscous fluids can be found in [FR10]. Two-sided estimates of modeling errors arising in homogenized models of elliptic problems with oscillating coefficients have been recently derived in [RSS12b].

Algorithm 6.1 Iteration method

Input: $x_0 \in X$ {initial approximation}, ε {accuracy}
$i = 1$
$x_1 = \mathcal{T}x_0 + g$
while $\|x_i - x_{i-1}\| > \varepsilon$ **do**
  $i = i + 1$
  $x_i = \mathcal{T}x_{i-1} + g$
end while
Output: x_i {approximate solution}

It is worth noting that simultaneous consideration of approximation and modeling errors generates conceptually new adaptive algorithms, in which values of these errors are compared on each step of refinement (related either to a mesh or to model description). Such combined modeling-discretization adaptive methods can be very efficient for complicated structures analyzed with the help of the *defeaturing* method (see, e.g., [RSS12a]).

6.7 Error Bounds for Iteration Methods

In this section, we consider iteration methods based on fixed point algorithms and the corresponding a posteriori error estimates, which are applied to algebraic problems as well as integral and differential equations.

6.7.1 General Iteration Algorithm

Consider the following general problem: Find x in a Hilbert space X such that

$$x = \mathcal{T}x + g, \tag{6.39}$$

where $\mathcal{T} : X \to X$ in a certain bounded operator and $g \in X$. The element $x_\odot \in X$ satisfying (6.39) is a *fixed point*. A natural way of solving (6.39) is to apply the iteration procedure

$$x_i = \mathcal{T}x_{i-1} + g, \quad i = 1, 2, \ldots, \tag{6.40}$$

which generates an infinite sequence $\{x_i\}_{i=1}^{\infty}$. The theory of fixed point iteration methods (see, e.g., [Col64, KF75, Zei86]) formulates conditions that guarantee convergence of this sequence to $x_\odot$.

The iteration process (6.40) generates Algorithm 6.1. If the sequence $\{x_i\}$ in (6.40) converges, (i.e., if $x_i \xrightarrow[i\to+\infty]{} x_\odot$ in X), then the sequence generated by Algorithm 6.1 is finite for any positive ε. In this case, it ends up with a certain approximation of $x_\odot$.

However, the fact that $\|x_i - x_{i-1}\| \le \varepsilon$ cannot guarantee that x_i is close to the fixed point (even if such a point indeed exists). Examples that demonstrate this fact are easy to construct.

Example 6.1 The iteration scheme

$$x_i = \sqrt{\delta + x_{i-1}^2}, \quad x_0 \in [0, +\infty), \quad i = 1, 2, \dots,$$

generates a sequence $\{x_i\}$ such that $|x_i - x_{i-1}| \to 0$. For example, if $\delta = 0.01$ and $x_0 = 0$, then after 50000 iterations we have $|x_i - x_{i-1}| = 0.000224$. However, this fact does not mean that we are close to a solution. Moreover, this sequence does not converge to any fixed point because the corresponding algebraic equation $x = \sqrt{\delta + x^2}$ has no real roots. This elementary example shows that heuristic iteration schemes similar to Algorithm 6.1 may lead to unreliable and wrong results.

6.7.2 A Priori Estimates of Errors

Reliable iteration algorithms can be constructed if the operator $\mathcal{T}$ possesses additional properties.

Assume that $\mathcal{T} : S \to S$, where S is a closed nonempty set in X and $\mathcal{T}$ is a q-*contractive* mapping, i.e.,

$$\|\mathcal{T}x - \mathcal{T}y\|_X \le q\|x - y\|_X, \quad \forall x, y \in X, \tag{6.41}$$

where $0 < q < 1$ (which is independent of x and y).

It is easy to show that the sequence $\{x_i\}$ converges to a unique fixed point $x_\odot$ (in the literature, this fact is known as the Banach fixed point theorem). We have

$$\begin{aligned}\|x_{i+1} - x_i\|_X &= \|\mathcal{T}x_i - \mathcal{T}x_{i-1}\|_X \\ &\le q\|x_i - x_{i-1}\|_X \le \dots \le q^i\|x_1 - x_0\|_X.\end{aligned} \tag{6.42}$$

For any $m > 1$, we have

$$\begin{aligned}\|x_{i+m} - x_i\|_X &\le \|x_{i+m} - x_{i+m-1}\|_X + \dots + \|x_{i+1} - x_i\|_X \\ &\le (q^{m-1} + q^{m-2} + \dots + 1)q^i\|x_1 - x_0\|_X \\ &\le \frac{q^i}{1-q}\|x_1 - x_0\|_X \xrightarrow[i\to+\infty]{} 0.\end{aligned} \tag{6.43}$$

In the view of (6.43), $\{x_i\}$ is a Cauchy sequence, which has a limit in the Banach space X.

A reliable algorithm must include a stopping criterion based on correct two-sided estimates of the error

$$e_i := \|x_i - x_\odot\|_X. \tag{6.44}$$

Taking (6.41) into account, we conclude that

$$e_i = \|\mathcal{T}x_{i-1} - \mathcal{T}x_\odot\|_X \le q e_{i-1} \le q^i e_0. \tag{6.45}$$

However, in general, e_0 is unknown. To get a computable bound, we use (6.43), tend m to infinity, and obtain an a priori estimate

$$e_i \le \frac{q^i}{1-q}\|x_1 - x_0\|_X =: \overline{\mathsf{M}}_i^0. \tag{6.46}$$

This error majorant is fully computable, but may seriously overestimate the error.

6.7.3 A Posteriori Estimates of Errors

In order to obtain a sharper majorant of the error, we set $i = 1$ in (6.43). Then, it has the form

$$\|x_{1+m} - x_1\|_X \le \frac{q}{1-q}\|x_1 - x_0\|_X, \quad \text{where } m > 1. \tag{6.47}$$

Since $x_{1+m} \xrightarrow[m\to+\infty]{} x_\odot$, we obtain

$$\|x_\odot - x_1\|_X \le \frac{q}{1-q}\|x_1 - x_0\|_X. \tag{6.48}$$

Let us consider x_{i-1} as the first element of the iteration sequence. Then, (6.48) implies the majorant

$$e_i \le \frac{q}{1-q}\|x_i - x_{i-1}\|_X =: \overline{\mathsf{M}}_i. \tag{6.49}$$

A lower bound of the error follows from the triangle inequality

$$\|x_i - x_{i+1}\|_X \le \|x_i - x_\odot\|_X + \|x_{i+1} - x_\odot\|_X \le (1+q)\|x_i - x_\odot\|_X. \tag{6.50}$$

Algorithm 6.2 Iteration algorithm with a posteriori error control

Input: $x_0 \in X$ {initial approximation}, ε {accuracy}
$i = 1,$
$x_1 = \mathcal{T}x_0 + g,$
$\overline{\mathsf{M}}_1 = \frac{q}{1-q}\|x_1 - x_0\|_X,$
$\underline{\mathsf{M}}_1 = 0$
while $\overline{\mathsf{M}}_i > \varepsilon$ **do**
  $x_i = \mathcal{T}x_{i-1} + g$
  $\overline{\mathsf{M}}_i = \frac{q}{1-q}\|x_i - x_{i-1}\|_X$
  $\underline{\mathsf{M}}_i = \frac{1}{1+q}\|x_i - x_{i+1}\|_X$
  $i = i + 1$
end while
Output: x_i {approximate solution}
  $\overline{\mathsf{M}}_i$ and $\underline{\mathsf{M}}_i$ {two-sided bounds of the error}

Thus,

$$e_i \geq \frac{1}{1+q}\|x_i - x_{i+1}\|_X =: \underline{\mathsf{M}}_i, \tag{6.51}$$

where $\underline{\mathsf{M}}_i$ is the minorant for the error. Now, (6.51) and (6.49) lead us to the following conclusion:

Two-sided bounds of the distance between x_i and $x_\odot$ can be computed using three neighboring elements of the iteration sequence, namely, x_{i-1}, x_i, and x_{i+1}.

The estimates (6.49) and (6.51) were derived by Ostrowski [Ost72]. By (6.51) and (6.49), we modify Algorithm 6.1 and obtain Algorithm 6.2, which includes two-sided estimates of the error on each step.

6.7.4 Advanced Forms of Error Bounds

The ratio $\overline{\mathsf{M}}_i/\underline{\mathsf{M}}_i$ exceeds $\frac{1+q}{1-q}$. Therefore, if q is close to 1, then the efficiency of the estimates may deteriorate. In order to compensate this effect, we can apply advanced error bounds, which use additional terms of the sequence $\{x_i\}_{i=1}^N$. For example, a simple modification of the majorant follows from the estimate

$$\|x_i - x_\odot\|_X \leq \|x_i - x_{i+1}\|_X + \|x_{i+1} - x_\odot\|_X \leq \|x_i - x_{i+1}\|_X + q\|x_i - x_\odot\|_X,$$

which yields

$$\|x_i - x_\odot\|_X \leq \frac{1}{1-q}\|x_i - x_{i+1}\|_X := \overline{\mathsf{M}}_i^1(x_i, x_{i+1}). \tag{6.52}$$

Since

$$\frac{1}{1-q}\|x_i - x_{i+1}\|_X \leq \frac{q}{1-q}\|x_{i-1} - x_i\|_X, \tag{6.53}$$

we see that $\overline{\mathsf{M}}_i^1(x_i, x_{i+1})$ is sharper than $\overline{\mathsf{M}}_i$. The same idea can be applied to subsequences of $\{x_i\}_{i=1}^N$, so that we obtain

$$\|x_i - x_\odot\|_X \leq \frac{1}{1-q^l}\|x_i - x_{i+l}\|_X := \overline{\mathsf{M}}_i^l(x_i, x_{i+l}), \quad l = 1, \ldots, L. \tag{6.54}$$

By means of three sequential elements x_i, x_{i+1}, and x_{i+2}, we deduce another estimate

$$\begin{aligned}\|x_i - x_\odot\|_X &\leq \|x_i - x_{i+1}\|_X + \|x_{i+1} - x_{i+2}\|_X + \|x_{i+2} - x_\odot\|_X \\ &\leq \|x_i - x_{i+1}\|_X + \frac{1}{1-q}\|x_{i+2} - x_{i+1}\|_X \\ &:= \overline{\mathsf{M}}_i^{1,2}(x_i, x_{i+1}, x_{i+2}). \end{aligned} \tag{6.55}$$

Error minorants can be improved by similar arguments, e.g.,

$$\|x_i - x_\odot\|_X \geq \frac{1}{1+q^l}\|x_i - x_{i+l}\|_X := \underline{\mathsf{M}}_i^l(x_i, x_{i+l}). \tag{6.56}$$

The ratio of advanced upper and lower bounds in (6.54) and (6.56) is $\frac{1+q^l}{1-q^l}$ so that they provide estimates better than (6.51) and (6.49) if q is close to 1.

These estimates and their modifications can be applied to linear and nonlinear algebraic systems, integral equations, and other problems solved by iteration methods. The major difficulty is that it is necessary to calculate a sharp upper bound of q and to prove that it is less than 1. Below we discuss several examples, which demonstrate typical behavior of error estimates.

Example 6.2 Solve the equation

$$x = \phi(x), \tag{6.57}$$

where

$$\phi(x) = \alpha x + \frac{1}{x^p},$$

$\alpha = 0.9$, and $p = 1$. The problem is to find the point of intersection of the two curves depicted in Fig. 6.1a. We apply the iteration method (with $x_0 = 1.0$) and compute

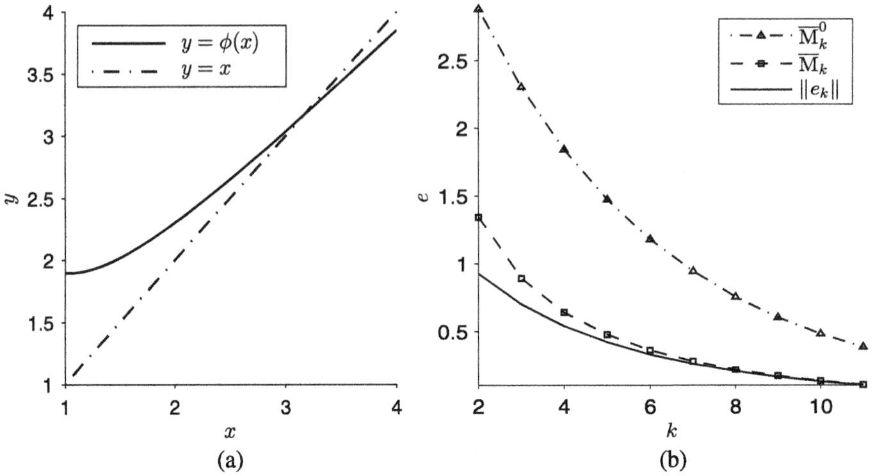

Fig. 6.1 Example 6.2. **a** Solution of equation $\phi(x) = x$. **b** A priori and a posteriori error majorants

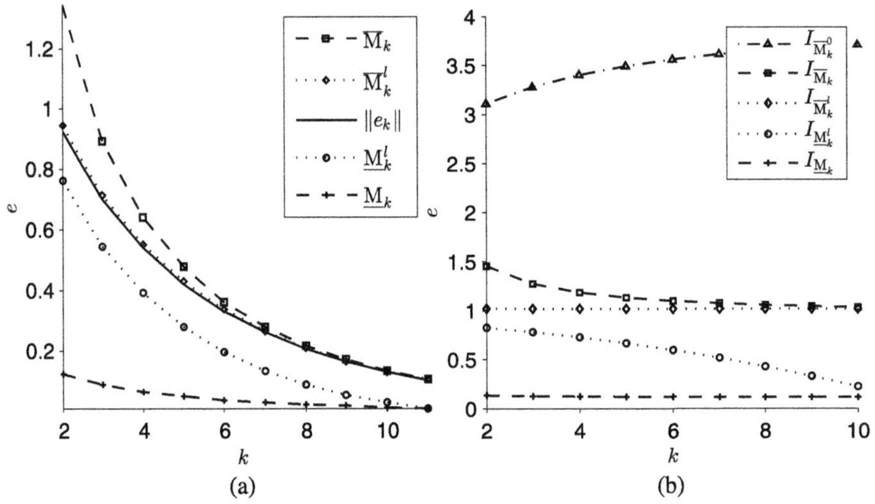

Fig. 6.2 Example 6.2. **a** The error and error bounds. **b** Efficiency indexes

different error estimates. In Fig. 6.1b, we compare a posteriori error majorant $\overline{\mathsf{M}}_k$ (6.49) with the priori error majorant $\overline{\mathsf{M}}_k^0$ (6.46). Clearly, $\overline{\mathsf{M}}_k$ provides more accurate error estimates than $\overline{\mathsf{M}}_k^0$.

Figure 6.2a illustrates the exact error together with upper and lower bounds (6.49) and (6.51), and their more advanced counterparts (6.55) and (6.56). In Fig. 6.2b, we present the corresponding efficiency indexes (i.e., values of minorants and majorants

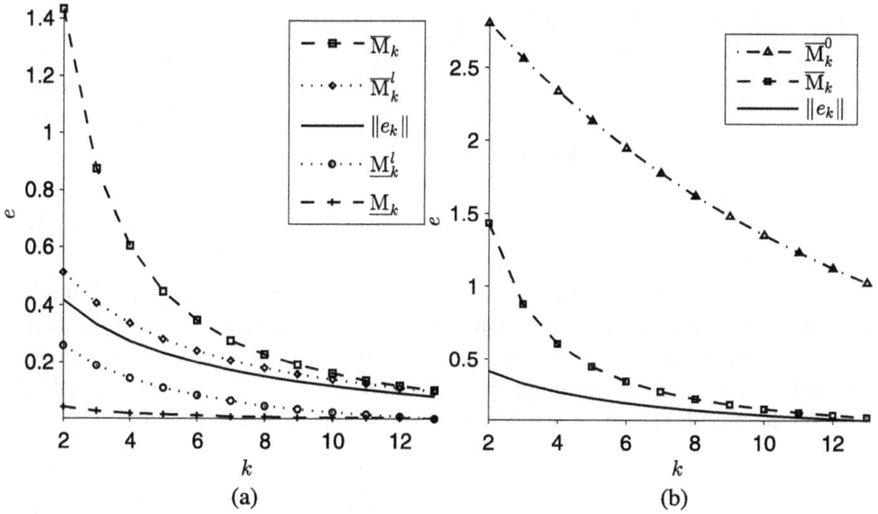

Fig. 6.3 Example 6.3. **a** The error and various a posteriori error bounds. **b** A priori and a posteriori error majorants

divided by the value of the exact error). Here, $I_{\overline{\mathsf{M}}_k^0} = \frac{\overline{\mathsf{M}}_k^0}{\|e_k\|}$, $I_{\overline{\mathsf{M}}_k} = \frac{\overline{\mathsf{M}}_k}{\|e_k\|}$, $I_{\overline{\mathsf{M}}_k^l} = \frac{\overline{\mathsf{M}}_k^l}{\|e_k\|}$, $I_{\underline{\mathsf{M}}_k^l} = \frac{\underline{\mathsf{M}}_k^l}{\|e_k\|}$, and $I_{\underline{\mathsf{M}}_k} = \frac{\underline{\mathsf{M}}_k}{\|e_k\|}$.

Example 6.3 In this example, we set

$$\phi(x) = 0.95\arctan(x) - 0.05x^2 + 0.01$$

in (6.57) and start from $x_0 = 0$. The results are shown in Fig. 6.3. As in the previous test, we see that a posteriori error bounds are more efficient than a priori bounds, and their advanced forms are even more efficient.

Example 6.4 Consider the equation

$$f(x) = 0.$$

In order to represent it in the form $x = \mathcal{T}x$, we set

$$\mathcal{T}x := x + \rho f(x), \tag{6.58}$$

where ρ is a certain parameter. Then, the iteration method has the form

$$x_i = x_{i-1} + \rho f(x_{i-1}). \tag{6.59}$$

Since $(\mathcal{T}x)' = 1 + \rho f'(x)$, we may try to provide q-contractivity of the operator $\mathcal{T}$ by selecting an appropriate ρ (for this purpose we must know two-sided bounds of the derivative in an interval containing the root).

If $\rho = -[f'(x)]^{-1}$, then we arrive at Newton's method

$$\mathcal{T}x = x - \frac{f(x)}{f'(x)}, \tag{6.60}$$

which implies the iteration scheme

$$x_i = x_{i-1} - \frac{f(x_{i-1})}{f'(x_{i-1})}. \tag{6.61}$$

It should be noted that such type methods need certain a priori assumptions. Namely, the root must be localized in some interval $[x_\ominus, x_\oplus]$, in which $f'(x) \neq 0$, starting point must satisfy the condition $f(x_0)f''(x_0) > 0$, and the corresponding operator $\mathcal{T}$ must perform a contractive mapping of this interval to itself. Sometimes, instead of (6.61) the modified Newton's method is used. It has the form

$$x_i = x_{i-1} - \frac{f(x_{i-1})}{f'(x_\sharp)}, \tag{6.62}$$

where $x_\sharp \in [x_\ominus, x_\oplus]$ is a fixed element.

Analogous methods are often used for systems of nonlinear algebraic equations and operator equations in Banach spaces. Let $A : X \to X$ be a continuous bounded operator and A' be the (Freshet) derivative of A. Then, the functional equation $Ax = 0$ can be solved by the iteration scheme $x_i = x_{i-1} - (A'(x_{i-1}))^{-1}Ax_{i-1}$, which is often called the Newton–Kantorovich method (concerning error estimates for this class of iteration methods, we refer to [Deu04, EW94, GT74, Mor89, Ort68, PP80] and the literature cited in these publications).

6.7.5 Systems of Linear Simultaneous Equations

Important applications of iteration methods are related to systems of linear simultaneous equations and other algebraic problems (see, e.g., [Axe94, Var62]).

Consider the system of linear simultaneous equations $Ax = f$ with $f \in \mathbb{R}^d$ and a non-degenerate matrix $A \in \mathbb{M}^{d \times d}$. Let

$$A = L + D + R, \tag{6.63}$$

where L and R are left and right triangular matrices, respectively, and D is a diagonal matrix. A linear iteration method has the form

$$x_{k+1} = \mathcal{L}x_k + b, \tag{6.64}$$

where $\mathcal{L}$ depends on L, D, R, and some parameters. In the examples below, we use standard schemes that can be found from [BBC94, GL96, Saa03, SG89, Ver00], and many other publications. Particular cases are the Jacobi scheme,

$$x_{k+1} = \mathcal{L}x_k + b, \quad \text{with } \mathcal{L} = D^{-1}(L + R), b = D^{-1}f, \tag{6.65}$$

the Gauss–Seidel scheme

$$\mathcal{L} = (L + D)^{-1} R, \qquad b = (L + D)^{-1} f, \tag{6.66}$$

and the so-called the SOR scheme

$$\mathcal{L} = I - \left(\frac{D}{\omega} + L\right)^{-1} A, \qquad b = \left(\frac{D}{\omega} + L\right)^{-1} f, \tag{6.67}$$

where $0 < \omega < 2$. Below, we discuss the applicability of the estimates (6.46) (6.49) to several iteration schemes of this type.

6.7.5.1 Stationary Methods

Usually, one-step iteration methods are presented in the canonical form

$$B^{k+1} \frac{x_{k+1} - x_k}{\tau_{k+1}} + A x_k = f, \quad k = 1, \ldots, n. \tag{6.68}$$

In stationary methods, $B_{k+1} = B$ and $\tau_{k+1} = \tau$. Then, (6.68) has the form

$$B \frac{x_{k+1} - x_k}{\tau} + A x_k = f \tag{6.69}$$

and the error $e_k = x_k - x_{\odot}$, $k = 1, 2, \ldots$ satisfies the equation

$$B \frac{e_k - e_{k-1}}{\tau} + A e_{k-1} = 0, \tag{6.70}$$

which yields the relation

$$e_k = \left(I - \tau B^{-1} A\right) e_{k-1}. \tag{6.71}$$

Hence, convergence of the iteration process depends on the properties of the transition matrix $\mathcal{L} := I - \tau B^{-1} A$. In view of the Banach theorem, the condition $\|\mathcal{L}\| < 1$ guarantees that the iteration sequence converges for any x_0.

Assume that A and B are positive definite and

$$\lambda_{\min}(Bx, x) \le (Ax, x) \le \lambda_{\max}(Bx, x), \quad \lambda_{\min}, \lambda_{\max} > 0, \forall x \in X. \tag{6.72}$$

Let $\kappa = \frac{\lambda_{\min}}{\lambda_{\max}}$, and $\tau = \frac{2}{\lambda_{\min} + \lambda_{\max}}$. Then, $q = \frac{1-\kappa}{1+\kappa}$ (see [BBC94, GL96, SG89, Saa03, Ver00])) and we can use the estimates

$$\|x_k - x_{\odot}\|_A \le q^k \|x_0 - x_{\odot}\|_A, \tag{6.73}$$

$$\|x_k - x_{\odot}\|_A = \|e_k\|_A \le \frac{q}{1-q} \|e_k - e_{k-1}\|_A = \frac{q}{1-q} \|x_k - x_{k-1}\|_A, \tag{6.74}$$

Algorithm 6.3 Stationary iteration method with a priori and a posteriori bounds

Input: $x_0 \in X$ {initial approximation}, ε {accuracy}
$\tau = \frac{2}{\lambda_{\max}(B^{-1}A)+\lambda_{\min}(B^{-1}A)}$
$\kappa = \frac{\lambda_{\min}(B^{-1}A)}{\lambda_{\max}(B^{-1}A)}$
$q = \frac{1-\kappa}{1+\kappa}$
$\mathcal{L} = I - \tau B^{-1}A$
$b = \tau B^{-1} f$
$k = 1$
$x_k = \mathcal{L}x_{k-1} + b$
$\overline{\mathsf{M}}_1 = \frac{q}{1-q}\|x_1 - x_0\|_A$
$\overline{\mathsf{M}}_1^0 = \frac{q^k}{1-q}\|x_1 - x_0\|_A$
while $\overline{\mathsf{M}}_k > \varepsilon$ **do**
  $k = k + 1$
  $x_k = \mathcal{L}x_{k-1} + b$
  $\overline{\mathsf{M}}_k^0 = \frac{q^k}{1-q}\|x_1 - x_0\|_A$ {a priori estimate for $\|e_k\|_A$}
  $\overline{\mathsf{M}}_k = \frac{q}{1-q}\|x_k - x_{k-1}\|_A$ {a posteriori estimate for $\|e_k\|_A$}
end while
Output: x_k {approximate solution}
  $\overline{\mathsf{M}}_k^0$, $\overline{\mathsf{M}}_k$ {a priori and a posteriori error bounds}

$$\|x_k - x_\odot\|_A \leq \frac{q^k}{1-q}\|x_1 - x_0\|_A, \tag{6.75}$$

where $\|x\|_A := \sqrt{(Ax, x)}$. The corresponding one-step stationary iteration scheme with guaranteed bounds is presented by Algorithm 6.3, and Example 6.5 is based on this algorithm.

Remark 6.1 If $\lambda_{\min}$ and $\lambda_{\max}$ are unknown, then we need computable estimates of them. Methods developed for this purpose can be found in [Kol11, Yam80, Yam82, Yam01, ZSM05].

Example 6.5 Let $A = c_1 Q^T DQ + c_2 D$, where c_1 and c_2 are constants, Q is a randomly generated unitary matrix, and $D = \{d_{ij}\}$, $d_{ij} = 0$ if $i \neq j$, $d_{ii} = i$. Typical results are presented in Table 6.1. Here $\overline{\mathsf{M}}_k^0$ is a priori error estimate, $\overline{\mathsf{M}}_k$ and $\underline{\mathsf{M}}_k$ are a posteriori error estimates, $\overline{\mathsf{M}}_k^l$ and $\underline{\mathsf{M}}_k^l$ are advanced a posteriori estimates), and, finally, $\|e_k\|$ is the exact error at step k of the iteration (for the case $c_1 = 1$, $c_2 = 10$, and $N = 10$). Efficiency indexes of the estimates are presented in Table 6.2. For $N = 100$ the results are presented in Table 6.3 and Table 6.4. They show that $\overline{\mathsf{M}}_k^l$ and $\underline{\mathsf{M}}_k^l$ efficiently estimate the error. We can not say the same about $\overline{\mathsf{M}}_k^0$.

Table 6.1 Example 6.5 ($N = 10$). Error estimates and exact error

k	$\underline{\mathsf{M}}_k$	$\underline{\mathsf{M}}_k^l$	$\|e_k\|$	$\overline{\mathsf{M}}_k^l$	$\overline{\mathsf{M}}_k$	$\overline{\mathsf{M}}_k^0$
2	5.67e-03	5.74e-02	5.74e-02	5.74e-02	1.10e-01	1.83e-01
4	3.93e-03	4.01e-02	4.01e-02	4.01e-02	6.74e-02	1.46e-01
8	2.03e-03	1.97e-02	1.97e-02	1.97e-02	3.39e-02	9.20e-02
16	5.87e-04	4.78e-03	4.78e-03	4.78e-03	9.63e-03	3.66e-02
32	6.71e-05	2.85e-04	2.85e-04	2.85e-04	1.12e-03	5.82e-03
64	1.53e-06	2.17e-06	2.17e-06	2.17e-06	2.71e-05	1.47e-04
128	9.69e-10	1.15e-09	1.31e-09	1.31e-09	1.73e-08	9.31e-08

Table 6.2 Example 6.5 ($N = 10$). Efficiency indexes

k	$I_{\underline{\mathsf{M}}_k}$	$I_{\underline{\mathsf{M}}_k^l}$	$I_{\overline{\mathsf{M}}_k^l}$	$I_{\overline{\mathsf{M}}_k}$	$I_{\overline{\mathsf{M}}_k^0}$
2	0.10	1.00	1.00	1.91	3.19
4	0.10	1.00	1.00	1.68	3.63
8	0.10	1.00	1.00	1.72	4.66
16	0.12	1.00	1.00	2.01	7.66
32	0.24	1.00	1.00	3.95	20.44
64	0.71	1.00	1.00	12.52	67.67
128	0.74	0.88	1.00	13.24	71.23

6.7.5.2 The Chebyshev Method

The Chebyshev method is one of the iteration schemes with variable steps. It has the form

$$\frac{x_{k+1} - x_k}{\tau_{k+1}} + Ax_k = f, \tag{6.76}$$

where

$$\tau_k = \frac{\tau_0}{1 + \rho_0 t_k}, \qquad k = 1, \ldots, n. \tag{6.77}$$

Here, n is the amount of Chebyshev parameters $t_k = \cos\frac{(2k-1)\pi}{2n}$ (which are introduced to minimize the error at every iteration step; see, e.g., p. 109 in [SG89] and [Saa03]),

$$\tau_0 = \frac{2}{\lambda_{\min} + \lambda_{\max}}, \qquad \rho_0 = \frac{1-\kappa}{1+\kappa}, \quad \text{and} \quad \kappa = \frac{\lambda_{\min}}{\lambda_{\max}}. \tag{6.78}$$

The iteration procedure can be represented in the form

$$x^{(n)}_{(l+1)} = \mathcal{L}x^{(n)}_l + b, \quad l = 1, 2, \ldots, \tag{6.79}$$

Table 6.3 Example 6.5 ($N = 100$). Error estimates and exact error

k	$\underline{\mathrm{M}}_k$	$\underline{\mathrm{M}}_k^l$	$\|e_k\|$	$\overline{\mathrm{M}}_k^l$	$\overline{\mathrm{M}}_k$	$\overline{\mathrm{M}}_k^0$
2	2.01e-03	4.34e-02	4.34e-02	4.34e-02	7.44e-01	8.28e-01
4	1.95e-03	4.19e-02	4.19e-02	4.19e-02	7.12e-01	8.18e-01
8	1.89e-03	3.96e-02	3.96e-02	3.96e-02	6.81e-01	8.00e-01
16	1.80e-03	3.60e-02	3.60e-02	3.60e-02	6.42e-01	7.65e-01
32	1.64e-03	3.06e-02	3.06e-02	3.06e-02	5.83e-01	7.00e-01
64	1.37e-03	2.30e-02	2.30e-02	2.30e-02	4.86e-01	5.85e-01
128	9.56e-04	1.39e-02	1.39e-02	1.39e-02	3.39e-01	4.09e-01
256	4.67e-04	5.59e-03	5.59e-03	5.59e-03	1.66e-01	2.00e-01
512	1.12e-04	1.09e-03	1.09e-03	1.09e-03	3.97e-02	4.78e-02
1024	6.37e-06	5.32e-05	5.32e-05	5.32e-05	2.27e-03	2.73e-03
2048	2.08e-08	1.64e-07	1.64e-07	1.64e-07	7.41e-06	8.93e-06

Table 6.4 Example 6.5 ($N = 100$). Efficiency indexes

k	$I_{\underline{\mathrm{M}}_k}$	$I_{\underline{\mathrm{M}}_k^l}$	$I_{\overline{\mathrm{M}}_k^l}$	$I_{\overline{\mathrm{M}}_k}$	$I_{\overline{\mathrm{M}}_k^0}$
2	0.05	1.00	1.00	17.15	19.07
4	0.05	1.00	1.00	17.00	19.53
8	0.05	1.00	1.00	17.22	20.23
16	0.05	1.00	1.00	17.86	21.28
32	0.05	1.00	1.00	19.07	22.89
64	0.06	1.00	1.00	21.11	25.43
128	0.07	1.00	1.00	24.44	29.48
256	0.08	1.00	1.00	29.69	35.82
512	0.10	1.00	1.00	36.51	44.04
1024	0.12	1.00	1.00	42.60	51.36
2048	0.13	1.00	1.00	45.15	54.42

where $x_l^{(n)}$ denotes the version obtained at the end of the cycle 'l', and

$$\mathcal{L} = (I - \tau_n A) \cdot \dots \cdot (I - \tau_1 A), \qquad b = \left(\sum_{i=2}^{n} \prod_{m=i}^{n} (I - \tau_m A) + 1 \right) f. \tag{6.80}$$

Here, n reminds about the amount of iterations on the internal subcycles. In this case, the estimate (6.46) reads as follows:

$$\left\| x_l^{(n)} - x_\odot \right\| \leq \left(q^{(n)} \right)^l \|x_0 - x_\odot\|, \tag{6.81}$$

Algorithm 6.4 Chebyshev iteration algorithm with a priori and a posteriori bounds

Input: $x_0 \in X$ {initial approximation}, ε {accuracy}, n {number of sub-cycles}
$\tau_0 = \frac{2}{\lambda_{\max}(A)+\lambda_{\min}(A)}$
$\kappa = \frac{\lambda_{\min}(A)}{\lambda_{\max}(A)}$
$\rho_0 = \frac{1-\kappa}{1+\kappa}$
$\rho_1 = \frac{1-\sqrt{\kappa}}{1+\sqrt{\kappa}}$
$q(n) = \frac{2\rho_1^n}{1+\rho_1^{2n}}$
$x_1 =$ chebyshev-subcycle($x_0, n, A, f, \tau_0, \rho_0$) {Apply Algorithm 6.5}
$i = 1$
$\overline{\mathsf{M}}_i = \frac{q}{1-q}\|x_i - x_{i-1}\|_A$ {a posteriori estimate}
$\overline{\mathsf{M}}_i^0 = \frac{q^i}{1-q}\|x_1 - x_0\|_A$ {a priori estimate}
while $\overline{\mathsf{M}}_i > \varepsilon$ **do**
  $i = i + 1$
  $x_i =$ chebyshev-subcycle ($x_{i-1}, n, A, f, \tau_0, \rho_0$) {Apply Algorithm 6.5}
  $\overline{\mathsf{M}}_i^0 = \frac{q^i}{1-q}\|x_1 - x_0\|_A$
  $\overline{\mathsf{M}}_i = \frac{q}{1-q}\|x_i - x_{i-1}\|_A$
end while
Output: x_i {approximate solution}
  $\overline{\mathsf{M}}_i^0, \overline{\mathsf{M}}_i$ {a priori and a posteriori error bounds}

Algorithm 6.5 Chebyshev sub-cycle algorithm

Input: x_0 {input vector}, n {number of sub-cycles}, A, f, τ_0, ρ_0
$\mathcal{L} = I$
$b = \bar{0}$
for $k = 1 : n$ **do**
  $t_k = \cos\frac{(2k-1)\pi}{2n}$
  $\tau_k = \frac{\tau_0}{1+\rho_0 t_k}$
  $\mathcal{L} = (I - \tau_k A)\mathcal{L}$
  $b = (I - \tau_k A)b + \tau_k f$
end for
$x = \mathcal{L}x_0 + b$
Output: x {output vector}

where

$$q^{(n)} = \frac{2\rho_1^n}{1+\rho_1^{2n}} \quad \text{and} \quad \rho_1 = \frac{1-\sqrt{\kappa}}{1+\sqrt{\kappa}}. \tag{6.82}$$

Algorithm 6.4 (together with Algorithm 6.5) formalizes the iteration schemes based on the Chebyshev method.

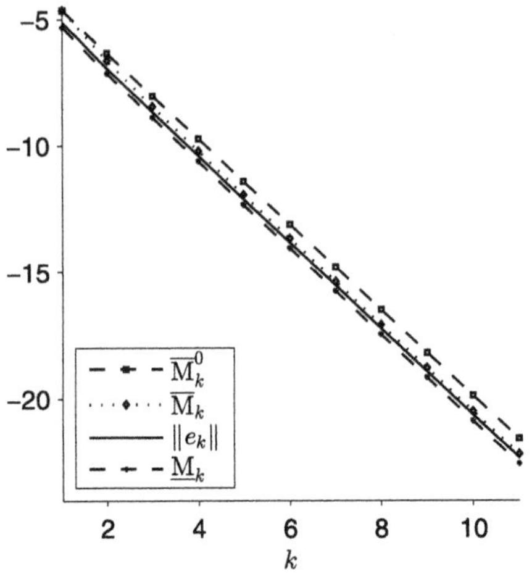

(a) N = 100

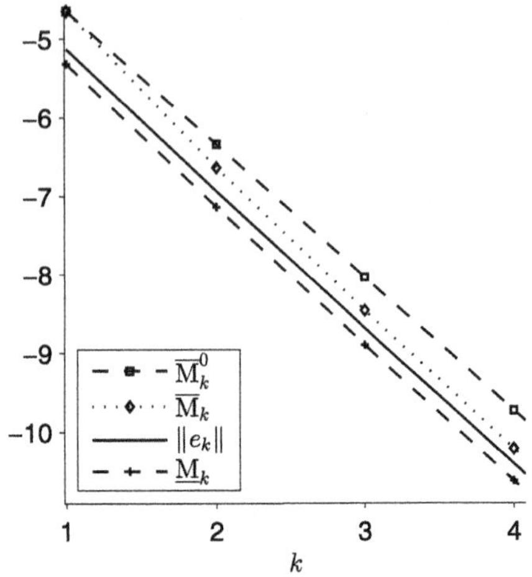

(b) N = 100

Fig. 6.4 Example 6.6. **a** The a priori estimate $\overline{M}_k^0$, the a posteriori estimate $\overline{M}_k$, the true error $\|e_k\|$, and the lower estimate $\underline{M}_k$. **b** Zoomed interval of a priori estimate $\overline{M}_k^0$, the a posteriori estimate $\overline{M}_k$, the true error $\|e_k\|$, and the lower estimate $\underline{M}_k$

Example 6.6 In this example, the set of matrixes was generated by the same procedure as in Example 6.5.

For each system, the set of accessorial parameters (i.e., the size of the Chebyshev subcycle) is set $n = 2k + 1$, where $k = 2, 3$, as before $N = 10^k$. The set of accessorial parameters for each system (the size of the Chebyshev sub-cycle) is set by the formula $n = 2k + 1$, where $k = 2, 3$, respectively. Systems with sizes $N = 10^k$, k defining the dimension $N = 10^k$, $k = 2, 3$.

In Figs. 6.4a and 6.5a, the estimates are depicted in the logarithmic scale. We see that the Chebychev method is more efficient than the stationary method considered in the previous example, and a posteriori majorants and minorants provide a correct presentation on the error.

Finally, we note that more information on a posteriori error control for various problems solved by iteration methods can be found in, e.g., [ES00, GTG76, Hay79, Leo94, Mey92, Pot85, Qn00, SO00, TO76], and in the literature cited therein.

6.7.6 Ordinary Differential Equations

Iteration methods can be used for computing approximate solutions of ordinary differential equations. In this case, a differential problem is presented in the form of an integral equation. If one can prove that the integral operator is subject to the conditions of the Banach theorem, then the results of the general theory (existence, convergence, and error estimates) are directly applicable.

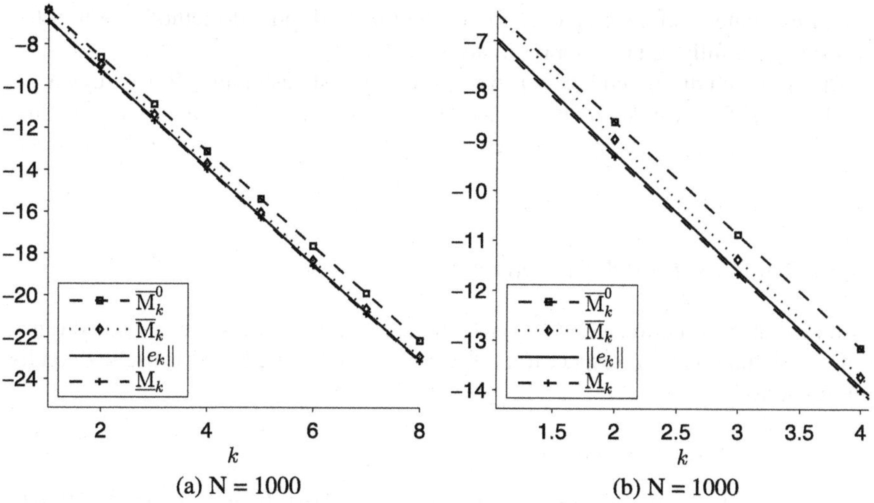

Fig. 6.5 Example 6.6. **a** The a priori estimate $\overline{\mathsf{M}}_k^0$, the a posteriori estimate $\overline{\mathsf{M}}_k$, the true error $\|e_k\|$, and the lower estimate $\underline{\mathsf{M}}_k$. **b** Zoomed interval of a priori estimate $\overline{\mathsf{M}}_k^0$, the a posteriori estimate $\overline{\mathsf{M}}_k$, the true error $\|e_k\|$, and the lower estimate $\underline{\mathsf{M}}_k$

In this section, we discuss a version of the Picard–Lindelöf method (see, e.g., [CL55, HNW93]), which is used to solve the Cauchy problem

$$\frac{\mathrm{d}u}{\mathrm{d}t} = \varphi\big(u(t), t\big), \qquad u(t_0) = a_0, \tag{6.83}$$

where the solution $u(t)$ (which may be a scalar or vector function) must be found in the interval $[t_0, t_K]$.

The Picard–Lindelöf method can be used not only for ODEs but also for time-dependent algebraic and functional equations (see, e.g., [Nev89a] and [Nev89b], where it was shown that the speed of convergence is independent of the step sizes). Numerical methods based on Picard–Lindelöf iterations for dynamical processes (the so-called waveform relaxation in the context of electrical networks) are discussed in [EKK+95]. A posteriori estimates and nodal superconvergence for time stepping methods were studied in [AMN11, MN06].

The existence and uniqueness of $u(t)$ follow from the Picard–Lindelöf theorem and the Picard's existence theorem or from the Cauchy–Lipschitz theorem (see pp. 1–15 in [CL55, Lin94]).

The problem (6.83) can be solved numerically by various well-known methods (e.g., by the methods of Runge–Kutta and Adams, see, e.g., [But03, PTVF07]). Typically, these methods are furnished by a priori asymptotic estimates, which show theoretical properties of the numerical algorithm. However, these estimates have mainly a qualitative meaning and do not provide information about the exact error bounds for particular numerical approximation. In this section, we deduce such type of estimates and discuss a version of the Picard–Lindelöf method as a tool for constructing a fully reliable approximation of (6.83).

The approach discussed below is based on Ostrowski estimates. It was suggested in [MNR12]. The corresponding numerical algorithm includes adaptation of the integration step and provides guaranteed bounds for the accuracy on the time interval $[t_0, t_K]$.

6.7.6.1 The Picard–Lindelöf Method

Assume that the function $\varphi(\xi(t), t)$ (which, in general, can be a vector-valued function) is continuous with respect to both variables and satisfies the Lipschitz condition in the form

$$\begin{aligned} &\left\|\varphi(u_2, t_2) - \varphi(u_1, t_1)\right\|_{C([t_1,t_2])} \\ &\quad \leq \mathrm{L}_1 \|u_2 - u_1\|_{C([t_1,t_2])} + \mathrm{L}_2 |t_2 - t_1|, \quad \forall (u_1, t_1), (u_2, t_2) \in Q, \end{aligned} \tag{6.84}$$

where $\mathrm{L}_1, \mathrm{L}_2$ are Lipschitz constants, and

$$Q := \left\{(\xi, t) \mid \xi \in \mathrm{U}, t_0 \leq t \leq t_N\right\}. \tag{6.85}$$

Here, U is the set of possible values of u (the information about this set comes from an a priori analysis of the problem).

In the Picard–Lindelöf method, we represent the differential equation in the integral form

$$u(t) = \int_{t_0}^{t} \varphi\big(u(s), s\big)\, \mathrm{d}s + a_0. \tag{6.86}$$

Now, the exact solution is a fixed point of (6.86), which can be found by the iteration method

$$u_j(t) = \int_{t_0}^{t} \varphi\big(u_{j-1}(s), s\big)\, \mathrm{d}s + a_0. \tag{6.87}$$

We write this relation in the form $u_j = \mathcal{T} u_{j-1} + a_0$, where $\mathcal{T} : X \to X$ is the integral operator. It is easy to see that the operator

$$\mathcal{T} u := \int_{t_k}^{t} \varphi\big(u(s), s\big)\, \mathrm{d}s \tag{6.88}$$

is q-contractive on $I_k = [t_k, t_{k+1}]$ (with respect to the uniform metric)

$$q := \mathrm{L}_1 (t_{k+1} - t_k) < 1. \tag{6.89}$$

Therefore, if the interval $[t_k, t_{k+1}]$ is sufficiently small, then the solution can be found by the iteration procedure (henceforth, it is called the Adaptive Picard–Lindelöf (APL) method) and the corresponding errors can be controlled by the Ostrowski estimates

$$
\begin{aligned}
\underline{\mathsf{M}}_j &:= \frac{1}{1+q}\|u_j - u_{j+1}\|_{C(I_k)} \leq \|u - u_j\|_{C(I_k)} \\
&\leq \frac{q}{1-q}\|u_j - u_{j-1}\|_{C(I_k)} =: \overline{\mathsf{M}}_j. \qquad (6.90)
\end{aligned}
$$

However, this theoretically simple scheme contains serious technical difficulties. Let $\mathcal{F}_K = \bigcup_{k=0}^{K-1}[t_k, t_{k+1}]$ be a mesh selected on $[t_0, t_K]$. Consider one step of the APL method. Assume that u_0 is the initial approximation defined as a piecewise affine continuous function on a certain sub-mesh $\Omega_{S_k} = \bigcup_{s=0}^{S_k-1}[z_s, z_{s+1}]$ induced on the interval $[t_k, t_{k+1}]$, where $z_0 = t_k$ and $z_{S_k} = t_{k+1}$. On the first interval, we have

$$u_1(t) = \int_{t_0}^{t} \varphi\big(u_0(s), s\big)\,\mathrm{d}s + a_0, \quad t \in [t_0, t_1]. \qquad (6.91)$$

If $q < 1$ and u_1 is computed exactly, then we can measure the distance between u_1 and the exact solution u by means of the estimate

$$\big\|u_1(t) - u(t)\big\|_{C([t_0,t_1])} \leq \frac{q}{1-q}\big\|u_1(t) - u_0(t)\big\|_{C([t_0,t_1])}. \qquad (6.92)$$

However, the integration operator in (6.91) does not transfer piecewise affine functions to piecewise affine functions, so that iterations lead to piecewise polynomial functions, the order of which increases from iteration to iteration and makes the exact integration more and more difficult. Very soon we will be forced to use approximate quadrature formulas. In this case, the estimate (6.92) cannot be applied. If we wish to perform iterations within the framework of a certain finite dimensional space X_h (e.g., the space of piecewise affine functions), then additional errors caused by integration and mapping of a function to this finite dimensional space must be taken into account. This situation is illustrated in Fig. 6.6.

The operator $\mathcal{T}$ maps X_h to $Z_h \subset X$ (i.e. $x_j = \mathcal{T}x_{j-1} \in Z_h$), where Z_h doesn't coincide with X_h. Therefore, if we wish to study the iteration process as a mapping of X_h to itself, then we need to apply a certain projection (interpolation) operator π and evaluate the corresponding error. Since in practice we use a numerical approximation of $\mathcal{T}$ (which is $\mathcal{T}_\Delta : X_h \to Z_h$), the function $\widehat{x}_j$ (which is computed numerically) also contains an integration error. Similar difficulties arise in many other iteration scheme based on integral type operators, where we need to estimate the interpolation and integration errors in order to provide guaranteed error estimates.

Henceforth, we consider the situation in which X_h is the space CP^1 of piecewise affine continuous functions constructed on a local mesh with nodes z_s ($s = 0, \ldots, S_k$, $z_0 = t_0$, $z_{S_k} = t_1$). In this case, on the first step of integration we construct

$$\bar{u}_1(t) = \pi u_1 \in CP^1\big([t_0, t_1]\big), \qquad (6.93)$$

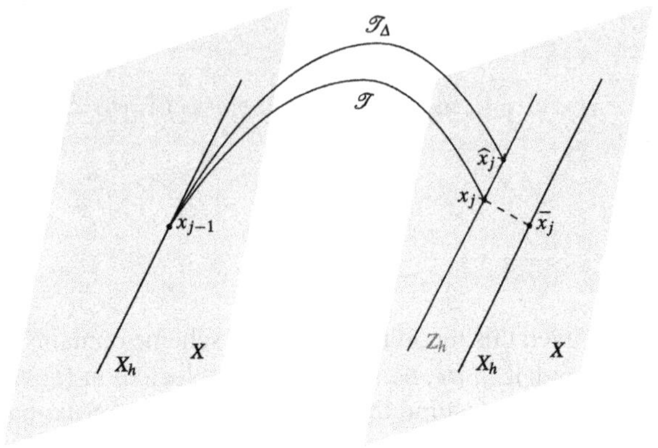

Fig. 6.6 Integration and interpolation errors generated by $\mathcal{T}$

where π is the projection operator $\pi : C \to CP^1([t_0, t_1])$ satisfying the condition $\pi u(z_s) = \bar{u}(z_s)$ at all points z_s. Thus, the norm on the right-hand side of (6.92) can be estimated as follows:

$$\begin{aligned}&\left\|u_1(t) - u_0(t)\right\|_{C([t_0,t_1])}\\ &\quad \le \left\|\bar{u}_1(t) - u_0(t)\right\|_{C([t_0,t_1])} + \left\|\bar{u}_1(t) - u_1(t)\right\|_{C([t_0 t_1])}.\end{aligned} \tag{6.94}$$

Here $\|\bar{u}_1(t) - u_1(t)\|_{C([t_0,t_1])} = \|\bar{e}_1\|_{C([t_0,t_1])}$ is the *interpolation error*. In general, this term is unknown, but we can estimate it.

Numerical integration generated errors have a different origin. In general, the values $\bar{u}(z_s)$, $s = 0, \ldots, S_k$ cannot be found exactly. Therefore, at every node z_s we have an approximate value $\widehat{u}_1(z_s)$ instead of $\bar{u}_1(z_s)$. In view of this fact, we reform (6.94) as follows:

$$\begin{aligned}\left\|u_1(t) - u_0(t)\right\|_{C([t_0,t_1])} &\le \left\|\widehat{u}_1(t) - u_0(t)\right\|_{C([t_0,t_1])} + \left\|\widehat{u}_1(t) - \bar{u}_1(t)\right\|_{C([t_0,t_1])}\\ &\quad + \left\|\bar{u}_1(t) - u_1(t)\right\|_{C([t_0,t_1])},\end{aligned} \tag{6.95}$$

where $\|\widehat{u}_1(t) - \bar{u}_1(t)\|_{C([t_0,t_1])} = \|\widehat{e}_1\|_{C([t_0,t_1])}$ is the *integration error*.

6.7.6.2 Estimates of Interpolation and Integration Errors

First, we study the difference between u_1 and $\bar{u}_1$, where $\bar{u}_1$ is the interpolant of u_1, which at $\{z_s\}_{s=0}^{S_k}$ is defined by the relation

$$\bar{u}_1(z_s) = u_1(z_s) = \int_{z_0}^{z_s} \varphi\big(u_0(t), t\big)\,\mathrm{d}t + a_0. \tag{6.96}$$

For any subinterval $z \in [z_s, z_{s+1}]$,

$$u_1(z) = u(z_s) + \int_{z_s}^{z} \varphi\big(u_0(t), t\big)\,\mathrm{d}t,$$

$$\bar{u}_1(z) = u_1(z_s) + \frac{u_1(z_{s+1}) - u_1(z_s)}{\Delta_s}(z - z_s),$$

where $\Delta_s = z_{s+1} - z_s$. Then, on this interval we have

$$\begin{aligned} \bar{e} &= \bar{u}_1(z) - u_1(z) \\ &= \frac{z - z_s}{\Delta_s} \int_{z_s}^{z_{s+1}} \varphi\big(u_0(t), t\big)\,\mathrm{d}t - \int_{z_s}^{z} \varphi\big(u_0(t), t\big)\,\mathrm{d}t. \end{aligned} \tag{6.97}$$

We recall that u_0 is an affine function, so that

$$\int_{z_s}^{z} \varphi\big(u_0(t), t\big)\,\mathrm{d}t = \int_{z_s}^{z} \varphi\left(u_{0,s} + \frac{u_{0,s+1} - u_{0,s}}{\Delta_s}(t - z_s), t\right)\mathrm{d}t. \tag{6.98}$$

Using these representations, the interpolation error can be estimated as follows (see [MNR12]):

$$\begin{aligned} &\big\|\bar{u}_1(z) - u_1(z)\big\|_{C([z_s, z_{s+1}])} \\ &\quad \le \frac{\varphi_{s+1} - \varphi_s}{8}\Delta_s + \frac{2}{3}\Delta_s\big(\mathrm{L}_{1,s}|u_{0,s+1} - u_{0,s}| + \mathrm{L}_{2,s}\Delta_s\big), \end{aligned} \tag{6.99}$$

where $\varphi_s = \varphi(u_{0,s}, z_s)$ and $\varphi_{s+1} = \varphi(u_{0,s+1}, z_{s+1})$.

Errors of numerical integration can be evaluated in a similar way. Assume that f is a Lipschitz function with the constant L. Then, we can easily find guaranteed bounds of integration errors associated with the well-known quadrature formula

$$S = \int_a^b f(x)\,\mathrm{d}x \simeq \widetilde{S} := \frac{f(a) + f(b)}{2}(b - a). \tag{6.100}$$

It is easy to see that the integration error is subject to the estimate

$$e_{int} = |S - \widetilde{S}| \le \frac{L}{4}(b - a)^2 - \frac{1}{4L}\big[f(b) - f(a)\big]^2.$$

In our case, $a = z_s$, $b = z_{s+1}$, $\mathrm{L}_s = \mathrm{L}_{1,s} l_s + \mathrm{L}_{2,s}$, and l_s is the slope of the piecewise affine function $\widehat{u}$ on every interval $[z_s, z_{s+1}]$, $s = 0, \ldots, S_{k-1}$. Thus,

$$\big\|\widehat{u}_1(t) - \bar{u}_1(t)\big\|_{C([z_s, z_{s+1}])} \le \frac{\mathrm{L}_s}{4}\Delta_s^2 - \frac{1}{4\mathrm{L}_s}[\varphi_{s+1} - \varphi_s]^2. \tag{6.101}$$

Interpolation and integration errors interact and we need to evaluate the overall impact of them accumulated in the process of integration.

On every subinterval $[z_s, z_{s+1}]$, the interpolation error can be estimated with the help of (6.99). Then, for the whole interval $[t_0, t_1] := \bigcup_{s=0}^{S_{k-1}} [z_s, z_{s+1}]$ we obtain

$$\begin{aligned}&\left\|\bar{u}_1(t)-u_1(t)\right\|_{C([t_0,t_1])}\\&\quad\le\sum_{s=0,\dots,S_k-1}\frac{\varphi_{s+1}-\varphi_s}{8}\Delta_s+\frac{2}{3}\big[\mathrm{L}_{1,s}|u_{0,s+1}-u_{0,s}|+\mathrm{L}_{2,s}\Delta_s\big]\Delta_s.\end{aligned}\tag{6.102}$$

For the error of integration, we have

$$\left\|\bar{u}_1(t)-\widehat{u}_1(t)\right\|_{C([t_0,t_1])}\le\sum_{s=0,\dots,S_{k-1}}\frac{\mathrm{L}_s}{2}\Delta_s^2-\frac{1}{2\mathrm{L}_s}[\varphi_{s+1}-\varphi_s]^2.\tag{6.103}$$

Then, (6.95) implies the estimate

$$\begin{aligned}&\left\|u_1(t)-u_0(t)\right\|_{C([t_0,t_1])}\\&\quad\le\left\|\widehat{u}_1(t)-u_0(t)\right\|_{C([t_0,t_1])}\\&\qquad+\sum_{s=0,\dots,S_{k-1}}\left(\frac{\varphi_{s+1}-\varphi_s}{8}\Delta_s+\frac{2}{3}\Delta_s\big[\mathrm{L}_{1,s}|u_{0,s+1}-u_{0,s}|+\mathrm{L}_{2,s}\Delta_s\big]\right)\\&\qquad+\sum_{s=0,\dots,S_k-1}\left(\frac{\mathrm{L}_s}{2}\Delta_s^2-\frac{1}{2\mathrm{L}_s}[\varphi_{s+1}-\varphi_s]^2\right),\end{aligned}\tag{6.104}$$

which right-hand side contains only known quantities and, therefore, can be used in the Ostrowski estimate. After j iterations we have

$$\begin{aligned}\left\|u_{j+1}(t)-u_j(t)\right\|_{C([t_0,t_1])}&\le\mathrm{M}^{\oplus,1}_{j+1}(\widehat{u}_j)\\&:=E^1_{iter}+E^1_{interp}+E^1_{integr},\end{aligned}\tag{6.105}$$

where

$$E^1_{iter}:=\left\|\widehat{u}_{j+1}(t)-\widehat{u}_j(t)\right\|_{C([t_0,t_1])},\tag{6.106}$$

$$\begin{aligned}E^1_{interp}:=\sum_{s=0,\dots,S_k-1}&\left(\frac{\varphi(\widehat{u}_{j,s+1},z_{s+1})-\varphi(\widehat{u}_{j,s},z_s)}{8}\Delta_s\right.\\&\left.+\frac{2}{3}\Delta_s\big[\mathrm{L}_{1,s}|\widehat{u}_{j,s+1}-\widehat{u}_{j,s}|+\mathrm{L}_{2,s}\Delta_s\big]\right),\end{aligned}\tag{6.107}$$

and

$$E^1_{integr}:=\sum_{s=0,\dots,S_k-1}\left(\frac{\mathrm{L}_s}{2}\Delta_s^2-\frac{1}{2\mathrm{L}_s}\big[\varphi(\widehat{u}_{j,s+1},z_{s+1})-\varphi(\widehat{u}_{j,s},z_s)\big]^2\right).\tag{6.108}$$

For $j = 0$ the function $\widehat{u}_j$ is taken as a piecewise affine interpolation of u_0, and for $j \geq 1$ it is taken from the previous iteration step. The quantity $\overline{\mathsf{M}}_j^{\oplus,1}$ is fully computable, and it shows the overall error associated with the step number j on the first interval.

Remark 6.2 The estimate of the overall error related to the interval $[t_0, t_K]$ includes all errors computed on the intervals. In other words, the error associated with $[t_0, t_{k-1}]$ is appended to the error on $[t_{k-1}, t_k]$ (formally, this rule follows from the fact that the initial condition on $[t_{k-1}, t_k]$ includes the errors on all previous intervals).

Thus, for the problem (6.83) (with the Lipschitz function φ) fully guaranteed and computable bounds of approximations can are presented by (6.105). Numerical tests discussed below were computed by S. Matculevich (see [MNR12]).

6.7.6.3 Adaptive Picard–Lindelöf Algorithm

The algorithm needs a suitable primal mesh $\mathcal{F}_K$ (should be generated a priori and changed in the iteration process). Here, we do not discuss this question in detail and only note that $\mathcal{F}_K$ should reflect the behavior of $\varphi(u(t), t)$ and requires information about U (see (6.85)). In practice, such information can be obtained in different ways (e.g., by solving the problem (6.83) numerically with the help of some other non-adaptive (e.g., Runge–Kutta) method on a coarse mesh or by an a priori analysis of the solution properties).

The APL algorithm is a cycle over all the intervals of the mesh $\mathcal{F}_K = \bigcup_{k=0}^{K-1} [t_k, t_{k+1}]$. On each subinterval, the algorithm is realized as a sub-cycle (the index of which is j). In the sub-cycle, we apply the PL method and try to find an approximation that meets the accuracy requirements (i.e., the accuracy must be smaller than ε^k). The initial data are taken from the previous step (for the first step, the initial condition is defined by a_0).

After computing an approximation on $[t_k, t_{k+1}]$, we use the majorant and find a guaranteed upper bound (which includes the interpolation and integration errors). Iterations are continued unless the required accuracy ε^k is achieved. After that we save the results and proceed to the next interval.

Note that in Algorithm 6.6, we do not discuss in detail the process of integration on an interval, which is performed on a local mesh with a certain amount of subintervals (which size is Δ_s). In principle, it may happen that the desired level of accuracy ε^k is not achieved with the Δ_s selected at some moment $t' < t_K$. This fact can be easily detected because the interpolation and integration errors will dominate and do not allow the overall error to decrease below ε^k. In this case, Δ_s must be reduced, and computations on the corresponding interval must be repeated.

Algorithm 6.6 The APL method

Input: ε {accuracy}, a_0 {initial condition}, q {contraction constant}
$\mathcal{F}_K = \bigcup_{k=0}^{K-1} [t_k, t_{k+1}]$ {initial mesh}
$\Omega_{S_k} = \bigcup_{s=0}^{S_k-1} [z_s, z_{s+1}]$ {initial submesh for each subinterval}
for $k = 0$ to $K - 1$ **do**
  $j = 1$
  do
    if $k = 0$
      $a = a_0$
    else
      $a = v^k(t_k)$
    endif
    $v_j^{k+1} = \mathit{Integration\ Procedure}(\varphi, v_{j-1}^{k+1}, S_k) + a$
    calculate E_{interp}^k and E_{integr}^k by using (6.107) and (6.108)
    $\overline{\mathsf{M}}_j^{\oplus,k+1} = E_{iter}^{k+1} + E_{interp}^{k+1} + E_{integr}^{k+1}$
    $e_j^{\oplus} = \frac{q}{1-q} \overline{\mathsf{M}}_j^{\oplus,k+1}$
    if $E_{interp}^{k+1} + E_{integr}^{k+1} > \varepsilon/K$
      $S_{k+1} = 2S_{k+1}$ {refine the mesh Ω_{S_k}}
    endif
    $j = j + 1$
  while $e_j^{\oplus} > \varepsilon/K$
  $v^{k+1} = v_j^{k+1}$ {approximate solution on $[t_{k-1}, t_k]$}
  $e^{\oplus,k+1} = e_j^{\oplus}$ {error bound for $[t_{k-1}, t_k]$}
end for
Output: $\{v^k\}_{k=0}^{K-1}$ {approximate solution}
$\{e^{\oplus,k}\}_{k=0}^{K-1}$ {error bounds on subintervals}

Algorithm 6.6 generates a piecewise linear approximation

$$v(t) := v^k(t), \quad \text{if } t \in [t_k, t_{k+1}], \quad k = \{0, 1, \dots, K-1\}$$

and a piecewise constant error bound

$$\overline{\mathsf{M}}(t) := e^{\oplus,k}, \quad \text{if } t \in [t_k, t_{k+1}], \quad k = \{0, 1, \dots, K-1\}. \tag{6.109}$$

The error bound is guaranteed. For the exact solution u holds

$$v(t) - \overline{\mathsf{M}}(t) \le u(t) \le v(t) + \overline{\mathsf{M}}(t), \quad t \in [t_0, t_K].$$

Moreover, we can improve the error bound further. If we perform l additional iterations after the stopping criteria $e_j^{\oplus} \le \varepsilon/K$ is satisfied, we can use an advanced form

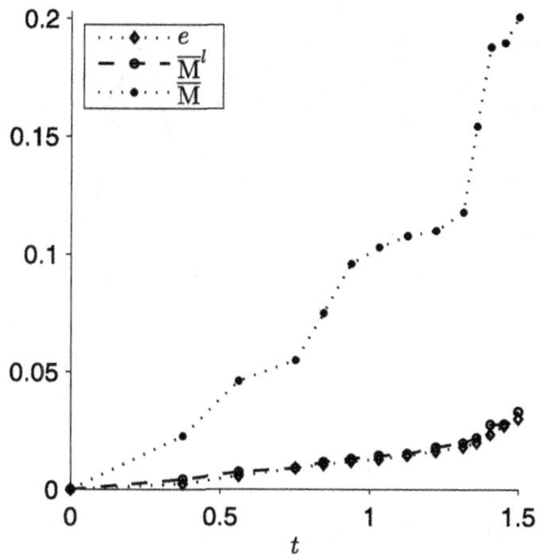

Fig. 6.7 Example 6.7. The error and the error majorants

of the estimate (6.54) instead of (6.49). We denote the error bound obtained in this manner by $\overline{\mathsf{M}}^l$. In the following examples, we select $l = 1$.

Example 6.7 Consider the problem

$$\frac{du}{dt} = 4ut\sin(8t), \quad t \in \left[0, \frac{3}{2}\right], \qquad u(0) = a_0 = 1. \tag{6.110}$$

The exact solution of this problem is $u = e^{(1/16)\sin(8t)-(1/2)t\cos(8t)}$.

In Fig. 6.7, we depict the error and error bounds generated by APL-method. In Figs. 6.8a and 6.9a, the results are presented in a different form. Here, we see the exact solution, the approximate solution computed by the APL method, and the bounds of possible variation of it. Therefore, we can guarantee that the exact solution belongs to the shadowed zone. Numerical results exposed in Figs. 6.8a and 6.9a show that the advanced majorant $\overline{\mathsf{M}}^l$ provides much sharper bounds of the deviation than the original Ostrowski estimate $\overline{\mathsf{M}}$.

Table 6.5 shows the true error, the majorant, and three components of the majorant on each time interval. In this example, the values of S_k are sufficient large, so that the interpolation and integration error estimates are insignificant with respect to the first term.

Example 6.8 This example is intended to show that the APL is applicable to stiff problems. We consider the classical stiff equation (see [HW96])

$$\frac{du}{dt} = 50\cos(t) - 50u, \quad t = [0, 1], \qquad u(0) = a_0 = 1 \tag{6.111}$$

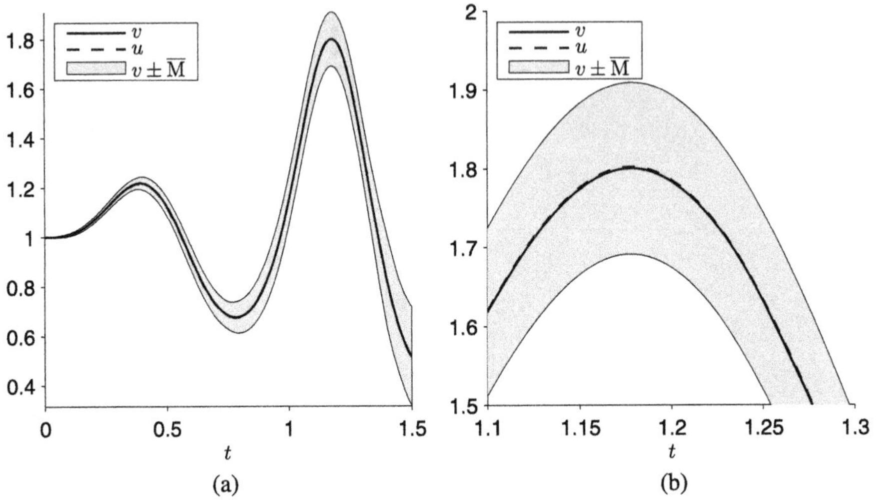

Fig. 6.8 Example 6.7. **a** The exact and approximate solutions with guaranteed bounds of the deviation computed by the Ostrowski estimate. **b** Zoomed interval with solutions and bounds of the deviation

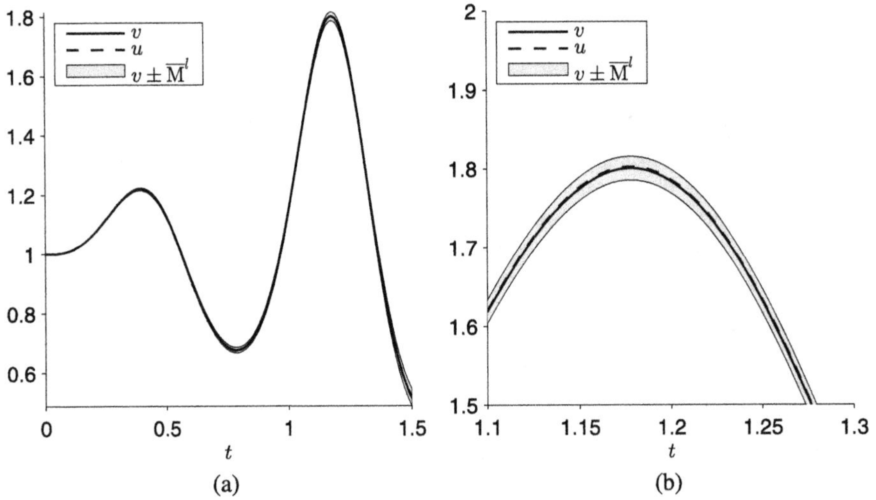

Fig. 6.9 Example 6.7. **a** The exact and approximate solutions with the guaranteed bounds of the deviation computed by the advanced error majorant. **b** Zoomed interval with solutions and bounds of the deviation

with the exact solution $u = \frac{1}{2501}e^{-50t} + \frac{2500}{2501}\cos(t) + \frac{50}{2501}\sin(t)$.

The results are exposed in Fig. 6.10a. The left picture shows the behavior of e and the corresponding majorants. We see that the usual Ostrowski estimate $\overline{\mathrm{M}}$ se-

Table 6.5 Example 6.7. The error, the majorant, and three components of the majorant related to the interval $[t_k, t_{k+1}]$, $k = \{0, 1, \ldots, K-1\}$

k	$\|e^{k+1}\|$	E^{k+1}_{iter}	E^{k+1}_{interp}	E^{k+1}_{integr}	$\overline{\mathrm{M}}^{\oplus,k+1}$
0	2.2841e-003	2.2658e-002	8.6160e-008	9.5725e-008	2.2658e-002
1	3.3368e-003	4.6095e-002	1.8847e-007	5.8148e-007	4.6095e-002
2	3.3368e-003	5.4949e-002	2.5299e-007	5.9301e-007	5.4949e-002
3	1.5150e-003	7.4818e-002	2.5768e-007	2.3618e-006	7.4818e-002
4	1.3213e-003	9.5993e-002	3.0190e-007	2.3699e-006	9.5993e-002
5	9.8338e-004	1.0302e-001	3.4216e-007	2.3807e-006	1.0302e-001
6	6.4687e-003	1.5427e-001	4.8963e-007	2.4320e-006	1.5427e-001
7	6.8425e-003	1.5647e-001	6.1877e-007	2.4999e-006	1.5647e-001
8	6.5957e-003	2.3495e-001	9.4891e-007	2.6183e-006	2.3495e-001
9	4.6256e-003	2.7145e-001	9.8935e-007	2.6328e-006	2.7145e-001
10	6.3005e-003	3.0533e-001	9.9923e-007	2.6373e-006	3.0533e-001
11	6.6933e-003	3.2838e-001	1.0158e-006	2.6404e-006	3.2838e-001
12	6.6933e-003	4.4629e-001	1.0182e-006	2.6517e-006	4.4629e-001

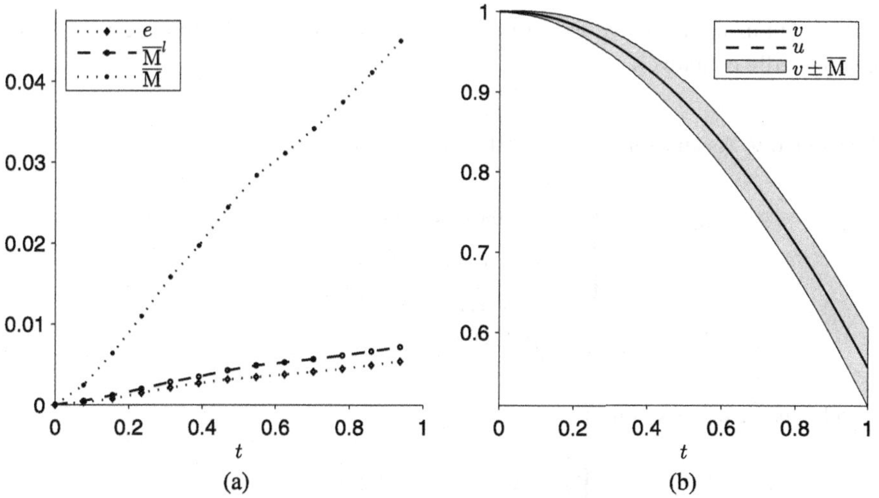

Fig. 6.10 Example 6.8. **a** The error and the error majorants. **b** The exact and the approximate solutions with guaranteed bounds of possible deviations

riously overestimates the error, but the advanced majorant $\overline{\mathrm{M}}^l$ provides the correct presentation on the error. The right picture shows the approximate solution v and the zone $v \pm \overline{\mathrm{M}}$, which contains the exact solution u. Certainly, in this test we know u and can check that it indeed belongs to this zone. However, in general u is un-

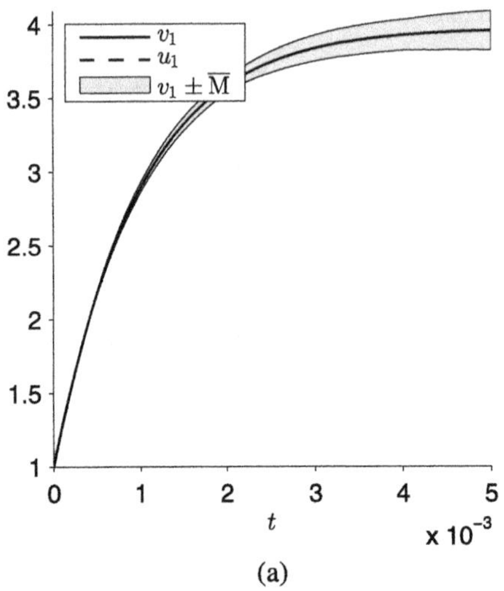

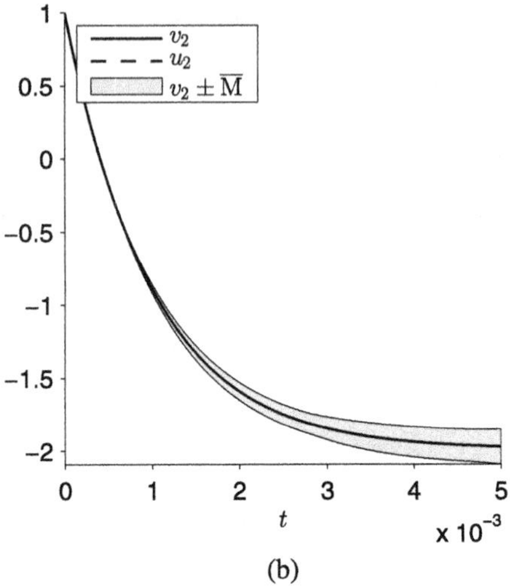

Fig. 6.11 Example 6.9. The exact and approximate solutions and guaranteed bounds of possible deviations

known, but using the above-described technique we can compute the interval which guaranteedly contains it.

Example 6.9 As an example of a stiff system, we consider

$$\begin{cases} \frac{du_1}{dt} = 998u_1 + 1998u_2, \\ \frac{du_2}{dt} = -999u_1 - 1999u_2, \\ u_1(t_0) = 1, u_2(t_0) = 1, \\ t \in [0, 5 \cdot 10^{-3}], \end{cases}$$

which has the exact solution

$$\begin{cases} u_1 = 4e^{-t} - 3e^{-1000t}, \\ u_2 = -2e^{-t} + 3e^{-1000t}. \end{cases}$$

In Figs. 6.11a, 6.11b, 6.12a and 6.12b, we present the corresponding results in the same form as in the previous example.

We note that for stiff equations getting an approximate solution with the guaranteed and sharp error bounds requires much larger expenditures than in relatively simple Examples 6.7 and 6.8. This fact is not surprising. It is clear that fully reliable computations for stiff models are much more expensive.

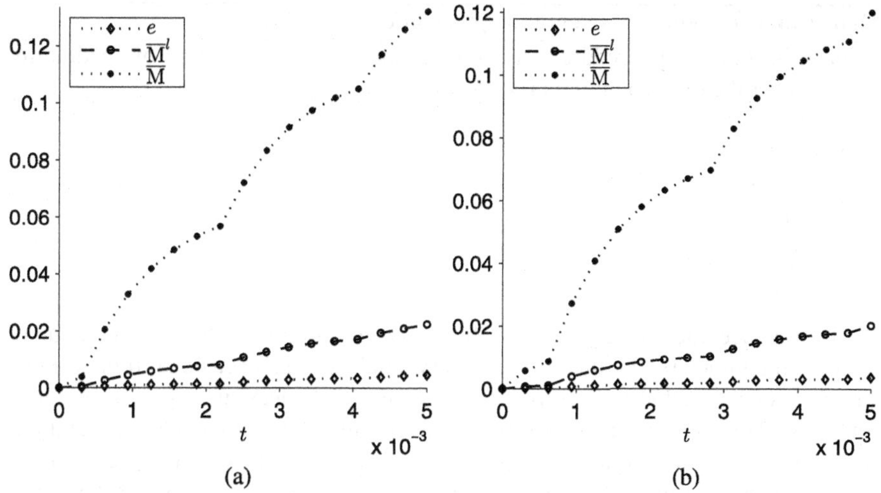

Fig. 6.12 Example 6.9. The errors and error majorants for the components u_1 (**a**) and u_2 (**b**)

6.8 Roundoff Errors

At the end of this section, we briefly discuss *roundoff errors*. They arise because numbers in computers are presented in the *floating point format*:

$$x = \pm\left(\frac{i_1}{q} + \frac{i_2}{q^2} + \cdots + \frac{i_k}{q^k}\right)q^l, \quad i_s < q.$$

These numbers form the set $\mathbb{R}_{qlk} \subset \mathbb{R}$, where q is the *base* of the representation and $l \in [l_1, l_2]$ is the *power*.

We outline that the set R_{qlk} is not closed with respect to elementary and advanced operations! For this reason, practically all operations performed on any computer generate roundoff errors. In general, the effects caused by such type errors are studied by interval analysis (see, e.g., [AH83, Var62]).

Example 6.10 (Summation of two drastically different real numbers) Assume that we use a rather primitive digital system with $k = 3$ and $q = 2$ to compute $a + b$, where $a = 1000$ and $b = 1$. In the process of summation, the smallest number must be normalized in such a way that both numbers have the same power l. Then, all nonzero digits of b will be lost and we formally obtain $a + b = a$. This example shows the difference between operations in $\mathbb{R}$ and $\mathbb{R}_{qlk}$. Certainly, modern computers operate with much better digital systems, but effects of this type may arise even in elementary operations.

Definition 6.1 The smallest floating point number, which when added to 1 gives a quantity other than 1, is the machine (computer) accuracy $\overline{h}$.

Algorithm 6.7 Numerical integration procedure

Input: a, b, n, f (procedure that computes the integrand), c (array of quadrature weights or the corresponding procedure)
$S = 0$
for $i = 1 : n$ **do**
  $S = S + c_i f(x_i)$
end for
Output: S

Machine accuracy depends on the digital system used in a particular computer. In modern computers, the quantity $\overline{h}$ is very small, so that occasional roundoff errors are usually unable to corrupt numerical results (at least in engineering computations). However, if such type errors are repeated many times (e.g., in iteration procedures), then the overall effect may be quite substantial. Numerical integration methods present examples in which roundoff errors may seriously affect the accuracy.

Example 6.11 Consider the standard numerical integration scheme

$$\int_b^a f(x)\,dx \cong \sum_{i=1}^{n} c_i f(x_i).$$

A simple method of numerical integration associated with this quadrature formula is presented in Algorithm 6.7.

Assume that we have made one half of the summations and now perform the next iteration step, which is to compute

$$\sum_{i=1}^{(1/2)n} c_i f(x_i) + c_{(1/2)n+1} f(x_{(1/2)n+1}),$$

where the first term is of the order 1 and the second one is a very small number ε. In the process of normalization, certain digits of the latter term may be lost. The smaller the integration step, the more valuable these roundoff effects could be, so that in practical computations it may happen that the overall accuracy deteriorates if the integration step goes to zero. In order to avoid such type effects, the integration process should be performed in a more sophisticated way, avoiding summation of drastically different real numbers.

6.9 Open Problems

At present, we know that the main classes of linear boundary value problems can be reliably modeled because for them we have computable and guaranteed error bounds (majorants and minorants). These two-sided estimates adequately reflect errors evaluated in terms of energy norms and other weaker measures. Moreover, they generate efficient and robust error indicators, which can be used in mesh-adaptive procedures. Also, majorants and minorants allow us to evaluate modeling errors and effects caused by incomplete knowledge of physical data. The corresponding theory has been successfully extended to nonconforming approximations of linear problems and to certain classes of convex nonlinear problems. However, in general, the theory of fully reliable computer simulation methods contains many open problems. Further development of this theory is faced with conceptual questions and challenging problems.

Below we formulate and briefly motivate eight problems. We hope that some of them will be studied and solved (at least partially) in the nearest future.

1. *Guaranteed and efficient error bounds in terms of pointwise norms.* It is well-known that a priori error estimates in terms of L^∞ type norms can be derived, provided that the exact solution possesses necessary regularity. Also, some error indicators valid for this case have been suggested. However, so far we do not know analogs of the estimates (3.28) and (3.35) (or other functional a posteriori estimates) which could present guaranteed and fully computable bounds of errors evaluated in terms of pointwise norms. A possible way of deriving such bounds may come from the theory of sub- and supersolutions, which are widely used in the theory of partial differential equations (see, e.g., [GT77]).
2. *Fully reliable numerical methods for coupled problems.* The majority of real life engineering problems are described by coupled models, which may include various differential, integral, and algebraical relations. They arise, e.g., in the theory of electro- or magneto-rheological fluids, piezoelectricity, poroelasticity, and in many other models. It is clear that trustable numerical simulations should be based on a careful analysis of interactions between different errors. This interaction together with effects caused by uncertain material data may lead to rather unpredictable results. The theory of fully reliable computer simulation for coupled models, which takes into account all these effects, is yet to be developed.
3. *Analysis of effects caused by uncertain data in nonlinear problems.* So far we have no systematic investigations of indeterminacy effects for strongly nonlinear problems. It is clear that in some cases these effects can be very significant, and reliable modeling of nonlinear models cannot be performed without quantitative analysis of indeterminacy errors.
4. *Efficient space-time error control methods for evolutionary models.* In spite of considerable progress in numerical analysis of evolutionary models, the majority of algorithms used in practice are based on heuristic grounds. The development of fully reliable integration schemes, which would be able to efficiently control accumulation of the overall error (and prevent an unacceptable growth of this error) is one of the most challenging tasks.
5. *Reliable modeling of nonlinear problems in fluid dynamics.* At this point, the Navier–Stokes (NS) equation should be mentioned first of all. Certainly, reliable numerical methods cannot be justified unless the existence and uniqueness of NS system is proved (which is one of the Millenium Problems stated by the Clay Mathematical Institute) or unless we properly define the meaning of the

word "solution". On the other hand, accurate and fully reliable numerical results performed for specially selected test problems may provide useful information for solving difficult theoretical questions. Similar questions arise in other models of viscous flow, including coupled models (e.g., in solid-fluid interaction models, generalized Newtonian fluids, phase transition models, etc.).

6. *Fully reliable methods for optimal control problems.* Partial differential equations often enter optimal control problems (in the form of state equations). It is clear that errors encompassed in solutions of state equations affect solutions of optimal control problems. This means that investigation of fully reliable methods (based on guaranteed and computable error majorants and minorants) is a challenging topic in the theory of optimal control and optimal shape design.
7. *Computable bounds of constants in embedding inequalities.* Error majorants contain different global constants, which come from embedding inequalities associated with the problem under consideration. Therefore, finding explicit bounds for such constants is an important problem. For some cases, such estimates are known (see, e.g., [AD03, Dob03, NR12, OC00, Pay07, PW60, Rep12]). However, many problems in this area are still unsolved. For example, there is no general method able to provide efficient estimates of the LBB constant for arbitrary Lipschitz domain. Such estimates are required if we wish to perform reliable modeling of problems defined on the subspace of divergence-free fields.
8. *Modeling errors and modeling-discretization adaptive methods.* It is clear that efficient and reliable numerical algorithms developed for engineering problems must be based on simultaneous analysis of approximation and modeling errors. Therefore, evaluation of modeling errors and validation of mathematical models is a fundamental problem, which nowadays is actively under study with the help of various methods and theoretical conceptions (see Sect. 6.6). We believe that the creation of new adaptive numerical methods balancing modeling errors and errors arising due to discretization, approximate integration, interpolation, etc. is a goal of utmost practical importance.

Appendix A
Mathematical Background

A.1 Vectors and Tensors

By $\mathbb{R}^d$ and $\mathbb{M}^{d\times d}$ we denote the spaces of real d-dimensional vectors and $d \times d$ matrices, respectively. The scalar product of vectors is denoted by $\cdot$, and for the product of tensors we use the symbol :, i.e.,

$$u \cdot v = u_i v_i, \qquad \tau : \sigma = \tau_{ij}\sigma_{ij},$$

where summation (from 1 to d) over repeated indices is implied. Symbol $\otimes$ stands for the dyad product of vectors $u \in \mathbb{R}^d$ and $v \in \mathbb{R}^d$, i.e.,

$$u \otimes v = u_i v_j \in \mathbb{R}^{d\times d}, \tag{A.1}$$

The norms of vectors and tensors are defined as follows:

$$|a| := \sqrt{a \cdot a}, \qquad |\sigma| := \sqrt{\sigma : \sigma}.$$

Henceforth, the symbol := means "equals by definition". The multiplication of a matrix $A \in \mathbb{M}^{d\times d}$ and a vector $b \in \mathbb{R}^d$ is a vector, which we denote by Ab. Matrices are usually denoted by capital letters (matrices associated with stresses and strains are denoted by Greek letters σ and ε). Any tensor τ is decomposed into a *deviatoric* part τ^{D} and a *trace* $\operatorname{tr}\tau := \tau_{ii}$, so that $\tau := \tau^{\mathrm{D}} + \frac{1}{d}\mathbb{I}\operatorname{tr}\tau$, where $\mathbb{I}$ is the unit tensor. It is easy to check that

$$\tau : \mathbb{I} = \operatorname{tr}\tau, \qquad \tau^{\mathrm{D}} : \mathbb{I} = 0, \tag{A.2}$$

$$|\tau|^2 = \left|\tau^{\mathrm{D}}\right|^2 + \frac{1}{d}\operatorname{tr}\tau^2, \tag{A.3}$$

so that τ is decomposed into two parts (which sometimes are called deviatorical and spherical).

For $a, b \in \mathbb{R}$ and any positive β, we have the algebraic Young's inequality

$$2ab \le \beta a^2 + \frac{1}{\beta}b^2. \tag{A.4}$$

O. Mali et al., *Accuracy Verification Methods*,
Computational Methods in Applied Sciences 32, DOI 10.1007/978-94-007-7581-7,

This inequality has a more general form

$$ab \leq \frac{1}{p}(\beta a)^p + \frac{1}{p'}\left(\frac{b}{\beta}\right)^{p'}, \quad p > 1, \frac{1}{p} + \frac{1}{p'} = 1. \tag{A.5}$$

With the help of (A.4) we find that

$$\frac{1}{1+\beta}|a|^2 - \frac{1}{\beta}|b|^2 \leq |a+b|^2 \leq (1+\beta)|a|^2 + \frac{1+\beta}{\beta}|b|^2. \tag{A.6}$$

Analogous relations hold for any pair of vectors (or tensors) a and b and any $\beta > 0$, i.e.,

$$2\tau : \sigma \leq \beta|\tau|^2 + \frac{1}{\beta}|\sigma|^2, \qquad |\tau + \sigma|^2 \leq (1+\beta)|\tau|^2 + \frac{1+\beta}{\beta}|\sigma|^2. \tag{A.7}$$

If U is a Hilbert space with scalar product $(\cdot,\cdot)$ and the respective norm $\|\cdot\|$ associated with the product, then it is easy to extend (A.4)–(A.7) to the elements of U.

Also, we use the inequality

$$(y,q) \leq \|y\|_{\mathcal{A}}\|q\|_{\mathcal{A}^{-1}}, \quad \forall y,q \in U, \tag{A.8}$$

where $\mathcal{A} : U \to U$ is a symmetric and positive definite operator,

$$\|y\|^2_{\mathcal{A}} := (\mathcal{A}y, y), \quad \text{and} \quad \|y\|^2_{\mathcal{A}^{-1}} := \left(\mathcal{A}^{-1}y, y\right).$$

Indeed, we have

$$0 \leq \left(\mathcal{A}(\mathcal{A}^{-1}y + \gamma q), \mathcal{A}^{-1}y + \gamma q\right) = \left(\mathcal{A}^{-1}y, y\right) - 2\gamma(y,q) + \gamma^2(\mathcal{A}q, q),$$

where we can set $\gamma := \frac{(\mathcal{A}^{-1}y,y)}{(y,q)}$ (we assume that $y \neq 0$ and $(y,q) \neq 0$, otherwise the inequality (A.8) is obvious) and arrive at (A.8).

A.2 Spaces of Functions

A.2.1 Lebesgue and Sobolev Spaces

We denote a bounded connected domain in $\mathbb{R}^d$ by Ω and its boundary by Γ. Everywhere in the book, we assume that Γ is Lipschitz continuous. This means that at any point $x \in \Gamma$, a certain local coordinate system can be introduced such that the boundary is presented by a one-valued function which is Lipschitz continuous at a vicinity of x. Usually, ω stands for an open subset of Ω. The closure of sets is denoted by a bar and the Lebesgue measure of a set ω by the symbol $|\omega|$.

By $L^{\alpha}(\omega)$ we denote the space of functions summable with power α endowed with norm

$$\|w\|_{\alpha,\omega} := \left(\int_{\omega} |w|^{\alpha} \, dx \right)^{1/\alpha}.$$

Also, we use the simplified notation

$$\|w\|_{\alpha} := \|w\|_{\alpha,\Omega}, \qquad \|w\| := \|w\|_{2,\Omega}.$$

The vector-valued functions with components that are square summable in Ω form the Hilbert space $L^2(\Omega, \mathbb{R}^d)$. Analogously, $L^2(\Omega, \mathbb{M}^{d\times d})$ is the Hilbert space of tensor-valued functions (sometimes we use the special notation Σ for this space). If tensor-valued functions are assumed to be symmetric, then we write $\mathbb{M}_s^{d\times d}$ (and Σ_s instead of $L^2(\Omega, \mathbb{M}_s^{d\times d})$). For $v \in L^2(\Omega, \mathbb{R}^d)$ and $\tau \in L^2(\Omega, \mathbb{M}^{d\times d})$, the norms are defined by the relations

$$\|v\|^2 := \int_{\Omega} |v|^2 \, dx \quad \text{and} \quad \|\tau\|^2 := \int_{\Omega} |\tau|^2 \, dx.$$

Similarly, for a vector-valued function $v \in L^{\alpha}(\omega)$,

$$\|v\|_{\alpha,\omega} := \left(\int_{\omega} |v|^{\alpha} \, dx \right)^{1/\alpha} \quad \text{and} \quad \|v\|_{\alpha,\Omega} = \|v\|_{\alpha} := \left(\int_{\Omega} |v|^{\alpha} \, dx \right)^{1/\alpha}. \tag{A.9}$$

The space of measurable essentially bounded functions is denoted by $L^{\infty}(\Omega)$. It is equipped with the norm

$$\|u\|_{\infty,\Omega} = \operatorname*{ess\,sup}_{x\in\Omega} |u(x)|.$$

By $\overset{\circ}{C}{}^{\infty}(\Omega)$ we denote the space of all infinitely differentiable functions with compact supports in Ω. The spaces of k-times differentiable scalar- and vector-valued functions are denoted by $C^k(\Omega)$ and $C^k(\Omega, \mathbb{R}^d)$, respectively; $\overset{\circ}{C}{}^k(\Omega)$ is the subspace of $C^k(\Omega)$ that contains functions vanishing on the boundary. $P^k(\Omega)$ denotes the set of polynomial functions defined in $\Omega \subset \mathbb{R}^d$, i.e., $v \in P^k(\Omega)$ if

$$v = \sum_{|\alpha|\le m} a_{\alpha} x^{\alpha}, \quad m \le k, a_{\alpha} = a_{\alpha_1,\dots,\alpha_d}, \tag{A.10}$$

where $\alpha := (\alpha_1, \dots, \alpha_d)$ is the so-called multi-index, $|\alpha| := \alpha_1 + \alpha_2 + \dots + \alpha_d$, and $x^{\alpha} = x^{\alpha_1} x^{\alpha_2} \cdots x^{\alpha_d}$. For partial derivatives we keep the standard notation and write $\frac{\partial f}{\partial x_i}$ or $f_{,i}$. Usually, we understood them in a generalized sense, namely, a function $g = f_{,i}$ is called a generalized derivative of $f \in L^1(\Omega)$ with respect to x_i if it satisfies the relation

$$\int_{\Omega} f w_{,i} \, dx = -\int_{\Omega} g w \, dx, \quad \forall w \in \overset{\circ}{C}{}^1(\Omega). \tag{A.11}$$

Generalized derivatives of higher orders are defined by similar integral relations.

By $\{\!|g|\!\}_S$ we denote the mean value of a function g on S, i.e.,

$$\{\!|g|\!\}_S := \frac{1}{|S|}\int_S g\,dx \quad \text{and} \quad \widetilde{g}_S := g - \{\!|g|\!\}_S.$$

Square-summable functions with zero mean form the space

$$\overset{\circ}{L}{}^2(\Omega) := \left\{q \in L^2(\Omega) \mid \{\!|q|\!\}_\Omega = 0\right\}.$$

The space $H(\Omega, \mathrm{div})$ is a subspace of $L^2(\Omega, \mathbb{R}^d)$ that contains vector-valued functions with square-summable divergence, and $H(\Omega, \mathrm{Div})$ is a subspace of Σ that contains tensor-valued functions with square-summable divergence, i.e.,

$$H(\Omega, \mathrm{div}) := \left\{v \in L^2\left(\Omega, \mathbb{R}^d\right) \mid \mathrm{div}\, v := \{v_{i,i}\} \in L^2(\Omega)\right\},$$
$$H(\Omega, \mathrm{Div}) := \left\{\tau \in L^2\left(\Omega, \mathbb{M}^{d\times d}\right) \mid \mathrm{Div}\,\tau := \{\tau_{ij,j}\} \in L^2\left(\Omega, \mathbb{R}^d\right)\right\}.$$

Both spaces $H(\Omega, \mathrm{div})$ and $H(\Omega, \mathrm{Div})$ are Hilbert spaces endowed with scalar products

$$(u, v)_{\mathrm{div}} := \int_\Omega (u \cdot v + \mathrm{div}\, u\, \mathrm{div}\, v)\, dx$$

and

$$(\sigma, \tau)_{\mathrm{Div}} := \int_\Omega (\sigma : \tau + \mathrm{Div}\,\sigma \cdot \mathrm{Div}\,\tau)\, dx,$$

respectively. The norms $\|\cdot\|_{\mathrm{div}}$ and $\|\cdot\|_{\mathrm{Div}}$ are associated with the above-defined scalar products.

Similarly, $H(\Omega, \mathrm{curl})$ is the Hilbert space of vector-valued functions having square-summable rotor, i.e.,

$$H(\Omega, \mathrm{curl}) := \left\{v \in L^2\left(\Omega, \mathbb{R}^3\right) \mid \mathrm{curl}\, v \in L^2\left(\Omega, \mathbb{R}^3\right)\right\},$$

where $\mathrm{curl}\, v := (v_{3,2} - v_{2,3};\, v_{1,3} - v_{3,1};\, v_{2,1} - v_{1,2})$. This space can be defined as the closure of smooth functions with respect to the norm

$$\|w\|_{\mathrm{curl}} := \left(\|w\|_\Omega^2 + \|\mathrm{curl}\, w\|_\Omega^2\right)^{1/2}.$$

The Sobolev spaces [Sob50] $W^{m,p}(\Omega)$ (where m and p are positive integers) contain functions summable with power p, its generalized derivatives up to order m belonging to L^p. For a function $f \in W^{m,p}(\Omega)$, the norm is defined as follows:

$$\|f\|_{m,p,\Omega} = \left(\int_\Omega \sum_{|\alpha|\le m} |D^\alpha f|^p\, dx\right)^{1/p},$$

where $D^\alpha v = \frac{\partial^{|\alpha|} v}{\partial x_1^{\alpha_1} \cdots \partial x_d^{\alpha_d}}$ is a derivative of the order $|\alpha| := \alpha_1 + \cdots + \alpha_d$.

Sobolev spaces with $p = 2$ are often denoted by the letter H (because they are Hilbert spaces), i.e.,

$$H^m(\Omega) := \{v \in L^2(\Omega) \mid D^\alpha v \in L^2(\Omega), \forall m : |\alpha| \le m\}.$$

A subset of $H^m(\Omega)$ formed by the functions vanishing on Γ is denoted by $\mathring{H}^m(\Omega)$. Relationships between the Sobolev spaces and other functional spaces (such as $L^p(\Omega)$ and $C^k(\Omega)$) are well studied. The following results (which are known in the literature as *embedding theorems*, see [GT77, Lad70, Sob50]) play an important role in analysis of PDEs. We recall that a space X is said to be continuously embedded in Y if $X \subset Y$ and a constant C exists such that

$$\|w\|_Y \le C\|w\|_X, \quad \forall w \in X.$$

This fact is usually denoted by $X \mapsto Y$. For example, it is easy to show that $L^p(\Omega) \mapsto L^q(\Omega)$, provided that $p \ge q \ge 1$.

We say that the embedding operator is compact if it maps bounded sets to compact sets (compact embedding is denoted by the symbol $\hookrightarrow$).

Theorem A.1 *Let Ω be a bounded domain in $\mathbb{R}^d$ with Lipschitz continuous boundary and $l \ge 0$ be an integer. Then,*

- *If $p, q \ge 1$, $p \le q$, and $l + \frac{n}{q} \ge \frac{n}{p}$, then $W^{l,p}(\Omega)$ is continuously embedded in $L^q(\Omega)$ provided that $l < \frac{n}{p}$. If $l = \frac{n}{p}$, then the above holds for $p < q < \infty$.*
- *if in the above $l + \frac{n}{q} > \frac{n}{p}$, then the embedding operator is compact.*
- *If $l - k > \frac{n}{p}$, then $W^{l,p}(\Omega)$ is compactly embedded in $C^k(\overline{\Omega})$.*

A.2.2 Boundary Traces

The functions in Sobolev spaces have counterparts on Γ (and on other manifolds of lower dimensions) that are associated with spaces of traces. Thus, there exist some bounded operators mapping the functions defined in Ω to functions defined on the boundary. For example, an operator $\gamma : H^1(\Omega) \to L^2(\Gamma)$ is called the *trace* operator if it satisfies the following conditions:

$$\gamma v = v|_\Gamma, \quad \forall v \in C^1(\Omega), \tag{A.12}$$

$$\|\gamma v\|_{2,\Gamma} \le c_{T\Gamma} \|v\|_{1,2,\Omega}, \tag{A.13}$$

where $c_{T\Gamma}$ is a positive constant independent of v. From these relations, we observe that γv is a natural generalization of the trace defined for a continuous function (in the pointwise sense). The image of γ is a subset of $L^2(\Gamma)$, which is the space $H^{1/2}(\Gamma)$. The functions from other Sobolev spaces are also known to have traces in Sobolev spaces with fractional indices. Thus, γ is a linear continuous operator from $H^1(\Omega)$ to $H^{1/2}(\Gamma)$ and the space $\mathring{H}^1(\Omega)$ is the kernel of γ. Also, for any $\phi \in H^{1/2}(\Gamma)$, one can define a continuation (lifting) operator $\mu : H^{1/2}(\Gamma) \to H^1(\Omega)$ such that

$$\mu\phi = w, \quad w \in H^1(\Omega), \qquad \gamma w = \phi \quad \text{on } \Gamma$$

and (e.g., see [LM68a, LM68b, LM68c])

$$\|\phi\|_{H^{1/2},\Gamma} \leq c_\gamma \|w\|_{1,2,\Omega}, \quad \|w\|_{1,2,\Omega} \leq c_\mu \|\phi\|_{H^{1/2},\Gamma}. \tag{A.14}$$

Using the operator γ, we define subspaces of functions vanishing on Γ or on some part Γ_1 of Γ. Such subspaces are marked by the zero subscript, e.g.,

$$V_0 := \{v \in V \parallel \gamma v = 0 \text{ a.e. on } \Gamma_1\}.$$

We understand the boundary values of functions in the sense of traces, so that "$u = \phi$ on Γ" means that the trace γu of a function u defined in Ω coincides with the given function ϕ defined on Γ. If for two functions u and v defined in Ω we say that u equals v on Γ, then we mean that $\gamma(u - v) = 0$ on Γ. For the sake of simplicity, we omit the trace symbol γ and write $u = v$.

Boundary conditions of boundary value problems are often written in the form

$$u = u_0 \quad \text{on } \Gamma,$$

where u_0 is a given function from V. This condition should be understood in the sense that the trace of u coincides with the trace of u_0.

A.2.3 Linear Functionals

Let V be a Banach space. A mapping $\ell : v \to \langle \ell, v \rangle \in \mathbb{R}$ (where $v \in V$) is called a linear functional if it is additive and $\langle \ell, \gamma v \rangle = \gamma \langle \ell, v \rangle$ for any $\gamma \in \mathbb{R}$. We define $|\!|\!|\ell|\!|\!| := \sup_{v \in V} \frac{\langle \ell, v \rangle}{\|v\|_V}$ and call the functional bounded if this quantity is finite. Any bounded linear functional is continuous. The set of all linear bounded functionals on V forms the dual space V^*. In view of the Riesz representation theorem, in a Hilbert space H with the scalar product $(\cdot, \cdot)$ any bounded linear functional has the form (v^*, v), where v^* is a unique element of H. Moreover, $|\!|\!|\ell|\!|\!| = \|v^*\|_H$. This result can be extended to a wide spectrum of reflexive metric spaces if we replace the scalar product by $\langle v^*, v \rangle$.

For $f \in L^2(\Omega)$, the functional

$$\langle f_{,i}, \varphi \rangle := -\int_\Omega f \frac{\partial \varphi}{\partial x_i} \, dx \tag{A.15}$$

is linear and continuous not only for functions in $\overset{\circ}{C}{}^\infty(\Omega)$ but also for all functions of the space $\overset{\circ}{H}{}^1(\Omega)$ (this fact follows from the density of smooth functions in $\overset{\circ}{H}{}^1(\Omega)$

and known theorems on the continuation of linear functionals). Such functionals can be viewed as generalized derivatives of square-summable functions. They form the space $H^{-1}(\Omega)$ dual to $\overset{\circ}{H}{}^{1}(\Omega)$. It is easy to see that the quantity

$$[\![f_{,i}]\!] := \sup_{\substack{\varphi \in \overset{\circ}{H}{}^{1}(\Omega) \\ \varphi \neq 0}} \frac{|\langle f_{,i}, \varphi \rangle|}{\|\nabla \varphi\|_{\Omega}} \tag{A.16}$$

is nonnegative and finite. It can be used to introduce the norm for $H^{-1}(\Omega)$. We can introduce other linear continuous functionals defined on $\overset{\circ}{H}{}^{1}(\Omega)$, e.g.,

$$\ell(w) = \int_{\Omega} \eta \cdot \nabla w \, \mathrm{d}x,$$

where η is a given function $\in L^2(\Omega, \mathbb{R}^d)$. It is easy to see that

$$[\![\ell]\!] \leq \|\eta\|_{\Omega}.$$

Example A.1 In a posteriori error analysis of the differential equation $\operatorname{div} A\nabla u + f = 0$, the residual functional

$$\ell_v(w) := \int_{\Omega} (f w - A\nabla v \cdot \nabla w) \, \mathrm{d}x$$

generated by an approximate solution v has an important role. Getting the computable upper bounds for such a functional (and for similar functionals arising in other boundary value problems) presents a problem studied in the theory of error estimation.

A.3 Inequalities

In this section, we recall several inequalities well-known in functional analysis (e.g., see [GT77, KF75, LU68, Sob50]) and discuss some implications that are employed in the book.

A.3.1 The Hölder Inequality

First, we recall the inequality

$$|a \cdot b| \leq \left(\sum_{i=1}^{d} |a_i|^{\alpha} \right)^{1/\alpha} \left(\sum_{i=1}^{d} |b_i|^{\alpha'} \right)^{1/\alpha'}, \tag{A.17}$$

where $\frac{1}{\alpha'} + \frac{1}{\alpha} = 1$ and $a, b \in \mathbb{R}^d$. It is known as the discrete Hölder inequality. The Hölder inequality for functions defined in a bounded Lipschitz domain ω reads

$$\int_\omega wv \,\mathrm{d}x \le \|w\|_{\alpha,\omega} \|v\|_{\alpha',\omega} \tag{A.18}$$

for $w \in L^\alpha(\omega)$ and $v \in L^{\alpha'}(\omega)$, $\alpha \in [1, +\infty)$.

A.3.2 The Poincaré and Friedrichs Inequalities

First, we recall the following well-known fact. Let $\ell(w)$ be a linear continuous functional defined for any $w \in W^{1,p}(\Omega)$, $p \ge 1$, such that if $\ell(w) = 0$ and $w \in P^0$, then $w = 0$. Then, the original norm of $W^{1,p}(\Omega)$ is equivalent to the norm $|\ell(w)| + \|\nabla w\|_{p,\Omega}$. The proof of this assertion is similar to the proof of Theorem A.2.

Since $W^{1,p}(\Omega)$ is embedded in $L^p(\Omega)$, we conclude that

$$\|w\|_{p,\Omega} \le C(p, \Omega, d)\big(|\ell(w)| + \|\nabla w\|_{p,\Omega}\big), \quad \forall w \in W^{1,p}(\Omega). \tag{A.19}$$

We can select $\ell(w)$ as

$$\ell(w) = \int_\Omega w \,\mathrm{d}x, \qquad \ell(w) = \int_\Gamma w \,\mathrm{d}s, \quad \text{or} \quad \ell(w) = \int_\omega w \,\mathrm{d}x,$$

where ω is a certain measurable subset of Ω.

Consider subspaces of $W^{1,p}(\Omega)$ defined by the condition $\ell(w) = 0$. Then, (A.19) reads

$$\|w\|_{p,\Omega} \le C(p, \Omega, d)\|\nabla w\|_{p,\Omega}, \quad \forall w \in \big\{W^{1,p}(\Omega) \mid w \in \ker \ell\big\}. \tag{A.20}$$

A.3.2.1 The Poincaré Inequality

If $\ell(w) = \int_\Omega w \,\mathrm{d}x$, then the condition $\ell(w) = 0$ is equivalent to $\{\!|w|\!\}_\Omega = 0$. In this case, we obtain the Poincaré inequality

$$\|w\|_{p,\Omega} \le C_P(p, \Omega, d)\|\nabla w\|_{p,\Omega}, \quad \forall w \in \big\{W^{1,p}(\Omega) \mid \{\!|w|\!\}_\Omega = 0\big\}. \tag{A.21}$$

Without the zero mean condition, the Poincaré inequality can also be represented in the form (for $p = 2$)

$$\|w\|^2_\Omega \le C_P^2(\Omega, d)\left(\|\nabla w\|^2_\Omega + \left(\int_\Omega w \,\mathrm{d}x\right)^2\right). \tag{A.22}$$

From (A.22) it follows that for $p = 2$

$$\|w\|_\Omega \leq C_P(\Omega, d)\|\nabla w\|_\Omega, \quad \forall w \in \overset{\circ}{L}{}^2(\Omega). \tag{A.23}$$

These results (and similar estimates associated with the spaces $W^{l,p}(\Omega)$) are proved with the help of the so-called "compactness method". Below we use this method to prove one more estimate, which has an exceptional importance for the approximation theory.

Theorem A.2 *Let $\Omega \in \mathbb{R}^d$ be a bounded domain with Lipschitz boundary Γ and $w \in W^{k+1,p}(\Omega)$. Then,*

$$\|w\|_{k,p,\Omega} \leq C|w|_{k+1,p,\Omega}, \quad \forall w \in V, \tag{A.24}$$

where $C(p, \Omega, d, k)$ is a positive constant independent of w and V is the space of functions orthogonal to all polynomials of degree less than or equal to k, i.e.,

$$V = \left\{ w \in W^{k+1,p}(\Omega) \;\middle|\; \int_\Omega w\eta \, dx = 0, \; \forall \eta \in P^k(\Omega) \right\}.$$

Proof Assume that (A.24) does not hold; then for any $i \in \mathbb{N}$, we can find $w^{(i)} \in V$ such that

$$\left\|w^{(i)}\right\|_{k,p,\Omega} > i\left|w^{(i)}\right|_{k+1,p,\Omega}. \tag{A.25}$$

Let $\bar{w}^{(i)} = w^{(i)}/\|w^{(i)}\|_{k,p,\Omega}$. We have

$$\left|\bar{w}^{(i)}\right|_{k+1,p,\Omega} < 1/i, \quad \left\|w^{(i)}\right\|_{k,p,\Omega} = 1. \tag{A.26}$$

Evidently the sequence $\bar{w}^{(i)}$ is bounded with respect to the norm $W^{k+1,p}(\Omega)$, which implies

$$\bar{w}^{(i)} \to \bar{w} \in V \quad \text{weakly in } W^{k+1,p}(\Omega).$$

In view of embedding theorems (cf. Theorem A.1),

$$\bar{w}^{(i)} \to \bar{w} \quad \text{strongly in } W^{k,p}(\Omega).$$

Moreover, $|\bar{w}^{(i)}|_{k+1,p,\Omega} \to 0$. Hence, $\bar{w} \in P^k(\Omega) \cap V$. Since

$$\int_\Omega \bar{w}\eta \, dx = 0, \quad \forall \eta \in P^k(\Omega),$$

we set $\eta = \bar{w}$ and conclude that $\bar{w} = 0$. Thus, $\|w^{(i)}\|_{k,p,\Omega} \to 0$ as $i \to \infty$. On the other hand, $\|w^{(i)}\|_{k,p,\Omega} = 1$ (cf. (A.26)). We arrive at a contradiction, which shows that our original assumption was wrong and the estimate (A.24) is true. □

In the book, we use (A.23) in the derivation of a posteriori estimates for elliptic partial differential equations. In this analysis, it is important to have computable estimates of the constant C_P.

If Ω is a convex domain and $p = 2$, then (for any d) $C_P(\Omega, d) \leq \frac{\operatorname{diam}\Omega}{\pi}$ (see [PW60]). Similar estimates for other $p \geq 1$ can be found in, e.g., Sect. 7.8 in [GT77] and [AD03].

A.3.2.2 The Friedrichs Inequality

For functions from $H^1(\Omega)$, the Friedrichs inequality

$$\|w\|_\Omega^2 \leq C_{F\Omega}^2 \left(\|\nabla w\|_\Omega^2 + \int_\Gamma |w|^2 \, ds \right) \tag{A.27}$$

holds, where $C_{F\Omega}^2$ does not depend on w. Similar inequalities hold for functions in $W^{1,p}(\Omega)$.

A particular form of (A.27) is

$$\|w\|_\Omega \leq C_F(\Omega, d) \|\nabla w\|_\Omega, \quad \forall w \in \mathring{H}^1(\Omega). \tag{A.28}$$

It also follows from (A.20) if $\ell(w) = \int_\Gamma w \, ds$ and $p = 2$.

For $p \in [1, +\infty)$ we have an analogous estimate (see, e.g., [GT77])

$$\|w\|_p \leq \left(\frac{|\Omega|}{\varpi_d} \right)^{1/d} \|\nabla w\|_{p,\Omega}. \tag{A.29}$$

Let $\Omega \subset \widehat{\Omega}$. For any $w \in \mathring{H}^1(\Omega)$, we can define $\widehat{w} \in \mathring{H}^1(\widehat{\Omega})$ by setting $\widehat{w} = w$ in Ω and $\widehat{w}(x) = 0$ for any $x \in \widehat{\Omega} \setminus \Omega$. For all $\widehat{w} \in \mathring{H}^1(\widehat{\Omega})$, we have the inequality

$$\|w\|_{\widehat{\Omega}} \leq C_F(\widehat{\Omega}, d) \|\nabla w\|_{\widehat{\Omega}}, \quad \forall w \in \mathring{H}^1(\widehat{\Omega}),$$

which means that $C_F(\Omega, d) \leq C_F(\widehat{\Omega}, d)$.

This fact opens a way of deriving simple upper bounds for the Friedrichs constant. Let

$$\Omega \subset \widehat{\Omega} := \left\{ x \in \mathbb{R}^d \mid a_i < x < b_i, b_i - a_i = l_i, i = 1, \ldots, d \right\}.$$

Since the constant $C_F(\widehat{\Omega}, d)$ is known, we find that

$$C_F(\Omega, d) \leq \frac{1}{\pi} \left(\sum_{i=1}^d \frac{1}{l_i^2} \right)^{-1}. \tag{A.30}$$

Finally, we note that the constant in (A.28) depends on the lowest eigenvalue of the operator Δ, which satisfies the Rayleigh relation

$$\frac{1}{C_F^2(\Omega, d)} = \lambda_\Omega := \inf_{\substack{w \in \overset{\circ}{H}{}^1(\Omega) \\ w \neq 0}} \frac{\|\nabla w\|^2}{\|w\|^2}. \tag{A.31}$$

Therefore, lower estimates of the minimal eigenvalue generate upper estimates of the Friedrichs constant.

Discrete versions of the Friedrichs and Poincaré inequalities valid for piecewise H^1 functions are established in [Bre03b]. They are used in error analysis of various nonconforming approximations.

A.3.3 Korn's Inequality

In continuum mechanics, of importance is the following assertion known as the Korn's inequality. Let Ω be an open, bounded domain with Lipschitz continuous boundary. Then,

$$\int_\Omega \left(|w|^2 + |\varepsilon(w)|^2\right) \mathrm{d}x \geq C_{K\Omega} \|w\|_{1,2,\Omega}^2, \quad \forall w \in H^1(\Omega, \mathbb{R}^d), \tag{A.32}$$

where $C_{K\Omega}$ is a positive constant independent of w and $\varepsilon(w)$ denotes the symmetric part of the tensor ∇w, i.e.,

$$\varepsilon_{ij}(w) = \frac{1}{2}\left(\frac{\partial w_i}{\partial x_j} + \frac{\partial w_j}{\partial x_i}\right). \tag{A.33}$$

It is not difficult to verify that the left-hand side of (A.32) is bounded from the above by the H^1-norm of w. Thus, it represents a norm equivalent to $\| \cdot \|_{1,2,\Omega}$. The kernel of $\varepsilon(w)$ is called the space of rigid deflections and is denoted by $\mathbf{R}(\Omega)$. If $w \in \mathbf{R}(\Omega)$, then it can be represented in the form $w = w_0 + \omega_0 x$, where w_0 is a vector independent of x and ω_0 is a skew-symmetric tensor with coefficients independent of x. It is easy to understand that the dimension of $\mathbf{R}(\Omega)$ is finite and equals $d + \frac{d(d-1)}{2}$.

For functions in $\overset{\circ}{H}{}^1(\Omega)$, the Korn's inequality is easy to prove. Indeed,

$$\begin{aligned}
|\varepsilon(w)|^2 &= \frac{1}{4}(w_{i,j} + w_{j,i})(w_{i,j} + w_{j,i}) \\
&= \frac{1}{4}(w_{i,j}w_{i,j} + w_{j,i}w_{j,i} + 2w_{i,j}w_{j,i}) = \frac{1}{2}\left(|\nabla w|^2 + w_{i,j}w_{j,i}\right),
\end{aligned}$$

where summation over repeated indices is implied. Therefore, for any $w \in \mathring{C}^2(\Omega)$ we have

$$\int_\Omega |\varepsilon(w)|^2 \,dx = \frac{1}{2}\int_\Omega \left(|\nabla w|^2 + w_{i,j}w_{j,i}\right) dx = \frac{1}{2}\int_\Omega \left(|\nabla w|^2 - w_i w_{j,ij}\right) dx$$
$$= \frac{1}{2}\int_\Omega \left(|\nabla w|^2 + w_{i,i}w_{j,j}\right) dx = \frac{1}{2}\int_\Omega \left(|\nabla w|^2 + |w_{i,i}|^2\right) dx$$
$$\geq \frac{1}{2}\|\nabla w\|^2.$$

Hence,

$$\|\nabla w\| \leq \sqrt{2}\,\|\varepsilon(w)\|, \quad \forall w \in \mathring{C}^2(\Omega). \tag{A.34}$$

Since $\mathring{C}^2(\Omega)$ is dense in $\mathring{H}^1(\Omega)$, this inequality is also valid for functions in $\mathring{H}^1(\Omega)$. Mathematical justifications of the inequality (A.32) are much more complicated (see, e.g., [DL72, Nit81]). Korn's inequalities for piecewise H^1 vector fields are established in [Bre03a]. Interesting generalizations of the Korn's inequality has been recently found and presented in [NPW12].

A.3.4 LBB Inequality

The following result plays an important role in analysis of boundary value problems related to the theory of viscous fluids.

Lemma A.1 *Let Ω be a bounded domain with Lipschitz continuous boundary. There exists a positive constant κ_Ω (which depends only on Ω) such that for any function $f \in \mathring{L}^2(\Omega)$ one can find a function $w \in V$ satisfying the relations* $\operatorname{div} w = f$ *and*

$$\|\nabla w\| \leq \kappa_\Omega \|f\|. \tag{A.35}$$

The reader will find the proof in [BA72, BF91, LS76]. Lemma A.1 implies several important results. First, it leads to the key condition in the mathematical theory of incompressible fluids known in the literature as the Inf–Sup (or *Ladyzhenskaya–Babuška–Brezzi* (LBB)) condition. The latter reads: there exists a positive constant c_{LBB} such that

$$\inf_{\substack{q \in \mathring{L}^2(\Omega)\\ q \neq 0}} \sup_{\substack{w \in V\\ w \neq 0}} \frac{\int_\Omega q \operatorname{div} w \,dx}{\|q\|\,\|\nabla w\|} \geq c_{LBB}. \tag{A.36}$$

By Lemma A.1, it is easy to show that (A.36) holds with $c_{LBB} = (\kappa_\Omega)^{-1}$. Indeed, for arbitrary $q \in \mathring{L}^2(\Omega)$, we can find w_q such that $\operatorname{div} w_q = q$ and $\|\nabla w_q\| \leq \kappa_\Omega \|q\|$, which implies the required result.

Inf–Sup condition (A.36) and its discrete analogs are used for proving the stability and convergence of numerical methods in various problems related to the theory of viscous incompressible fluids. In [Bab73b] and [Bre74], this condition was proved and used to justify the convergence of the so-called *mixed* methods, in which a boundary-value problem is reduced to a saddle-point problem for a certain Lagrangian. Also, (A.36) can be justified by the Nečas inequality (for domains with Lipschitz boundaries a simple proof is presented in [Bra03]). Estimates of c_{LBB} for various domains are discussed, e.g., in [Dob03, OC00, Pay07].

A.4 Convex Functionals

A consequent exposition of convex analysis can be found, e.g., in [ET76, Roc70]. Here, we make a brief summary of the most commonly used notions.

Definition A.1 A functional $J : V \to \mathbb{R}$ is called convex if

$$J(\lambda_1 v_1 + \lambda_2 v_2) \leq \lambda_1 J(v_1) + \lambda_2 J(v_2)$$

for any nonnegative λ_1 and λ_2 such that $\lambda_1 + \lambda_2 = 1$.

The convexity of quadratic type functionals is easy to show, e.g., for any v_1 and v_2,

$$\begin{aligned}(\lambda_1 v_1 + \lambda_2 v_2)^2 &= \lambda_1^2 v_1^2 + 2\lambda_1\lambda_2 v_1 v_2 + \lambda_2^2 v_2^2 \\ &\leq \lambda_1^2 v_1^2 + \lambda_1\lambda_2 v_1^2 + \lambda_1\lambda_2 v_2^2 + \lambda_2^2 v_2^2 = \lambda_1 v_1^2 + \lambda_2 v_2^2.\end{aligned}$$

Important subsets of the set of convex functionals are formed by *strictly convex* and *uniformly convex* functionals.

Definition A.2 A functional $J : V \to \mathbb{R}$ is called strictly convex if

$$J(\lambda_1 v_1 + \lambda_2 v_2) < \lambda_1 J(v_1) + \lambda_2 J(v_2)$$

for any positive λ_1 and λ_2 such that $\lambda_1 + \lambda_2 = 1$.

Uniform convexity is a property which is important in the theory of nonlinear convex variational problems related to a posteriori error estimation.

Definition A.3 A convex functional $J : V \to \mathbb{R}$ is called uniformly convex in a ball $B(0_V, \rho)$ if there exists a nonnegative functional $\Upsilon_\rho \not\equiv 0$ such that for all $v_1, v_2 \in B(0_V, \rho)$ the following inequality holds:

$$J\left(\frac{v_1 + v_2}{2}\right) + \Upsilon_\rho(v_1 - v_2) \leq \frac{1}{2}\big(J(v_1) + J(v_2)\big). \tag{A.37}$$

For many convex functionals, the estimate (A.37) holds with

$$\Upsilon_\rho = \Upsilon_\rho\big(\|v_1 - v_2\|_V\big),$$

i.e., with a monotone function dependent on the norm and vanishing only at zero (see, e.g., [NR04, Rep97b]).

Let V be a reflexive Banach space (see Sect. A.2.3). The functional $J^* : V^* \to \mathbb{R}$ defined by the relation

$$J^*(v^*) = \sup_{v \in V}\big\{\langle v^*, v\rangle - J(v)\big\} \tag{A.38}$$

is said to be *dual* (or *conjugate*) to J.

Remark A.1 If J is a smooth function that increases at infinity faster than any linear function, then J^* is the Legendre transform of J. The dual functionals were studied by Young, Fenchel, Moreau, and Rockafellar (e.g., see [ET76, Fei93, Roc70]). The functional J^* is also called *polar* to J.

The functional

$$J^{**}(v) = \sup_{v^* \in V^*}\big\{\langle v^*, v\rangle - J^*(v^*)\big\} \tag{A.39}$$

is called the *second conjugate* to J (or *bipolar*). If J is a convex functional attaining finite values, then J coincides with J^{**}.

To illustrate the definitions of conjugate functionals, consider functionals defined on the Euclidean space E^d. In this case, V and V^* consist of the same elements: d-dimensional vectors (denoted by ξ and ξ^*, respectively) and the quantity $\langle \xi^*, \xi\rangle$ is presented by the scalar product $\xi^* \cdot \xi$.

Let $A = \{a_{ij}\}$ be a positive definite matrix. We have the following pair of mutually conjugate functionals:

$$J(\xi) = \frac{1}{2}A\xi \cdot \xi \quad \text{and} \quad J^*(\xi^*) = \frac{1}{2}A^{-1}\xi^* \cdot \xi^*. \tag{A.40}$$

Another example is given by the functionals

$$J(\xi) = \frac{1}{\alpha}|\xi|^\alpha \quad \text{and} \quad J^*(\xi^*) = \frac{1}{\alpha'}|\xi^*|^{\alpha'}, \tag{A.41}$$

where $\frac{1}{\alpha} + \frac{1}{\alpha'} = 1$. If φ is an odd convex function, then $(\varphi(\|u\|_V))^* = \varphi^*(\|u^*\|_{V^*})$.

Let a functional $J : V \to \mathbb{R}$ be finite at $v_0 \in V$. The functional J is called subdifferentiable at v_0 if there exists an affine minorant l such that $J(v_0) = l(v_0)$. The element v^* is called a *subgradient* of J at v_0. The set of all subgradients of J at v_0 forms a *subdifferential*, which is usually denoted by $\partial J(v_0)$. It may be empty, or it may contain one element or infinitely many elements.

An important property of convex functionals follows directly from the fact that they have an exact affine minorant

$$J(v) - J(v_0) \geq \langle v^*, v - v_0\rangle, \tag{A.42}$$

where $v^* \in \partial J(v_0)$. The inequality (A.42) represents the basic incremental relation for convex functionals. For proper convex functionals, there exists a simple criterion that enables one to verify whether or not an element v^* belongs to the set $\partial J(v)$.

Lemma A.2 *The following two statements are equivalent*:

$$J(v) + J^*(v^*) - \langle v^*, v\rangle = 0, \tag{A.43}$$

$$v^* \in \partial J(v), \tag{A.44}$$

$$v \in \partial J^*(v^*). \tag{A.45}$$

Proof Assume that $v^* \in \partial J(v)$. In accordance with (A.42), we have

$$J(w) \geq J(v) + \langle v^*, w - v\rangle, \quad \forall w \in V.$$

Hence,

$$\langle v^*, v\rangle - J(v) \geq \langle v^*, w\rangle - J(w), \quad \forall w \in V$$

and, consequently,

$$\langle v^*, v\rangle - J(v) \geq \sup_{w\in V}\{\langle v^*, w\rangle - J(w)\} = J^*(v^*). \tag{A.46}$$

However, by the definition of J^*, we know that for any v and v^*

$$J^*(v^*) \geq \langle v^*, v\rangle - J(v). \tag{A.47}$$

We observe that (A.46) and (A.47) imply (A.43).

Assume that $v \in \partial J^*(v^*)$. Then, $J^*(w^*) \geq J^*(v^*) + \langle w^* - v^*, v\rangle$, so that

$$\langle v^*, v\rangle - J^*(v^*) \geq \sup_{w^*\in V^*}\{\langle w^*, v\rangle - J^*(w^*)\} = J^{**}(v).$$

On the other hand,

$$\langle v^*, v\rangle - J^*(v^*) \geq J^{**}(v) = J(v),$$

and we again arrive at (A.43).

Assume that (A.43) holds. By the definition of J^*, we obtain

$$0 = J(v) + J^*(v^*) - \langle v^*, v\rangle \geq J(v) - J(w) - \langle v^*, v - w\rangle,$$

where w is an arbitrary element of V. Thus,

$$J(w) - J(v) \geq \langle v^*, w - v\rangle, \quad \forall w \in V,$$

which means that $J(v) + \langle v^*, v - w\rangle$ is an exact affine minorant of J (at v) and, consequently, (A.44) holds. The proof of (A.45) is quite similar. □

Let J and J^* be a pair of conjugate functionals. The functional

$$D_J : V \times V^* \to \mathbb{R}_+$$
$$D_J(v, v^*) := J(v) + J^*(v^*) - \langle v^*, v \rangle$$

is called the *compound* functional.

In view of Lemma A.2, $D_J(v, v^*) \geq 0$. Moreover, $D_J(v, v^*) = 0$ if and only if the arguments satisfy (A.44) and (A.45), which are also called the *duality relations* and very often represent the constitutive relations of a physical model. Compound functionals play an important role in the a posteriori error estimation of nonlinear variational problems.

Finally, we recall some basic notions related to the differentiation of convex functionals. We say that J has a weak derivative $J'(v_0) \in V^*$ (at the point v_0) in the sense of Gâteaux if

$$\lim_{\lambda \to +0} \frac{J(v_0 + \lambda w) - J(v_0)}{\lambda} = \langle J'(v_0), w \rangle, \quad \forall w \in V. \tag{A.48}$$

Assume that J is differentiable in the above sense and $v^* \in \partial J(v_0)$. Then, for any $v \in V$ we know that $J(v) - J(v_0) \geq \langle v^*, v - v_0 \rangle$. Set $v = v_0 + \lambda w$, where $\lambda > 0$. Now, we have $J(v_0 + \lambda w) - J(v_0) \geq \lambda \langle v^*, w \rangle$. Therefore,

$$\langle J'(v_0), w \rangle = \lim_{\lambda \to +0} \frac{J(v_0 + \lambda w) - J(v_0)}{\lambda} \geq \langle v^*, w \rangle,$$

and $\langle J'(v_0) - v^*, w \rangle \geq 0$ for any $w \in V$. This inequality means that, in such a case, the Gâteaux derivative coincides with v^*.

Appendix B
Boundary Value Problems

This chapter briefly discusses the main approaches to quantitative analysis of elliptic boundary value problems. They are based on classical and generalized settings of a BVP and on representation of a problem in the variational or saddle point forms. Each of the approaches generates the corresponding approximation procedures and numerical methods, which are considered in the last part of the chapter.

B.1 Generalized Solutions of Boundary Value Problems

Solutions of boundary value problems are usually considered in a generalized sense. The definition of a solution to BVP is connected with approximation methods and, therefore, it is impossible to discuss accuracy and error estimation methods without addressing these questions. In this chapter we briefly recall basic facts from the corresponding mathematical theory with the example of the problem

$$\Delta u + f = 0 \quad \text{in } \Omega, \tag{B.1}$$

$$u = u_0 \quad \text{on } \Gamma. \tag{B.2}$$

Originally, this problem was understood in the classical sense: Find $u \in C^2(\Omega) \cap C(\bar{\Omega})$ such that the boundary condition (B.2) is satisfied and

$$\frac{\partial^2 u}{\partial^2 x_1} + \frac{\partial^2 u}{\partial^2 x_2} + f = 0, \quad \forall (x_1, x_2) \in \Omega,$$

where $\frac{\partial^2 u}{\partial^2 x_k}$, $k = 1, 2$, are the classical derivatives.

However, the question as to how one can guarantee that a solution of (B.1)–(B.2) does exist must be answered. This question happened to be very difficult, and the answer has been found only after about one hundred years of studies, which completely reconstructed the theory of partial differential equations. The modern theory is based on the works of D. Hilbert, H. Poincaré, S. Sobolev, R. Courant,

O. Mali et al., *Accuracy Verification Methods*,
Computational Methods in Applied Sciences 32, DOI 10.1007/978-94-007-7581-7,

O. Ladyzhenskaya, and many others mathematicians who contributed to the conception of a *generalized* or *weak* solution. These solutions are closely related to the *Petrov–Bubnov–Galerkin method* [Gal15]. The idea of this method is to find $u_N = \sum_{i=1}^{N} \alpha_i w_i$ such that

$$\int_\Omega (\Delta u_N + f) w_i \,\mathrm{d}x = 0, \quad \forall w_i, i = 1, 2, \ldots, N.$$

This means that the residual of the differential equation generated by u_N is *orthogonal* to the finite dimensional space V_N formed by linearly independent trial functions w_i.

The conception of generalized solutions naturally extends this idea. Let us find a function that makes the residual orthogonal to all the functions in w from a proper functional space V, i.e.,

$$\int_\Omega (\Delta u + f) w \,\mathrm{d}x = 0, \quad \forall w \in V.$$

Integration by parts leads to the *generalized statement* of (B.1)–(B.2): Find $u \in \overset{\circ}{H}{}^1(\Omega) + u_0$ such that

$$\int_\Omega \nabla u \cdot \nabla w \,\mathrm{d}x = \int_\Omega f w \,\mathrm{d}x, \quad \forall w \in V = \overset{\circ}{H}{}^1(\Omega). \tag{B.3}$$

It is easy to see that if (B.1)–(B.2) has a classical solution, then it automatically satisfies (B.3). However, some boundary value problems may have generalized solutions but do not have classical solutions.

The idea used in (B.3) admits wide extensions (see, e.g., [Lad85]) to various differential equations. In particular, it is easy to generalize it to the class of elliptic problems generated by V-elliptic bilinear forms, defined on a Hilbert space V. We recall that a symmetric bilinear form $a : V \times V \to R$ is called V-elliptic if two positive constants c_1 and c_2 exists such that

$$a(u, u) \geq c_1 \|u\|_V^2, \qquad \forall u \in V, \tag{B.4}$$

$$\left|a(u, v)\right| \leq c_2 \|u\|_V \|v\|_V, \quad \forall u, v \in V. \tag{B.5}$$

In particular, the left-hand side of (B.3) is presented by $a(u, v) = \int_\Omega \nabla u \cdot \nabla v \,\mathrm{d}x$.

Let $\ell : V \to \mathbb{R}$ be a bounded linear functional. The problem of finding u in a Hilbert space V such that

$$a(u, w) = \ell(w), \quad \forall w \in V, \tag{B.6}$$

is a generalization of (B.3). It is not difficult to prove that (B.6) has a solution. The proof of this fact is based on the following theorem (often called the Lax–Milgram lemma).

Lemma B.1 *For a V-elliptic bilinear form a there exists a linear bounded operator $A \in \mathcal{L}(V, V)$ such that*

$$a(u, v) = (Au, v), \quad \forall u, v \in V.$$

It has an inverse $A^{-1} \in \mathcal{L}(V, V)$. The norms of the operators A and A^{-1} are bounded by the constants c_2 and $\frac{1}{c_1}$, respectively.

Lemma B.1 is sufficient to prove that the problem (B.6) is well defined and to deduce the first a priori estimate that bounds $\|u\|_V$ through the norm of ℓ (such estimates are often called "energy estimates").

Theorem B.1 *Let a be a V-elliptic bilinear form, then* (B.6) *has a unique solution and*

$$\|u\|_V \leq \frac{1}{c_1}\|\ell\|. \tag{B.7}$$

Proof In view of the Reisz theorem, there exists $w \in V$ such that

$$\ell(v) = (v, w), \quad \forall v \in V$$

and $\|w\|_V = \|\ell\|$. By Lemma B.1, we know that $a(u, v) = (Au, v)$ and, therefore, (B.6) reads

$$(Au, v) = (w, v), \quad \forall v \in V,$$

which is equivalent to $Au = w$. Since the inverse operator exists, we conclude that $u = A^{-1}w$. Hence, u exists and is unique. By Lemma B.1, we also conclude that

$$\|u\|_V = \left\|A^{-1}w\right\|_V \leq \left\|A^{-1}\right\| \|w\|_V \leq \frac{1}{c_1}\|w\|_V = \frac{1}{c_1}\|\ell\|. \qquad \square$$

Example B.1 Consider the problem (B.1)–(B.2) and reformulate it in the following form: find $\bar{u} := u - u_0 \in V_0 := \overset{\circ}{H}{}^1(\Omega)$ such that

$$\int_\Omega \nabla\bar{u} \cdot \nabla w \, dx = \ell(w), \quad \forall w \in V_0,$$

where $\ell(w) := \int_\Omega (fw - \nabla u_0 \cdot \nabla w) \, dx$. We set $V = V_0$ and endow it with the norm $\|\nabla v\|$. Since the corresponding bilinear form is $a(\bar{u}, w) = \int_\Omega \nabla\bar{u} \cdot \nabla w \, dx$, we see that $c_1 = c_2 = 1$. Then (B.7) implies the energy estimate

$$\|\nabla\bar{u}\| \leq \|\ell\| = \sup_{w \in V} \frac{\int_\Omega (fw - \nabla u_0 \cdot \nabla w) \, dx}{\|\nabla w\|}. \tag{B.8}$$

Note that

$$\int_\Omega fw \, dx \leq \|f\|\|w\| \leq \|f\| C_{F\Omega} \|\nabla w\|,$$

where $C_{F\Omega}$ is the constant in the Friedrichs inequality. Therefore, (B.8) yields the desired estimate

$$\|\nabla \bar{u}\| \leq C_{F\Omega}\|f\| + \|\nabla u_0\|,$$

which means that $\|\nabla u\| \leq C(\|f\| + \|\nabla u_0\|)$.

B.2 Variational Statements of Elliptic Boundary Value Problems

The variational approach arose in the 19th century shortly after the first PDE's had been presented. They are motivated by physical "minimal energy" principles in mechanics. In fact, the variational method originates from the famous Fermat theorem: *If f is a differentiable function that attains the minimum at $\bar{x}$, then $f'(\bar{x}) = 0$.*

Later, L. Euler and J. L. Lagrange created the calculus of variations and generalized this principle to one-dimensional variational problems. Moreover, it was proved that a minimizer of the functional $\int_0^T g(t, y, \dot{y})\,dt$ must satisfy a certain ODE (which is often named the *Euler–Lagrange equation*). A further development of the variational approach was made by K. Weierstrass, A. Clebsch, D. Hilbert, E. Noether, H. Lebesgue, J. Hadamard, L. Pontryagin, J. Moreau, C. B. Morrey, T. Rockafellar, and many others. It has been shown that variational statements can be extended to various multidimensional problems. In the general form, such a problem consists of finding $u(x)$ such that $u = u_0$ on Γ and

$$J(u) = \inf_v J(v), \tag{B.9}$$

where the infimum is taken over all admissible functions v (i.e., all such functions that $J(v)$ is finite and $v = u_0$ on Γ).

For example, it is easy to see that the functional generating the problem (B.1)–(B.2) is

$$J(v) = \int_\Omega \left(\frac{1}{2}|\nabla v|^2 - f v\right) dx.$$

Indeed, let w be an admissible (smooth) function vanishing on the boundary. Then for any $\lambda > 0$ we have

$$J(u) \leq J(u + \lambda w) = \int_\Omega \left(\frac{1}{2}\left|\nabla(u + \lambda w)\right|^2 - f(u + \lambda w)\right) dx. \tag{B.10}$$

This inequality is equivalent to

$$\int_\Omega (\nabla u \cdot \nabla w - f w)\,dx \geq \int_\Omega -\frac{\lambda}{2}|\nabla w|^2\,dx.$$

Since λ is an arbitrary real number, we find that (B.10) can be true only if the left-hand side is nonnegative. Obviously, w can be replaced by $-w$ and, therefore, for any w

$$\int_{\Omega} (\nabla u \cdot \nabla w - f w)\,\mathrm{d}x = 0. \tag{B.11}$$

Since smooth functions are dense in $\overset{\circ}{H}{}^1(\Omega)$, we conclude that (B.11) also holds for any function from this space. Thus, we have derived the generalized statement (B.3) using variational arguments. If u possesses second derivatives, then we can integrate by parts in the above integral and use the *Du–Bois–Reymond Lemma*, which says that if g is a locally integrable function defined on an open set Ω and satisfies the condition

$$\int_{\Omega} g w\,\mathrm{d}x = 0, \quad \forall w \in \overset{\circ}{C}{}^{\infty}(\Omega),$$

then $g = 0$ for almost all points in Ω. We conclude that u is the classical solution, i.e.,

$$\Delta u + f = 0, \quad \text{in } \Omega, \tag{B.12}$$

$$u = u_0 \quad \text{on } \Gamma. \tag{B.13}$$

Now, it is necessary to discuss how the existence and uniqueness theorems can be proved within the framework of the variational method. Certainly, we can refer to previous results (based on the Lax–Milgram lemma) and say that since (B.11) has a solution, a minimizer exists. However, there exists another quite different approach (which comes from the works of K. Weierstrass). It is applicable to a much wider class of boundary value problems than those generated by V-elliptic bilinear forms.

Theorem B.2 *If K is a closed bounded set in $\mathbb{R}^d$, and J is a continuous functional defined on K, then the problem*

$$\inf_{v \in K} J(v), \tag{B.14}$$

has a minimizer $u \in K$.

Proof Let $\{v_k\}$ be a minimizing sequence, i.e., $J(v_k) \to \inf_{v \in K} J$. We can extract a converging subsequence out of it (by the Boltzano–Weierstrass Lemma). Denote this subsequence by $\{v_{k_s}\}$. Since K is closed, we know that the limit of this sequence (we denote it by u) belongs to K. Since J is continuous, we find that

$$\inf_{v \in K} J = \lim_{s \to \infty} J(v_{k_s}) = J(u).$$

Thus, u is a minimizer. □

It is not difficult to extend this method to the case of functional spaces and lower semicontinuous functionals. We recall that the functional $J : V \to \mathbb{R}$ is called lower semicontinuous at $\widehat{v}$ if $\lim_{v_s \to \widehat{v}} J(v_s) \geq J(\widehat{v})$ for any sequence v_s converging to $\widehat{v}$. The functional is called lower semicontinuous if it possesses this property at any point of the set on which it is defined.

Theorem B.3 *Let V be a full metric space, $K \subset V$ be a compact set, and J be a lower semicontinuous functional defined and finite on K. Then, the problem* $\inf_{v \in K} J(v)$ *has a solution (minimizer) $u \in K$.*

Proof Let $\{v_k\}$ be a minimizing sequence, i.e., $J(v_k) \to \inf_K J$. Since K is compact, we can extract a convergent subsequence out of it, which we denote $\{v_{k_s}\}$. K is a closed set; therefore, the limit of this sequence (we denote it by u) belongs to K. Finally, we recall that J is lower semicontinuous and conclude that

$$\inf_K J = \lim_{s \to \infty} J(v_{k_s}) \geq J(u).$$

We conclude that u is a minimizer. □

Regrettably, this theorem cannot be directly applied to functionals minimized on the whole space $\overset{\circ}{H}{}^1(\Omega)$ (and other functional spaces). It is necessary to reduce the requirement imposed on K (compactness) and compensate this reduction by strengthening conditions imposed on J. More precisely, we replace *compactness* by *weak compactness*, which means that if K is bounded then a weakly convergent sequence can be extracted. This change would be very convenient, because any closed bounded subset of a Hilbert space is weakly compact. On the other hand, we also change the lower semicontinuity of J to weak lower semicontinuity (which means that the functional is lower semicontinuous for all weakly convergent sequences).

Theorem B.4 *Let K be weakly compact and J be a weakly lower semicontinuous functional defined on K. Then, the problem* $\inf_{v \in K} J(v)$ *has a minimizer $u \in K$.*

Proof Let $\{v_k\}$ be a minimizing sequence, i.e., $J(v_k) \to \inf_K J$. We can extract a weakly convergent subsequence $\{v_{k_s}\} \rightharpoonup u \in K$. Since J is weakly lower semicontinuous, we find that

$$\inf_K J = \lim_{s \to \infty} J(v_{k_s}) \geq J(u).$$

Thus, u is a minimizer. □

Theorem B.4 opens a way of proving the existence of a minimizer for the functional (B.9). Indeed, the set $K := \{w \in V_0 \mid J(w) \leq J(v_1)\}$ that contains a minimizing sequence is bounded. This fact follows from the relation

$$\frac{1}{2}\|\nabla w\|^2 - \int_\Omega f w \, dx \leq J(v_1),$$

where v_1 is the first element of a minimizing sequence. This relation infers the estimate (for $w \in K$)

$$\frac{1}{2}\|\nabla w\|^2 \leq J(v_1) + \|f\|\|w\| \leq J(v_1) + \|f\|C_F\|\nabla w\|$$
$$\leq \left(J(v_1) + \frac{1}{2}C_F^2\|f\|^2\right) + \frac{1}{2}\|\nabla w\|^2,$$

which shows that $\|\nabla w\|$ is bounded.

It remains to verify the weak lower semicontinuity of J. Fortunately, for functionals defined on reflexive spaces there is a simple criterion: *a convex lower semicontinuous functional is weakly lower semicontinuous*. Typically, the functionals arising in variational statements of boundary-value problems are continuous on reflexive spaces and the convexity of them is easy to check.

Thus, Theorem B.4 guarantees the existence of a minimizer and, consequently, proves the existence of the generalized solution of the corresponding differential equation.

It is worth remarking that the variational method is extendable to a much wider class of problems, which includes many nonlinear boundary value problems. To show this, we need one more definition.

Definition B.1 The functional J is called *coercive* on V if

$$J(v_k) \to +\infty$$

for any sequence $\{v_k\} \in V$ such that $\|v_k\|_V \to +\infty$.

Coercivity plays an important role in establishing the existence.

Lemma B.2 *Let J be coercive; then the set*

$$V_\alpha := \left\{v \in V \mid J(v) \leq \alpha\right\}$$

is bounded.

Proof Assume the contrary, i.e., V_α is unbounded and it is not contained in any ball

$$B(0,d) = \left\{v \in V \mid \|v\|_V \leq d\right\}.$$

This means that for any integer k, one can find $v_k \in V_\alpha$ such that $\|v_k\|_V > k$. By coercivity, we conclude that

$$J(v_k) \to +\infty \quad \text{as } k \to +\infty.$$

But this is impossible because elements of V_α are such that the functional does not exceed α. □

As a result, we arrive at the following theorem, which provides easily verifiable conditions sufficient to guarantee the existence of a minimizer for a wide class of convex variational problems (see, e.g., [ET76])

Theorem B.5 *Let $J : K \to R$ be a convex, continuous, and coercive functional, and let K be a nonempty, convex, and closed subset of a Hilbert space V. Then, the problem $\inf_{w \in K} J(w)$ has a minimizer u. If J is strictly convex, then the minimizer is unique.*

Proof Let $\{v_k\}$ be a minimizing sequence, i.e., $J(v_k) \to \inf_K J$. The set

$$K_1 := \left\{ v \in K \mid J(v) \le J(v_1) \right\}$$

is bounded (by Lemma B.2). Obviously, it is also closed. In a Hilbert space all closed bounded sets are weakly compact. Therefore, we can extract a weakly convergent subsequence $\{v_{k_s}\} \rightharpoonup u \in K_1$. Since J is convex and continuous, it is weakly lower semicontinuous, and we find that

$$\inf_K J = \lim_{s \to \infty} J(v_{k_s}) \ge J(u),$$

which means that u is a minimizer.

Assume that J is strictly convex, i.e.,

$$J(\lambda_1 v_1 + \lambda_2 v_2) < \lambda_1 J(v_1) + \lambda_2 J(v_2), \quad \lambda_1 + \lambda_2 = 1, \lambda_i > 0.$$

If u_1 and u_2 are two different minimizers, then we immediately arrive at a contradiction, because

$$J(\lambda_1 u_1 + \lambda_2 u_2) < \lambda_1 J(u_1) + \lambda_2 J(u_2) = \inf_K J.$$

Hence, we conclude that the minimizer is unique. □

At the end of this section, we discuss several examples showing how to apply Theorem B.5 to various variational problems.

Example B.2 Consider the functional $J(w) = \frac{1}{2} a(w, w) - \langle \ell, w \rangle$, where a is the bilinear form used in the definition of problem (B.6). It is easy to verify that the minimizer of $J(v)$ on V satisfies (B.6). We set $K = V$ and apply Theorem B.5.

Since

$$\frac{1}{2} a(w, w) \ge c_1 \|w\|_V^2, \qquad \left| \langle \ell, w \rangle \right| \le \|\ell\| \|w\|_V,$$

we see that

$$J(w) \ge c_1 \|w\|_V^2 - \|\ell\| \|w\|_V \to +\infty \quad \text{as } \|w\|_V \to +\infty.$$

Hence, J is coercive on V. Since J is strictly convex and continuous we apply Theorem B.5 and establish the existence and uniqueness of the minimizer without Lemma B.1.

Example B.3 One of the simplest models in the theory of nonlinear fluids is related to the functional

$$J(w) = \int_\Omega \left(\frac{\nu}{2} |\nabla w|^2 + k_* |\nabla w| - f w \right) \mathrm{d}x, \tag{B.15}$$

which must be minimized on the set $K = \overset{\circ}{H}{}^1(\Omega)$. Here, μ and k_* are positive constants dependent on the viscosity and plasticity properties of the so-called Bingham fluid and $f \in L^2(\Omega)$. Since

$$-\int_\Omega f w \,\mathrm{d}x \geq -\|f\| C_F(\Omega, d) \|\nabla w\|,$$

the functional (B.15) is coercive. It is also strictly convex and continuous on V. Therefore, the solution of this problem exists and is unique.

Remark B.1 Finally, we discuss applicability limits of the method based on lower semicontinuity. There exist many practically interesting problems in which conditions of Theorem B.5 are not satisfied. They are related to (a) nonconvexity of the functional J, (b) nonconvexity of the set K, or (c) nonreflexivity of the space V.

Nonconvex variational problems often arise, e.g., in the theory of phase transitions in solids. In these problems, the energy functional contains two (or more) branches generated by different phases, so that the whole energy is not convex. A simple example presents the functional

$$\int_\Omega \big(g(\nabla w) - f w\big) \,\mathrm{d}x, \quad g = \min\{g_1, g_2\},$$

where g_1 and g_2 are related to two different phases. In these problems, a minimizing sequence may have no strong convergence and "solutions" are presented by structures with rapidly oscillating layers (see [BJ87]).

Optimal control problems with control η in the main part of the operator A, e.g., problems of type

$$\inf_{(\eta, w) \in K} J(\eta, w), \quad K := \big\{(\eta, w) \mid A(\eta) w + f = 0\big\}, \eta \in C,$$

generate a class of problems with nonconvex K (even if the set of admissible control functions C is convex and closed). The nonexistence of a minimizer often arises in the form of the so-called "sliding regimes". Mathematically, these problems require the so-called G-closure of the operator set.

Problems defined on nonreflexive spaces arise if the energy has linear growth with respect to the differential operator. Typical example is presented by the *non-parametric minimal surface problem*

$$J(v) = \int_\Omega \sqrt{1 + |\nabla w|^2} \,\mathrm{d}x. \tag{B.16}$$

The functional J is defined on the Sobolev space $V := W^{1,1}(\Omega)$ and

$$K := \{w \in V \mid w = u_0 \text{ on } \Gamma\},$$

where u_0 is a given function defining the boundary condition. This functional is *convex* and continuous on V. Since $J(w) \geq \|\nabla w\|$, it is coercive on K. Also, K is convex and closed (in V).

However, the variational problem associated with the functional (B.15) may have no solution because $W^{1,1}(\Omega)$ is a nonreflexive space. For such spaces, convexity and boundedness do not imply weak compactness (so that we cannot apply the above-presented method).

Practically important classes of engineering problems related to such type phenomena arise in the theory of capillary surfaces and in perfect plasticity. In these models, a minimizing sequence may converge to a discontinuous function. Therefore, special approximation methods are required (see, e.g., [BMR12, NW03, Rep94, ST87] and references therein).

Numerical methods generated by variational statements use a natural idea of minimizing $J(v)$ on a certain finite dimensional subspace of V. We consider some of them in Sect. B.4.2.

B.3 Saddle Point Statements of Elliptic Boundary Value Problems

B.3.1 Introduction to the Theory of Saddle Points

Minimax approaches to elliptic partial differential equations and the corresponding numerical methods are based on the theory of saddle points, which play an important role in mathematical analysis of boundary value problems. Saddle points are often considered as corresponding solutions. We refer, e.g., to [ET76, Roc70] for a systematic exposition of the saddle point theory. Concerning numerical methods for saddle point problems we refer to, e.g., [Glo84, GLT76] and the literature cited therein. Below, the reader can find a collection of results used in the book.

Saddle point theory operates with functionals defined on a pair of elements (functions), which are called Lagrangians. Let $L : (V \times Q) \to \mathbb{R}$ be such a Lagrangian, V and Q be two Banach spaces, and $K \subset V$ and $M \subset Q$ be two closed subsets in these spaces.

Saddle point problem is to find $(u, p) \in K \times M$ such that

$$L(u, q) \leq L(u, p) \leq L(v, p), \quad \forall q \in M, \forall v \in K. \tag{B.17}$$

It is easy to see that regardless of the structure of the Lagrangian and the nature of nonempty sets K and M,

$$\sup_{q \in M} \inf_{v \in K} L(v, q) \leq \inf_{v \in K} \sup_{q \in M} L(v, q). \tag{B.18}$$

Our first goal is to present conditions that guarantee the existence of a saddle point. First, we mention the following simple criterion.

Lemma B.3 *If there exist a constant α and two elements $u \in K$ and $p \in M$ such that*

$$L(u, q) \leq \alpha, \quad \forall q \in M, \tag{B.19}$$

and

$$L(v, p) \geq \alpha, \quad \forall v \in K, \tag{B.20}$$

then (u, p) is a saddle point. Moreover, we have the relation

$$\alpha = \inf_{v \in K} \sup_{q \in M} L(v, q) = \sup_{q \in M} \inf_{v \in K} L(v, q). \tag{B.21}$$

Proof From (B.19) and (B.20), we obtain

$$L(u, p) \leq \alpha \leq L(u, p).$$

Therefore, $L(u, p) = \alpha$ and

$$L(u, q) \leq L(u, p) \leq L(v, p), \quad \forall v \in K, \forall y \in M,$$

which means that (u, p) is a saddle point. Since

$$\sup_{q \in M} L(u, q) = L(u, p) = \alpha \quad \text{and} \quad \inf_{v \in K} L(v, p) = L(u, p) = \alpha,$$

we have

$$\inf_{v \in K} \sup_{q \in M} L(v, q) \leq \sup_{q \in M} L(u, q) = \alpha,$$

$$\sup_{q \in M} \inf_{v \in K} L(v, q) \geq \inf_{v \in K} L(v, p) = \alpha.$$

In view of (B.18), we arrive at (B.21). □

Regrettably, in many cases it is not easy to justify the conditions (B.20) and (B.21), so that other existence criteria are required. We will discuss them later, but first prove one general result about the structure of saddle points. Henceforth, we assume that

- V and Q are reflexive Banach spaces (e.g., Hilbert spaces) and K and M are convex and closed subsets of V and M, respectively.
- The functional $v \mapsto L(v,q)$ is *convex and lower semicontinuous* for any $q \in M$.
- The functional $q \mapsto L(v,q)$ is *concave and upper semicontinuous* for any $v \in K$.

These conditions are easily verifiable and hold for many practically interesting problems. However, they are not sufficient to guarantee the existence of a saddle point.

Lemma B.4 *All saddle points of L form a set $K_0 \times M_0$, where K_0 and M_0 are convex subsets of K and M, respectively.*

Proof Assume that (u_1, p_1) and (u_2, p_2) are two different saddle points. Then,

$$L(u_1,q) \le L(u_1,p_1) = \alpha = L(u_2,p_2) \le L(v,p_1),$$
$$L(u_2,q) \le L(u_1,p_1) = \alpha = L(u_2,p_2) \le L(v,p_2),$$

where v and q are arbitrary elements of the sets K and M, respectively. From the first relation, we obtain $L(u_2,p_1) \ge \alpha$, and from the second one we have $L(u_2,p_1) \le \alpha$.

Now, Lemma B.3 implies that (u_2,p_1) is a saddle point. The same conclusion is obviously true for (u_1,p_2). Let u_1 and u_2 be two different elements of K_0. Then,

$$L(u_1,q) \le L(u_1,p_1) = \alpha, \quad \forall q \in M,$$
$$L(u_2,q) \le L(u_2,p_1) = \alpha, \quad \forall q \in M.$$

Since $v \mapsto L(v,q)$ is a convex mapping, we have

$$L(\lambda_1 u_1 + \lambda_2 u_2, q) \le \lambda_1 L(u_1,q) + \lambda_2 L(u_2,q) \le \alpha, \quad \lambda_1 + \lambda_2 = 1.$$

In particular, $L(\lambda_1 u_1 + \lambda_2 u_2, p_1) \le \alpha$. Since $L(v,p_1) \ge \alpha$ for all $v \in K$, we obtain the opposite inequality

$$L(\lambda_1 u_1 + \lambda_2 u_2, p_1) \ge \alpha$$

and conclude that $L(\lambda_1 u_1 + \lambda_2 u_2, p_1) = \alpha$.

Hence, $\lambda_1 u_1 + \lambda_2 u_2 \in K_0$, where $\lambda_1 + \lambda_2 = 1$. Hence, K_0 is a convex set. The convexity of M is proved analogously. □

Using the Lagrangian L, we define two functionals:

$$J(v) := \sup_{q \in M} L(v,q) \quad \text{and} \quad I^*(q) := \inf_{v \in K} L(v,q),$$

which generate two variational problems.

Problem $\mathcal{P}$. Find $u \in K$ such that

$$J(u) = \inf \mathcal{P} := \inf_{v \in K} J(v). \tag{B.22}$$

Problem $\mathcal{P}^*$. Find $p \in M$ such that

$$I^*(p) = \sup \mathcal{P}^* := \sup_{q \in M} I^*(q). \tag{B.23}$$

Henceforth, Problems $\mathcal{P}$ and $\mathcal{P}^*$ are called *primal* and *dual*, respectively. Their solutions are closely related. Solutions of these problems (if they exist) form the corresponding saddle point of (u, p).

By (B.18), we see that

$$\sup \mathcal{P}^* \leq \inf \mathcal{P}. \tag{B.24}$$

However, in many cases a stronger relation $\sup \mathcal{P}^* = \inf \mathcal{P}$. Below we state the main theorem, which establishes a link between the solutions of Problems $\mathcal{P}$ and $\mathcal{P}^*$ and the saddle points of Problems $\mathcal{L}$.

Theorem B.6 *The following two statements are equivalent:*

1. *there exists a pair of elements $u \in K$ and $p \in M$ such that*

$$J(u) = \inf \mathcal{P}, \tag{B.25}$$

$$I^*(p) = \sup \mathcal{P}^*, \tag{B.26}$$

$$\inf \mathcal{P} = \sup \mathcal{P}^*. \tag{B.27}$$

2. *(u, p) is a saddle point of the Lagrangian L on $K \times M$.*

Moreover, any of the above two assertions implies the principal relation

$$I^*(p) = L(u, p) = J(u). \tag{B.28}$$

Proof Let the first assumption be true. We set $\alpha = \inf \mathcal{P} = \sup \mathcal{P}^*$. Then,

$$L(u, q) \leq \sup_{q \in M} L(u, q) = J(u) = \alpha, \quad \forall q \in M,$$

$$L(v, p) \geq \inf_{v \in K} L(v, p) = I^*(p) = \alpha, \quad \forall v \in K.$$

According to Lemma B.3, (u, p) is a saddle point.

Let (u, p) be a saddle point, i.e.,

$$L(u, q) \leq L(u, p) \leq L(v, p), \quad \forall v \in K, q \in M.$$

From this double inequality we obtain

$$J(u) = \sup_{q \in M} L(u, q) \leq L(u, p) \leq L(v, p) \leq \sup_{q \in M} L(v, q) = J(v), \quad \forall v \in K.$$

Hence, $u \in K$ is a minimizer. Analogously,

$$I^*(p) = \inf_{v \in K} L(v, p) \geq L(u, p) \geq L(u, q) \geq \inf_{v \in K} L(v, q) = I^*(q), \quad \forall q \in M$$

and $p \in M$ is a maximizer. Furthermore,

$$L(u,p) \leq \sup_{q \in M} L(u,q) = J(u) \leq L(u,p),$$

$$L(u,p) \geq \inf_{v \in K} L(v,p) = I^*(p) \geq L(u,p),$$

and the relation (B.28) follows. □

At the end of this concise overview of the saddle point theory, we expose sufficient conditions, which are useful in checking the correctness of particular saddle point problems.

Theorem B.7 *Let the assumptions imposed on L and the sets K hold. Assume that the sets K and M are bounded or there exist elements $p_0 \in M$ and $u_0 \in K$ such that*

$$L(v_k, p_0) \to +\infty \quad \text{for any sequence } \{v_k\} \in K \text{ such that } \|v_k\|_V \to +\infty, \tag{B.29}$$

$$L(u_0, q_k) \to -\infty \quad \text{for any sequence } \{q_k\} \in M \text{ such that } \|q_k\|_Q \to +\infty. \tag{B.30}$$

Then, the Lagrangian L has at least one saddle point on $K \times M$.

Remark B.2 It is possible to prove that a saddle point exists if K is bounded and the coercivity condition for q holds (or M is bounded and the coercivity condition for v holds). It is also worth noting that the basic relation $\inf \mathcal{P} = \sup \mathcal{P}^*$ is true even if only one of the above coercivity conditions holds. Proofs and a more detailed exposition of the saddle point theory can be found in [ET76].

B.3.2 Saddle Point Statements of Linear Elliptic Problems

We discuss mathematical statements of elliptic boundary value problems, which generate mixed finite element approximations. All the main principles of this approach can be demonstrated with the paradigm of the problem

$$\operatorname{div} A\nabla u + f = 0 \quad \text{in } \Omega, \tag{B.31}$$

$$u = u_0 \quad \text{on } \Gamma_D, \tag{B.32}$$

$$A\nabla u \cdot n = F \quad \text{on } \Gamma_N, \tag{B.33}$$

where $A \in \mathbb{M}^{d\times d}$ is a matrix with bounded entries, which satisfies the condition

$$c_1^2|\xi|^2 \leq A(x)\xi\cdot\xi \leq c_2^2|\xi|^2, \quad \forall \xi \in \mathbb{R}^d, \text{ for a.e. } x \in \Omega. \tag{B.34}$$

We assume that $u_0 \in V = H^1(\Omega)$, $f \in L_2(\Omega)$, $F \in L_2(\Gamma_N)$. In mixed methods, the system (B.31)–(B.33) is represented with the help of two functions: u and p (this vector-valued function is often called a "flux"). Now the system reads

$$\operatorname{div} p + f = 0 \quad \text{in } \Omega, \tag{B.35}$$

$$p = A\nabla u \quad \text{in } \Omega, \tag{B.36}$$

$$u = u_0 \quad \text{on } \Gamma_D, \tag{B.37}$$

$$p \cdot n = F \quad \text{on } \Gamma_N, \tag{B.38}$$

and the "solution" is a pair of functions (u, p) satisfying (B.35)–(B.38) in a generalized sense, which is explained below. We assume that

$$u \in V_0 + u_0 := \{v \in V \mid v = w + u_0,\, w \in V_0\}, \quad V_0 := \{v \in V \mid v = 0 \text{ on } \Gamma_D\},$$

and $p \in Q := L_2(\Omega, \mathbb{R}^d)$. By

$$\|q\|_A^2 := \int_\Omega Aq \cdot q \,\mathrm{d}x \quad \text{and} \quad \|q\|_{A^{-1}}^2 := \int_\Omega A^{-1} q \cdot q \,\mathrm{d}x$$

we denote norms generated by A and A^{-1} (for which a two-sided estimate similar to (B.34) holds with constants $\bar{c}_1$ and $\bar{c}_2$). It is clear that these norms are equivalent to the usual norm of $L^2(\Omega, \mathbb{R}^d)$. The Lagrangian

$$L(v, q) = \int_\Omega \left(\nabla v \cdot q - \frac{1}{2} A^{-1} q \cdot q - f v \right) \mathrm{d}x - \int_{\Gamma_N} F v \,\mathrm{d}s$$

generates two functionals. The first one is

$$J(v) := \sup_{q \in Q} L(v, q) = \frac{1}{2} \|\nabla v\|_A^2 - \ell(v), \quad \ell(v) := \int_\Omega f v \,\mathrm{d}x + \int_{\Gamma_N} F v \,\mathrm{d}s. \tag{B.39}$$

It leads to Problem $\mathcal{P}$: Find $u \in V_0 + u_0$ such that

$$\inf_{w \in V_0 + u_0} J(w) = \inf \mathcal{P} = J(u).$$

It is clear that $J(v)$ is convex, continuous, and coercive on $V_0 + u_0$, so that the minimizer u exists.

Another (dual) functional is defined by the relation

$$I^*(q) = \inf_{v \in V_0 + u_0} \left\{ \int_\Omega \left(\nabla v \cdot q - \frac{1}{2} A^{-1} q \cdot q \right) \mathrm{d}x - \ell(v) \right\}.$$

We represent v as $u_0 + w$, where w is an element of the linear subspace V_0 and see that the infimum is finite if and only if q belongs to a special set, namely,

$$q \in Q_\ell := \left\{ q \in Q \,\middle|\, \int_\Omega (\nabla w \cdot q - f w) \,\mathrm{d}x - \int_{\Gamma_N} F w \,\mathrm{d}s = 0,\ \forall w \in V_0 \right\}.$$

For any $q \in Q_\ell$, we have

$$I^*(q) := -\frac{1}{2}\|q\|^2_{A^{-1}} - \ell(u_0) + \int_\Omega \nabla u_0 \cdot q \, \mathrm{d}x. \tag{B.40}$$

Hence, the dual Problem $\mathcal{P}^*$ reads as follows:

$$\sup_{q \in Q_\ell} \left(-\frac{1}{2}\|q\|^2_{A^{-1}} - \ell(u_0) + \int_\Omega \nabla u_0 \cdot q \, \mathrm{d}x \right). \tag{B.41}$$

The functional $-I^*$ is convex, continuous, and coercive. The set Q_ℓ is an affine manifold, so that it is convex. It is easy to see that it is closed in Q. Hence, Theorem B.5 guarantees the existence of a maximizer p.

It is easy to show that $J(u) = I^*(p)$. We know that $J(u) \geq I^*(p)$. Take $\bar{q} := A\nabla u$ and note that

$$\int_\Omega A\nabla u \cdot \nabla w \, \mathrm{d}x = \ell(w), \quad \forall w \in V_0.$$

Thus, $\bar{q} \in Q_\ell$ and

$$\begin{aligned}\int_\Omega \nabla u_0 \cdot \bar{q} \, \mathrm{d}x - \ell(u_0) &= \int_\Omega \nabla u \cdot \bar{q} \, \mathrm{d}x - \ell(u) + \int_\Omega \nabla(u_0 - u) \cdot \bar{q} \, \mathrm{d}x - \ell(u_0 - u) \\ &= \int_\Omega \nabla u \cdot A\nabla u \, \mathrm{d}x - \ell(u).\end{aligned}$$

Also

$$-\frac{1}{2}\|\bar{q}\|^2_{A^{-1}} = -\frac{1}{2}\int_\Omega A^{-1}(A\nabla u) \cdot (A\nabla u) \, \mathrm{d}x = -\frac{1}{2}\int_\Omega A\nabla u \cdot \nabla u \, \mathrm{d}x,$$

and we find that

$$I^*(\bar{q}) = \frac{1}{2}\|\nabla u\|^2_A - \ell(u) = J(u).$$

By Theorem B.6, we conclude that the saddle point (u, p) exists and satisfies the relations

$$\inf_{v \in V_0 + u_0} J(v) := \inf \mathcal{P} = L(u, p) = \sup \mathcal{P}^* := \sup_{q \in Q_\ell} I^*(q). \tag{B.42}$$

The problem of finding $(u, p) \in (V_0 + u_0) \times Q$ such that

$$L(u, q) \leq L(u, p) \leq L(v, p), \quad \forall q \in Q, \forall v \in V_0 + u_0, \tag{B.43}$$

leads to the *Primal Mixed Form* of the problem (B.31)–(B.33).

What relations follow from (B.43)? Consider the left-hand inequality. For any $\lambda > 0$, we have

$$L(u, p+\lambda\eta) \leq L(u, p), \quad \forall \eta \in Q,$$

which implies the relation

$$\int_\Omega \left(\lambda \nabla u \cdot \eta - \frac{\lambda^2}{2} A^{-1}\eta \cdot \eta - \lambda A^{-1} p \cdot \eta\right) \mathrm{d}x \leq 0.$$

Dividing by λ, we obtain

$$\int_\Omega (\nabla u \cdot \eta - A^{-1} p \cdot \eta)\, \mathrm{d}x \leq \int_\Omega \frac{\lambda}{2} A^{-1}\eta \cdot \eta\, \mathrm{d}x, \quad \forall \lambda > 0.$$

This inequality holds if and only if

$$\int_\Omega (\nabla u \cdot \eta - A^{-1} p \cdot \eta)\, \mathrm{d}x = 0, \quad \forall \eta \in Q.$$

Thus, $\nabla u = A^{-1} p$ and (B.36) holds at almost all points of Ω.

By the right-hand inequality, we have

$$L(u, p) \leq L(u+w, p), \quad \forall w \in V_0.$$

Hence,

$$\int_\Omega \nabla w \cdot p\, \mathrm{d}x - \ell(w) \geq 0, \quad \forall w \in V_0.$$

Since V_0 is a linear manifold, this relation holds as equality.

We see that the saddle point $(u, p) \in (V_0 + u_0) \times Q$ satisfies the integral relations

$$\int_\Omega (A^{-1} p - \nabla u) \cdot \eta\, \mathrm{d}x = 0, \quad \forall \eta \in Q, \tag{B.44}$$

$$\int_\Omega p \cdot \nabla w\, \mathrm{d}x - \ell(w) = 0, \quad \forall w \in V_0, \tag{B.45}$$

which define the generalized solution of the above mixed problem. Substituting $p = A\nabla u$ (this relation follows from (B.44)) into (B.45), we arrive at the integral identity

$$\int_\Omega A\nabla u \cdot \nabla w\, \mathrm{d}x = \ell(w), \quad \forall w \in V_0, \tag{B.46}$$

which defines the generalized solution of (B.31)–(B.33) and is a particular form of (B.6).

Remark B.3 It is worth noting that, in the primal mixed problem, the *constitutive relation* (i.e., a physical law postulated for the model) $p = A\nabla u$ is satisfied in the sense of $L_2(\Omega)$ and the *conservation law* (the equation of balance) $\operatorname{div} p + f = 0$ and the boundary condition $p \cdot n = F$ on Γ_N are satisfied in the sense of integral relation (B.45), which involves trial functions from a narrower space V_0. If p is sufficiently regular, then the relations (B.35) and (B.38) are satisfied in the classical sense.

Another mixed form of the boundary value problem (B.31)–(B.33) arises if we represent L in a somewhat different form and define the saddle point in $\widehat{Q} \times \widehat{V}$, where $\widehat{Q} := H(\Omega; \operatorname{div})$ and $\widehat{V} := L^2(\Omega)$. It is called the *dual mixed statement*.

First, we introduce a functional $g : (V_0 + u_0) \times \widehat{Q} \to \mathbb{R}$ by the relation

$$g(v, q) := \int_\Omega (\nabla v \cdot q + v \operatorname{div} q)\, dx.$$

In fact, we select a flux q even in a narrower set

$$\widehat{Q}_F := \left\{ q \in \widehat{Q} \;\middle|\; g(w, q) = \int_{\Gamma_N} F w \, ds, \ \forall w \in V_0 \right\}.$$

Then, using integration by parts, we obtain

$$\begin{aligned} L(v, q) &= \int_\Omega \left(\nabla v \cdot q - \frac{1}{2} A^{-1} q \cdot q \right) dx - \ell(v) \\ &= g(v, q) - \int_\Omega v \operatorname{div} q \, dx - \frac{1}{2} \|q\|^2_{A^{-1}} - \ell(v), \end{aligned}$$

For any $q \in \widehat{Q}_F$ and $w \in V_0$, we have

$$g(v, q) = g(w + u_0, q) = g(w, q) + g(u_0, q) = \int_{\Gamma_N} F w \, ds + g(u_0, q).$$

Therefore, if the variable q is taken not in Q (as in the primal mixed form) but in the narrower set $\widehat{Q}_F$, then the Lagrangian can be represented in a different form:

$$\begin{aligned} \widehat{L}(v, q) &= g(v, q) - \int_\Omega (v \operatorname{div} q + f v)\, dx - \frac{1}{2} \|q\|^2_{A^{-1}} - \int_{\Gamma_N} F v \, ds \\ &= -\frac{1}{2} \|q\|^2_{A^{-1}} - \int_\Omega v \operatorname{div} q \, dx - \int_\Omega f v \, dx - \int_{\Gamma_N} F u_0 \, ds + g(u_0, q). \end{aligned}$$

We see that $\widehat{L}$ is defined on a much wider set of primal functions $v \in \widehat{V}$.

The problem of finding $(\widehat{u}, \widehat{p}) \in \widehat{V} \times \widehat{Q}_F$ such that

$$\widehat{L}(\widehat{u}, \widehat{q}) \le \widehat{L}(\widehat{u}, \widehat{p}) \le \widehat{L}(\widehat{v}, \widehat{p}), \quad \forall \widehat{q} \in \widehat{Q}_F, \forall \widehat{v} \in \widehat{V} \tag{B.47}$$

is called the *Dual Mixed Form* of the problem (B.31)–(B.33).

From (B.47) we obtain necessary conditions for the dual mixed statement. Since

$$\widehat{L}(\widehat{u},\widehat{q}) \le \widehat{L}(\widehat{u},\widehat{p}), \quad \forall \widehat{q} \in \widehat{Q}_F,$$

we have

$$-\frac{1}{2}\|\widehat{p}+\lambda\eta\|^2_{A^{-1}} - \int_\Omega (\widehat{u}\,\mathrm{div}\,\widehat{p}+\lambda\eta + f\widehat{u})\,\mathrm{d}x - \int_{\Gamma_N} Fu_0\,\mathrm{d}s + g(u_0,\widehat{p}+\lambda\eta)$$
$$\le -\frac{1}{2}\|\widehat{p}\|^2_{A^{-1}} - \int_\Omega \widehat{u}\,\mathrm{div}\,\widehat{p}\,\mathrm{d}x - \int_\Omega f\widehat{u}\,\mathrm{d}x - \int_{\Gamma_N} Fu_0\,\mathrm{d}s + g(u_0,\widehat{p}),$$

where λ is a real number and η is a function in $\widehat{Q}_0 := \widehat{Q}_F$ with $F = 0$. Now, we arrive at the relation

$$-\lambda\int_\Omega \left(A^{-1}\widehat{p}\cdot\eta + \widehat{u}\,\mathrm{div}\,\eta\right)\mathrm{d}x + \lambda g(u_0,\eta) \le \frac{\lambda^2}{2}\int_\Omega A^{-1}\eta\cdot\eta\,\mathrm{d}x,$$

which is equivalent to

$$\int_\Omega \left(A^{-1}\widehat{p}\cdot\eta + \widehat{u}\,\mathrm{div}\,\eta\right)\mathrm{d}x - g(u_0,\eta) \ge \frac{\lambda}{2}\int_\Omega A^{-1}\eta\cdot\eta\,\mathrm{d}x.$$

Since $\lambda > 0$ can be taken arbitrarily small and $\widehat{Q}_0$ is a linear manifold, the latter relation holds if and only if

$$\int_\Omega \left(A^{-1}\widehat{p}\cdot\eta + \widehat{u}\,\mathrm{div}\,\eta\right)\mathrm{d}x - g(u_0,\eta) = 0, \quad \forall \eta \in \widehat{Q}_0.$$

Another saddle point inequality

$$\widehat{L}(\widehat{u},\widehat{p}) \le \widehat{L}(\widehat{u}+\widehat{v},\widehat{p}), \quad \forall \widehat{v} \in \widehat{V} := L^2(\Omega)$$

yields

$$\int_\Omega (\widehat{v}\,\mathrm{div}\,\widehat{p} + f\widehat{v})\,\mathrm{d}x = 0.$$

Thus, we arrive at the system

$$\int_\Omega \left(A^{-1}\widehat{p}\cdot\eta + \mathrm{div}\,\eta\widehat{u}\right)\mathrm{d}x = g(u_0,\eta), \quad \forall \eta \in \widehat{Q}_0, \tag{B.48}$$

$$\int_\Omega (\mathrm{div}\,\widehat{p} + f)\widehat{v}\,\mathrm{d}x = 0, \quad \forall \widehat{v} \in \widehat{V}. \tag{B.49}$$

Remark B.4 Now the condition $\mathrm{div}\,\widehat{p} + f = 0$ is satisfied in a "strong" sense (for all trial functions $L^2(\Omega)$), the Neumann boundary condition is viewed as the essential boundary condition, and the relation $\widehat{p} = A\nabla\widehat{u}$ and a Dirichlet type boundary condition are satisfied in a weaker sense.

These properties of the dual mixed statement are employed in the corresponding dual mixed finite element method (see the next section). This method generates approximations, which satisfy equilibrium type relations much better than the standard finite element approximations. This fact is important in many applications, where strict satisfaction of the equilibrium relations is indeed required.

The Lagrangian $\widehat{L}$ also generates two functionals

$$\widehat{J}(\widehat{v}) := \sup_{\widehat{q} \in \widehat{Q}_F} \widehat{L}(\widehat{v}, \widehat{q}) \quad \text{and} \quad \widehat{I}^*(\widehat{q}) := \inf_{\widehat{v} \in \widehat{V}} \widehat{L}(\widehat{v}, \widehat{q}).$$

The corresponding two variational problems are

$$\widehat{\mathcal{P}}: \inf_{\widehat{v} \in \widehat{V}} \widehat{J}(\widehat{v}) \quad \text{and} \quad \widehat{\mathcal{P}}^*: \sup_{\widehat{q} \in \widehat{Q}_F} \widehat{I}^*(\widehat{q}).$$

Note that the functional $\widehat{J}$ (unlike J) does not have a simple explicit form. However, we can prove the solvability of Problem $\widehat{\mathcal{P}}$ by the following Lemma.

Lemma B.5 *For any $\widehat{v} \in \widehat{V}$ and $F \in L_2(\Gamma_N)$, there exists $p^v \in \widehat{Q}_F$ such that*

$$\operatorname{div} p^v + \widehat{v} = 0 \quad \text{in } \Omega, \tag{B.50}$$

$$\|p^v\|_{A^{-1}} \le C_\Omega \big(\|\widehat{v}\| + \|F\|_{\Gamma_N} \big). \tag{B.51}$$

Proof We know that the problem

$$\begin{aligned} \operatorname{div} A\nabla u^v + \widehat{v} &= 0 \quad \text{in } \Omega, \\ u^v &= 0 \quad \text{on } \Gamma_D, \\ A\nabla u^v \cdot n &= F \quad \text{on } \Gamma_N \end{aligned}$$

has a (unique) solution $u^v \in V_0$, which satisfies the energy estimate

$$\|\nabla u^v\|_A \le C_\Omega \big(\|\widehat{v}\| + \|F\|_{\Gamma_N} \big).$$

Let $p^v := A\nabla u^v$. We have $\operatorname{div} p^v + \widehat{v} = 0$. Obviously, $p^v \in \widehat{Q}_F$ and, since

$$\|p^v\|_{A^{-1}}^2 = \int_\Omega A^{-1}(A\nabla u^v) \cdot (A\nabla u^v)\, dx = \|\nabla u^v\|_A^2,$$

we find that (B.51) also holds. □

By the Lemma B.5, we can easily prove the coercivity of $\widehat{J}$ on $\widehat{V}$. Indeed,

$$\begin{aligned}\widehat{J}(\widehat{v}) &\geq \widehat{L}(\widehat{v}, \alpha p^v)\\ &= -\frac{1}{2}\|\alpha p^v\|_{A^{-1}}^2 - \alpha\int_\Omega \widehat{v}(\operatorname{div} p^v)\,dx - \int_\Omega f\widehat{v}\,dx - \int_{\Gamma_N} Fu_0\,ds + g(u_0, \alpha p^v)\\ &= -\frac{1}{2}\alpha^2\|p^v\|_{A^{-1}}^2 + \alpha\|\widehat{v}\|^2 - \|f\|\|\widehat{v}\| + g(u_0, \alpha p^v) - \int_{\Gamma_N} Fu_0\,ds.\end{aligned}$$

Here $|g(u_0, \alpha p^v)| \leq \alpha\|p^v\|_{\operatorname{div}}\|u_0\|_{1,2,\Omega}$ and

$$\begin{aligned}\|p^v\|_{\operatorname{div}}^2 = \|p^v\|^2 + \|\operatorname{div} p^v\|^2 &\leq \frac{1}{\bar{c}_1}\|p^v\|_{A^{-1}}^2 + \|\widehat{v}\|^2\\ &\leq \frac{1}{\bar{c}_1}C_\Omega^2\left(\|\widehat{v}\| + \|F\|_{\Gamma_N}\right)^2 + \|\widehat{v}\|^2,\end{aligned}$$

where $\bar{c}_1$ is the smallest eigenvalue of A^{-1}.

Therefore

$$\widehat{J}(\widehat{v}) \geq -\frac{1}{2}\alpha^2 C_\Omega^2\|\widehat{v}\|^2 + \alpha\|\widehat{v}\|^2 + \Theta(\|\widehat{v}\|) + \Theta_0,$$

where $\Theta(\|\widehat{v}\|)$ contains terms linear with respect to $\|\widehat{v}\|$ and Θ_0 does not depend on $\widehat{v}$. Take $\alpha = 1/C_\Omega^2$. Then,

$$\widehat{J}(\widehat{v}) \geq \frac{1}{2C_\Omega^2}\|\widehat{v}\|^2 + \Theta(\|\widehat{v}\|) + \Theta_0 \to +\infty \quad \text{as } \|\widehat{v}\| \to \infty.$$

It is not difficult to prove that the functional $\widehat{J}$ is convex and lower semicontinuous. Therefore, Problem $\widehat{\mathcal{P}}$ has a solution $\widehat{u}$.

Corollary B.1 *Lemma* B.5 *implies the Inf–Sup condition for the dual mixed form*:

$$\inf_{\substack{\phi\in L^2(\Omega)\\ \psi\in L^2(\Gamma_N)}} \sup_{q\in\widehat{Q}_F} \frac{\int_\Omega \phi \operatorname{div} q\,dx + \int_{\Gamma_N}\psi q\cdot n\,ds}{\|q\|_{\operatorname{div}}(\|\phi\|^2 + \|\psi\|_{\Gamma_N}^2)^{1/2}} \geq C_0 > 0.$$

Now construct the dual functional $\widehat{I}^*$. It is easy to see that if $\operatorname{div}\widehat{q} + f = 0$ (in the L_2-sense), then

$$\begin{aligned}\widehat{I}^*(\widehat{q}) &= \inf_{\widehat{v}} \widehat{L}(\widehat{v}, \widehat{q})\\ &= \inf_{\widehat{v}}\left\{-\frac{1}{2}\|\widehat{q}\|_{A^{-1}}^2 - \int_\Omega v(\operatorname{div}\widehat{q})\,dx - \int_\Omega fv\,dx - \int_{\Gamma_N} Fu_0\,ds + g(u_0, \widehat{q})\right\}\\ &= -\frac{1}{2}\|\widehat{q}\|_{A^{-1}}^2 + g(u_0, \widehat{q}) - \int_{\Gamma_N} Fu_0\,ds\end{aligned}$$

$$= -\frac{1}{2}\|\widehat{q}\|_{A^{-1}}^2 + \int_\Omega (\nabla u_0 \cdot \widehat{q} - f u_0)\,\mathrm{d}x - \int_{\Gamma_N} F u_0\,\mathrm{d}s$$

$$= \int_\Omega \nabla u_0 \cdot \widehat{q}\,\mathrm{d}x - \frac{1}{2}\|\widehat{q}\|_{A^{-1}}^2 - \ell(u_0).$$

In all other cases, $\widehat{I}^*(\widehat{q}) = -\infty$.

In the dual mixed form $\widehat{q} \in \widehat{Q}_F$. As we have seen, I^* attains finite values only if $\operatorname{div}\widehat{q} + f = 0$. This means that

$$\int_\Omega \nabla w \cdot \widehat{q}\,\mathrm{d}x = -\int_\Omega (\operatorname{div}\widehat{q}) w\,\mathrm{d}x + \int_{\Gamma_N} F w\,\mathrm{d}s = \ell(w), \qquad \forall w \in V_0$$

and $\widehat{q} \in Q_\ell$.

Thus, Problems $\mathcal{P}^*$ and $\widehat{\mathcal{P}}^*$ coincide and are reduced to the maximization of I^* on the set Q_ℓ, which means that

$$\sup \mathcal{P}^* = \sup \widehat{\mathcal{P}}^*.$$

We know that $\sup \widehat{\mathcal{P}}^* \leq \inf \widehat{\mathcal{P}}$. Set $q = p = A\nabla u$. Then,

$$\sup_{q \in \widehat{Q}_F} \inf_{\hat{v} \in \hat{V}} \widehat{L}(\hat{v}, q) \geq \inf_{\hat{v} \in \hat{V}} \widehat{L}(\hat{v}, p) = -\frac{1}{2}\|A\nabla u\|_{A^{-1}}^2 - \int_{\Gamma_N} F u_0\,\mathrm{d}s + g(u_0, p)$$

$$= -\frac{1}{2}\|\nabla u\|_A^2 - \int_{\Gamma_N} F u_0\,\mathrm{d}s + \int_\Omega (\nabla u_0 \cdot p - f u_0)\,\mathrm{d}x.$$

Here u_0 is a function satisfying the prescribed boundary conditions (we can set $u_0 = u$). We see that $\sup\inf \widehat{L}(\hat{v}, q) \geq J(u)$. By Theorem B.6, we conclude that a saddle point of $\widehat{L}$ exists and

$$\widehat{L}(\widehat{u}, \widehat{p}) = \inf \widehat{\mathcal{P}} = \sup \widehat{\mathcal{P}}^*.$$

On the other hand

$$\sup \widehat{\mathcal{P}}^* = \sup \mathcal{P}^* = \inf \mathcal{P}.$$

We infer that $\inf \widehat{\mathcal{P}} = \inf \mathcal{P}$ and $u \in V_0 + u_0$ (minimizer of the Problem $\mathcal{P}$) also minimizes $\widehat{J}$ on $\widehat{V}$.

Analogously, if $p \in Q_\ell$ is the maximizer of Problem $\mathcal{P}^*$, then

$$\int_\Omega \nabla w \cdot p\,\mathrm{d}x = \int_\Omega f w\,\mathrm{d}x + \int_{\Gamma_N} F w\,\mathrm{d}s, \qquad \forall w \in V_0.$$

From here we see that $\operatorname{div} p + f = 0$ a.e. in Ω and, hence,

$$\int_\Omega (\nabla w \cdot p + (\operatorname{div} p) w)\, dx = \int_{\Gamma_N} F w \, ds, \quad \forall w \in V_0,$$

that is, $p \in \widehat{Q}_F$. Thus, p is also the maximizer of Problem $\widehat{\mathcal{P}}^*$. We conclude that

both mixed statements have the same solution (u, p), which is in fact a generalized solution of our problem.

B.3.3 Saddle Point Statements of Nonlinear Variational Problems

Finally, we comment on minimax statements of nonlinear elliptic problems, which serve as a basis for minimax approximations and saddle point algorithms.

Let the Lagrangian L be defined on a pair of Hilbert spaces V and Y with scalar products $(\cdot, \cdot)$ and $((\cdot, \cdot))$, respectively. As before, we assume that $K \subset V$ and $\Lambda \subset Y$ are convex and closed sets.

Consider the functional

$$J(v) = \frac{1}{2} a(v, v) - (f, v) + \sup_{y \in \Lambda} \big((y, \Phi(v)) \big) \tag{B.52}$$

and the corresponding convex variational problem $\mathcal{P}$: Find $u \in K$ such that

$$J(u) = \inf_{v \in K} J(v). \tag{B.53}$$

This problem has a solution and can be represented in a somewhat different form

$$\inf_{v \in K} \sup_{y \in \Lambda} L(v, y), \tag{B.54}$$

where

$$L(v, y) := \frac{1}{2} a(v, v) - (f, v) + \big((y, \Phi(v)) \big),$$

$a(v, v) : V \times V \to \mathbb{R}$ is a bilinear form, which satisfies the inequality

$$a(v, v) \geq c_1 \|v\|_V^2, c_1 > 0, \quad \forall v \in V, \tag{B.55}$$

and $\Phi : V \to Y$ is a given functional. Henceforth, we assume that

$$\big\| \Phi(u) - \Phi(v) \big\|_Y \leq C \|u - v\|_V, \quad \forall u, v \in K. \tag{B.56}$$

In order to guarantee the existence of the minimizer u, we need to assume additionally that the functional $v \to ((y, \Phi(v)))$ is convex and l.s.c. for any y. In this case, the functional $v \to \sup_{y\in\Lambda}((y, \Phi(v)))$ is convex and l.s.c. as the supremum of convex l.s.c. functionals.

Example B.4 Consider the problem

$$\inf_{v\in V} J(v), \quad J(v) = \int_\Omega \big(g(v, \nabla v) + \alpha|v|\big)\,\mathrm{d}x, \alpha \geq 0.$$

We assume that the integrand g is convex, continuous, and differentiable. Also, we assume that it is coercive on V so that the problem $\inf_{v\in V} J(v)$ is correctly stated. However, it is generated by a non-differentiable functional, which may lead to known difficulties in the process of minimization. We can avoid them if we reformulate the problem and represent it in the minimax form.

Define the set

$$K = \big\{y \in L_\infty(\Omega) \mid |y(x)| \leq 1 \text{ in } \Omega\big\}.$$

It is not difficult to show that

$$\int_\Omega |v|\,\mathrm{d}x = \sup_{y\in K} \int_\Omega yv\,\mathrm{d}x.$$

Then, the Lagrangian associated with the functional J is

$$L(v, y) = \int_\Omega \big(g(v, \nabla v) + \alpha yv\big)\,\mathrm{d}x.$$

We note that L is differentiable with respect to both variables v and y. Thus, passing to a minimax setting allows us to exclude non-differentiable terms, which may cause serious technical (computational) difficulties.

Example B.5 Another example is presented by the classical Stokes problem. In this case,

$$J(v) = \int_\Omega \big(\nu|\varepsilon(v)|^2 - fv\big)\,\mathrm{d}x, \quad \nu > 0,$$

where $v \in V := \overset{\circ}{H}{}^1(\Omega, \mathbb{R}^d)$ and $\varepsilon(v)$ is the tensor of small strains. The problem is to minimize $J(v)$ on a subset that contains only solenoidal (divergence free) fields (we denote this set by $\overset{\circ}{\mathrm{S}}(\Omega)$). We set $\Lambda = \overset{\circ}{L}(\Omega, \mathbb{R}^d)$, $Y = L^2(\Omega)$, and

$$L(v, q) = \frac{1}{2}a(v, v) - (f, v) + \int_\Omega q\,\mathrm{div}(v)\,\mathrm{d}x,$$

where $a(u, v) = \int_\Omega \nu\varepsilon(u) : \varepsilon(v)\,\mathrm{d}x$. It is easy to see that

$$J(v) = \sup_{q\in\Lambda} L(v, q).$$

Hence, finding a saddle point of $L(v, q)$ on $V \times \Lambda$ means solving the Stokes problem.

Example B.6 Consider the problem of minimizing a convex functional J on a convex closed set $K \in V$. It can be written in the minimax form with the help of the characteristic functional

$$\chi_K(v) = \begin{cases} 0, & v \in K, \\ +\infty, & v \notin K. \end{cases}$$

It is easy to see that

$$\inf_{v \in K} J(v) = \inf_{v \in V} \big(J(v) + \chi_K(v)\big).$$

Let $\chi_K^*(v^*) : V^* \to \mathbb{R}$ denote the Fenchel conjugate to χ_K. Here V^* is the space topologically dual to V, i.e., the space of linear continuous functionals $v^*(v)$ represented in the form $\langle v^*, v\rangle$. If V is a Hilbert space, then we write (v^*, v) instead of $\langle v^*, v\rangle$. By the definition (see Sect. A.4)

$$\chi_K^*(v^*) = \sup_{v \in V} \big(\langle v^*, v\rangle - \chi_K(v)\big).$$

With the help of χ_K^* we construct the Lagrangian

$$L(v, v^*) = \langle v^*, v\rangle + J(v) - \chi_K^*(v^*),$$

which gives the minimax form of this constrain minimization problem. In particular, if $K = \{v \in V \mid \|v\|_V \leq 1\}$, then χ_K^* has a simple form $\chi_K^*(v^*) = \|v^*\|_{V^*}$.

Many other nonlinear problems in mechanics, physics, economy, biology and other sciences cannot be stated in the form of identities (equations). Two main reasons for this are that either the energy functional J is non-differentiable or that the set of admissible functions K is not a linear manifold. Problems of such a type lead us to new mathematical objects called *variational inequalities* (see, e.g., [DL72, Glo84, GLT76, Pan85]). Let

$$J(v) = J_0(v) + j(v),$$

where $J_0(v) : V \to \mathbb{R}$ is a convex, coercive, and continuous functional having the Gateaux derivative G' (see Sect. A.4). It is assumed that $j(v) : V \to \mathbb{R}$ is convex and continuous but not necessarily differentiable. The problem $\mathcal{P}$ is to minimize J over a convex closed set $K \subset V$.

The following theorem presents one of the basic facts in the theory of convex nonlinear variational problems (often called Lions–Stampacchia lemma).

Lemma B.6 *A function u is a minimizer of the problem $\mathcal{P}$ if and only if it is a solution of the variational inequality*

$$\langle J_0'(u), w - u\rangle + j(w) - j(u) \geq 0, \quad \forall w \in K. \tag{B.57}$$

Proof We show that the minimizer of the variational problem satisfies (B.57). Assume that $u \in K$ realizes the lowest value of the functional, i.e., $J(u) = \inf \mathcal{P}$. Then,

$$J(v) \geq J(u), \quad \forall v \in K.$$

Since K is a convex set, we know that $v = \lambda u + (1 - \lambda)w$ belongs to K, provided that $w \in K$ and $\lambda \in (0, 1)$. Consequently, $J(w + \lambda(u - w)) \geq J(u)$ and

$$J_0\big(w + \lambda(u - w)\big) + j\big(w + \lambda(u - w)\big) \geq J_0(u) + j(u).$$

We recall that the functional j is convex, i.e.,

$$j\big(w + \lambda(u - w)\big) = j\big(\lambda u + (1 - \lambda)w\big) \leq \lambda j(u) + (1 - \lambda) j(w).$$

Therefore

$$J_0\big(w + \lambda(u - w)\big) - J_0(u) + (1 - \lambda)\big(j(w) - j(u)\big) \geq 0.$$

We set $\mu = 1 - \lambda$, and obtain

$$J_0\big(w + (1 - \mu)(u - w)\big) - J_0(u) + \mu\big(j(w) - j(u)\big) \geq 0, \quad \forall w \in K,$$

which implies

$$\frac{J_0(u + \mu(w - u)) - J_0(u)}{\mu} + j(w) - j(u) \geq 0.$$

Passing to the limit as $\mu \to +0$, we find that the minimizer u satisfies the variational inequality

$$\big\langle J_0'(u), w - u\big\rangle + j(w) - j(u) \geq 0, \quad \forall w \in K.$$

Now, we show that (2) implies (1). Let u be a solution of the variational inequality

$$\big\langle J_0'(u), w - u\big\rangle + j(w) - j(u) \geq 0, \quad \forall w \in K.$$

By convexity of J_0, we have $J_0(w) \geq J_0(u) + \langle J_0'(u), w - u\rangle$. Hence,

$$J_0(w) \geq J_0(u) + j(u) - j(w), \quad \forall w \in K,$$

i.e.,

$$J(u) \leq J(w), \quad \forall w \in K.$$

In other words, u is a minimizer of the problem P. □

Relation (B.57) is called an *elliptic variational inequality*. In view of the above-proved theorem, it is equivalent to the variational Problem $\mathcal{P}$.

Corollary B.2 *If* $J_0(v) = \frac{1}{2}a(v, v) - (f, v)$, *then*

$$J(v) = \frac{1}{2}a(v, v) - (f, v) + j(v)$$

and we minimize it over the set K. *Then, the corresponding variational inequality is as follows*:

$$a(u, v - u) - (f, v - u) + j(v) - j(u) \geq 0, \quad \forall v \in K. \tag{B.58}$$

We note that $J(v)$ *is continuous and strictly convex and* K *is a convex closed set. Then, by the general existence Theorem* B.5 *we conclude the minimizer of* $J(v)$ *on* K *exists and it is unique. Hence, the problem* (B.58) *has a unique solution.*

If $j(v) \equiv 0$, *then we arrive at the elliptic variational inequality of the first kind*:

$$a(u, v - u) \geq (f, v - u), \quad \forall v \in K. \tag{B.59}$$

Problems generated by (B.58) *with* $K = V$ *form another special class of elliptic variational inequalities, which are often called the variational inequalities of the second kind.*

B.4 Numerical Methods for Boundary Value Problems

Almost all approximation methods developed for boundary value problems are based upon the idea to approximate the exact solution by a certain combination of known functions ϕ, which are usually called *trial* functions. This idea probably has its origin in [Rit09]. Approximation methods can be classified by

- types of the trial functions (finite dimensional spaces),
- the mathematical statement used,
- the residual form, which the method aims to minimize.

If the trial functions belong to the energy space V, then the approximations are called *conforming* or *internal*. If they belong to a wider space, then they are *non-conforming* or *external*. Also, trial functions can be classified as *globally* or *locally supported*. In the first case, trial functions do not vanish at almost all points of Ω. Locally supported functions (which are widely used in finite element approximations) vanish everywhere except for a small supporting domain.

From the viewpoint of the mathematical origin, the methods can be generated by *classical* or *generalized (integral)*, *variational* (or *dual variational*), mixed (or dual mixed) statements of a boundary value problem.

By the "residual type" they are classified into the methods minimizing the residual in a pointwise (strong) sense (as, e.g., in the collocation method) and those that minimize it in L^2 or even in a weaker sense.

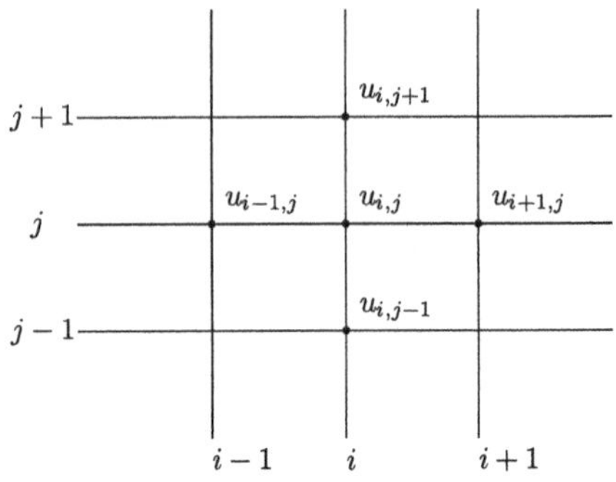

Fig. B.1 Regular finite difference grid Ω_h

B.4.1 Finite Difference Methods

The finite difference (FD) method is one of the first methods developed for quantitative analysis of PDE's. For a systematic study, we refer the reader to the books [FW60, SG89, MA81] and numerous publications cited therein.

The FD method is based on the *classical statement* of a boundary value problem and belongs to the class of *nonconforming* methods.

In the FD method, Ω is replaced by Ω_h, which is a collection of mesh nodes forming a mesh (grid). Instead of u, a mesh-function u_h defined only at nodes is considered. Let Ω be a plain domain covered by a regular mesh with nodes (ih, jh), where i and j are natural numbers and h is the mesh size (see Fig. B.1). An approximate solution to be found is a mesh function v_{ij}, which is a set of real numbers associated with mesh nodes.

The FD method exploits approximations of classical derivatives, generated by well-known incremental relations, e.g.,

$$v(x_1 + h, x_2) = v(x_1, x_2) + \frac{\partial v}{\partial x_1}(x_1, x_2)h + \frac{1}{2}\frac{\partial^2 v}{\partial^2 x_1}(x_1 + \xi, x_2)h^2,$$

where $\xi \in [x_1, x_1 + h]$. Assuming that the function v is sufficiently smooth and the last term on the right-hand side is small, we conclude that

$$\frac{\partial v}{\partial x_1}(x_1, x_2) \approx \frac{v_h(i+1, j) - v_h(i, j)}{h},$$

where i is the index related to x_1, j is related to x_2, and $v_h(i, j)$ denotes the value of the mesh function v_h at the node (ih, jh). Analogously,

$$\frac{\partial v}{\partial x_1}(x_1 - h, x_2) \approx \frac{v_h(i, j) - v_h(i-1, j)}{h}.$$

For the second derivatives, we have similar relations

$$\frac{\partial^2 v}{\partial^2 x_1}(x_1, x_2) \approx \frac{1}{h}\left(\frac{\partial v}{\partial x_1}(x_1, x_2) - \frac{\partial v}{\partial x_1}(x_1 - h, x_2)\right)$$
$$\approx \frac{v_h(i+1, j) - 2v_h(i, j) + v_h(i-1, j)}{h^2} = \Lambda_{11} v_h,$$
$$\frac{\partial^2 v}{\partial^2 x_2}(x_1, x_2) \approx \frac{v_h(i, j+1) - 2v_h(i, j) + v_h(i, j-1)}{h^2} = \Lambda_{22} v_h.$$

Hence, the operator Δ is replaced by the following difference operator

$$\Delta v \approx \Delta_h v_h := \Lambda_{11} v_h + \Lambda_{22} v_h$$
$$= \frac{v_h(i+1, j) + v_h(i, j+1) - 2v_h(i, j) + v_h(i-1, j) + v_h(i, j-1)}{h^2} \tag{B.60}$$

and the boundary value problem $\Delta u + f = 0$, $u = u_0$ on Γ (we assume that Γ is approximated by the boundary nodes of Ω_h) is replaced by the finite difference equation

$$\Delta_h u_h = f_h, \quad f_h(i, j) = f(x_i, x_j),$$
$$u_h(i, j) = u_0(x_i, x_j), \quad (x_i, x_j) \in \Gamma.$$

The latter relations lead to a system of linear simultaneous equations with respect to the unknown nodal values $u_h(i, j)$ associated with the internal nodes. This is the main idea of the FD method, which suggests to replace the differential problem

$$\mathcal{L}u = f \tag{B.61}$$

associated with the operator $\mathcal{L}$ by the finite difference equation

$$\mathcal{L}_h u_h = f_h, \tag{B.62}$$

where $\mathcal{L}_h$ is the *mesh-operator*, $u_h \in V_h$, and V_h is the space of *mesh-functions*. The first condition, which must be satisfied, is that all discrete problems $\mathcal{L}_h u_h = f_h$ are solvable.

Let $\pi_h : V \to V_h$ be an interpolation operator which transforms functions into mesh functions. Then, the mesh function $e_h = u_h - \pi_h(u)$ is the error of the FD scheme, and the quantity $\|e_h\|_h$ (which is computed with the help of a mesh norm $\|\cdot\|_h$) is an error measure. If $\mathcal{L}$ is a linear operator, then

$$\mathcal{L}_h e_h = \mathcal{L}_h u_h - \mathcal{L}_h(\pi_h u) = f_h - \mathcal{L}_h(\pi_h u).$$

Note that $\pi_h(\mathcal{L}u) = \pi_h f$. Therefore, we obtain the following relation of importance for the error analysis of FD approximations:

$$\mathcal{L}_h e_h = f_h - \pi_h f + \pi_h(\mathcal{L}u) - \mathcal{L}_h(\pi_h u). \tag{B.63}$$

Here $f_h - \pi_h f$ is the approximation of the right-hand side (and of the boundary conditions) and $\pi_h(\mathcal{L}u) - \mathcal{L}_h(\pi_h u)$ is the error arising owing to the approximation of the differential operator $\mathcal{L}$ by a mesh-operator $\mathcal{L}_h$.

For example, let $\mathcal{L} = \Delta$ and π_h be defined in the simplest way, i.e.,

$$(\pi_h w)(i, j) = w(x_i, x_j).$$

Then, $\pi_h(\mathcal{L}u) = \Delta u(ih, jh)$. Using the Lagrange formula, it is not difficult to show that for smooth functions $\Delta_h(\pi_h u) = \Delta u + ch^2$, where the constant c depends on the values of fourth order derivatives of u. Thus, under these (rather restrictive) assumptions the difference operator Δ_h approximates Δ with the accuracy h^2. Regrettably, this fact it is not sufficient to prove that $e_h \to 0$.

In order to guarantee that errors tend to zero as h tends to zero, we must justify another condition (*stability*).

Definition B.2 A finite deference scheme is correct if (a) for any f_h there exists a unique mesh-solution u_h and (b) a constant $c > 0$ exists such that

$$\|u_h\|_h \leq c\|f_h\|_h, \quad \forall f_h \in V_h, \tag{B.64}$$

where c does not depend on f_h.

The stability condition means that mesh-norms of discrete solutions are controlled by the mesh-norm $\|f_h\|_h$. We note that stability is a crucial property not only for the FD method, but also for many other numerical methods. The notion of stability can be formulated in a very general form. Consider an abstract problem $\mathcal{L}u_f = f$, where $\mathcal{L} : X \to Y$. If we can prove that

$$\|u_f\|_X \leq c\|f\|_Y, \quad \forall f \in Y, \tag{B.65}$$

with the positive constant c independent of f, then the problem associated with the operator $\mathcal{L}$ is stable.

For example, if $\mathcal{L} = \Delta$ is understood as a differentiable operator that acts from $X = \{u \in C^2(\Omega), u = u_0 \text{ on } \Gamma\}$ to $Y = C(\Omega)$ (or from $X = \{u \in H^2(\Omega), u = u_0 \text{ on } \Gamma\}$ to L^2), then, in general, we cannot prove (B.65).

However, if we consider a generalized solution (for $f \in L^2$) and understand $\mathcal{L}$ as an operator from H^1 to L^2, then by the relation

$$\int_\Omega \nabla u_f \cdot \nabla w \,\mathrm{d}x = \int_\Omega f w \,\mathrm{d}x,$$

we find that

$$\left\|\nabla(u_f - u_0)\right\|^2 \leq \|f\| \|u_f - u_0\| \leq \|f\| C_{F\Omega} \left\|\nabla(u_f - u_0)\right\|$$

and easily obtain

$$\left\|\nabla(u_f - u_0)\right\| \leq C_{F\Omega}\|f\|.$$

Since u_0 is a given function, we deduce (B.65) by the triangle inequality. Moreover, such an estimate can be proved for a wider class of right-hand sides, namely, for

$$\ell(w) = \int_\Omega (gw + \tau \cdot \nabla w)\,\mathrm{d}x, \quad g \in L^2(\Omega), \tau \in L^2(\Omega, \mathbb{R}^d).$$

Linear functionals of this type belong to the space $H^{-1}(\Omega)$, so that Δ possesses stability as an operator $V_0 \to H^{-1}(\Omega)$.

Let us now formulate and prove a theorem that is central in error analysis of FD methods.

Theorem B.8 *Let a solution of the differential problem exist and possess necessary differentiability properties. If the FD scheme approximates the differential operator and the function* f *with power* h^k *and it is stable, then*

$$\|e_h\|_h \le Ch^k. \tag{B.66}$$

Proof In view of the approximation property, we have

$$\left\|\pi_h(\mathcal{L}u) - \mathcal{L}_h(\pi_h u)\right\|_h \le ch^k,$$
$$\|f - f_h\|_h \le ch^k.$$

Then, from (B.63) if follows that $\|g_h\|_h \le 2ch^k$, where

$$\mathcal{L}_h e_h = g_h := f_h - \pi_h f + \pi_h(\mathcal{L}u) - \mathcal{L}_h(\pi_h u).$$

In view of the stability property,

$$\|e_h\|_h \le C\|g_h\|_h \le 2cCh^k. \qquad \square$$

We end up this section with comments related to practical applications of classical FD schemes. The FD methods have clear advantages: they are relatively simple and the derivation of a discrete problem is very transparent. They are often used in the analysis of evolutionary problems of type

$$\frac{\partial u}{\partial t} + Au = f,$$

where $u = u(x, t)$, $t \in (0, T)$. If time derivatives are replaced by the finite difference relation, then the following two numerical schemes arise

$$u^{k+1} = (I - \tau A)u^k + \tau f^k, \quad 0 \le \tau k \le T \text{ (explicit)}, \tag{B.67}$$
$$u^{k+1} = (I + \tau A)^{-1}u^k + \tau f^k, \quad 0 \le \tau k \le T \text{ (implicit)}. \tag{B.68}$$

It is well-known that the first scheme is stable only for sufficiently small τ, unlike the second one which is unconditionally stable (provided that A possesses a positive discrete spectrum). Such schemes are efficiently used in computer simulation methods.

However, for elliptic problems the FD methods are used not very often. This is due to several properties of the method, namely

- Strong assumptions on the differentiability of exact solutions, which may not hold in many practically interesting problems.
- Difficulties in approximation of complicated boundaries.
- Verification of stability (especially for nonlinear problems) may be a very difficult task.

Practical implementation of FD methods is often based on heuristic grounds. In engineering and scientific computations, analysts typically justify the results using model problem(s) (where solutions are known). A suitable value of h usually comes not from theoretical estimates, but is due to comparison of results computed for meshes with various h. Reliable justification of numerical results obtained by FD method may be a difficult problem, which however can be solved by means of the technique discussed in Sect. 6.1.

B.4.2 Variational Difference Methods

Various variational methods originate from of the *Ritz method* [Rit09], which is based on the variational statement and minimizes the corresponding energy functional J on a certain finite dimensional space V_n formed by a collection of linearly independent functions $\{v_i\}$, $i = 1, 2, \ldots, n$. In the Ritz method, an approximation is sought in the form

$$u_n = \sum_{i=1}^{n} \alpha_i v_i. \tag{B.69}$$

Example B.7 If $J(v) = \frac{1}{2}a(v, v) - (f, v)$, then we arrive at the problem

$$\min_{\alpha_i} \sum_i \sum_j \left(\frac{1}{2} a_{ij} \alpha_i \alpha_j - f_i \alpha_i \right), \quad a_{ij} = a(v_i, v_j),\ f_i = (f, v_i),$$

which is reduced to a system of linear simultaneous equations with respect to the unknown coefficients α.

Definition B.3 We say that a collection of spaces $\{V_n\}$ is *limit dense* in V if for any $v \in V$ and small $\varepsilon > 0$, we can find a natural number $n^*(v)$ such that

$$\text{dist}\{v, V_n\} := \inf_{w \in V_n} \|v - w\|_V \le \varepsilon, \quad \forall n \ge n^*(v). \tag{B.70}$$

In other words, if $\{V_n\}$ is limit dense in V, then for any $v \in V$ one can find a sequence v_k such that $v_k \in V_k$ and $v_k \to v$ strongly in V.

Let $J : V \to \mathbb{R}$ be convex, continuous, and coercive on V, and the collection of spaces $\{V_n\}$ be *limit dense* in V. Then, it is easy to prove that the sequence u_n constructed by the variational method tends to the exact minimizer u as $n \to +\infty$. Assume the opposite, i.e.,

$$\lim_{n\to+\infty} \inf J(u_n) > J(u) = \inf_{w\in V} J(w).$$

In view of the limit density property, we can find a sequence of elements $v_n \in V_n$ convergent to u in V. Then,

$$\lim_{n\to+\infty} \inf J(v_n) = J(u).$$

On the other hand, $J(v_n) \geq J(u_n) = \inf_{w_n\in V_n} J(w_n)$, and we arrive at a contradiction. Hence, $\{u_n\}$ is a minimizing sequence. Since J is coercive, we conclude that the sequence is bounded and contains a weakly convergent subsequence (which we denote by the same letters): $u_n \rightharpoonup \widetilde{u}$. Since J is weakly lower semicontinuous, we conclude that

$$\inf J = \lim_{n\to+\infty} \inf J(v_n) \geq J(\widetilde{u}),$$

whence it follows that $\widetilde{u}$ is a minimizer (if J is strictly convex, we conclude that $\widetilde{u} = u$). Thus, approximations constructed by the variational method weakly converge to the exact solution under very general assumptions on J and finite dimensional subspaces used.

In order to prove strong convergence, we need to strengthen the conditions imposed on J and assume that it is uniformly convex (see Sect. A.4), i.e., there exists a nonnegative monotone function $\Upsilon_\rho(\|u_n - u\|_V) \not\equiv 0$, $\Upsilon_\rho(0) = 0$ such that for all,

$$J\left(\frac{v_1 + v_2}{2}\right) + \Upsilon_\rho\big(\|u_n - u\|_V\big) \leq \frac{J(v_1) + J(v_2)}{2}, \quad \forall v_1, v_2 \in B(0_V, \rho).$$

If the functional J is coercive, then the minimizing sequence is bounded and, therefore, it is contained in a ball. We select ρ (radius of the ball) accordingly, set $v_1 = u_n$, $v_2 = u$, and see that

$$\Upsilon_\rho\big(\|u_n - u\|_V\big) \leq \frac{1}{2}J(u_n) + \frac{1}{2}J(u) - J\left(\frac{u_n + u}{2}\right) \leq \frac{1}{2}\big(J(u_n) - J(u)\big). \quad \text{(B.71)}$$

Hence, for such type functionals any minimizing sequence tends to the minimizer in the norm of V.

For a special but important case $J(v) = \frac{1}{2}a(v, v) - (f, v)$, we have the relation (see [Mik64])

$$\frac{1}{2}a(u - v, u - v) = J(v) - J(u). \quad \text{(B.72)}$$

In view of this relation, for quadratic type variational problems strong convergence of a minimizing sequence is guaranteed.

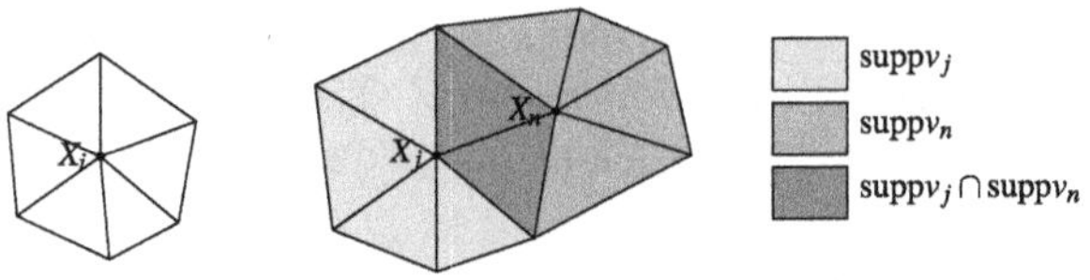

Fig. B.2 Patch (supporting set of v_i) associated with the node X_i, and the intersection of two adjacent patches associated with X_j and X_n

For a wide set of convex variational problems, the convergence of approximate solutions (conforming approximations) constructed by the variational method is easy to prove. The proof is based upon the limit density of approximation spaces and very general properties of the energy functional.

B.4.3 Petrov–Galerkin Methods

The Petrov–Galerkin method is based on the generalized statement of a boundary value problem. For example, assume that $u \in V$ is defined by the integral identity (B.6). In particular, if $\ell(v) = (f, v)$, then the problem reads

$$a(u, v) = (f, v), \quad \forall v \in V. \tag{B.73}$$

We can seek an approximate solution u_n in the form (B.69) and select the trial functions in the same subspace V_n (or in another subspace $V_{n'}$). Certainly, the selection of v_i is an important question. If all v_i belong to V, then we say that the method is conforming. In conforming approximation methods, approximate solutions are represented by functions (not mesh-functions). The function u_n satisfying the relation

$$a(u_n, v_i) = \ell(v_i), \quad \forall i = 1, 2, \ldots, n \tag{B.74}$$

is called a *Galerkin* approximation of u.

Methods using global trial functions (such as, e.g., the Ritz method) have serious technical drawbacks. First, for domains with complicated boundaries, a proper collection of trial functions v_i may be difficult to construct. Moreover, the integration of complicated global functions generates an error, and the corresponding matrix $\{a_{ij}\}$ does not have a special (sparse, n-diagonal) structure.

The idea of finite element approximations (which is contained in [Cou43]) is to use trial functions made of lower order polynomials having small supporting sets $\mathrm{supp}\, v_i = \omega_i \subset \Omega$. The set ω_i of elements associated with the node i (see Fig. B.2) is often called a *patch* of elements. Traditionally, the diameter of $\mathrm{supp}\, v_i$ is denoted by h. The subspace formed by such type functions is called V_h, and we denote the respective Galerkin solution by u_h, i.e.,

$$a(u_h, w_h) = \ell(w_h), \quad \forall w_h \in V_h. \tag{B.75}$$

In the asymptotic analysis of the finite element method (FEM) (see, e.g., [Bra07, Cia78a, SF73]), it is assumed that

$$u_h \in V_h \subset V, \quad \dim V_h = n(h) < +\infty, \quad \text{and} \quad n(h) \to +\infty \quad \text{as } h \to 0.$$

Such type conforming approximations generate dispersed matrices $\{a_{ij}\}$ whose entries are exactly computed by simple integration formulas. Moreover, these approximations are able to approximate solutions of boundary value problems in domains having complicated boundaries. Another attractive property is that the stability and convergence of the scheme is easily proved.

The convergence of a sequence of conforming approximations to the exact solution of an elliptic boundary value problem is usually proved by the method, the main idea of which is easy to demonstrate with the example of the problem (B.6).

Theorem B.9 *Let $a(\cdot,\cdot)$ be a V-elliptic bilinear form and let finite dimensional subspaces V_h be limit dense in V. Then, the sequence of Galerkin approximations u_h tends to u in V as $h \to 0$.*

Proof In view of (B.75), we have

$$c_1 \|u_h\|^2 \leq a(u_h, u_h) = \ell(u_h) \leq |\ell| \|u_h\|.$$

Whence

$$\|u_h\| \leq \frac{1}{c_1} |\ell|.$$

The sequence u_h is bounded and contains a weakly convergent subsequence (for the sake of simplicity, we denote it by u_h as well), i.e.,

$$u_h \rightharpoonup \widetilde{u} \quad \text{in } V \text{ as } h \to 0.$$

Let w be an arbitrary element of V. Since the collection of V_h is limit dense in V, we know that a sequence $w_h \in V_h$ exists such that $w_h \to v$ in V. Then,

$$a(u_h, w_h) \to a(\widetilde{u}, w), \qquad \ell(w_h) \to \ell(w),$$

and, consequently,

$$a(\widetilde{u}, w) = \ell(w), \quad \forall w \in V.$$

We conclude that $\widetilde{u}$ satisfies (B.6). We know that the generalized solution is unique. Therefore, $\widetilde{u} = u$. Now, we pass to the limit in the relation

$$\begin{aligned} a(u_h - u, u_h - u) &= a(u_h, u_h) - a(u, u_h - u) - a(u_h, u) \\ &= \ell(u_h) - a(u, u_h - u) - a(u_h, u) \end{aligned}$$

and find that

Table B.1 Different types of mixed approximations

Primal variable	Dual variable	Method
$u_h \in V_0 + u_0$	$p_h \in L^2(\Omega, \mathbb{R}^d)$	Primal Mixed (PM) Method
$u_h \in L^2(\Omega)$	$p_h \in H(\Omega, \mathrm{div})$	Dual Mixed (DM) Method
$u_h \in V_0 + u_0$	$p_h \in H(\Omega, \mathrm{div})$	Least Squares Mixed (LSM) Method

$$c_1 \|u_h - u\|^2 \le a(u_h - u, u_h - u) \to \ell(u) - a(u,u) = 0 \quad \text{as } h \to 0.$$

Thus, strong convergence of u_h to u is established. □

Nowadays, finite element methods are widely used in engineering and scientific computations. A systematic discussion of the finite element method (and other methods) for problems in structural mechanics can be found in [BLM00, Oña09].

B.4.4 Mixed Finite Element Methods

Mixed methods are based on saddle point statements of elliptic problems, in which a solution is understood as a pair of variables satisfying two integral type relations defined on a proper set of trial functions.

Consider again the problem (B.35). We know that the generalized solution to this problem u belongs to the space $H^1(\Omega)$, and the respective flux is $p = A\nabla u$ belongs to $H(\Omega, \mathrm{div})$. Different approximations of the primal and dual variables lead to different versions of mixed methods. The main three cases are presented in Table B.1.

B.4.4.1 The Primal Mixed Method

This method is based on the statement

$$\int_\Omega \left(A^{-1}p - \nabla u\right) \cdot q \,\mathrm{d}x = 0, \quad \forall q \in Q = L^2(\Omega), \tag{B.76}$$

$$\int_\Omega p \cdot \nabla w \,\mathrm{d}x - \ell(w) = 0, \quad \forall w \in V_0, \tag{B.77}$$

which defines the pair $(u, p) \in \{V_0(\Omega) + u_0\} \times L^2(\Omega)$. By $Q_h \subset Q$ and $V_{0h} \subset V_0$ we denote subspaces constructed by the FE approximation. Then, a discrete analog of this system is the Primal Mixed Finite Element Method (PMM).

In the PMM, we need to find a pair of functions $(u_h, p_h) \in (V_{0h} + u_0) \times Q_h$, where $Q_h \subset L^2(\Omega)$, such that

$$\int_\Omega \left(A^{-1}p_h - \nabla u_h\right) \cdot q_h \,\mathrm{d}x = 0, \quad \forall q_h \in Q_h, \tag{B.78}$$

$$\int_\Omega p_h \cdot \nabla w_h \,\mathrm{d}x - \ell(w_h) = 0, \quad \forall w_h \in V_{0h}. \tag{B.79}$$

In the simplest case, u_h is constructed by means of piecewise affine (C^0) elements, and p_h uses piecewise constant functions (they should satisfy the compatibility condition $A\nabla u_h \in Q_h$). With respect to the pair of spaces $(V_0 + u_0) \times Q$, this method operates with conforming approximations, but with respect to $(V_0 + u_0) \times H(\Omega, \mathrm{div})$ it should be viewed as a nonconforming method.

If p_h is a piecewise constant function, then the relation $p = A\nabla u$ is satisfied on any element T in the integral sense

$$\int_T (p_h - A\nabla u_h)\,\mathrm{d}x = 0.$$

If A is a matrix with constant entries, then

$$p_h = A\nabla u_h \quad \text{in } \Omega. \tag{B.80}$$

On the other hand, the equation of balance $\operatorname{div} p + f = 0$ is satisfied in a weaker sense: namely, the residual of this equation is orthogonal to a certain amount of trial functions in V_{0h}. From the physical point of view, this fact means that, generally speaking, these approximations are more focused on the relation $p = A\nabla u$ than on the balance equation. However, the constitutive relation $p = A\nabla u$ is often known with some precision only, unlike the balance equation representing the basic energy conservation principle, which must be exactly satisfied. In other words, we are more interested in keeping the balance relation (at least integrally) on every element than making accurate satisfaction of the constitutive relation. A way of doing this is considered below.

Finally, we note that the corresponding u_h is the usual Galerkin approximation. Indeed, by (B.80) we exclude p_h and find that

$$\int_\Omega A\nabla u_h \cdot \nabla w_h \,\mathrm{d}x = \ell(w_h), \quad \forall w_h \in V_h.$$

Therefore, all the results of the approximation theory for Galerkin solutions can be used in the analysis of PM approximations.

B.4.4.2 The Dual Mixed Method

Conforming variational approximations and similar approximations generated by the PM method have an essential drawback, namely, the respective flux p_h does not satisfy the equilibrium equation even in an integral sense. For this reason, nowadays the "classical" FEM schemes (using *nodal* approximations) are often replaced by approximations based upon *edge*-type elements, which are natural for the dual mixed statement. In these approximations, the major attention is focused on the conservation (balance) relations. The corresponding theory is systematically exposed in [BF91, Bra07, RT91] and other publications.

In our simple example, the dual mixed mathematical statement is defined on the pair of spaces $L^2(\Omega) \times H(\Omega, \mathrm{div})$. As in Sect. B.3.2, the spaces and functions used in dual mixed approximations are denoted by hats. In particular, $\widehat{V} = L^2(\Omega)$. Let

$$\widehat{V}_h \subset \widehat{V}, \qquad \widehat{Q}_{0h} \subset \widehat{Q}_0, \qquad \widehat{Q}_{Fh} \subset \widehat{Q}_F.$$

A discrete analog of the dual mixed problem is as follows: Find $(\widehat{u}_h, \widehat{p}_h) \in \widehat{V}_h \times \widehat{Q}_{Fh}$ such that

$$\int_\Omega \left(A^{-1}\widehat{p}_h \cdot \widehat{q}_h + \widehat{u}_h \operatorname{div} \widehat{q}_h\right) \mathrm{d}x = g(u_0, \widehat{q}_h), \quad \forall \widehat{q}_h \in \widehat{Q}_{0h}, \tag{B.81}$$

$$\int_\Omega (\operatorname{div} \widehat{p}_h + f)\widehat{v}_h \, \mathrm{d}x = 0, \qquad \forall \widehat{v}_h \in \widehat{V}_h. \tag{B.82}$$

Approximations of the primal and dual variables must satisfy a discrete analog of the "infsup" condition (see Lemma B.5). This question is discussed in Sect. C.3.

B.4.4.3 Least Squares Mixed Method

We end up this concise overview of mixed finite elements methods with a short comment on the Least Squares Mixed FEM. If this method is applied to a second order elliptic problem (see, e.g., [CLMM94, PCL94]), then the problem is reduced to minimization of the quadratic functional

$$J(v, q) := \| \operatorname{div} q + f \|^2 + \|q - A\nabla v\|^2$$

on the product space $(V_0 + u_0) \times Q_{\Gamma_N}$, where Q_{Γ_N} contains vector- valued functions from $H(\Omega, \mathrm{div})$ satisfying the Neumann boundary conditions (the latter requirement is not indeed important and can be avoided by adding the term $\|q \cdot n - F\|^2_{\Gamma_n}$). In [PCL94]), the reader will find a discussion of a mixed FEM method that uses conforming approximations u_h and p_h in the corresponding spaces and a priori rate convergence estimates.

B.4.5 Trefftz Methods

The classical Trefftz method [Tre26] was suggested for elliptic problems with constant coefficients, such as, e.g.,

$$\Delta u = 0 \quad \text{in } \Omega, \qquad u = u_0 \quad \text{on } \Gamma.$$

Let $\phi_i, i = 1, 2, \dots, N$, be a set of linearly independent harmonic functions (i.e., they satisfy the equation $\Delta\phi_i = 0$). We define an approximate solution (see, e.g., [Mik64]) as a function

$$\widetilde{u} = \sum_{i=1}^{N} \alpha_i \phi_i$$

that minimizes the energy norm $\|\nabla(u - \widetilde{u})\|^2$. Since $\widetilde{u}$ does not satisfy the prescribed boundary condition, the Trefftz method operates with nonconforming approximations. In this method, we select the coefficients α_i in such a way that the functional

$$J(\widetilde{u}) := \int_\Omega \left(|\nabla\widetilde{u}|^2 - 2\nabla\widetilde{u} \cdot \nabla u\right) \mathrm{d}x$$

attains the minimum. Since $\widetilde{u}$ is a harmonic function, we have

$$J(\widetilde{u}) = \sum_{i,j=1}^{N} \alpha_i \alpha_j \int_\Omega \nabla\phi_i \cdot \nabla\phi_j \,\mathrm{d}x - 2 \int_{\partial\Omega} \sum_{i=1}^{N} \alpha_i \frac{\partial\phi_i}{\partial\nu} u_0 \,\mathrm{d}s.$$

The minimization of the above functional leads to a linear system with respect to α_i.

It is convenient to rewrite the functional in the form that does not include the normal derivatives of ϕ_i on the boundary. Since

$$\int_\Omega \nabla\widetilde{u} \cdot \nabla(u - u_0) \,\mathrm{d}x = 0,$$

we can rewrite the functional as

$$J(\widetilde{u}) := \int_\Omega \left(|\nabla\widetilde{u}|^2 - 2\nabla\widetilde{u} \cdot \nabla u_0\right) \mathrm{d}x$$

and define α_i by minimizing

$$\frac{1}{2} \sum_{i=1}^{N} \sum_{j=1}^{N} \alpha_i \alpha_j \int_\Omega \nabla\phi_i \cdot \nabla\phi_j \,\mathrm{d}x - \sum_{i=1}^{N} \alpha_i \int_\Omega \nabla\phi_i \cdot \nabla u_0 \,\mathrm{d}x.$$

For the equation $\operatorname{div} A\nabla u = 0$ and other elliptic equations, this method can be presented quite analogously. The classical Trefftz method admits various generalizations (e.g., see [CLP09, Her84, Jir78]).

B.4.6 Finite Volume Methods

The finite volume (FV) method was first developed for evolutionary models in order to generate approximations, which are first of all oriented toward the proper satisfaction of conservation laws (this idea was suggested in [God59]; a consequent exposition is presented in [LeV02]). Nowadays various modifications are used for solving partial differential equations of all types. The FV method is very popular in engineering computations. Mathematical studies of the method can be found in, e.g., [ABC03, EGH00, Nic06] and many other publications. Below we briefly discuss one version, which is often applied to elliptic problems. To this end, we select the problem

$$-\operatorname{div}\big(\alpha(x)\nabla u\big) = f \quad \text{in } \Omega, \tag{B.83}$$

$$u = u_0 \quad \text{on } \Gamma_D, \tag{B.84}$$

$$\alpha\nabla u \cdot \nu = g_N \quad \text{on } \Gamma_N, \tag{B.85}$$

where Ω is a bounded Lipschitz domain in $\mathbb{R}^2$, $u_0 \in H^1(\Omega)$, $f \in L^2(\Omega)$, $g_N \in L^2(\Gamma_N)$, and

$$\alpha \in L^\infty(\Omega), \quad 0 < \alpha_\ominus \leq \alpha(x) \leq \alpha_\oplus < +\infty, \quad \forall x \in \Omega, \tag{B.86}$$

are given data. The solution $u \in V_0 + u_0$ satisfies the integral relation

$$\int_\Omega \alpha\nabla u \cdot \nabla w \, \mathrm{d}x = \int_\Omega f w \, \mathrm{d}x + \int_{\Gamma_N} g_N w \, \mathrm{d}s, \quad \forall w \in V_0. \tag{B.87}$$

Let Ω be divided into a collection of simplicial cells T_i, $i = 1, 2, \ldots, N$ (e.g., triangles). We denote by Γ_i the boundary of an element T_i and by ν_i its outward normal vectors. The relation

$$-\int_{T_i} \operatorname{div} p \, \mathrm{d}x = -\int_{\Gamma_i} p \cdot \nu_i \, \mathrm{d}s = \int_{T_i} f \, \mathrm{d}x \tag{B.88}$$

reflects the *conservation law principle*. It holds for the true flux $p = \alpha\nabla u$ on each cell that has no common boundary with Γ_N. The relation

$$-\int_{\Gamma_i} p \cdot \nu_i \, \mathrm{d}s = \int_{T_i} f \, \mathrm{d}x + \int_{\Gamma_{Ni}} g_N \, \mathrm{d}s \tag{B.89}$$

presents the same law for a cell having common boundary Γ_{Ni} with Γ_N. The relations (B.88) and (B.89) form a basis of the FV method. Moreover, we use a special representation of normal fluxes in terms of the values of the approximation u_h on the cells. In the simplest case, we assume that $u_h \in P^0(T_i)$. For the control volume T_i, we select a certain point x_i called "cell center" (see Fig. B.3), where u_i denotes $u_h(x_i)$.

The flux along an edge of the triangulation is approximated by using the so-called cell-centered scheme of the finite volume method. Let $\alpha_i = \{\!|\alpha|\!\}_{T_i}$, E_{ij} be an interior edge of the triangulation, shared by two elements T_i and T_j of respective centers x_i and x_j. If we suppose that this edge E_{ij} is orthogonal to the straight line joining x_i

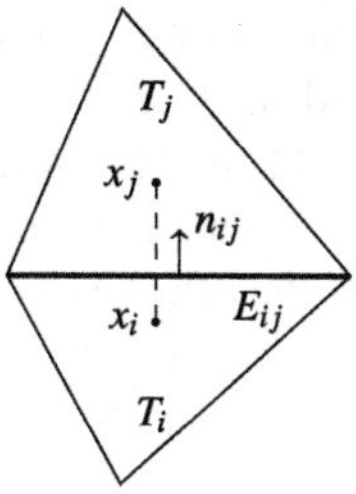

Fig. B.3 Two adjacent cells T_i and T_j, and the flux associated with the edge E_{ij}

and x_j, then

$$-\int_{E_{ij}} \alpha \nabla u \cdot \nu_i \, ds \cong -\tau_{E_{ij}}(u_j - u_i) = F_{T_i,E_{ij}}, \tag{B.90}$$

where the factor is defined as a weighted harmonic mean

$$\tau_{E_{ij}} := |E_{ij}| \frac{\alpha_i \alpha_j}{\alpha_i d_{T_j,E_{ij}} + \alpha_j d_{T_i,E_{ij}}},$$

and $d_{T_i,E_{ij}}$ denotes the minimal distance between the center x_i and the edge E_{ij}.

If an edge E_{Di} of an element T_i belongs to the Dirichlet boundary of the domain, then we denote by $d_{T_i,E_{Di}}$ the distance between the center x_i and $\bar{x}_i$, where $\bar{x}_i$ is defined by the relation $|x_i - \bar{x}_i| = \inf_{x \in E_{Di}} |x_i - x|$. Along this edge, the flux is approximated by

$$-\int_{E_{Di}} \alpha \nabla u \cdot \nu_i \, ds \cong -\tau_{E_{Di}}(u_0 - u_i) = F_{T_i,E_{Di}}, \tag{B.91}$$

where $\tau_{E_{Di}} = |E_{Di}| \frac{\alpha_i}{d_{T_i,E_{Di}}}$.

If an edge E_{Ni} belongs to the Neumann boundary, then

$$-\int_{E_{Ni}} \alpha \nabla u \cdot \nu_i \, ds = -\int_{E_{Ni}} g_N \, ds, \tag{B.92}$$

and the corresponding flux is given by the relation

$$F_{T_i,E_{Ni}} = -\int_{E_{Ni}} g_N \, ds. \tag{B.93}$$

Equations (B.88) and (B.89) then lead to the following finite volume scheme:

$$-\sum_{E_{ij} \subset \Gamma_i} F_{T_i,E_{ij}} = \int_{T_i} f \, dx, \tag{B.94}$$

$$-\sum_{E_{ij} \subset \Gamma_i} F_{T_i,E_{ij}} = \int_{T_i} f \, dx + \int_{E_{Ni}} g_N \, ds. \tag{B.95}$$

This system of equations is considered together with the relations (B.90), (B.91), and (B.93).

Then, the numerical flux associated with the internal edge E_{ij} common to T_i and T_j is defined by the relation $q_{E_{ij}} := -F_{T_i,E_{ij}}$. On Γ_{Ni} and Γ_D, the fluxes are defined by the relations $q_{E_{Ni}} = \int_{E_{Ni}} g_N \, ds$ and $q_{E_{Di}} := -F_{T_i,E_{Di}}$, respectively.

Thus, the approximations produced by the finite volume method are presented by a set of piecewise constant functions

$$u_h(x) = u_i \quad \text{for } x \in T_i$$

and the set

$$\mathfrak{Q}_h := \{q_{ij} \mid q_{ij} \in P^0(E_{ij})\}$$

of normal fluxes on the edges.

Remark B.5 Certainly, the approximations u_i and q_{ij} generated by the FV method are nonconforming (mesh) approximations. However, by these data we can construct a pair of functions

$$(\widetilde{u}_h, \widetilde{q}_h) \in H^1(\Omega) \times H(\Omega, \mathrm{div}),$$

viewed as conforming approximations of u and p, respectively. For example, we can do this as follows. Let $\mathfrak{P}_k$ be the patch related to a common node k. We define the value of $\widetilde{u}_h$ at the interior node k as an averaged value

$$\widetilde{u}_h(x_k) := \frac{\sum_{s=i_1,\dots,i_{m_k}} |T_s| u_s}{\sum_{s=i_1,\dots,i_{m_k}} |T_s|},$$

where m_k is the number of elements in the patch $\mathfrak{P}_k$. Inside T_i the function $\widetilde{u}_h$ is defined as the affine function having the above-defined values at all the nodes. For a boundary node k, we take

$$\widetilde{u}_h(x_k) := u_0(x_k),$$

assuming that $u_0|_\Gamma \in C(\Gamma)$.

The function $\widetilde{q}_h$ is defined by the extension of edge fluxes inside T_i with the help of Raviart–Thomas elements of the lowest order (RT^0 elements).

Remark B.6 A posteriori error estimates for finite volume approximations were obtained by different methods. For example, in [Zou10] a hierarchical approach was used. Papers [HO02, LT02, Nic06, Voh07a, Voh07b, Ye11] use various versions of the residual approach. Functional type a posteriori estimates for FV approximations has been suggested and tested in [CDNR09]. Modifications of a posteriori estimates adapted to singularly perturbed convection diffusion problems were studied in [Ang95, Ohl01].

B.4.7 Discontinuous Galerkin Methods

The discontinuous Galerkin (DG) method belongs to the class of *nonconforming methods*, based on weakened mixed statements. It was initially proposed in the 1970s/1980s (see, e.g., [Arn82, Bab73a, Bab73b, BZ73] and the references in [ABCM02]). Usually it uses piecewise polynomial functions that do not preserve continuity along faces of finite elements (subdomains). To discuss the basic ideas, we consider the simplest elliptic problem $\Delta u + f = 0$ with homogeneous boundary conditions on Γ.

As before, we split the equation into two relations:

$$p = \nabla u, \tag{B.96}$$

$$\operatorname{div} p + f = 0. \tag{B.97}$$

Let $\omega \subset \Omega$, $q \in H(\omega, \operatorname{div})$, and $w \in H^1(\omega)$. Since

$$\int_\omega (p \cdot q - \nabla u \cdot q)\, \mathrm{d}x = \int_\omega (p \cdot q + u \operatorname{div} q)\, \mathrm{d}x - \int_{\partial\omega} u(q \cdot n)\, \mathrm{d}s,$$

$$\int_\omega (\operatorname{div} p + f) w\, \mathrm{d}x = \int_\omega (-p \cdot \nabla w + f w)\, \mathrm{d}x + \int_{\partial\omega} (p \cdot n) w\, \mathrm{d}s,$$

we find that

$$\int_\omega p \cdot q\, \mathrm{d}x = \int_\omega -u \operatorname{div} q\, \mathrm{d}x + \int_{\partial\omega} u(q \cdot n)\, \mathrm{d}s, \tag{B.98}$$

$$\int_\omega p \cdot \nabla w\, \mathrm{d}x = \int_\omega f w\, \mathrm{d}x + \int_{\partial\omega} (p \cdot n) w\, \mathrm{d}s. \tag{B.99}$$

Assume that Ω is decomposed into a collection of subdomains Ω_i and Γ_{ij} denotes the common boundary of Ω_i and Ω_j. We define the so-called "broken" spaces

$$\widehat{V} := \{ w = \{w^i\}, w^i \in H^1(\Omega_i), i = 1, 2, \ldots, N \}$$

$$\widehat{Q} := \{ q = \{q^i\}, q^i \in H(\Omega_i, \operatorname{div}), q \cdot n \in L^2(\partial\Omega_i) \},$$

use (B.98) and (B.99), and arrive at the relations, which generate the following statement: Find $u, p \in (\widehat{V}, \widehat{Q})$ such that

$$\int_{\Omega_i} p \cdot q^i\, \mathrm{d}x = \int_{\Omega_i} -u \operatorname{div} q^i\, \mathrm{d}x + \int_{\partial\Omega_i} u\left(q^i \cdot n\right) \mathrm{d}s, \quad \forall q^i, \tag{B.100}$$

$$\int_{\Omega_i} p \cdot \nabla w^i\, \mathrm{d}x = \int_{\Omega_i} f w\, \mathrm{d}x + \int_{\partial\Omega_i} (p \cdot n) w^i\, \mathrm{d}s, \quad \forall w^i. \tag{B.101}$$

In the DG method, $\widehat{V}$ and $\widehat{Q}$ are replaced by finite dimensional spaces $\widehat{V}_h$ and $\widehat{Q}_h$, and the system (B.100)–(B.101) is replaced by a close (but different) system

$$\int_{\Omega_i} p_h \cdot q_h^i \,\mathrm{d}x = -\int_{\Omega_i} u_h \operatorname{div} q_h^i \,\mathrm{d}x + \int_{\partial\Omega_i} \widetilde{u}(q_h^i \cdot n)\,\mathrm{d}s, \quad \forall q^i \tag{B.102}$$

$$\int_{\Omega_i} p_h \cdot \nabla w_h^i \,\mathrm{d}x = \int_{\Omega_i} f w \,\mathrm{d}x + \int_{\partial\Omega_i} (\widetilde{p}\cdot n) w_h^i \,\mathrm{d}s, \quad \forall w^i, \tag{B.103}$$

where $\widetilde{u}$ and $\widetilde{p}$ are the so-called *numerical fluxes*, which are approximations of ∇u and u on the boundary $\partial\Omega_i$, respectively, and w_h^i and p_h^i are polynomials on Ω_i. Typically, Ω_i are assumed to be of size h. Unknown numerical fluxes are expressed in terms of u_h and p_h, i.e.,

$$\widetilde{u} = G_1(u_h, p_h), \qquad \widetilde{p} = G_2(u_h, p_h). \tag{B.104}$$

Different versions of the DG method make use of different forms of G_1 and G_2. We rewrite the above relations in the form

$$\int_{\Omega} p_h \cdot q_h \,\mathrm{d}x = -\int_{\Omega} u_h \widehat{\operatorname{div}} q_h \,\mathrm{d}x + \sum_{i=1}^{N} \int_{\partial\Omega_i} \widetilde{u}(q_h \cdot n)\,\mathrm{d}s, \quad \forall q_h \in \widehat{Q}_h, \tag{B.105}$$

$$\int_{\Omega} p_h \cdot \widehat{\nabla} w_h \,\mathrm{d}x = \int_{\Omega} f w \,\mathrm{d}x + \int_{\Gamma_0} (\widetilde{p}\cdot n) w_h \,\mathrm{d}s + \int_{\Gamma} (\widetilde{p}\cdot n) w_h \,\mathrm{d}s, \quad \forall w_h \in \widehat{V}_h, \tag{B.106}$$

where Γ_0 is the set of internal faces and $\widehat{\nabla}$ and $\widehat{\operatorname{div}}$ are (generalized) differential operators defined on each Ω_i.

Set $w_h = 1$ on Ω_i. Then, from (B.103) it follows that

$$\int_{\Omega_i} f \,\mathrm{d}x + \int_{\partial\Omega_i} \widetilde{p}\cdot n \,\mathrm{d}s = 0. \tag{B.107}$$

This fact means that the scheme is *conservative* on any subdomain.

Consider the face Γ_{ij}. We introduce the *mean* values of w and q on Γ_{ij}

$$\{\!|w|\!\} := \frac{1}{2}(w^i + w^j), \qquad \{\!|q|\!\} := \frac{1}{2}(q^i + q^j)$$

and *jumps*

$$[w] := w^i n_{ij} + w^j n_{ji}, \qquad [q] := q^i \cdot n_{ij} + q^j \cdot n_{ji}.$$

On Γ, we set $[w] = wn$ and $\{\!|q|\!\} = q$. Note that

$$\begin{aligned}
&\sum_{i,j=1}^{N} \int_{\Gamma_{ij}} (w_h^i q_h^i \cdot n_{ij} - w_h^j q_h^j \cdot n_{ij})\, \mathrm{d}s \\
&\quad = \sum_{i,j=1}^{N} \int_{\Gamma_{ij}} \frac{1}{2}(w_h^i n_{ij} - w_h^j n_{ij})(q_h^i + q_h^j)\, \mathrm{d}s \\
&\qquad + \sum_{i,j=1}^{N} \int_{\Gamma_{ij}} \frac{1}{2}(w_h^i + w_h^j)(q_h^i \cdot n_{ij} - q_h^j \cdot n_{ij})\, \mathrm{d}s \\
&\quad = \sum_{i,j=1}^{N} \int_{\Gamma_{ij}} (\{\!|q_h|\!\} \cdot [w_h] + \{\!|w_h|\!\}[q_h])\, \mathrm{d}s.
\end{aligned} \tag{B.108}$$

Here an integral over Γ_{ij} is zero if $\Gamma_{ij} = \emptyset$. On the external boundary, we have

$$\int_{\Gamma} w_h (q_h \cdot n)\, \mathrm{d}s = \int_{\Gamma} [w_h] \cdot \{\!|q_h|\!\}\, \mathrm{d}s. \tag{B.109}$$

Then,

$$\begin{aligned}
&\sum_i \int_{\partial\Omega_i} w_h (q_h \cdot n_i)\, \mathrm{d}s \\
&\quad = \sum_{i,j=1}^{N} \int_{\Gamma_{ij}} (\{\!|q_h|\!\} \cdot [w_h] + \{\!|w_h|\!\}[q_h])\, \mathrm{d}s + \int_{\Gamma} [w_h] \cdot \{\!|q_h|\!\}\, \mathrm{d}s
\end{aligned} \tag{B.110}$$

and

$$\begin{aligned}
&\int_{\Omega} (\widehat{\nabla} w_h \cdot q_h + w_h \widehat{\mathrm{div}} q_h)\, \mathrm{d}x \\
&\quad = \sum_i \int_{\partial\Omega_i} w_h (q_h \cdot n_i)\, \mathrm{d}s \\
&\quad = \sum_{i,j=1}^{N} \int_{\Gamma_{ij}} (\{\!|q_h|\!\} \cdot [w_h] + \{\!|w_h|\!\}[q_h])\, \mathrm{d}s + \int_{\Gamma} [w_h] \cdot \{\!|q_h|\!\}\, \mathrm{d}s.
\end{aligned} \tag{B.111}$$

We reform (B.105) and (B.106) by the above relations and arrive at the system

$$\begin{aligned}
\int_{\Omega} p_h \cdot q_h\, \mathrm{d}x &= \int_{\Omega} -u_h \widehat{\mathrm{div}} q_h\, \mathrm{d}x + \sum_{i,j=1}^{N} \int_{\Gamma_{ij}} (\{\!|q_h|\!\} \cdot [\widetilde{u}] + \{\!|\widetilde{u}|\!\}[q_h])\, \mathrm{d}s \\
&\quad + \int_{\Gamma} [\widetilde{u}]\{\!|q_h|\!\}\, \mathrm{d}s, \quad \forall q_h \in \widehat{Q}_h,
\end{aligned} \tag{B.112}$$

$$\int_{\Omega} p_h \cdot \widehat{\nabla} w_h \, \mathrm{d}x = \int_{\Omega} f w \, \mathrm{d}x + \sum_{i,j=1}^{N} \int_{\Gamma_{ij}} \left(\{\!|\widetilde{p}_h|\!\} \cdot [w_h] + \{\!|u_h|\!\}[\widetilde{p}_h] \right) \mathrm{d}s$$
$$+ \int_{\Gamma} [w_h] \cdot \{\!|p_h|\!\} \, \mathrm{d}s, \quad \forall w_h \in \widehat{V}_h. \tag{B.113}$$

Now, we set $w_h = u_h$ in (B.111) and obtain

$$\int_{\Omega} u_h \widehat{\mathrm{div}} q_h \, \mathrm{d}x = \sum_{i,j=1}^{N} \int_{\Gamma_{ij}} \left(\{\!|q_h|\!\} \cdot [u_h] + \{\!|u_h|\!\}[q_h] \right) \mathrm{d}s + \int_{\Gamma} [u_h] \cdot \{\!|q_h|\!\} \, \mathrm{d}s$$
$$- \int_{\Omega} \widehat{\nabla} u_h \cdot q_h \, \mathrm{d}x. \tag{B.114}$$

Use this relation in (B.112). Then, this equation comes in the form

$$\int_{\Omega} p_h \cdot q_h \, \mathrm{d}x = \int_{\Omega} \widehat{\nabla} u_h \cdot q_h \, \mathrm{d}x$$
$$+ \sum_{i,j=1}^{N} \int_{\Gamma_{ij}} \left(\{\!|q_h|\!\} \cdot [\widetilde{u} - u_h] + \{\!|\widetilde{u} - u_h|\!\}[q_h] \right) \mathrm{d}s$$
$$+ \int_{\Gamma} [\widetilde{u} - u_h] \cdot \{\!|q_h|\!\} \, \mathrm{d}s, \tag{B.115}$$

where q_h is an arbitrary trial function in $\widehat{Q}_h$. (B.115) is an important relation. First, using it we can express p_h in terms of u_h. Second, it defines a bilinear form B associated with the DG method.

Now, we need to introduce certain trace and lifting operators associated with the DG method. We recall that the operator γ that assigns the boundary data related to a function in ω is called the *trace operator* and the operator extending the boundary data inside ω is called the *lifting operator* μ (cf. Sect. A.2.2). We can define $\mu(q)$ in such a way that the relation

$$\int_{T} \mu(q) \cdot q \, \mathrm{d}x = - \sum_{E_{st} \in \partial T} \int_{E_{st}} q \cdot \{\!|q|\!\} \, \mathrm{d}s$$

holds for a certain set of admissible q. Let us consider the simplest example. Assume that we consider only constant q. Then, we can define

$$\mu(q) = \alpha \left\{ \int_{\partial T} q_1 \, \mathrm{d}s, \int_{\partial T} q_2 \, \mathrm{d}s \right\}.$$

From the above-defined conservation principle, it follows that α should be selected as $\alpha = -\frac{1}{|T|}$. Indeed,

$$\int_T \mu(q) \cdot q \, \mathrm{d}x = |T|\alpha\Big(q_1 \int_{\partial T} q_1 \, \mathrm{d}s + q_2 \int_{\partial T} q_2 \, \mathrm{d}s\Big)$$
$$= |T|\alpha \int_{\partial T} q \cdot q \, \mathrm{d}s \quad \Rightarrow \quad \alpha = -\frac{1}{|T|}.$$

In the DG scheme, we need two lifting operators defined for the *whole sampling*

$$\mu_1 : L^2(\Gamma_0 + \Gamma, R^2) \to \widehat{Q}_h, \qquad \mu_2 : L^2(\Gamma_0) \to \widehat{Q}_h.$$

They are defined by the relations

$$\int_\Omega \mu_1(q) \cdot q \, \mathrm{d}x = -\int_{\Gamma_0 + \Gamma} q \cdot \{\!|q|\!\} \, \mathrm{d}s,$$
$$\int_\Omega \mu_2(w) \cdot q \, \mathrm{d}x = -\int_{\Gamma_0} w[q] \, \mathrm{d}s.$$

Now we return to (B.115) and replace the terms in the right hand side, using lifting operators. We have

$$\int_{\Gamma_0 + \Gamma} \{\!|q_h|\!\} \cdot [\widetilde{u} - u_h] \, \mathrm{d}s = -\int_\Omega \mu_1\big([\widetilde{u} - u_h]\big) \cdot q_h \, \mathrm{d}x,$$
$$\int_{\Gamma_0} [q_h]\{\!|\widetilde{u} - u_h|\!\} \, \mathrm{d}s = -\int_\Omega \mu_2\big(\{\!|\widetilde{u} - u_h|\!\}\big) \cdot q_h \, \mathrm{d}x,$$

and (B.115) implies that

$$\int_\Omega p_h \cdot q_h \, \mathrm{d}x = \int_\Omega \big(\widehat{\nabla} u_h - \mu_1\big([\widetilde{u} - u_h]\big) - \mu_2\big(\{\!|\widetilde{u} - u_h|\!\}\big)\big) \cdot q_h \, \mathrm{d}x.$$

Instead of the relation $p_h = \nabla u_h$ (in the conforming FEM) we have a more sophisticated relation $p_h = \widehat{\nabla} u_h - \mu_1([\widetilde{u} - u_h]) - \mu_2(\{\!|\widetilde{u} - u_h|\!\})$, which includes additional terms depending on discontinuities.

Take (B.115) and set $q_h = \widehat{\nabla} w_h$, where w_h is an arbitrary function in the broken space $\widehat{V}_h$. We have

$$\begin{aligned}\int_\Omega p_h \cdot \widehat{\nabla} w_h \, \mathrm{d}x = &\int_\Omega \widehat{\nabla} u_h \cdot \widehat{\nabla} w_h \, \mathrm{d}x \\ &+ \int_{\Gamma_0 + \Gamma} \{\!|\widehat{\nabla} w_h|\!\} \cdot [\widetilde{u} - u_h] \, \mathrm{d}s \\ &+ \int_{\Gamma_0} \{\!|u - u_h|\!\}[\widehat{\nabla} w_h] \, \mathrm{d}s.\end{aligned} \tag{B.116}$$

Recall (B.113)

$$\int_{\Omega} p_h \cdot \widehat{\nabla} w_h \, dx = \int_{\Omega} f w_h \, dx + \int_{\Gamma_0+\Gamma} \{\!|\widetilde{p}_h|\!\} \cdot [w_h] \, ds + \int_{\Gamma_0} \{\!|u_h|\!\}[\widetilde{p}_h] \, ds.$$

We rewrite (B.116) in the standard form

$$B_h(u_h, w_h) = \int_{\Omega} f w_h \, dx, \quad \forall w_h \in \widehat{V}_h, \tag{B.117}$$

where

$$B_h(u_h, w_h) := \int_{\Omega} \widehat{\nabla} u_h \cdot \widehat{\nabla} w_h \, dx + \int_{\Gamma_0+\Gamma} \{\!|\widehat{\nabla} w_h|\!\} \cdot [\widetilde{u} - u_h] - \{\!|\widetilde{p}_h|\!\}[w_h] \, ds$$
$$+ \int_{\Gamma_0} \{\!|u - u_h|\!\} \cdot [\widehat{\nabla} w_h] - \{\!|u_h|\!\}[\widetilde{p}_h] \, ds$$

is the bilinear form of the DG method.

Example B.8 An example is given by the Bassi–Rebay method. Here

$$\widetilde{u} = \{\!|u_h|\!\} \quad \text{on } \Gamma_0, \qquad \widetilde{u} = 0 \quad \text{on } \Gamma, \qquad \widetilde{p} = \{\!|p_h|\!\} \quad \text{on } \Gamma_0 + \Gamma. \tag{B.118}$$

In this case, $\{\!|\widetilde{u} - u_h|\!\} = 0$, $[\widetilde{u} - u_h] = [u_h]$, and we have

$$p_h = \widehat{\nabla} u_h + \mu_1([u_h]),$$
$$B_h(u_h, v_h) = \int_{\Omega} \left(\widehat{\nabla} u_h \cdot \widehat{\nabla} v_h + \mu_1([u_h])\mu_1([v_h])\right) dx$$
$$- \int_{\Gamma_0+\Gamma} \left(\{\!|\widehat{\nabla} u_h|\!\} \cdot [v_h] + [u_h] \cdot \{\!|\widehat{\nabla} v_h|\!\}\right) ds.$$

Further mathematical analysis of DG schemes is mainly based on proving two properties: boundedness and stability, i.e.,

$$B_h(v_h, v_h) \geq c_1 \big|[v_h]\big|^2, \qquad \forall v_h \in V_h \tag{B.119}$$
$$B_h(u_h, v_h) \leq c_2 \big|[u_h]\big|\big|[v_h]\big|, \quad \forall u_h, v_h \in V_h, \tag{B.120}$$

where c_1 and c_2 are positive constants independent of h and $|[v_h]|$ is a suitable "broken" energy norm.

A priori rate convergence estimates for DG approximations are derived with the help of these properties and interpolation-type estimates. A systematic discussion of these questions can be found in [ABCM02] and literature cited in this paper.

A posteriori estimates for DG approximations of elliptic type equations were investigated by many authors. In [Ain07, BGC05, BHL03, EP05, ESV10, HSW07, KP03, YC06], residual type error indicators for the energy norm were suggested

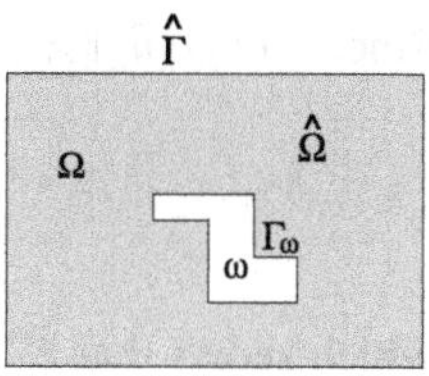

Fig. B.4 The domains Ω, ω, and $\widehat{\Omega}$

and, in [SXZ06], the authors considered a posteriori estimates based on local solutions and on gradient recovery. [Kim66, Kim07] are devoted to a posteriori error analysis for locally conservative mixed methods, with applications to P^1 nonconforming FEM and interior penalty DG (IPDG) methods and mixed FEM. A posteriori estimates in terms of L^2 norm were derived in [Cas05] for the so-called "local DG method". Methods based on equilibration of residuals are discussed in [BS08, Sch08]. In [EP05, HRS00, JS95, SW05, YC06] time-dependent (transport) equations are considered. Functional a posteriori estimates have been applied for DG approximations in [LRT09, RT11, TR09].

B.4.8 Fictitious Domain Methods

The fictitious domain method is often used for problems associated with complicated geometry. In general terms, the idea of this method is to get an approximate solution by means of a problem defined in a simpler domain. Closeness of two solutions can be proved if the coefficients of the latter problem are selected in a special way. Let us discuss this idea, using the basic elliptic problem: Find $u \in H^1(\Omega)$ such that

$$\operatorname{div} A\nabla u + f = 0 \quad \text{in } \Omega, \tag{B.121}$$

$$u = u_0 \quad \text{on } \Gamma = \widehat{\Gamma} + \Gamma_\omega, \tag{B.122}$$

where $\Omega = \widehat{\Omega} \setminus \omega$, $\widehat{\Omega}$ is a "simple" domain with the boundary $\widehat{\Gamma}$ (e.g., rectangular domain in Fig. B.4), and ω is a hole (holes). We consider the following modified problem: Find $u_\varepsilon \in H_0^1(\widehat{\Omega})$ such that

$$\int_{\widehat{\Omega}} \widehat{A}_\varepsilon \nabla \widehat{u}_\varepsilon \cdot \nabla \widehat{w}\, dx + \int_{\widehat{\Omega}} \widehat{b}_\varepsilon \widehat{u}_\varepsilon \widehat{w}_\varepsilon\, dx = \int_{\widehat{\Omega}} \widehat{f} \widehat{w}\, dx, \quad \forall \widehat{w} \in H_0^1(\widehat{\Omega}), \tag{B.123}$$

where $\widehat{A}_\varepsilon$, $\widehat{b}_\varepsilon$, and $\widehat{f}$ are selected such that $\widehat{u}_\varepsilon$ tends to u in Ω. For example, we can set $\widehat{A}_\varepsilon = A$ in Ω, $\widehat{A}_\varepsilon = \frac{1}{\varepsilon} I$ in ω, $b_\varepsilon = 0$ in Ω, $b_\varepsilon = \frac{1}{\varepsilon}$ in ω, and somehow extend f to ω. In the simplest case, penalization is applied only to the second term. Then, (B.123) infers the estimate

$$c_1 \|\nabla \widehat{u}_\varepsilon\|_\Omega^2 + \frac{1}{\varepsilon} \|\widehat{u}_\varepsilon\|_\omega^2 \le \int_{\widehat{\Omega}} \widehat{f} \widehat{u}_\varepsilon\, dx \le \|\widehat{f}\|_{\widehat{\Omega}} \|\widehat{u}_\varepsilon\|_{\widehat{\Omega}} \le C_{F\widehat{\Omega}} \|\widehat{f}\|_{\widehat{\Omega}} \|\nabla \widehat{u}_\varepsilon\|_{\widehat{\Omega}}.$$

Since $\|\widehat{f}\|_{\Omega}\|\widehat{u}_\varepsilon\|_{\widehat{\Omega}} \le \frac{c_1}{2}\|\nabla\widehat{u}_\varepsilon\|^2_{\widehat{\Omega}} + \frac{1}{2c_1}\|\widehat{f}\|^2_{\widehat{\Omega}}$, we have the estimate

$$\frac{c_1}{2}\|\nabla\widehat{u}_\varepsilon\|^2_{\Omega} + \frac{1}{\varepsilon}\|\widehat{u}_\varepsilon\|^2_{\omega} \le \frac{C_{F\widehat{\Omega}}}{2c_1}\|\widehat{f}\|^2_{\widehat{\Omega}}, \tag{B.124}$$

which shows that $\|\widehat{u}_\varepsilon\|^2_{\omega} \to 0$ as $\varepsilon \to 0$.

Also, (B.125) shows that $\|\nabla\widehat{u}_\varepsilon\|^2_{\Omega}$ is bounded and, therefore, contains a subsequence tending to a function $\widetilde{u}$ weakly in $H^1(\Omega)$.

Let $\widehat{w}$ in (B.123) be a test function supported in Ω (therefore, we denote it by w). Then, we pass to the limit in the relation

$$\int_{\Omega} A\nabla\widehat{u}_\varepsilon \cdot \nabla w\,\mathrm{d}x = \int_{\Omega} f w\,\mathrm{d}x \tag{B.125}$$

and see that in Ω the function $\widetilde{u}$ satisfies (in a generalized sense) the differential equation $\operatorname{div} A\nabla\widetilde{u} + f = 0$. This fact suggests the idea to use (B.123) instead of (B.121). Convergence of $\widehat{u}_\varepsilon$ to u can be strictly proved and qualified in terms of ε for various schemes of the method (see, e.g., [BGH$^+$01, GPP94b, GPP06, Kop68] and the literature cited therein). Applications of the method to viscous flow problems can be found in, e.g., [Ang99, GPP94a, GGP99] and to free boundary problems in [NK81]. We note that fictitious domain method can be considered as a "domain imbedding method" (see [BDGG71]).

Finally, we show that functional a posteriori estimates considered in Chap. 3 infer computable upper bounds of the error associated with the fictitious domain method. Let $\phi \in H^1(\widehat{\Omega})$ be a correction function such that

$$\phi(x) = 0 \quad \text{on } \widehat{\Gamma}, \qquad \phi(x) = -u_\varepsilon \quad \text{on } \Gamma_\omega. \tag{B.126}$$

Then, $v = \widehat{u}_\varepsilon + \phi$ can be viewed as a conforming approximation of u. We use the estimate (3.38)

$$|\!|\!| u - v |\!|\!|_{\Omega} \le \|A\nabla v - y\|_{A^{-1},\Omega} + \mathrm{C}\|f + \operatorname{div} y\|_{\Omega}, \tag{B.127}$$

where $y = p_\varepsilon := A\nabla\widehat{u}_\varepsilon$ in Ω. In this case, $\operatorname{div} y + f = 0$ in Ω and the majorant contains only one term. Note that $A\nabla v - p_\varepsilon = A\nabla\phi$. Thus, we obtain

$$|\!|\!| u - v |\!|\!|^2_{\Omega} \le \|A\nabla\phi\|^2_{A^{-1},\Omega} = \int_{\Omega} A\nabla\phi \cdot \nabla\phi\,\mathrm{d}x = e^2(\phi, \widehat{u}_\varepsilon|_{\Gamma_\omega}). \tag{B.128}$$

This estimate has a clear sense: the best possible error bound is obtained if the correction function $\phi \in H^1(\Omega)$ minimizes the quadratic integral on the right-hand side of (B.128) over the set of functions satisfying (B.126). It is clear that $e(\bar{\phi}, \widehat{u}_\varepsilon|_{\Gamma_\omega})$ represents the "nonconformity" error caused by an inexactness in the boundary condition on Γ_ω. Since $\widehat{u}_\varepsilon$ tends to zero in ω, this quantity tends to zero.

In [RSS03], this error estimate was applied to approximations computed with the help of a fictitious domain method. Numerical tests have shown its high efficiency and robustness.

Appendix C
A Priori Verification of Accuracy

The a priori convergence of approximate solutions and the corresponding error estimates provide the first and the most general information on the accuracy of numerical results. Methods of a priori error estimation for linear partial differential equations were developed in the 1950s/1960s years. Subsequently, they were extended to practically all classes of boundary (initial boundary) value problems including nonlinear ones. A priori rate of convergence estimates establish error bounds in terms of the mesh size parameter (and other parameters characterizing finite dimensional subspaces). The derivation of them is based upon two keystones: projection type error estimates and interpolation theory for functions in Sobolev spaces. The goal of this section is to demonstrate the main ideas within the paradigm of a linear elliptic problem. The reader will find a detailed discussion of the theory in [BS94, Cia78a, SF73] and other publications.

C.1 Projection Error Estimate

We consider the boundary value problem presented in the generalized form

$$a(u, w) = \ell(w), \quad \forall w \in V, \tag{C.1}$$

where the bilinear form a satisfies the conditions (B.4) and (B.5). We recall that the solution u minimizes the functional

$$J(v) := \frac{1}{2}a(v, v) - (f, v)$$

on a Banach space V. Let $\widehat{V}$ be a subspace of V and $\widehat{u} \in \widehat{V}$ be such that $J(\widehat{u}) \leq J(\widehat{v})$ for any $\widehat{v} \in \widehat{V}$. Then, $\widehat{u}$ satisfies the relation

$$a(\widehat{u}, \widehat{w}) = \ell(\widehat{w}), \quad \forall \widehat{w} \in \widehat{V}, \tag{C.2}$$

and for any $\widehat{v} \in \widehat{V}$ we have (cf. (B.72))

O. Mali et al., *Accuracy Verification Methods*,
Computational Methods in Applied Sciences 32, DOI 10.1007/978-94-007-7581-7,

$$\frac{1}{2}a(u-\widehat{u},u-\widehat{u}) = J(\widehat{u}) - J(u) \le J(\widehat{v}) - J(u) = \frac{1}{2}a(u-\widehat{v},u-\widehat{v}).$$

Thus,

$$a(u-\widehat{u},u-\widehat{u}) = \inf_{\widehat{v}\in\widehat{V}} a(u-\widehat{v},u-\widehat{v}). \tag{C.3}$$

The right-hand side of the *projection error estimate* (C.3) can be viewed as the distance from $u \in V$ to the subspace $\widehat{V}$ computed in terms of the norm generated by the form a. If $u \in \widehat{V}$, then (C.3) shows that $\widehat{u} = u$.

By (B.4) and (B.5), we obtain the estimate

$$\|u-\widehat{u}\|_V^2 \le \mathbb{C}_a \inf_{\widehat{v}\in\widehat{V}} \|u-\widehat{v}\|_V^2, \tag{C.4}$$

where $\mathbb{C}_a := \frac{c_2}{c_1}$ is the "condition number" of the bilinear form a.

We see that the error $e = u - \widehat{u}$ is bounded from the above by the distance between the exact solution u and the subspace $\widehat{V}$.

Approximations of partial differential equations are usually constructed with the help of finite dimensional subspaces (i.e., $\dim \widehat{V} = N < +\infty$). In the finite element method (cf. Sect. B.4.3), the basis of $\widehat{V}$ is created with the help of piecewise polynomial and locally supported trial functions. Assume that all the support domains of trial functions have character size h and denote the corresponding $\widehat{V}$ by V_h. Then, the relation (C.2) reads

$$a(u_h, w_h) = \ell(w_h), \quad \forall w_h \in V_h. \tag{C.5}$$

By (C.4), we conclude that the *Galerkin approximation* u_h satisfies the estimate

$$\|u-u_h\|_V^2 \le \mathbb{C}_a \inf_{v_h\in V_h} \|u-v_h\|_V^2. \tag{C.6}$$

This result is often called the Cea's lemma. The estimate (C.6) plays an important role in error analysis. It serves as a basis for deriving a priori convergence rate estimates expressed in terms of the parameter h.

Generalized versions of (C.6) (known as lemmas of Strang) extend Cea's lemma to nonconforming approximations (see, e.g., [Bra07, Cia78a]).

C.2 Interpolation Theory in Sobolev Spaces

Interpolation theory investigates the difference between a function in a Sobolev space and a suitable counterpart of it in some finite dimensional subspace. A natural way of defining such a counterpart is to project the function to the corresponding subspace. However, exact (orthogonal) projections may be difficult to construct, so that much simpler interpolation procedures are used instead. For example, if $V = \overset{\circ}{H}{}^1(0,1)$ and V_h is the subspace of piecewise affine continuous functions with N nodes, then the interpolant of $v \in V$ is constructed by assigning $v_h(x_k) = v(x_k)$ at each node x_k, $k = 1, 2, \dots, N$.

We begin with two lemmas (the proofs of which can be found in, e.g., [Cia78a]). They establish important facts related to approximations generated by piecewise polynomial functions.

Lemma C.1 *Let $\Omega \in \mathbb{R}^d$ be a connected bounded domain with Lipschitz boundary, $1 \le q \le +\infty$, and $k \ge 0$. There exists a constant $C > 0$ such that for any $v \in W^{k+1,q}(\Omega)$*

$$\inf_{p \in P^k(\Omega)} \|v - p\|_{k+1,q,\Omega} \le C(\Omega)|v|_{k+1,q,\Omega}. \tag{C.7}$$

An advanced version of this inequality reads as follows:

$$\inf_{p \in P^k(\Omega)} \|v - p\|_{t,q,\Omega} \le C(\operatorname{diam}\Omega)^{k+1-t}|v|_{k+1,q,\Omega}, \quad t = 0, 1, \dots, k+1,$$

where the constant C depends on d, k, q and on the aspect ratio of Ω.

Lemma C.2 *Let the conditions of Lemma* C.1 *hold and let $\ell : V^* \to \mathbb{R}$ be a linear continuous functional vanishing on $P^k(\Omega)$, where V^* is the space dual to $W^{k+1,q}$. Then, there exists a constant $c > 0$ such that for any $v \in W^{k+1,q}(\Omega)$*

$$\left|\ell(v)\right| \le c\|\ell\|_{V^*}|v|_{k+1,q,\Omega}. \tag{C.8}$$

In the literature, these results are often called Deny–Lions and Bramble–Hilbert lemmas ([BH70, DL55]), respectively.

Definition C.1 We say that domains Ω and $\widehat{\Omega}$ are affine equivalent if there exists an affine non-degenerate mapping $F(\widehat{x}) = B\widehat{x} + b$, where $B \in \mathbb{M}^{d\times d}$, $\det B > 0$ and $b \in \mathbb{R}^d$, which maps $\widehat{\Omega}$ to Ω.

Since

$$|\Omega| = \int_\Omega \mathrm{d}x = \int_{\widehat{\Omega}} \det\left(\frac{\partial x}{\partial \widehat{x}}\right) \mathrm{d}\widehat{x} = \int_{\widehat{\Omega}} \det Bd\,\mathrm{d}\widehat{x} = |\widehat{\Omega}| \det B,$$

we see that

$$\det B = \frac{|\Omega|}{|\widehat{\Omega}|}. \tag{C.9}$$

Let $v \in W^{m,p}(\Omega)$ and $\widehat{V}(\widehat{x}) = v(B\widehat{x} + b)$, where $\widehat{x} \in \widehat{\Omega}$. It is clear that $\widehat{v} \in W^{m,p}(\widehat{\Omega})$. Moreover, the constants $C_1(m, d, p)$ and $C_2(m, d, p) > 0$ exist such that

$$|\widehat{v}|_{m,p,\widehat{\Omega}} \le C_1(m, d, p, \Omega)\|B\|^m(\det B)^{-1/p}|v|_{m,p,\Omega}, \tag{C.10}$$

and

$$|v|_{m,p,\Omega} \leq C_2(m,d,p,\Omega)\left\|B^{-1}\right\|^m (\det B)^{1/p} |\widehat{v}|_{m,p,\Omega}. \tag{C.11}$$

The corresponding proofs can be found, e.g., in [Cia78a]. The quantities $\|B\|$ and $\det B$ can be estimated throughout geometrical characteristics of Ω and $\widehat{\Omega}$. Let

$$h = \operatorname{diam} \Omega := \sup\left\{|x-y| \mid x, y \in \Omega\right\}, \tag{C.12}$$

$$\widehat{h} = \operatorname{diam} \widehat{\Omega} := \sup\left\{|\widehat{x}-\widehat{y}| \mid \widehat{x}, \widehat{y} \in \widehat{\Omega}\right\}, \tag{C.13}$$

$\mathcal{B}$ denote a ball in Ω, and $\widehat{\mathcal{B}}$ denote a ball in $\widehat{\Omega}$. Define the numbers

$$\rho = \sup\{\operatorname{diam} \mathcal{B} \mid \mathcal{B} \subset \Omega\}, \tag{C.14}$$

$$\widehat{\rho} = \sup\{\operatorname{diam} \widehat{\mathcal{B}} \mid \widehat{\mathcal{B}} \subset \widehat{\Omega}\}. \tag{C.15}$$

Since

$$\|B\| = \sup_{|\widehat{x}|=1} |B\widehat{x}| = \frac{1}{\widehat{r}} \sup_{|\widehat{x}|=\widehat{r}} |B\widehat{x}|, \quad \widehat{r} > 0, \tag{C.16}$$

we estimate $\|B\|$ if the quantity $\sup\{|B\widehat{x}| \mid |\widehat{x}| = \widehat{r}\}$ is estimated from the above for some $\widehat{r}$. In accordance with (C.15), for any small $\varepsilon > 0$ there exist $\widehat{z}, \widehat{y} \in \widehat{\Omega}$ such that

$$\widehat{r} = |\widehat{z} - \widehat{y}| = \widehat{\rho} - \varepsilon.$$

Denote $\widehat{x} = \widehat{z} - \widehat{y}$. It is clear that $x = F(\widehat{x})$ and $y = F(\widehat{y})$ belong to Ω. It is easy to see that

$$F(\widehat{z}) - F(\widehat{y}) = B\widehat{z} + b - B\widehat{y} - b = B\widehat{x}.$$

On the other hand,

$$F(\widehat{z}) - F(\widehat{y}) = z - y, \quad \forall z, y \in \Omega.$$

By (C.12), we conclude that $|B\widehat{x}| \leq h$. Thus,

$$\|B\| \leq \frac{1}{\widehat{\rho} - \varepsilon} h, \quad \forall \varepsilon > 0,$$

and we obtain

$$\|B\| \leq h/\widehat{\rho}. \tag{C.17}$$

The inverse estimate

$$\|B^{-1}\| \leq \rho/\widehat{h}$$

follows from similar arguments.

C.2.1 Interpolation Operators

Assume that the numbers $k, t, m, s, d \in \mathbb{N}$ are such that the space $W^{k+1,s}(\widehat{\Omega})$ is continuously embedded in $W_t^m(\widehat{\Omega})$, i.e, there exits C_1, dependent on $\widehat{\Omega}$, k, m, d, t, s, such that

$$\|\widehat{v}\|_{m,t,\widehat{\Omega}} \leq C_1 \|\widehat{v}\|_{k+1,s,\widehat{\Omega}}, \quad \forall \widehat{v} \in W^{k+1,s}(\widehat{\Omega}). \tag{C.18}$$

Let

$$\widehat{\Pi} : W^{k+1,s}(\widehat{\Omega}) \to W^{m,t}(\widehat{\Omega})$$

be a continuous operator, i.e., there exists constant $c_2(\widehat{\Omega}, k, s, d, t, m) > 0$ such that

$$\|\widehat{\Pi}\widehat{v}\|_{m,t,\widehat{\Omega}} \leq C_2 \|\widehat{v}\|_{k+1,s,\widehat{\Omega}}, \quad \forall \widehat{v} \in W^{k+1,s}(\widehat{\Omega}). \tag{C.19}$$

In further analysis, it is required that $\widehat{\Pi}$ do not change polynomials, i.e.,

$$\widehat{\Pi}\widehat{p} = \widehat{p}, \quad \forall \widehat{p} \in P^k(\widehat{\Omega}), \tag{C.20}$$

and that Ω and $\widehat{\Omega}$ be affine equivalent. Now we can define the operator $\Pi_\Omega v = \widehat{\Pi}\widehat{v}$, where $v(x) := \widehat{v}(\widehat{x})$. The operator Π_Ω is called the *interpolation operator* and the function $\Pi_\Omega v$ is called the *interpolant* of v. It is clear that the operator Π_Ω possesses the same property:

$$\Pi_\Omega p(x) = \widehat{\Pi}\widehat{p}(x) = \widehat{p}(\widehat{x}) = p(x), \quad \forall p \in P^k(\Omega).$$

Moreover, $p \in P^k(\Omega)$ if and only if $\widehat{p} \in P^k(\widehat{\Omega})$.

Theorem C.10 *Let the conditions* (C.18)–(C.20) *be satisfied. Then,*

$$|v - \Pi_\Omega v|_{m,t,\Omega} \leq C(\widehat{\Pi}, \widehat{\Omega}) \left(\frac{h}{\rho}\right)^m h^{k+1-m} (\det B) |v|_{k+1,s,\Omega}. \tag{C.21}$$

Proof Take a polynomial $\widehat{p} \in P^k(\widehat{\Omega})$, then

$$\widehat{v} - \widehat{\Pi}\widehat{v} = \widehat{v} - \widehat{p} - \widehat{\Pi}(\widehat{v} - \widehat{p}) = (I - \widehat{\Pi})(\widehat{v} - \widehat{p}).$$

From here, it follows that

$$|\widehat{v} - \widehat{\Pi}\widehat{v}|_{m,t,\widehat{\Omega}} \leq C(\widehat{\Pi}) \|\widehat{v} - \widehat{p}\|_{m,t,\widehat{\Omega}}, \quad \forall \widehat{p} \in P^k(\widehat{\Omega}).$$

We take the infimum with respect to $\widehat{p} \in P^k(\widehat{\Omega})$ and obtain

$$|\widehat{v} - \widehat{\Pi}\widehat{v}|_{m,t,\widehat{\Omega}} \leq C(\widehat{\Pi}) \inf_{\widehat{p} \in P^k(\widehat{\Omega})} \|\widehat{v} - \widehat{p}\|_{m,t,\widehat{\Omega}} \leq C(\widehat{\Pi}) C_1 \inf_{\widehat{p} \in P^k(\widehat{\Omega})} \|\widehat{v} - \widehat{p}\|_{k+1,s,\widehat{\Omega}}.$$

Now, (C.7) implies the estimate

$$|\widehat{v} - \widehat{\Pi}\widehat{v}|_{m,t,\widehat{\Omega}} \leq C(\widehat{\Pi}, \widehat{\Omega}) |\widehat{v}|_{k+1,s,\widehat{\Omega}}.$$

By (C.10)–(C.11), we have

$$|v - \Pi v|_{m,t,\Omega} \le C \| B^{-1} \|^m (\det B)^{1/t} |\widehat{v} - \widehat{\Pi}\widehat{v}|_{m,t,\widehat{\Omega}}.$$

Therefore,

$$\begin{aligned} |v - \Pi v|_{m,t,\Omega} &\le C \| B^{-1} \|^m (\det B)^{1/t} |\widehat{v}|_{k+1,s,\widehat{\Omega}} \\ &\le C \| B \|^{k+1} (\det B)^{-1/s} \| B^{-1} \|^m (\det B)^{1/t} |v|_{k+1,s,\Omega}. \end{aligned}$$ □

C.2.2 Interpolation on Polygonal Sets

In the majority of approximation methods, meshes are formed by rather simple cells (simplexes, quadrilaterals, polygons). Consider the simplest case, in which Ω is a simplex and Π_Ω maps functions from $W^{l,p}(\Omega)$ to $P^k(\Omega)$.

Definition C.2 Let $x_1, x_2, \dots, x_d$ be elements of $\mathbb{R}^d$, which do not belong to a common $\mathbb{R}^{d-1}$ plane. The set

$$T := \left\{ y \in \mathbb{R}^d \;\middle|\; y = \sum_{i=1}^{d} \lambda_i x_i,\ \lambda_i \in (0,1),\ \sum_{i=1}^{d} \lambda_i = 1 \right\}.$$

In other words, simplex is a convex envelope of elements $x_1, x_2, \dots, x_d$, which are called "nodes".

If $d = 2$, then $T = (x_1, x_2)$; if $d = 3$, then simplex is a tetrahedron. Faces of a simplex are simplexes of lower dimension. Two simplexes of one dimension are affine equivalent (see Definition C.1).

Usually, interpolation estimates are first studied on the basic ("etalon") simplex (Fig. C.1). Then, by affine mappings they can be easily extended to any (non-degenerate) simplex. Since polygonal sets are representable as unions of simplexes, this method opens a way of deriving estimates of the difference between a function and an interpolant of it for any domain of such a type.

In $\mathbb{R}^d$, the basic (etalon) simplex $\widehat{T}$ has the nodes

$$\begin{aligned} &\widehat{X}_0 = (0,0,\dots,0), \qquad \widehat{X}_1 = (1,0,0,\dots,0), \\ &\widehat{X}_2 = (0,1,0,\dots,0), \qquad \dots, \qquad \widehat{X}_d = (0,0,\dots,1). \end{aligned}$$

Henceforth, we consider the simplest case, in which $m = 0$ or 1 ($m = 0$ corresponds to approximation of functions and $m = 1$ corresponds to approximation

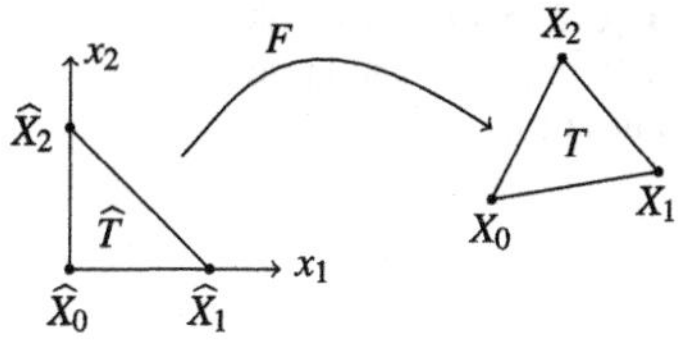

Fig. C.1 Affine mapping of $\widehat{T}$ to T

of derivatives), $k = 1$ and $t = s > 1$. This situation arises if we are interested in approximation of solutions in $W^{2,t}$ using the simplest polynomial approximations with $k = 1$.

In view of Theorem A.1, $W^{2,t}(\widehat{T})$ is continuously embedded in $C(\widehat{T})$ if $2 > d/t$. In particular, if $d = 2$, $t = 2$, then $H^2(\widehat{T})$ is continuously embedded in $C(\widehat{T})$ and, therefore, a simple interpolation operator can be constructed, namely,

$$\widehat{\Pi}\widehat{v} \in P^1(\widehat{T}), \qquad \widehat{\Pi}\widehat{v}(\widehat{X}_i) = \widehat{v}(\widehat{X}_i), \quad i = 0, 1, \ldots, d. \tag{C.22}$$

It is easy to see that the function $\widehat{\Pi}\widehat{v}$ is uniquely determined and the operator $\widehat{\Pi}$ does not change polynomials of order 1. Let us check that the condition (C.19) holds. In our case, it has the form

$$\|\widehat{\Pi}\widehat{v}\|_{m,t,\widehat{T}} \leq C\|\widehat{v}\|_{2,t,\widehat{T}}, \tag{C.23}$$

Indeed, on a finite dimensional space all the norms are equivalent, i.e.,

$$\|\widehat{\Pi}\widehat{v}\|_{m,t,\widehat{T}} \leq C\|\widehat{\Pi}\widehat{v}\|_{C(\widehat{T})}. \tag{C.24}$$

On the other hand,

$$\|\widehat{\Pi}\widehat{v}\|_{C(\widehat{T})} = \max_{i=0,1,\ldots,d} \left|\widehat{\Pi}\left(\widehat{v}(\widehat{X}_i)\right)\right| = \max_{i=0,1,\ldots,d} \left|\widehat{v}(\widehat{X}_i)\right| \leq C\|\widehat{v}\|_{C(\widehat{T})}. \tag{C.25}$$

Since (by the embedding theorem)

$$\|\widehat{v}\|_{C(\widehat{T})} \leq C\|\widehat{v}\|_{2,t,\widehat{T}}, \tag{C.26}$$

we find that (C.23) follows from (C.24)–(C.26).

Now we apply previously derived estimates. Let h denote the size (length of the largest edge) of the simplex T_h and ρ denote the radius of the largest inscribed ball. Then, (C.21) yields

$$|v - \Pi_h v|_{m,t,T_h} \leq C(m,d,t)\left(\frac{h}{\rho}\right)^m h^{2-m}|v|_{2,t,T_h}, \tag{C.27}$$

where Π_h is the interpolation operator associated with T_h.

Definition C.3 The aspect ratio $\wp(T)$ of a simplex T is the ratio of the length of the largest edge E to the diameter of the largest ball B inscribed in $\overline{T}$ i.e.,

$$\wp(T) := \frac{\max_{E \in \mathcal{E}(T)} |E|}{\max_{B \in \overline{T}} \operatorname{diam}(B)}, \tag{C.28}$$

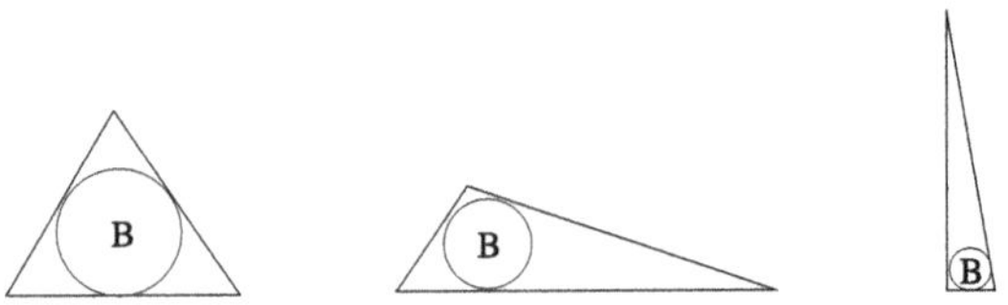

Fig. C.2 Simplexes with small, medium, and large values of $\wp$

where $\mathcal{E}(T)$ denotes the set of edges of T (see Fig. C.2 for illustrations).

Now, we rewrite (C.27) in the form

$$|v - \Pi_h v|_{m,t,T_h} \leq C(m,d,t)\big(\wp(T_h)\big)^m h^{2-m}|v|_{2,t,T_h}. \tag{C.29}$$

This estimate can be extended to functions defined in Ω, provided that this domain is a union of simplexes and the parameter h denotes the maximum over all the simplexes.

Let $\{\mathcal{T}_h\}$ be a sequence of simplicial triangulations of Ω ($\mathcal{T}_h = \{T_h^i\}_{i=1,2,\dots,N(h)}$) such that the number of simplexes N increases and the parameter h decreases. In order to guarantee that the corresponding approximations converge with the optimal rate, we need to impose additional conditions.

Definition C.4 We say that the sequence $\{\mathcal{T}_h\}$ is regular if

(a) $T_h^i \cap T_h^j = 0$ if $i \neq j$.
(b) Each edge of T_h^i is either a part of Γ or coincides with an edge of some other simplex in $\mathcal{T}_h$.
(c) All the quantities $\wp(T_i)$ associated with different simplexes $T^i \in \mathcal{T}_h$ in all triangulations $\mathcal{T}_h$ are uniformly bounded with respect to h, i.e.,

$$\wp(T_i) < C_\wp, \quad i = 1, 2, \dots, N. \tag{C.30}$$

Requirements (a) and (b) of Definition C.4 are necessary to guarantee that the functions constructed belong to the energy space V, so that they belong to the class of *conforming approximations*. The condition (c) is important for a priori error estimates: it allows one to guarantee that constants in interpolation estimates do not degenerate as h tends to zero (see below).

Let $\Pi_h^i v$ be the interpolant of v on T_h^i. We define the global interpolation operator Π_h by setting

$$\Pi_h v(x) = \Pi_h^i v(x) \quad \text{if } x \in T_h^i. \tag{C.31}$$

$\Pi_h v$ is an affine function on any simplex and coincides with v at the nodes. Moreover, $\Pi_h v \in C(\bar{\Omega})$ and, in addition,

$$\Pi_h v \in W^{1,t}(\Omega) \cap C(\bar{\Omega}). \tag{C.32}$$

In view of (C.29), on T_h^i we have:

$$|v - \Pi_h v|^t_{m,t,T_h^i} \leq \big(C(m,d,t)h^{2-m}\big)^t |v|^t_{2,t,T_h^i}. \tag{C.33}$$

We sum over all $i = 1, \ldots, N$ and obtain

$$|v - \Pi_h v|^t_{m,t,\Omega} \leq \left(C(m,d,t)h^{2-m}\right)^t |v|^t_{2,t,\Omega}, \tag{C.34}$$

whence the general interpolation estimate follows (where $m = 0$ or 1)

$$|v - \Pi_h v|_{m,t,\Omega} \leq C_{int} h^{2-m} |v|_{2,t,\Omega}, \tag{C.35}$$

where the *interpolation constant* C_{int} depends on the numbers m and t, the dimension d, and on the constant $C_{\wp}$ associated with the selected type of simplexes (or other elements) forming $\mathfrak{T}_h$. This estimate allows us to obtain qualified convergence estimates for finite element approximations.

Assume that Ω is a polygonal domain (which can be exactly approximated by a simplicial mesh) and $u \in H^2(\Omega)$. We know that

$$a(u - u_h, u - u_h) \leq a(u - \Pi_h u, u - \Pi_h u) \leq c_2 \|u - \Pi_h u\|^2.$$

For $m = 1$ and $t = 2$ the estimate (C.35) yields

$$|v - \Pi_h v|_{m,t,\Omega_h} \leq C_{int} h^{2-m} \|v\|_{2,t,\Omega_h}.$$

Hence, we conclude that the error is subject to the estimate

$$\|\nabla(u - u_h)\| \leq C_{int} \mathbb{C}_a h |u|_{2,2,\Omega}.$$

However, exact solutions in polygonal domains may not have H^2 regularity, so that this estimate is conditional so far. Below we consider one class of problems, which possess H^2 regularity and, therefore, the corresponding convergence results are guaranteed.

Remark C.1 Interpolation operators for functions in $H^1(\Omega)$ cannot be based on pointwise relations (such as (C.22)). They use more complicated constructions based upon integral type relations. Clement's interpolation operator considered in Sect. 2.2.1 belongs to this class. Various modifications of Clement's interpolation operator are suggested in [Ber89, BG98, Car99, SZ90].

C.3 A Priori Convergence Rate Estimates

Projection type and interpolation error estimates yield rate convergence estimates. We discuss this method for the problem

$$\operatorname{div} A\nabla u = -g \quad \text{in } \Omega, \qquad u = u_0 \quad \text{on } \Gamma.$$

We assume that Ω is a convex domain, $d = 2$, Γ is a smooth boundary, $g \in L^2(\Omega)$, $A \in C^1(\overline{\Omega}, \mathbb{M}^{d\times d})$ and $u_0 \in H^2(\Omega)$. In this case (owing to known results for elliptic

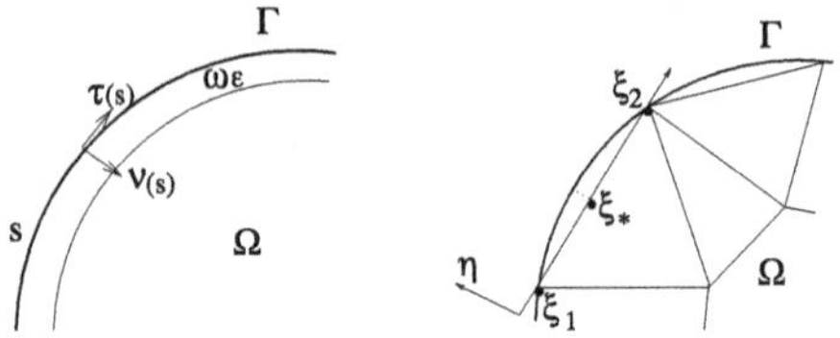

Fig. C.3 Boundary strip ω_ε supplied with the curvilinear coordinate system (*left*) and local coordinates associated with the element of the triangulation (*right*)

partial differential equations, see, e.g., [GT77, Lad85]), $u \in H^2(\Omega, \mathbb{R}^2)$, and the following *regularity* estimate holds:

$$\|u\|_{2,2,\Omega} \le c_r\big(\|g\|_{2,\Omega} + \|u_0\|_{2,2,\Omega}\big),$$

where c_r is a positive constant.

Let $\widetilde{u} = u - u_0$. Then, the problem is restated as a problem with homogeneous boundary conditions and a modified right-hand side

$$\begin{aligned} \operatorname{div} A\nabla\widetilde{u} &= -\big(g + \operatorname{div} A(\nabla u_0)\big) = -f \quad \text{in } \Omega, \\ \widetilde{u} &= 0 \quad \text{on } \Gamma. \end{aligned}$$

For $\widetilde{u}$ we have the regularity estimate

$$\|\widetilde{u}\|_{2,2,\Omega} \le c_r \|f\|_{2,\Omega}. \tag{C.36}$$

Consider a thin strip around the boundary. Henceforth, we assume that its width does not exceed ε and call it ε-strip:

$$\omega_\varepsilon = \big\{x \in \Omega \mid \operatorname{dist}(x, \Gamma) < \varepsilon\big\}.$$

If ε is sufficiently small ($\varepsilon \le \varepsilon_0$, where ε_0 depends on the maximal curvature of Γ), then we can introduce a local coordinate system, which uniquely determines any point of the strip. In Fig. C.3, these coordinates are denoted by s (length of the curve) and t (distance from Γ computed along the normal n oriented inside Ω). Let $l := |\Gamma|$ denote the length of Γ. Then, for any $s \in [0, l)$ and $t \in [0, \varepsilon)$ we have the representation in terms of local coordinates

$$\boldsymbol{x}(s, t) = \boldsymbol{x}(s) + t\boldsymbol{\nu}(s), \tag{C.37}$$

where $\boldsymbol{x}(s) = \boldsymbol{x}(s, 0)$ is the vector associated with the point on Γ, the distance of which to the reference point on Γ is s, and $\boldsymbol{\tau}(s)$ and $\boldsymbol{\nu}(s)$ are the corresponding unit vectors satisfying the relations

$$\boldsymbol{\tau}(s) = \frac{d\boldsymbol{x}(s)}{ds}, \qquad \frac{d\boldsymbol{\nu}(s)}{ds} = -k(s)\boldsymbol{\tau}(s).$$

It is not difficult to prove that the Jacobian of this curvilinear coordinate system is positive in ω_ε. Indeed,

$$\frac{\partial x(s,t)}{\partial s} = \tau(s) - k(s)t\tau(s),$$

$$x_1(s,t) = x_1(s) + t\nu_1(s), \qquad x_2(s,t) = x_2(s) + t\nu_2(s).$$

Hence,

$$\begin{vmatrix} \frac{\partial x_1}{\partial s} & \frac{\partial x_2}{\partial s} \\ \frac{\partial x_1}{\partial t} & \frac{\partial x_2}{\partial t} \end{vmatrix} = \begin{vmatrix} \tau_1(s)(1-tk(s)) & \tau_2(s)(1-tk(s)) \\ \nu_1(s) & \nu_2(s) \end{vmatrix}, \tag{C.38}$$

and we conclude that

$$|J| = \big|1-tk(s)\big||\tau_1\nu_2 - \tau_2\nu_1| = \big|1-tk(s)\big|\begin{vmatrix} \tau_1 & \tau_2 \\ \nu_1 & \nu_2 \end{vmatrix}. \tag{C.39}$$

Since $\tau_1 = -\nu_2$ and $\tau_2 = \nu_1$, we obtain

$$|J| = \big|1-tk(s)\big| \geq 1 - |t|\big|k(s)\big|. \tag{C.40}$$

Assume that ε_0 is sufficiently small, i.e., $\varepsilon_0|k(\gamma)| \leq \theta < 1$ for any point $\gamma \in \Gamma$. Then, $|J| \geq 1-\theta > 0$.

Now we can prove an auxiliary estimate, which is needed for deducing estimates of norms associated with the strip ω_ε (provided that $\varepsilon < \varepsilon_0$).

Lemma C.3 *For any $w \in H^1(\omega_\varepsilon)$, we have the estimate*

$$\|w\|^2_{\omega_\varepsilon} \leq \beta\varepsilon\big(\|w\|^2_{2,\Gamma} + \varepsilon\|\nabla w\|^2_{\omega_\varepsilon}\big), \tag{C.41}$$

where β does not depend on w and ε.

Proof We pass to the coordinate system (s,t) (see (C.37)) and define $w^{(\varepsilon)}(s,t) = w(x(s,t))$ in ω_ε. By the integral representation formula, we have

$$w^{(\varepsilon)}(s,t) = w^{(\varepsilon)}(s,0) + \int_0^t \frac{\partial w^{(\varepsilon)}(s,\mu)}{\partial \mu}\,\mathrm{d}\mu.$$

From here

$$\begin{aligned} \big|w^{(\varepsilon)}(s,t)\big| &\leq \big|w^{(\varepsilon)}(s,0)\big| + \int_0^t \big|\nabla w^{(\varepsilon)}(s,\mu)\big|\,\mathrm{d}\mu \\ &\leq \big|w^{(\varepsilon)}(s,0)\big| + \int_0^\varepsilon \big|\nabla w^{(\varepsilon)}(s,\mu)\big|\,\mathrm{d}\mu. \end{aligned}$$

We square both parts and obtain

$$\begin{aligned} \big|w^{(\varepsilon)}(s,t)\big|^2 &\leq 2\left\{\big|w^{(\varepsilon)}(s,0)\big|^2 + \left|\int_0^\varepsilon \big|\nabla w^{(\varepsilon)}(s,\mu)\big|\,\mathrm{d}\mu\right|^2\right\} \\ &\leq 2\left\{\big|w^{(\varepsilon)}(s,0)\big|^2 + \varepsilon\int_0^\varepsilon \big|\nabla w^{(\varepsilon)}(s,\mu)\big|^2\,\mathrm{d}\mu\right\}. \end{aligned}$$

Then,

$$\int_0^l \left|w^{(\varepsilon)}(s,t)\right|^2 \mathrm{d}s \le 2\left\{\int_0^l \left|w^{(\varepsilon)}(s,0)\right|^2 \mathrm{d}s + \varepsilon \int_0^l \int_0^\varepsilon \left|\nabla w^{(\varepsilon)}(s,t)\right|^2 \mathrm{d}s\, \mathrm{d}t\right\}.$$

We integrate both parts again over t, apply the Fubini theorem, and arrive at the estimate

$$\int_0^l \int_0^\varepsilon \left|w^{(\varepsilon)}(s,t)\right|^2 \mathrm{d}s\, \mathrm{d}t \le 2\varepsilon\left\{\int_0^l \left|w^{(\varepsilon)}(s,0)\right|^2 \mathrm{d}s + \varepsilon \int_0^l \int_0^\varepsilon \left|\nabla w^{(\varepsilon)}(s,t)\right|^2 \mathrm{d}s\, \mathrm{d}t\right\}.$$

Now we return to the original coordinate system (x_1, x_2), invoke the fact that the corresponding Jacobian is uniformly bounded and obtain the required estimate. □

Corollary C.3 *If* $w \in \overset{\circ}{H}{}^1(\Omega)$, *then*

$$\|w\|^2_{\omega_\varepsilon} \le \beta\varepsilon^2 \|\nabla w\|^2_{\omega_\varepsilon}. \tag{C.42}$$

If $w \in H^1(\Omega)$, *then*

$$\|w\|^2_{\omega_\varepsilon} \le \beta\varepsilon \|w\|^2_{1,2,\Omega}. \tag{C.43}$$

Assume that Ω is a convex domain which contains a regular triangulation $\mathfrak{T}_h$. Elements of $\mathfrak{T}_h$ form a polygonal domain $\Omega_h \subset \Omega$. Let

$$V_h = \left\{v_h \in C(\Omega_h) \mid v_h \text{ is affine on any simplex } T \in \mathfrak{T}_h,\, v_h = 0 \text{ on } \Gamma_h := \partial\Omega_h\right\}$$

and $\widetilde{u}_h \in V_h$ be defined by the relation

$$\int_{\Omega_h} A(x)\nabla\widetilde{u}_h \cdot \nabla w_h \,\mathrm{d}x = \int_{\Omega_h} f w_h \,\mathrm{d}x, \quad \forall w_h \in V_h.$$

Since $w_h = 0$ on Γ_h, we can extend $\widetilde{u}_h$ and w_h by zero to $\Omega \setminus \Omega_h$, and $u_h = \widetilde{u}_h + u_0$ can be considered as an approximation of u. The corresponding finite dimensional set of extended functions $\widetilde{V}_h$ is a subset of $V = \overset{\circ}{H}{}^1(\Omega)$. We see that the finite dimensional problem is equivalent to finding $\widetilde{u}_h \in \widetilde{V}_h$ such that

$$J(\widetilde{u}_h) = \inf_{\widetilde{w} \in \widetilde{V}_h} J(\widetilde{w}). \tag{C.44}$$

We apply the estimate (C.4) to this case and obtain

$$\left\|\nabla(\widetilde{u} - \widetilde{u}_h)\right\|_{2,\Omega} \le \mathbb{C}_a \left\|\nabla(\widetilde{u} - \tilde{v}_h)\right\|_{2,\Omega}, \quad \forall \tilde{v}_h \in \widetilde{V}_h. \tag{C.45}$$

Now, the goal is to construct a suitable interpolant of $\widetilde{u}$ and to derive the corresponding estimate of the difference $\widetilde{u} - \Pi_h\widetilde{u}$, which gives an upper bound of the error qualified in terms of the parameter h.

For $d = 2$, the space $H^2(\Omega)$ is continuously embedded in $C(\overline{\Omega})$, and we can construct the interpolant $\Pi_h \widetilde{u}$ in Ω_h by the above-discussed method, exploiting the fact that $\widetilde{u} \in H^2(\Omega)$. Then, we deduce the estimates

$$\left\| \nabla(\widetilde{u} - \Pi_h \widetilde{u}) \right\|_{\Omega_h} \leq C_{int} h |\widetilde{u}|_{2,2,\Omega_h}, \tag{C.46}$$

$$\|\widetilde{u} - \Pi_h \widetilde{u}\|_{\Omega_h} \leq C_{int} h^2 |\widetilde{u}|_{2,2,\Omega_h}. \tag{C.47}$$

Let $\Pi_h \widetilde{u}$ be zero in $\Omega \setminus \Omega_h$. We need to extend the estimates (C.46) and (C.47) from Ω_h to Ω.

First, we prove that $\Omega \setminus \Omega_h \subset \omega_\varepsilon$, where $\varepsilon = \kappa h^2$ and κ is a constant independent of h. Indeed, consider one edge of the triangulation having two common points with Γ (Fig. C.3, right). We introduce local coordinates ξ and η associated with the edge. Since the boundary is smooth, the function $\eta = \phi(\xi)$ that determines the boundary in the local coordinate system is also smooth. The distance between Γ and the edge is maximal at a point ξ_* where $\phi'(\xi_*) = 0$. We use the representation

$$\begin{aligned} \phi(\xi_2) &= 0 \\ &= \phi(\xi_*) + \phi'(\xi_*)(\xi_2 - \xi_*) + \frac{1}{2}\phi''\big(\xi_* + \theta(\xi_2 - \xi_*)\big)(\xi_2 - \xi_*)^2, \quad \theta \in [0, 1], \end{aligned}$$

and arrive at the estimate

$$\left|\phi(\xi_*)\right| \leq \frac{1}{2}\left|\phi''\big(\xi_* + \theta(\xi_2 - \xi_*)\big)\right|\left|(\xi_* - \xi_2)^2\right| \leq \kappa h^2.$$

Since the curvature is bounded, the constant κ is uniformly bounded for all edges that have two common points with Γ. Thus, we conclude that the distance between Γ and Γ_h does not exceed κh^2. This means that $\Omega \setminus \Omega_h \subset \omega_\varepsilon$, where $\varepsilon = \kappa h^2$.

Note that

$$\|\widetilde{u} - \Pi_h \widetilde{u}\|^2_{2,\Omega/\Omega_h} = \|\widetilde{u}\|^2_{2,\Omega/\Omega_h} \leq \|\widetilde{u}\|^2_{\omega_\varepsilon}. \tag{C.48}$$

We use (C.42) (since $\widetilde{u} = 0$ on Γ) and obtain

$$\|\widetilde{u}\|_{\omega_\varepsilon} \leq \beta\varepsilon \|\nabla \widetilde{u}\|_{\omega_\varepsilon} \leq \beta\kappa h^2 \|\nabla \widetilde{u}\|_{\omega_\varepsilon}. \tag{C.49}$$

In order to estimate the right-hand side of this inequality, we use (C.43) and recall that $\widetilde{u} \in H^2(\Omega)$. Hence, any generalized derivative of $\widetilde{u}$ belongs to $H^1(\Omega)$ and on ω_ε we can estimate the gradient as follows:

$$\|\nabla \widetilde{u}\|_{\omega_\varepsilon} \leq \beta\kappa^{1/2} h \|\widetilde{u}\|_{2,2,\Omega}. \tag{C.50}$$

From (C.49) and (C.50), it follows that

$$\|\widetilde{u}\|_{\omega_\varepsilon} \leq C^2 \kappa^{3/2} h^3 \|\widetilde{u}\|_{2,2,\Omega} = C_\kappa h^3 \|\widetilde{u}\|_{2,2,\Omega},$$

where C_κ is another constant (independent of h). We have

$$\|\widetilde{u}-\Pi_h\widetilde{u}\|^2_{\Omega}=\|\widetilde{u}-\Pi_h\widetilde{u}\|^2_{\Omega_h}+\|\widetilde{u}-\Pi_h\widetilde{u}\|^2_{\Omega/\Omega_h}$$
$$\leq C_{int}h^4\|\widetilde{u}\|^2_{2,2,\Omega_h}+\|\widetilde{u}\|^2_{\omega_\varepsilon}\leq C_{int}h^4\|\widetilde{u}\|^2_{2,2,\Omega_h}+C^2_\kappa h^6\|\widetilde{u}\|^2_{2,2,\Omega}.$$

Thus, for sufficiently small h

$$\|\widetilde{u}-\Pi_h\widetilde{u}\|_{\Omega}\leq\widehat{C}h^2\|\widetilde{u}\|_{2,2,\Omega},\tag{C.51}$$

where $\widehat{C}$ is a constant independent of h.

Estimates for the derivatives are derived analogously. We use the identity

$$\left\|\nabla(\widetilde{u}-\Pi_h\widetilde{u})\right\|^2_{\Omega}=\left\|\nabla(\widetilde{u}-\Pi_h\widetilde{u})\right\|^2_{\Omega_h}+\left\|\nabla(\widetilde{u}-\Pi_h\widetilde{u})\right\|^2_{\Omega/\Omega_h}.\tag{C.52}$$

By (C.46), the first norm is estimated as follows:

$$\left\|\nabla(\widetilde{u}-\Pi_h\widetilde{u})\right\|_{\Omega_h}\leq Ch\|\widetilde{u}\|_{2,2,\Omega_h}\leq Ch|\widetilde{u}|_{2,2,\Omega}.\tag{C.53}$$

Since

$$\left\|\nabla(\widetilde{u}-\Pi_h\widetilde{u})\right\|^2_{\Omega/\Omega_h}=\|\nabla\widetilde{u}\|^2_{\Omega/\Omega_h}\leq\|\nabla\widetilde{u}\|^2_{\omega_\varepsilon},$$

we apply (C.43) again and deduce estimates for the derivatives of $\widetilde{u}$ in the strip, which imply

$$\|\nabla\widetilde{u}\|^2_{\Omega/\Omega_h}\leq C\kappa h^2\|\widetilde{u}\|^2_{2,2,\Omega}.\tag{C.54}$$

By (C.52)–(C.54), we obtain

$$\left\|\nabla(\widetilde{u}-\Pi_h\widetilde{u})\right\|^2_{\Omega}\leq C_3h^2\|\widetilde{u}\|^2_{2,2,\Omega}.\tag{C.55}$$

Estimates (C.51) and (C.55) yield the estimate

$$|\widetilde{u}-\Pi_h\widetilde{u}|^2_{i,2,\Omega}\leq Ch^{2-i}\|\widetilde{u}\|^2_{2,2,\Omega},\quad i=0,1.\tag{C.56}$$

Now we use (C.45) and set $\tilde{v}_h=\Pi_h\widetilde{u}$. Then,

$$\left\|\nabla(\widetilde{u}-\widetilde{u}_h)\right\|_{2,\Omega}=|\widetilde{u}-\widetilde{u}_h|_{1,2,\Omega}$$
$$\leq\mathbb{C}_a|\widetilde{u}-\Pi_h\widetilde{u}|_{1,2,\Omega}\leq\mathbb{C}_aCh\|\widetilde{u}\|_{2,2,\Omega}\leq\mathbb{C}_aCc_rh\|f\|_{\Omega}.$$

Since $u=\widetilde{u}+u_0$ and $u_h=\widetilde{u}_h+u_0$, we arrive at the estimate

$$\left\|\nabla(u-u_h)\right\|_{\Omega}\leq\mathrm{C}_1h\|f\|_{\Omega},\quad\mathrm{C}_1=\mathbb{C}_aCc_r.\tag{C.57}$$

This estimate (and similar estimates derived for many other boundary value problems) represents the main result of a priori asymptotic error analysis. It shows that if all the computations are performed exactly, then the error encompassed in Galerkin solutions decreases with the same rate as the parameter h of regular triangulations. We note that the corresponding constant depends on the interpolation constant C, the ellipticity constants c_1 and c_2 and the regularity constant c_r.

Interpolation estimates (C.46) and (C.47) suggest the idea that the convergence rate in terms of the weak (L^2) norm may be better than h. To prove this fact, we use the so-called *Aubin–Nitsche estimate*. Let u_g be the solution of the adjoint problem

$$a^*(u_g, v) = \int_\Omega g \cdot v \, dx, \quad \forall v \in V, \tag{C.58}$$

which is generated by the adjoint matrix A^*. If the matrix is selfadjoint, then the problem reads

$$\int_\Omega A(x)\nabla u_g \cdot \nabla w \, dx = \int_\Omega g w \, dx, \quad \forall w \in \overset{\circ}{H}{}^1(\Omega). \tag{C.59}$$

For any $g \in L_2(\Omega)$ there exists a unique solution $u_g \in \overset{\circ}{H}{}^1(\Omega)$. Moreover, under the above-made assumptions on the problem data $u_g \in H^2(\Omega)$ and

$$\|u_g\|_{2,2,\Omega} \leq c_r^* \|g\|_\Omega. \tag{C.60}$$

It is easy to see that

$$\|u - u_h\|_{2,\Omega} = \sup_{\substack{g \in L^2(\Omega) \\ g \neq 0}} \frac{\int_\Omega (u - u_h) \cdot g \, dx}{\|g\|_{2,\Omega}}.$$

By (C.59), we rewrite this relation in the form

$$\|u - u_h\|_\Omega = \sup_{\substack{g \in L^2(\Omega) \\ g \neq 0}} \frac{\int_\Omega A(x)\nabla u_g \cdot \nabla(u - u_h) \, dx}{\|g\|_{2,\Omega}}.$$

We use the Galerkin orthogonality and insert the function $\Pi_h u_g \in V_h$ in the above quotient. Then,

$$\begin{aligned} \|u - u_h\|_\Omega &= \sup_{\substack{g \in L^2(\Omega) \\ g \neq 0}} \frac{\int_\Omega A(x)\nabla(u_g - \Pi_h u_g) \cdot \nabla(u - u_h) \, dx}{\|g\|_\Omega} \\ &\leq c_2 \sup_{\substack{g \in L^2(\Omega) \\ g \neq 0}} \frac{\|\nabla(u_g - \Pi_h u_g)\|_\Omega}{\|g\|_\Omega} \big\|\nabla(u - u_h)\big\|_\Omega. \end{aligned}$$

By the interpolation estimate and (C.60),

$$\big\|\nabla(u_g - \Pi_h u_g)\big\|_\Omega \leq C h \|u_g\|_{2,2,\Omega} \leq C c_r^* h \|g\|_\Omega,$$

and we arrive at the estimate

$$\|u - u_h\|_\Omega \leq C c_r^* c_2 h \left\| \nabla(u - u_h) \right\|_\Omega \leq \mathrm{C}_0 h^2 \|f\|_\Omega,$$

where $\mathrm{C}_0 = C^2 c_r c_r^* c_2 \mathbb{C}_a$. It shows that the convergence in a weaker norm has a higher rate. It should be noted that this conclusion is justified only if all the solutions of the auxiliary problem (C.59) are H^2-regular.

C.4 A Priori Error Estimates for Mixed FEM

Mixed methods are based on saddle point statements of elliptic problems (see, e.g., [BF91, Bra07, RT91] and Sect. B.3). Consider again the problem

$$\begin{aligned} \operatorname{div} A\nabla u + f &= 0 \quad \text{in } \Omega, \\ u &= u_0 \quad \text{on } \Gamma_D, \\ A\nabla u \cdot n &= F \quad \text{on } \Gamma_N, \end{aligned}$$

where $f \in L^2(\Omega)$ and $F \in L^2(\Gamma_N)$.

Our goal is to discuss specific problems arising in the a priori error analysis of dual mixed approximations (for primal mixed approximations we can, in principle, use methods very close to that considered in Sect. C.3). We derive computable upper bounds for the quantities

$$\left\| \nabla(u - u_h) \right\|_A, \qquad \|p - p_h\|_{A^{-1}}, \qquad \|p - p_h\|_{\mathrm{div}}.$$

Here, the main difference (with respect to the previous section) is in the derivation of projection-type error estimates. Combining them with standard interpolation results, one can obtain known a priori estimate of convergence rate. Below we present a simplified version of the corresponding analysis, which, however, contains the main ideas usually used for the dual mixed approximations. A detailed exposition of this subject can be found in the above-cited books.

For the sake of simplicity, we consider the case of uniform Dirichlét boundary conditions and a constant matrix A. The dual mixed setting is presented by the relations

$$\int_\Omega \left(A^{-1}\widehat{p} \cdot \widehat{q} + (\operatorname{div} \widehat{q})\widehat{u} \right) \mathrm{d}x = 0, \quad \forall \widehat{q} \in \widehat{Q}_0,$$

$$\int_\Omega (\operatorname{div} \widehat{p} + f)\widehat{v} \, \mathrm{d}x = 0, \quad \forall \widehat{v} \in \widehat{V}.$$

Since there is no Neumann part of the boundary, the sets $\widehat{Q}_F$ and $\widehat{Q}_0$ coincide with $\widehat{Q} := H(\Omega, \mathrm{div})$.

In the case considered, the system of dual mixed finite element approximations satisfies the following system:

$$\int_\Omega \left(A^{-1}\widehat{p}_h \cdot \widehat{q}_h + \widehat{u}_h \operatorname{div}\widehat{q}_h\right) \mathrm{d}x = 0, \quad \forall \widehat{q}_h \in \widehat{Q}_h,$$

$$\int_\Omega (\operatorname{div}\widehat{p}_h + f)\widehat{v}_h \,\mathrm{d}x = 0, \quad \forall \widehat{v}_h \in \widehat{V}_h.$$

Assume that

$$\mathfrak{T}_h \text{ is a regular triangulation of a polygonal domain } \Omega, \tag{C.61}$$

$$\widehat{V}_h = \left\{v_h \in L^2 \mid v_h \in P^0(T), \ \forall T \in \mathfrak{T}_h\right\}, \tag{C.62}$$

$$\widehat{Q}_h = \left\{q_h \in H(\Omega, \operatorname{div}) \mid q_h \in RT^0(T), \ \forall T \in \mathfrak{T}_h\right\}, \tag{C.63}$$

$$f \in P^0(T), \quad \forall T \in \mathfrak{T}_h. \tag{C.64}$$

Note that under the assumptions made

$$\operatorname{div}\widehat{p}_h + f = 0 \quad \text{on any } T. \tag{C.65}$$

This fact directly follows from the relation

$$\int_\Omega (\operatorname{div}\widehat{p}_h + f)\widehat{v}_h \,\mathrm{d}x = 0, \quad \forall \widehat{v}_h \in \widehat{V}_h.$$

In view of (C.65), $\widehat{p}_h \in Q_f$.

C.4.1 Compatibility and Stability Conditions

In order to provide the stability of the discrete DM problem, we need to further restrict conditions imposed on approximation spaces. Let $\widehat{V}_h$, $\widehat{Q}_h$ satisfy the following condition:

For any $v_h \in \widehat{V}_h$ exists $q_h^v \in \widehat{Q}_h$ such that

$$\operatorname{div} q_h^v = v_h \quad \text{(compatibility)}, \tag{C.66}$$

$$\left\|q_h^v\right\| \le C\|v_h\| \quad \text{(stability)}. \tag{C.67}$$

We show that the above two conditions are sufficient conditions for proving that a discrete DM problem is *correct* (e.g., has a solution), *stable* and *has a projection-type error estimate*.

From (C.66) and (C.67), it follows that

$$\inf_{v_h \in \widehat{V}_h} \sup_{q_h \in \widehat{Q}_h} \frac{\int_\Omega v_h \operatorname{div} q_h \,\mathrm{d}x}{\|v_h\| \|q_h\|_{\mathrm{div}}} \ge \mathbb{C} > 0,$$

which is called the *discrete Inf–Sup condition*. Indeed,

$$\sup_{q_h \in \widehat{Q}_h} \frac{\int_\Omega v_h \operatorname{div} q_h \, dx}{\|v_h\| \|q_h\|_{\text{div}}} \geq \frac{\int_\Omega v_h \operatorname{div} q_h^v \, dx}{\|v_h\| \|q_h^v\|_{\text{div}}} = \frac{\|v_h\|}{\|q_h\|_{\text{div}}} \geq \frac{1}{\sqrt{1+C^2}}.$$

Proving the solvability of the discrete dual mixed problem is based on (C.66) and (C.67) in the same way as the solvability of the dual mixed problem is based on (B.50) and (B.51). The corresponding proof is quite analogous and leads to the following conclusion: if the triangulations are "regular" and the discrete Inf–Sup condition holds, then the discrete problem has a unique solution.

Now, we deduce a projection type error estimate for dual mixed approximations. Since p is a maximizer, i.e.,

$$-\frac{1}{2}\|q\|^2_{A^{-1}} \leq -\frac{1}{2}\|p\|^2_{A^{-1}}, \quad \forall q \in Q_f,$$

we find that

$$\int_\Omega A^{-1} p \cdot q \, dx = 0, \quad \forall q \in Q_0,$$

where Q_0 is the space of solenoidal functions. Therefore, for any $q \in Q_f$,

$$\begin{aligned} \frac{1}{2}\|q - p\|^2_{A^{-1}} &= \frac{1}{2}\|q\|^2_{A^{-1}} - \frac{1}{2}\|p\|^2_{A^{-1}} + \int_\Omega A^{-1} p \cdot (p - q) \, dx \\ &= \frac{1}{2}\|q\|^2_{A^{-1}} - \frac{1}{2}\|p\|^2_{A^{-1}}. \end{aligned}$$

Let $Q_{fh} = Q_f \cap \widehat{Q}_h$. Note that $p_h \in Q_{fh}$ is also the maximizer of $-\frac{1}{2}\|q_{fh}\|^2_{A^{-1}}$ on Q_{fh}, so that

$$\begin{aligned} \frac{1}{2}\|p_h - p\|^2_{A^{-1}} &= \frac{1}{2}\|p_h\|^2_{A^{-1}} - \frac{1}{2}\|p\|^2_{A^{-1}} \leq \frac{1}{2}\|q_{fh}\|^2_{A^{-1}} - \frac{1}{2}\|p\|^2_{A^{-1}} \\ &= \frac{1}{2}\|q_{fh} - p\|^2_{A^{-1}}, \quad \forall q_{fh} \in Q_{fh}. \end{aligned}$$

Thus, we arrive at the first projection estimate

$$\|p - p_h\|_{A^{-1}} \leq \inf_{q_{fh} \in Q_{fh}} \|p - q_{fh}\|_{A^{-1}}. \tag{C.68}$$

However, this projection error estimate has an obvious drawback. It is applicable only for a very narrow class of approximations: conforming (internal) approximations of the set Q_f.

To obtain an estimate for a wider class, we first derive one auxiliary result. Let us consider a *Modified DM problem*. Let $\widetilde{f} = \operatorname{div}(\widehat{q}_h - p)$ where $\widehat{q}_h \in \widehat{Q}_h$. The modified DM problem is to find $\widehat{p}_h^f$ and $\widehat{u}_h^f$ such that

$$\int_{\Omega} (A^{-1}\widehat{p}_h^f \cdot \widehat{q}_h + \widehat{u}_h^f \operatorname{div} \widehat{q}_h)\, dx = 0, \quad \forall \widehat{q}_h \in \widehat{Q}_{0h}, \tag{C.69}$$

$$\int_{\Omega} (\operatorname{div} \widehat{p}_h^f + \widetilde{f})\widehat{v}_h\, dx = 0, \quad \forall \widehat{v}_h \in \widehat{V}_h. \tag{C.70}$$

Under the assumptions made $\widetilde{f} \in P^0(T)$, the above DM problem is solvable, and

$$\|\widehat{p}_h^f\|_{A^{-1}}^2 + \int_{\Omega} \widehat{u}_h^f \operatorname{div} \widehat{p}_h^f\, dx = 0,$$

$$\|\widehat{p}_h^f\|_{A^{-1}}^2 \leq \|\widehat{u}_h^f\| \|\operatorname{div} \widehat{p}_h^f\| = \|\widehat{u}_h^f\| \|\widetilde{f}\|.$$

From here, we observe that

$$\bar{c}_1 \|\widehat{p}_h^f\|^2 \leq \|\widehat{p}_h^f\|_{A^{-1}}^2 \leq \|\widehat{u}_h^f\| \|\widetilde{f}\|. \tag{C.71}$$

By (C.66) and (C.67) we conclude that for $\widehat{u}_h^f$ we can find $\bar{q}_h$ in $\widehat{Q}_h$ such that

$$\operatorname{div} \bar{q}_h + \widehat{u}_h^f = 0 \quad \text{and} \quad \|\bar{q}_h\| \leq C\|\widehat{u}_h^f\|.$$

Use $\bar{q}_h$ in the first identity (C.69). We have,

$$\int_{\Omega} (A^{-1}\widehat{p}_h^f \cdot \bar{q}_h + \widehat{u}_h^f \operatorname{div} \bar{q}_h)\, dx = 0.$$

Thus,

$$\begin{aligned}\|\widehat{u}_h^f\|^2 &= \int_{\Omega} \widehat{u}_h^f \operatorname{div} \bar{q}_h \leq \|\widehat{p}_h^f\|_{A^{-1}} \|\bar{q}_h\|_{A^{-1}} \\ &\leq \bar{c}_2 \|\widehat{p}_h^f\|_{A^{-1}} \|\bar{q}_h\| \leq \bar{c}_2 C \|\widehat{p}_h^f\|_{A^{-1}} \|\widehat{u}_h^f\|.\end{aligned}$$

We conclude that

$$\|\widehat{u}_h^f\| \leq \bar{c}_2 C \|\widehat{p}_h^f\|_{A^{-1}}. \tag{C.72}$$

Now, we use (C.71) and obtain

$$\|\widehat{p}_h^f\|_{A^{-1}}^2 \leq \|\widehat{u}_h^f\| \|\widetilde{f}\| \leq \bar{c}_2 C \|\widehat{p}_h^f\|_{A^{-1}} \|\widetilde{f}\|.$$

Thus,

$$\bar{c}_1 \|\widehat{p}_h^f\| \leq \|\widehat{p}_h^f\|_{A^{-1}} \leq \bar{c}_2 C \|\widetilde{f}\| \tag{C.73}$$

and

$$\|\widehat{p}_h^f\|_{\text{div}}^2 = \|\widehat{p}_h^f\|^2 + \|\operatorname{div} \widehat{p}_h^f\|^2 \leq \left(1 + \frac{c_2^2}{c_1^2} C^2\right) \|\widetilde{f}\|^2. \tag{C.74}$$

We note that the estimates (C.72), (C.73), and (C.74) show that the modified DM problem is stable, i.e. its solutions $(\widehat{p}_h^f, \widehat{u}_h^f)$ are bounded by the problem data uniformly with respect to h.

If we replace $\widetilde{f}$ by f, then we can derive the same stability estimate for the functions $(\widehat{p}_h, \widehat{u}_h)$ that present an approximate solution of the original DM problem.

C.4.2 Projection Estimates for Fluxes

Now, we return to the projection error estimates. As we have seen

$$\|p - p_h\|_{A^{-1}} \leq \inf_{q_{fh} \in Q_{fh}} \|p - q_{fh}\|.$$

Let $\eta_h = \widehat{p}_h^f + \widehat{q}_h$, where $\widehat{q}_h$ is *an arbitrary element of* $\widehat{Q}_h$. We have,

$$\begin{aligned} \operatorname{div} \eta_h &= \operatorname{div} \widehat{p}_h^f + \operatorname{div} \widehat{q}_h = -\widetilde{f} + \operatorname{div} \widehat{q}_h \\ &= \operatorname{div}(p - \widehat{q}_h) + \operatorname{div} \widehat{q}_h = \operatorname{div} p = -f. \end{aligned}$$

Therefore, $\eta_h \in Q_f$. Now, we use the projection inequality with η_h

$$\begin{aligned} \|p - p_h\|_{A^{-1}} &\leq \|p - \eta_h\|_{A^{-1}} = \left\|p - \widehat{p}_h^f - \widehat{q}_h\right\|_{A^{-1}} \\ &\leq \|p - \widehat{q}_h\|_{A^{-1}} + \left\|\widehat{p}_h^f\right\|_{A^{-1}}. \end{aligned} \tag{C.75}$$

Note that in the case considered $\operatorname{div}(p - p_h) = 0$, so that

$$\|p - p_h\|_{\operatorname{div}} = \|p - p_h\| \leq \frac{1}{\bar{c}_1} \|p - p_h\|_{A^{-1}}.$$

With the help of (C.73), we find that

$$\|p - p_h\|_{\operatorname{div}} \leq \frac{1}{\bar{c}_1} \left(\|p - \widehat{q}_h\|_{A^{-1}} + \left\|\widehat{p}_h^f\right\|_{A^{-1}}\right) \leq \frac{1}{\bar{c}_1} \left(\|p - \widehat{q}_h\|_{A^{-1}} + \bar{c}_2 C \|\widetilde{f}\|\right).$$

Now, we recall that $\widetilde{f} = \operatorname{div}(p - \widehat{q}_h)$ and arrive at the estimate

$$\|p - p_h\|_{\operatorname{div}} \leq \frac{1}{\bar{c}_1} \left(\|p - \widehat{q}_h\|_{A^{-1}} + \bar{c}_2 C \left\|\operatorname{div}(p - \widehat{q}_h)\right\|\right), \quad \forall \widehat{q}_h \in \widehat{Q}_h.$$

Hence,

$$\|p - p_h\|_{\operatorname{div}} \leq \bar{C}_p \inf_{\widehat{q}_h \in \widehat{Q}_h} \left\{\|p - \widehat{q}_h\|_{A^{-1}} + \left\|\operatorname{div}(p - \widehat{q}_h)\right\|\right\}, \tag{C.76}$$

where $\bar{C}_p$ depends on C, $\bar{c}_1$, and $\bar{c}_2$ and does not depend on h.

Error estimates for $\widehat{u} - \widehat{u}_h$ can be derived analogously. We have

$$\int_\Omega \left(A^{-1}\widehat{p}_h \cdot \widehat{q}_h + \widehat{u}_h \operatorname{div} \widehat{q}_h\right) \mathrm{d}x = 0, \quad \forall \widehat{q}_h \in \widehat{Q}_h.$$

Since $\widehat{Q}_h \subset Q$, we also have

$$\int_\Omega \left(A^{-1} p \cdot \widehat{q}_h + u \operatorname{div} \widehat{q}_h\right) \mathrm{d}x = 0.$$

From here, we see that

$$\int_\Omega \left(A^{-1}(\widehat{p}_h - p) \cdot \widehat{q}_h + (\widehat{u}_h - u) \operatorname{div} \widehat{q}_h\right) \mathrm{d}x = 0, \quad \forall \widehat{q}_h \in \widehat{Q}_h.$$

Denote

$$\{\!|u|\!\}_T = \frac{1}{|T|} \int_T u \,\mathrm{d}x, \qquad \{\!|u|\!\}_h(x) = \{\!|u|\!\}_{T_i} \quad \text{if } x \in T_i.$$

Since $\operatorname{div} \widehat{q}_h$ is constant on each T_i, we rewrite the relation as follows:

$$\int_\Omega \left(A^{-1}(\widehat{p}_h - p) \cdot \widehat{q}_h + \left(\widehat{u}_h - \{\!|u|\!\}_h\right) \operatorname{div} \widehat{q}_h\right) \mathrm{d}x = 0, \quad \forall \widehat{q}_h \in \widehat{Q}_h. \tag{C.77}$$

Note that $\{\!|u|\!\}_h \in \widehat{V}_h$ and therefore $\bar{u}_h := \widehat{u}_h - \{\!|u|\!\}_h \in \widehat{V}_h$ Now, we exploit the compatibility and stability conditions (C.66) and (C.67) again. For $\bar{u}_h$ one can find $q_h' \in \widehat{Q}_h$ such that

$$\operatorname{div} q_h' + \bar{u}_h = 0 \quad \text{and} \quad \|q_h'\| \leq C\|\bar{u}_h\|.$$

Let us use this function q_h' in the integral relation (C.77). We have

$$\int_\Omega \left(A^{-1}(\widehat{p}_h - p) \cdot q_h' + \bar{u}_h \operatorname{div} q_h'\right) \mathrm{d}x = 0.$$

From here, we conclude that

$$\begin{aligned} \|\bar{u}_h\|^2 &= \left| \int_\Omega A^{-1}(\widehat{p}_h - p) \cdot q_h' \,\mathrm{d}x \right| \\ &\leq \|\widehat{p}_h - p\|_{A^{-1}} \|q_h'\|_{A^{-1}} \leq C\bar{c}_2 \|\widehat{p}_h - p\|_{A^{-1}} \|\bar{u}_h\|. \end{aligned}$$

Thus,

$$\|\bar{u}_h\| \leq C\bar{c}_2 \|\widehat{p}_h - p\|_{A^{-1}}.$$

We have

$$\begin{aligned} \|u - \widehat{u}_h\| &\leq \left\|u - \{\!|u|\!\}_h\right\| + \left\|\{\!|u|\!\}_h - \widehat{u}_h\right\| = \left\|u - \{\!|u|\!\}_h\right\| + \|\bar{u}_h\| \\ &\leq \left\|u - \{\!|u|\!\}_h\right\| + C\bar{c}_2 \|\widehat{p}_h - p\|_{A^{-1}}. \end{aligned}$$

Note that by the definition of $\{\!|u|\!\}_h$

$$\left\| u - \{\!|u|\!\}_h \right\| \leq \|u - v_h\|, \quad \forall v_h \in \widehat{V}_h.$$

Thus,

$$\|u - \widehat{u}_h\| \leq C\bar{c}_2 \|\widehat{p}_h - p\|_{A^{-1}} + \inf_{v_h \in \widehat{V}_h} \|u - v_h\|$$

We recall that

$$\|p - p_h\|_{A^{-1}} \leq \|p - \widehat{q}_h\|_{A^{-1}} + \left\|\widehat{p}_h^f\right\|_{A^{-1}} \leq \bar{c}_2 C \left\|\mathrm{div}(p - \widehat{q}_h)\right\| + \|p - \widehat{q}_h\|_{A^{-1}}$$

and arrive at the projection type error estimate for the primal variable

$$\|u - \widehat{u}_h\| \leq C_u \Big(\inf_{\widehat{q}_h \in \widehat{Q}_h} \left\{ \|p - \widehat{q}_h\|_{A^{-1}} + \left\|\mathrm{div}(p - \widehat{q}_h)\right\| \right\} + \inf_{v_h \in \widehat{V}_h} \|u - v_h\| \Big), \quad \text{(C.78)}$$

where C_u depends on C, $\bar{c}_1$, and $\bar{c}_2$ and does not depend on h. Estimates (C.76) and (C.78) lead to a qualified a priori convergence estimates, provided that the solution possesses proper regularity.

It is worth concluding this brief overview of a priori error estimation methods by the following comment. As we have seen mathematical justifications of a priori convergence rate estimates are based on several assumptions, namely,

- the exact solution of a boundary value problem possesses an extra regularity (e.g., the generalized solution of a second order differential equation belongs to the space H^2);
- u_h is the Galerkin approximation, i.e., it is the exact solution of the corresponding finite dimensional problem;
- all the meshes $\mathcal{T}_h$ are regular, i.e., the elements do not "degenerate" in the refinement process.

In real life computations, it is difficult to guarantee that all these assumptions do hold. Even if we can ensure this, the a priori convergence estimates cannot guarantee that the error monotonically decreases as $h \to 0$ (this can be easily proved only for nested meshes). Moreover, in practice we are interested in the error of a *concrete approximation on a particular mesh*. A priori asymptotic estimates could hardly be efficient in such a context, because they are derived for the whole class of approximate solutions of a particular type, which encompasses all possible cases. Therefore, a priori convergence estimates have mainly a theoretical value: they show that an approximation method is correct "in principle". Efficient quantitative analysis of approximation errors associated with a particular approximate solution is performed by a posteriori estimates.

References

[ABC03] Y. Achdou, C. Bernardi, F. Coquel, A priori and a posteriori analysis of finite volume discretizations of Darcy's equations. Numer. Math. **96**(1), 17–42 (2003)

[ABCM02] D. Arnold, F. Brezzi, B. Cockburn, L. Marini, Unified analysis of discontinuous Galerkin methods for elliptic problems. SIAM J. Numer. Anal. **39**(5), 1749–1779 (2002)

[AD03] G. Acosta, R. Duran, An optimal Poincaré inequality in L^1 for convex domains. Proc. Am. Math. Soc. **132**(1), 195–202 (2003)

[AF89] D.N. Arnold, R.S. Falk, A uniformly accurate finite element method for the Reissner-Mindlin plate. SIAM J. Numer. Anal. **26**(6), 1276–1290 (1989)

[Ago02] A. Agouzal, On the saturation assumption and hierarchical a posteriori error estimator. Comput. Methods Appl. Math. **2**(2), 125–131 (2002)

[AH83] G. Alefeld, J. Herzberger, *Introduction to Interval Computations*. Computer Science and Applied Mathematics (Academic Press, New York, 1983). Translated from the German by Jon Rokne

[Ain98] M. Ainsworth, A posteriori error estimation for fully discrete hierarchic models of elliptic boundary-value problems on thin domains. Numer. Math. **80**(3), 325–362 (1998)

[Ain07] M. Ainsworth, A posteriori error estimation for discontinuous Galerkin finite element approximation. SIAM J. Numer. Anal. **45**(4), 1777–1798 (2007)

[AMM$^+$09] I. Anjam, O. Mali, A. Muzalevsky, P. Neittaanmäki, S. Repin, A posteriori error estimates for a Maxwell type problem. Russ. J. Numer. Anal. Math. Model. **24**(5), 395–408 (2009)

[AMN11] G. Akrivis, C. Makridakis, R.H. Nochetto, Galerkin and Runge-Kutta methods: unified formulation, a posteriori error estimates and nodal superconvergence. Numer. Math. **118**(3), 429–456 (2011)

[AMNR12] I. Anjam, O. Mali, P. Neittaanmäki, S. Repin, On the reliability of error indication methods for problems with uncertain data. in *Numerical Mathematics and Advanced Applications*, ed. by A. Cangiani, R.L. Davidchack, E. Georgoulis, A.N. Gorban, J. Levesley, M.V. Tretyakov. ENUMATH 2011 (Springer, Berlin, 2012)

[AMZ02] D. Arnold, A.L. Madureira, S. Zhang, On the range of applicability of the Reissner-Mindlin and Kirchhoff-Love plate bending models. J. Elast. **67**(3), 171–185 (2002)

[Ang95] L. Angermann, Balanced a posteriori error estimates for finite-volume type discretizations of convection-dominated elliptic problems. Computing **55**(4), 305–323 (1995)

[Ang99] Ph. Angot, Analysis of singular perturbations on the Brinkman problem for fictitious domain models of viscous flows. Math. Methods Appl. Sci. **22**(16), 1395–1412 (1999)

O. Mali et al., *Accuracy Verification Methods*,
Computational Methods in Applied Sciences 32, DOI 10.1007/978-94-007-7581-7,

[AO92] M. Ainsworth, J.T. Oden, A procedure for a posteriori error estimation for h-p finite element methods. Comput. Methods Appl. Mech. Eng. **101**(1–3), 73–96 (1992). Reliability in computational mechanics (Kraków, 1991)

[AO00] M. Ainsworth, J.T. Oden, *A Posteriori Error Estimation in Finite Element Analysis* (Wiley, New York, 2000)

[AR10] M. Ainsworth, R. Rankin, Fully computable error bounds for discontinuous Galerkin finite element approximations on meshes with an arbitrary number of levels of hanging nodes. SIAM J. Numer. Anal. **47**(6), 4112–4141 (2010)

[Arn82] D.N. Arnold, An interior penalty finite element method with discontinuous elements. SIAM J. Numer. Anal. **19**(4), 742–760 (1982)

[AS99] M. Ainsworth, B. Senior, hp-finite element procedures on non-uniform geometric meshes: adaptivity and constrained approximation, in *Grid Generation and Adaptive Algorithms*, ed. by M.W. Bern, J.E. Flaherty, M. Luskin. Minneapolis, MN, 1997. IMA Vol. Math. Appl., vol. 113 (Springer, New York, 1999), pp. 1–27

[Axe94] O. Axelsson, *Iterative Solution Methods* (Cambridge University Press, Cambridge, 1994)

[BA72] I. Babuška, A.K. Aziz, Survey lectures on the mathematical foundations of the finite element method, in *The Mathematical Foundations of the Finite Element Method with Applications to Partial Differential Equations*, ed. by A.K. Aziz. Univ. Maryland, Baltimore, MD, 1972 (Academic Press, New York, 1972), pp. 1–359. With the collaboration of G. Fix and R.B. Kellogg

[Bab61] I. Babuška, On randomized solutions of Laplace's equation. Čas. Pěst. Mat. **86**(3), 269–276 (1961)

[Bab73a] I. Babuška, The finite element method with penalty. Math. Comput. **27**, 221–228 (1973)

[Bab73b] I. Babuška, The finite element method with Lagrangian multipliers. Numer. Math. **20**(3), 179–192 (1973)

[BBC94] R. Barrett, M. Berry, T.F. Chan et al., *Templates for the Solution of Linear Systems: Building Blocks for Iterative Methods* (Society for Industrial and Applied Mathematics, Philadelphia, 1994)

[BBRR07] R. Becker, M. Braack, R. Rannacher, T. Richter, Mesh and model adaptivity for flow problems, in *Reactive Flows, Diffusion and Transport*, ed. by W. Jager, R. Rannacher, J. Warnatz (Springer, Berlin, 2007), pp. 47–75

[BC02] S. Bartels, C. Carstensen, Each averaging technique yields reliable a posteriori error control in FEM on unstructured grids. Part II: higher order FEM. Math. Comput. **71**(239), 971–994 (2002)

[BCH09] D. Braess, C. Carstensen, R.W.H. Hoppe, Error reduction in adaptive finite element approximations of elliptic obstacle problems. J. Comput. Math. **27**(2–3), 148–169 (2009)

[BDGG71] B.L. Buzbee, F.W. Dorr, G.A. George, G.H. Golub, The direct solution of the discrete Poisson equation on irregular regions. SIAM J. Numer. Anal. **8**(4), 722–736 (1971)

[BE03] M. Braack, A. Ern, A posteriori control of modeling errors and discretization errors. Multiscale Model. Simul. **1**(2), 221–238 (2003)

[Ber89] C. Bernardi, Optimal finite-element interpolation on curved domains. SIAM J. Numer. Anal. **26**(5), 1212–1240 (1989)

[Ber99] A. Bernardini, What are random and fuzzy sets and how to use them for uncertainty modelling in engineering systems? in *Whys and Hows in Uncertainty Modelling, Probability, Fuzziness and Anti-optimization*, ed. by I. Elishakoff. CISM Courses and Lectures, vol. 388 (Springer, Berlin, 1999)

[BF86] F. Brezzi, M. Fortin, Numerical approximation of Mindlin-Reissner plates. Math. Comput. **47**(175), 151–158 (1986)

[BF91] F. Brezzi, M. Fortin, *Mixed and Hybrid Finite Element Methods*. Springer Series in Computational Mathematics, vol. 15 (Springer, New York, 1991)

[BG98] C. Bernardi, V. Girault, A local regularization operator for triangular and quadrilateral finite elements. SIAM J. Numer. Anal. **35**(5), 1893–1916 (1998)

[BGC05] R. Bustinza, G.N. Gatica, B. Cockburn, An a posteriori error estimate for the local discontinuous Galerkin method applied to linear and nonlinear diffusion problems. J. Sci. Comput. **22–23**(1–3), 147–185 (2005)

[BGH⁺01] C. Bernardi, V. Girault, F. Hecht, H. Kawarada, O. Pironneau, A finite element problem issued from fictitious domain techniques. East-West J. Numer. Math. **9**(4), 253–263 (2001)

[BGP89] I. Babuška, M. Griebel, J. Pitkäranta, The problem of selecting the shape functions for a p-type finite element. Int. J. Numer. Methods Eng. **28**(8), 1891–1908 (1989)

[BH70] J.H. Bramble, S.R. Hilbert, Estimation of linear functionals on Sobolev spaces with application to Fourier transforms and spline interpolation. SIAM J. Numer. Anal. **7**, 112–124 (1970)

[BH81] M. Bernadou, K. Hassan, Basis functions for general Hsieh-Clough-Tocher triangles, complete or reduced. Int. J. Numer. Methods Eng. **17**(5), 784–789 (1981)

[BHHW00] R. Beck, R. Hiptmair, R. Hoppe, B. Wohlmuth, Residual based a posteriori error estimators for eddy current computation. M2AN Math. Model. Numer. Anal. **34**(1), 159–182 (2000)

[BHL03] R. Becker, P. Hansbo, M.G. Larson, Energy norm a posteriori error estimation for discontinuous Galerkin methods. Comput. Methods Appl. Mech. Eng. **192**(5–6), 723–733 (2003)

[BHS08] D. Braess, R.W.H. Hoppe, J. Schöberl, A posteriori estimators for obstacle problems by the hypercircle method. Comput. Vis. Sci. **11**(4–6), 351–362 (2008)

[BJ87] J.M. Ball, R.D. James, Fine phase mixtures as minimizers of energy. Arch. Ration. Mech. Anal. **100**(1), 13–52 (1987)

[BKR00] R. Becker, H. Kapp, R. Rannacher, Adaptive finite element methods for optimal control of partial differential equations: basic concept. SIAM J. Control Optim. **39**(1), 113–132 (2000)

[BLM00] T. Belytschko, W.K. Liu, B. Moran, *Nonlinear Finite Elements for Continua and Structures* (Wiley, Chichester, 2000)

[BMP09] A.N. Bogolyubov, M.D. Malykh, A.A. Panin, The dependence of the effectiveness of a posteriori estimation of an approximate solution of an elliptic boundary value problem on the input data and the algorithm parameters. Vestnik Moskov. Univ. Ser. III Fiz. Astronom **2009**(1), 18–22 (2009)

[BMR08] M. Bildhauer, M. Fuchs, S. Repin, Duality based a posteriori error estimates for higher order variational inequalities with power growth functionals. Ann. Acad. Sci. Fenn., Ser. A 1 Math. **33**(2), 475–490 (2008)

[BMR12] S. Bartels, A. Mielke, T. Roubíček, Quasi-static small-strain plasticity in the limit of vanishing hardening and its numerical approximation. SIAM J. Numer. Anal. **50**(2), 951–976 (2012)

[BNP10] A. Bonito, R.H. Nochetto, M.S. Pauletti, Geometrically consistent mesh modification. SIAM J. Numer. Anal. **48**(5), 1877–1899 (2010)

[BR78a] I. Babuška, W.C. Rheinboldt, Error estimates for adaptive finite element computations. SIAM J. Numer. Anal. **15**(4), 736–754 (1978)

[BR78b] I. Babuška, W.C. Rheinboldt, A-posteriori error estimates for the finite element method. Int. J. Numer. Methods Eng. **12**(10), 1597–1615 (1978)

[BR93] I.M. Babuška, R. Rodríguez, The problem of the selection of an a posteriori error indicator based on smoothening techniques. Int. J. Numer. Methods Eng. **36**(4), 539–567 (1993)

[BR96] R. Becker, R. Rannacher, A feed-back approach to error control in finite element methods: basic analysis and examples. East-West J. Numer. Math. **4**(4), 237–264 (1996)

[BR00] H.M. Buss, S.I. Repin, A posteriori error estimates for boundary value problems with obstacles, in *Numerical Mathematics and Advanced Applications*, ed. by P. Neit-

taanmäki, T. Tiihonen, P. Tarvainen. Jyväskylä, 1999 (World Scientific, River Edge, 2000), pp. 162–170
[BR01] R. Becker, R. Rannacher, An optimal control approach to a posteriori error estimation in finite element methods. Acta Numer. **10**, 1–102 (2001)
[BR03] W. Bangerth, R. Rannacher, *Adaptive Finite Element Methods for Differential Equations* (Birkhäuser, Basel, 2003)
[BR07] M. Bildhauer, S. Repin, Estimates of the deviation from the minimizer for variational problems with power growth functionals. J. Math. Sci. (N.Y.) **143**(2), 2845–2856 (2007)
[BR12] M. Besier, R. Rannacher, Goal-oriented space-time adaptivity in the finite element Galerkin method for the computation of nonstationary incompressible flow. Int. J. Numer. Methods Fluids **70**(9), 1139–1166 (2012)
[Bra03] J. Bramble, A proof of the inf-sup condition for the Stokes equations on Lipschitz domains. Math. Models Methods Appl. Sci. **13**(3), 361–371 (2003)
[Bra05] D. Braess, A posteriori error estimators for obstacle problems—another look. Numer. Math. **101**(3), 421–451 (2005)
[Bra07] D. Braess, *Finite Elements* (Cambridge University Press, Cambridge, 2007)
[Bre74] F. Brezzi, On the existence, uniqueness and approximation of saddle-point problems arising from Lagrange multipliers. Rev. Française Automat. Informat. Recherche Opérationnelle Sér. Rouge **8**(R-2), 129–151 (1974)
[Bre03a] S. Brenner, Korn's inequalities for piecewise h^1 vector fields. Math. Comput. **78**(247), 1067–1087 (2003)
[Bre03b] S. Brenner, Poincaré-Friedrichs inequalities for piecewise h^1 functions. SIAM J. Numer. Anal. **41**(1), 306–324 (2003)
[BS94] S. Brenner, R.L. Scott, *The Mathematical Theory of Finite Element Methods* (Springer, New York, 1994)
[BS98] J. Bramble, T. Sun, A negative-norm least squares method for Reissner-Mindlin plates. Math. Comput. **67**(223), 901–916 (1998)
[BS01] I. Babuška, T. Strouboulis, *The Finite Element Method and Its Reliability. Numerical Mathematics and Scientific Computation* (Oxford University Press, New York, 2001)
[BS08] D. Braess, J. Schöberl, Equilibrated residual error estimator for Maxwell's equation. Math. Comput. **77**(262), 651–672 (2008)
[BS12] S. Bartels, P. Schreier, Local coarsening of simplicial finite element meshes generated by bisections. BIT Numer. Math. **52**(3), 559–569 (2012)
[BSS11] D. Braess, S. Sauter, C. Schwab, On the justification of plate models. J. Elast. **103**(1), 53–71 (2011)
[But03] J.C. Butcher, *Numerical Methods for Ordinary Differential Equations* (Wiley, Chichester, 2003)
[BW91] R.E. Bank, B.D. Welfert, A posteriori error estimates for the Stokes problem. SIAM J. Numer. Anal. **28**(3), 591–623 (1991)
[BWS11] I. Babuška, J.R. Whiteman, T. Strouboulis, *Finite Elements, an Introduction to the Method and Error Estimation* (Oxford University Press, New York, 2011)
[BZ73] I. Babuška, M. Zlamal, Nonconforming elements in the finite element method with penalty. SIAM J. Numer. Anal. **10**(5), 863–875 (1973)
[Car99] C. Carstensen, Quasi-interpolation and a posteriori error analysis of finite element methods. M2AN Math. Model. Numer. Anal. **6**(33), 1187–1202 (1999)
[Cas05] P. Castillo, An a posteriori error estimate for the local discontinuous Galerkin method. J. Sci. Comput. **22/23**(1–3), 187–204 (2005)
[CB02] C. Carstensen, S. Bartels, Each averaging technique yields reliable a-posteriori error control in fem on unstructured grids. I: low order conforming, nonconforming, and mixed FEM. Math. Comput. **239**(71), 945–969 (2002)
[CDNR09] S. Cochez-Dhondt, S. Nicaise, S. Repin, A posteriori error estimates for finite volume approximations. Math. Model. Nat. Phenom. **4**(1), 106–122 (2009)

[CDP97] A. Charbonneau, K. Dossou, R. Pierre, A residual-based a posteriori error estimator for Ciarlet-Raviart formulation of the first biharmonic problem. Numer. Methods Partial Differ. Equ. **13**(1), 93–111 (1997)

[CF00a] C. Carstensen, S.A. Funken, Constants in Clément-interpolation error and residual based a posteriori error estimates in Finite Element Methods. East-West J. Numer. Math. **8**(3), 153–175 (2000)

[CF00b] C. Carstensen, S.A. Funken, Fully reliable localized error control in the FEM. SIAM J. Sci. Comput. **21**(4), 1465–1484 (2000)

[Cia78a] P.G. Ciarlet, *The Finite Element Method for Elliptic Problems* (North-Holland, New York, 1978)

[Cia78b] P.G. Ciarlet, Interpolation error estimates for the reduced Hsieh-Clough-Tocher triangle. Math. Comput. **32**(142), 335–344 (1978)

[CKP11] C. Carstensen, D. Kim, E.-J. Park, A priori and a posteriori pseudostress-velocity mixed finite element error analysis for the Stokes problem. SIAM J. Numer. Anal. **49**(6), 2501–2523 (2011)

[CL55] E.A. Coddington, N. Levinson, *Theory of Ordinary Differential Equations* (McGraw-Hill, New York, 1955)

[Clé75] Ph. Clément, Approximation by finite element functions using local regularization. Rev. Fr. Autom. Inform. Rech. Opér., Anal. Numér. **9**(R-2), 77–84 (1975)

[CLMM94] Z. Cai, R. Lazarov, T.A. Manteuffel, S.F. McCormick, First-order system least squares for second-order partial differential equations: part I. SIAM J. Numer. Anal. **31**(6), 1785–1799 (1994)

[CLRS01] T.H. Cormen, C.E. Leiserson, R.L. Rivest, C. Stein, *Introduction to Algorithms*, 2nd edn. (MIT Press, Cambridge, 2001)

[CN00] Z. Chen, R.H. Nochetto, Rezidual type a posteriori error estimates for elliptic obstacle problems. Numer. Math. **84**(4), 527–548 (2000)

[Col64] L. Collatz, *Funktionanalysis und Numerische Mathematik* (Springer, Berlin, 1964)

[Cou43] R. Courant, Variational methods for some problems of equilibrium and vibration. Bull. Am. Math. Soc. **49**, 1–23 (1943)

[CLP09] D. Copeland, U. Langer, D. Pusch, From the boundary element domain decomposition methods to local Trefftz finite element methods on polyhedral meshes, in *Domain Decomposition Methods in Science and Engineering XVIII*. Lect. Notes Comput. Sci. Eng., vol. 70 (Springer, Berlin, 2009), pp. 315–322

[CV99] C. Carstensen, R. Verfürth, Edge residuals dominate a posteriori error estimates for low order finite element methods. SIAM J. Numer. Anal. **36**(5), 1571–1587 (1999)

[DA05] W. Dörfler, M. Ainsworth, Reliable a posteriori error control for nonconformal finite element approximation of Stokes flow. Math. Comput. **74**(252), 1599–1619 (2005)

[Dem67] A.P. Dempster, Upper and lower probabilities induced by a multi-valued mapping. Ann. Math. Stat. **38**, 325–339 (1967)

[Dem07] L. Demkowicz, *Computing with hp Finite Elements. I. One- and Two-Dimensional Elliptic and Maxwell Problems* (Chapman & Hall/CRC Press, Boca Raton, 2007)

[Deu04] P. Deuflhard, *Newton Methods for Nonlinear Problems: Affine Invariance and Adaptive Algorithms* (Springer, Berlin, 2004)

[DL55] J. Deny, J.L. Lions, Les espaces du type Beppo-Levi. Ann. Inst. Fourier (Grenoble) **5**, 305–370 (1955)

[DL72] G. Duvaut, J.-L. Lions, *Les Inéquations en Mécanique et en Physique*. Travaux et Recherches Mathématiques, vol. 21 (Dunod, Paris, 1972)

[DLY89] P. Deuflhard, P. Leinen, H. Yserentant, Concept of an adaptive hierarchical finite element code. Impact Comput. Sci. Eng. **1**(1), 3–35 (1989)

[DMR91] R. Duran, M.A. Muschietti, R. Rodriguez, On the asymptotic exactness of error estimators for linear triangle elements. Numer. Math. **59**(2), 107–127 (1991)

[DN02] W. Dörfler, R.H. Nochetto, Small data oscillation implies the saturation assumption. Numer. Math. **91**(1), 1–12 (2002)

[Dob03] M. Dobrowolski, On the LBB constant on stretched domains. Math. Nachr. **254/255**, 64–67 (2003)

[Dör96] W. Dörfler, A convergent adaptive algorithm for Poisson's equation. SIAM J. Numer. Anal. **33**(3), 1106–1124 (1996)

[DR98] W. Dörfler, M. Rumpf, An adaptive strategy for elliptic problems including a posteriori controlled boundary approximation. Math. Comput. **67**(224), 1361–1382 (1998)

[EGH00] R. Eymard, T. Gallouët, R. Herbin, Finite volume methods, in *Handbook of Numerical Analysis*, vol. VII, ed. by P.G. Ciarlet, J.L. Lions (North-Holland, Amsterdam, 2000), pp. 713–1020

[EJ88] K. Eriksson, C. Johnson, An adaptive finite element method for linear elliptic problems. Math. Comput. **50**(182), 361–383 (1988)

[EKK+95] T. Eirola, A.M. Krasnosel'skii, M.A. Krasnosel'skii, N.A. Kuznersov, O. Nevanlinna, Incomplete corrections in nonlinear problems. Nonlinear Anal. **25**(7), 717–728 (1995)

[Eli83] I. Elishakoff, *Probabilistic Methods in the Theory of Structures* (Wiley, New York, 1983)

[EP05] A. Ern, J. Proft, A posteriori discontinuous Galerkin error estimates for transient convection-diffusion equations. Appl. Math. Lett. **18**(7), 833–841 (2005)

[ES00] H.W. Engl, O. Scherzer, Convergence rates results for iterative methods for solving nonlinear ill-posed problems, in *Surveys on Solution Methods for Inverse Problems*, ed. by D. Colton, H.W. Engl, A.K. Louis, J.R. McLaughlin, W. Rundell (Springer, Vienna, 2000), pp. 7–34

[ESV10] A. Ern, A.F. Stephansen, M. Vohralik, Guaranteed and robust discontinuous Galerkin a posteriori error estimates for convection-diffusion-reaction problems. J. Comput. Appl. Math. **234**(1), 114–130 (2010)

[ET76] I. Ekeland, R. Temam, *Convex Analysis and Variational Problems* (North-Holland, New York, 1976)

[EW94] S.C. Eisenstat, H.F. Walker, Globally convergent inexact Newton methods. SIAM J. Optim. **4**(2), 393–422 (1994)

[Fal74] R.S. Falk, Error estimates for the approximation of a class of variational inequalities. Math. Comput. **28**, 963–971 (1974)

[Fei93] M. Feistauer, *Mathematical Methods in Fluid Dynamics* (Longman, Harlow, 1993)

[FNP09] M. Farhloul, S. Nicaise, L. Paquet, A priori and a posteriori error estimations for the dual mixed finite element method of the Navier-Stokes problem. Numer. Methods Partial Differ. Equ. **25**(4), 843–869 (2009)

[FNR02] M. Frolov, P. Neittaanmäki, S. Repin, On the reliability, effectivity and robustness of a posteriori error estimation methods. Technical report B14, Department of Mathematical Information Technology, University of Jyväskylä, 2002

[FNR03] M. Frolov, P. Neittaanmäki, S. Repin, On computational properties of a posteriori error estimates based upon the method of duality error majorant, in *Proc. 5th European Conference on Numerical Mathematics and Applications*, ed. by M. Feistauer, V. Dolejš, P. Knobloch, K. Najzar. Prague (2003)

[FNR06] M.E. Frolov, P. Neittaanmäki, S. Repin, Guaranteed functional error estimates for the Reissner-Mindlin plate problem. J. Math. Sci. (N.Y.) **132**(4), 553–561 (2006)

[FR06] M. Fuchs, S. Repin, A posteriori error estimates of functional type for variational problems related to generalized Newtonian fluids. Math. Methods Appl. Sci. **29**(18), 2225–2244 (2006)

[FR10] M. Fuchs, S. Repin, Estimates of the deviations from the exact solutions for variational inequalities describing the stationary flow of certain viscous incompressible fluids. Math. Methods Appl. Sci. **33**(9), 1136–1147 (2010)

[Fro04a] M. Frolov, On efficiency of the dual majorant method for the quality estimation of approximate solutions of fourth-order elliptic boundary value problems. Russ. J. Numer. Anal. Math. Model. **19**(5), 407–418 (2004)

[Fro04b] M. Frolov, Reliable control over approximation errors by functional type a posteriori estimates. Ph.D. thesis, Jyväskylä Studies in Computing 44, University of Jyväskylä, 2004

[FW60] G. Forsythe, W. Wasow, *Finite Difference Methods for Partial Differential Equations* (Wiley, New York, 1960)

[Gal15] B.G. Galerkin, Beams and plates. Series in some questions of elastic equilibrium of beams and plates. Vestn. Ingenerov, St. Petersbg. **19**, 897–908 (1915). Approximate translation of the title from Russian

[GGP99] V. Girault, R. Glowinski, T.W. Pan, A fictitious-domain method with distributed multiplier for the Stokes problem, in *Applied Nonlinear Analysis*, ed. by A. Sequera, H. Berao da Vega, J. Videman (Kluwer/Plenum, New York, 1999), pp. 159–174

[GH08] A. Günther, M. Hinze, A posteriori error control of a state constrained elliptic control problem. J. Numer. Math. **16**(4), 307–322 (2008)

[GHR07] A. Gaevskaya, R.W.H. Hoppe, S. Repin, Functional approach to a posteriori error estimation for elliptic optimal control problems with distributed control. J. Math. Sci. (N.Y.) **144**(6), 4535–4547 (2007)

[GK90] D.H. Greene, D.E. Knuth, *Mathematics for the Analysis of Algorithms*, 3rd edn. (Birkhäuser, Boston, 1990)

[GL96] G.H. Golub, C.F. Van Loan, *Matrix Computations*, 3rd edn. (Johns Hopkins University Press, Baltimore, 1996)

[Glo84] R. Glowinski, *Numerical Methods for Nonlinear Variational Problems* (Springer, New York, 1984)

[GLT76] R. Glowinski, J.L. Lions, R. Trémolierés, *Analyse Numérique des Inéquations Variationnelles* (Dunod, Paris, 1976)

[GMNR07] E. Gorshkova, A. Mahalov, P. Neittaanmäki, S. Repin, A posteriori error estimates for viscous flow problems with rotation. J. Math. Sci. (N.Y.) **142**(1), 1749–1762 (2007)

[GNR06] E. Gorshkova, P. Neittaanmäki, S. Repin, Comparative study of the a posteriori error estimators for the Stokes problem, in *Numerical Mathematics and Advanced Applications*, ed. by M. Feistauer, V. Dolejší, P. Knobloch, K. Najzar. ENUMATH 2003 (Springer, Berlin, 2006), pp. 252–259

[God59] S. Godunov, A difference method for numerical calculation of discontinuous solutions of the equations of hydrodynamics. Mat. Sb. **47**(89), 271–306 (1959). In Russian

[Gor07] E. Gorshkova, A posteriori error estimates and adaptive methods for incompressible viscous flow problem. Ph.D. thesis, Jyväskylä Studies in Computing 86, University of Jyväskylä, 2007

[Gou57] S.H. Gould, *Variational Methods for Eigenvalue Problems. An Introduction to the Methods of Rayleigh, Ritz, Weinstein, and Aronszajn* (University of Toronto Press, Toronto, 1957)

[GPP94a] R. Glowinski, T.-W. Pan, J. Périaux, A fictitious domain method for external incompressible viscous flow modeled by Navier-Stokes equations. Comput. Methods Appl. Mech. Eng. **112**(1–4), 133–148 (1994). Finite element methods in large-scale computational fluid dynamics (Minneapolis, MN, 1992)

[GPP94b] R. Glowinski, T.-W. Pan, J. Périaux, A fictitious domain method for Dirichlet problem and applications. Comput. Methods Appl. Mech. Eng. **111**(3–4), 283–303 (1994)

[GPP06] R. Glowinski, T.W. Pan, J. Periaux, Numerical simulation of a multi-store separation phenomenon: a fictitious domain approach. Comput. Methods Appl. Mech. Eng. **195**(41–43), 5566–5581 (2006)

[GR86] V. Girault, P.A. Raviart, *Finite Element Approximation of the Navier-Stokes Equations: Theory and Algorithms.* (Springer, Berlin, 1986)

[GR05] A. Gaevskaya, S. Repin, A posteriori error estimates for approximate solutions of linear parabolic problems. Differ. Equ. **41**(7), 970–983 (2005)

[GT74] W.B. Gragg, R.A. Tapia, Optimal error bounds for the Newton-Kantorovich theorem. SIAM J. Numer. Anal. **11**, 10–13 (1974)

[GT77] D. Gilbarg, N.S. Trudinger, *Elliptic Partial Differential Equations of Second Order* (Springer, Berlin, 1977)

[GTG76] E.G. Geisler, A.A. Tal, D.P. Garg, On a-posteriori error bounds for the solution of ordinary nonlinear differential equations, in *Computers and Mathematics with Applications*, ed. by E.Y. Rodin (Pergamon, Oxford, 1976), pp. 407–416

[Han08] A. Hannukainen, Functional type a posteriori error estimates for Maxwell's equations, in *Numerical Mathematics and Advanced Applications*, ed. by K. Kunisch, G. Of, O. Steinbach. ENUMATH 2007 (Springer, Berlin, 2008), pp. 41–48

[Hay79] Y. Hayashi, On a posteriori error estimation in the numerical solution of systems of ordinary differential equations. Hiroshima Math. J. **9**(1), 201–243 (1979)

[HCB04] I. Hlaváček, J. Chleboun, I. Babuška, *Uncertain Input Data Problems and the Worst Scenario Method* (Elsevier, Amsterdam, 2004)

[Her84] I. Herrera, Trefftz method, in *Basic Principles and Applications*, ed. by C.A. Brebbia. Topics in Boundary Element Research, vol. 1 (Springer, Berlin, 1984), pp. 225–253

[HH08] M. Hintermüller, R.H.W. Hoppe, Goal-oriented adaptivity in control constrained optimal control of partial differential equations. SIAM J. Control Optim. **47**(4), 1721–1743 (2008)

[HHIK08] M. Hintermüller, R.H.W. Hoppe, Y. Iliash, M. Kieweg, An a posteriori error analysis of adaptive finite element methods for distributed elliptic control problems with control constraints. ESAIM Control Optim. Calc. Var. **14**(3), 540–560 (2008)

[HK87] I. Hlaváček, M. Křížek, On a superconvergent finite element scheme for elliptic systems. I. Dirichlet boundary condition. Apl. Mat. **32**(2), 131–154 (1987)

[HK94] R.H.W. Hoppe, R. Kornhuber, Adaptive multilevel methods for obstacle problems. SIAM J. Numer. Anal. **31**(2), 301–323 (1994)

[HK10] R.H.W. Hoppe, M. Kieweg, Adaptive finite element methods for mixed control-state constrained optimal control problems for elliptic boundary value problems. Comput. Optim. Appl. **46**(3), 511–533 (2010)

[HM03] J. Haslinger, R.A.E. Mäkinen, *Introduction to Shape Optimization: Theory, Approximation, and Computation*. Advances in Design and Control, vol. 7 (Society for Industrial and Applied Mathematics, Philadelphia, 2003)

[HN96] J. Haslinger, P. Neittaanmäki, *Finite Element Approximation for Optimal Shape, Material and Topology Design*, 2nd edn. (Wiley, Chichester, 1996)

[HNW93] E. Hairer, S.P. Nørsett, G. Wanner, *Solving Ordinary Differential Equations. I: Nonstiff Problems*, 2nd edn. Springer Series in Computational Mathematics, vol. 8 (Springer, Berlin, 1993)

[HO02] R. Herbin, M. Ohlberger, A posteriori error estimate for finite volume approximations of convection diffusion problems, in *Finite Volumes for Complex Applications, III*. Porquerolles, 2002 (Hermes Sci. Publ., Paris, 2002), pp. 739–746

[HPS07] P. Houston, I. Perugia, D. Schötzau, An a posteriori error indicator for discontinuous Galerkin discretizations of $H(\mathrm{curl})$-elliptic partial differential equations. IMA J. Numer. Anal. **27**(1), 122–150 (2007)

[HRS00] P. Houston, R. Rannacher, E. Süli, A posteriori error analysis for stabilised finite element approximations of transport problems. Comput. Methods Appl. Mech. Eng. **190**(11–12), 1483–1508 (2000)

[HSV12] A. Hannukainen, R. Stenberg, M. Vohralík, A unified framework for a posteriori error estimation for the Stokes problem. Numer. Math. **122**(4), 725–769 (2012)

[HSW07] P. Houston, D. Schötzau, T.P. Wihler, Energy norm a posteriori error estimation of hp-adaptive discontinuous Galerkin methods for elliptic problems. Math. Models Methods Appl. Sci. **17**(1), 33–62 (2007)

[HTW02] B.-O. Heimsund, X.-C. Tai, J. Wang, Superconvergence for the gradient of finite element approximations by L^2 projections. SIAM J. Numer. Anal. **40**(4), 1263–1280 (2002)

[HW96] E. Hairer, G. Wanner, *Solving Ordinary Differential Equations II: Stiff and Differential-Algebraic Problems*, 2nd edn. (Springer, Berlin, 1996)

[HWZ10] R.H.W. Hoppe, H. Wu, Z. Zhang, Adaptive finite element methods for the Laplace eigenvalue problem. J. Numer. Math. **18**(4), 281–302 (2010)

[JH92] C. Johnson, P. Hansbo, Adaptive finite elements in computational mechanics. Comput. Methods Appl. Mech. Eng. **101**(1–3), 143–181 (1992)

[Jir78] J. Jirousek, Basis for development of large finite elements locally satisfying all field equations. Comput. Methods Appl. Mech. Eng. **14**, 65–92 (1978)

[JS95] C. Johnson, A. Szepessy, Adaptive finite element methods for conservation laws based on a posteriori error estimates. Commun. Pure Appl. Math. **48**(3), 199–234 (1995)

[JSV10] P. Jiranek, Z. Strakos, M. Vohralik, A posteriori error estimates including algebraic error and stopping criteria for iterative solvers. SIAM J. Sci. Comput. **32**(3), 1567–1590 (2010)

[KC67] J.B. Keller, D.S. Cohen, Some positone problems suggested by nonlinear heat generation. J. Math. Mech. **16**, 1361–1376 (1967)

[KF75] A.N. Kolmogorov, S.V. Fomin, *Introductory Real Analysis* (Dover, New York, 1975)

[KH99] J.P.C. Kleijnen, J.C. Helton, Statistical analyses of scatter plots to identify important factors in large-scale simulations, 1: review and comparison of techniques. Reliab. Eng. Syst. Saf. **65**, 147–185 (1999)

[Kim66] K.Y. Kim, A posteriori error analysis for locally conservative mixed methods. Math. Comput. **76**(257), 43–66 (2007)

[Kim07] K.Y. Kim, A posteriori error estimators for locally conservative methods of nonlinear elliptic problems. Appl. Numer. Math. **57**(9), 1065–1080 (2007)

[KM10] D. Kuzmin, M. Möller, Goal-oriented mesh adaptation for flux-limited approximations to steady hyperbolic problems. J. Comput. Appl. Math. **233**(12), 3113–3120 (2010)

[KN84] M. Křížek, P. Neittaanmäki, Superconvergence phenomenon in the finite element method arising from averaging gradients. Numer. Math. **45**(1), 105–116 (1984)

[KN87] M. Křížek, P. Neittaanmäki, On superconvergence techniques. Acta Appl. Math. **9**(3), 175–198 (1987)

[KNR03] S. Korotov, P. Neittaanmäki, S. Repin, A posteriori error estimation of goal-oriented quantities by the superconvergence patch recovery. J. Numer. Math. **11**(1), 33–59 (2003)

[KNS98] M. Křížek, P. Neittaanmäki, R. Stenberg (eds.), *Finite Element Methods: Superconvergence, Post-processing and a Posteriori Error Estimates*. Lecture Notes in Pure and Applied Mathematics, vol. 196 (Marcel Dekker, New York, 1998)

[Knu97] D.E. Knuth, *The Art of Computer Programming. Vol. 1: Fundamental Algorithms*, 2nd edn. (Addison-Wesley, Reading, 1969)

[Kol11] L.Yu. Kolotilina, On Ostrowski's disk theorem and lower bounds for the smallest eigenvalues and singular values. J. Math. Sci. (N.Y.) **176**(1), 68–77 (2011)

[Kop68] V.D. Kopchenov, Approximate solution of the Dirichlet problem by, fictitious domain method. Differ. Equ. **4**(1), 151–164 (1968). In Russian

[Kor96] R. Kornhuber, A posteriori error estimates for elliptic variational inequalities. Comput. Math. Appl. **31**(1), 49–60 (1996)

[KP03] O.A. Karakashian, F. Pascal, A posteriori error estimates for a discontinuous Galerkin approximation of second-order elliptic problems. SIAM J. Numer. Anal. **41**(6), 2374–2399 (2003)

[KR13] Yu. Kuznetsov, S. Repin, Guaranteed lower bounds of the smallest eigenvalues of elliptic differential operators. J. Numer. Math. **21**(2), 135–156 (2013)

[KRW07] A. Klawonn, O. Rheinbach, B. Wohlmuth, Dual-primal iterative substructuring for almost incompressible elasticity, in *Domain Decomposition Methods in Science and Engineering XVI*, ed. by O. Widlund, D.E. Keyes. Lect. Notes Comput. Sci. Eng., vol. 55 (Springer, Berlin, 2007), pp. 397–404

[KS78] J.R. Kuttler, V.G. Sigillito, Bounding eigenvalues of elliptic operators. SIAM J. Math. Anal. **9**(4), 767–773 (1978)

[KT13] S.K. Kleiss, S. Tomar, Guaranteed and sharp a posteriori error estimates in isogeometric analysis. Technical report, Johann Radon Institute for Computational and Applied Mathematics (RICAM), Austrian Academy of Sciences, 2013

[KS84] J.R. Kuttler, V.G. Sigillito, Eigenvalues of the Laplacian in two dimensions. SIAM Rev. **26**(2), 163–193 (1984)

[Kuz09] Yu.A. Kuznetsov, Lower bounds for eigenvalues of elliptic operators by overlapping domain decomposition, in *Domain Decomposition Methods in Science and Engineering XVIII*, ed. by M. Bercovier, M. Gander, R. Kornhuber, O. Widlund. Lect. Notes Comput. Sci. Eng., vol. 70 (Springer, Berlin, 2009), pp. 307–314

[Lad70] O.A. Ladyzhenskaya, *Mathematical Problems in the Dynamics of a Viscous Incompressible Fluid* (Nauka, Moscow, 1970). In Russian

[Lad85] O.A. Ladyzhenskaya, *The Boundary Value Problems of Mathematical Physics* (Springer, New York, 1985)

[Leo94] A.S. Leonov, Some a posteriori stopping rules for iterative methods for solving linear ill-posed problems. Comput. Math. Math. Phys. **34**(1), 121–126 (1994)

[LeV02] R. LeVeque, *Finite Volume Methods for Hyperbolic Problems* (Cambridge University Press, Cambridge, 2002)

[Lin94] E. Lindelöf, Sur l'application de la méthode des approximations successives aux équations différentielles ordinaires du premier ordre, in *Comptes Rendus Hebdomadaires des Séances de l'Académie des Sciences*, vol. 116, ed. by M.J. Bertrand (Impremerie Gauthier-Villars et Fils, Paris, 1894), pp. 454–457

[Lit00] V.G. Litvinov, *Optimization in Elliptic Boundary Value Problems with Applications to Mechanics of Deformable Bodies and Fluid Mechanics* (Birkhäuser, Basel, 2000)

[LL83] P. Ladevèze, D. Leguillon, Error estimate procedure in the finite element method and applications. SIAM J. Numer. Anal. **20**(3), 485–509 (1983)

[LM68a] J.L. Lions, E. Magenes, *Problèmes aux limites non homogènes et applications*, vol. 1 (Dunod, Paris, 1968)

[LM68b] J.L. Lions, E. Magenes, *Problèmes aux limites non homogènes et applications*, vol. 2 (Dunod, Paris, 1968)

[LM68c] J.L. Lions, E. Magenes, *Problèmes aux limites non homogènes et applications*, vol. 3 (Dunod, Paris, 1970)

[LNS07] M. Lyly, J. Niiranen, R. Stenberg, Superconvergence and postprocessing of MITC plate elements. Comput. Methods Appl. Mech. Eng. **196**(33–34), 3110–3126 (2007)

[LRT09] R. Lazarov, S. Repin, S. Tomar, Functional a posteriori error estimates for discontinuous Galerkin method. Numer. Methods Partial Differ. Equ. **25**(4), 952–971 (2009)

[LS76] O. Ladyzhenskaya, V. Solonnikov, Some problems of vector analysis, and generalized formulations of boundary value problems for the Navier-Stokes equation. Zap. Nauč. Semin. POMI **59**, 81–116 (1976). In Russian

[LT02] R. Lazarov, S. Tomov, A posteriori error estimates for finite volume element approximations of convection-diffusion-reaction equations. Comput. Geosci. **6**(3–4), 483–503 (2002). Locally conservative numerical methods for flow in porous media

[LU68] O.A. Ladyzhenskaya, N.N. Uraltseva, *Linear and Quasilinear Elliptic Equations* (Academic Press, New York, 1968)

[MA81] G. Marchuk, V. Agoshkov, *Introduction to Projection-Grid Methods* (Nauka, Moscow, 1981)

[Mal09] O. Mali, A posteriori error estimates for the Kirchhoff-Love arch model, in *Proceedings of the 10th Finnish Mechanics Days, Number 1 in A*, ed. by R. Mäkinen, P. Neittaanmäki, T. Tuovinen, K. Valpe. University of Jyväskylä (2009), pp. 315–323

[Mal11] O. Mali, Analysis of errors caused by incomplete knowledge of material data in mathematical models of elastic media. Ph.D. thesis, Jyväskylä Studies in Computing 132, University of Jyväskylä, 2011

[Mey92] P. Meyer, A unifying theorem on Newton's method. Numer. Funct. Anal. Optim. **13**(5–6), 463–473 (1992)

[Mik64] S.G. Mikhlin, *Variational Methods in Mathematical Physics* (Macmillan Co., New York, 1964). In Russian

[MM80] P. Mosolov, P. Myasnikov, *Mechanics of Rigid Plastic Bodies* (Nauka, Moscow, 1980). In Russian

[MN06] C. Makridakis, R.H. Nochetto, A posteriori error analysis for higher order dissipative methods for evolution problems. Numer. Math. **104**(4), 489–514 (2006)

[MNR12] S. Matculevich, P. Neittaanmäki, S. Repin, Guaranteed error bounds for a class of Picard-Lindelof iteration methods, in *Numerical Methods for Differential Equations, Optimization, and Technological Problems*, ed. by S. Repin, T. Tiihonen, T. Tuovinen (Springer, Berlin, 2012), pp. 151–168

[Mon98] P. Monk, A posteriori error indicators for Maxwell's equations. J. Comput. Appl. Math. **100**(2), 173–190 (1998)

[Mon03] P. Monk, *Finite Element Methods for Maxwell's Equations* (Oxford University Press, Oxford, 2003)

[Mor89] I. Moret, A Kantorovich-type theorem for inexact Newton methods. Numer. Funct. Anal. Optim. **10**(3–4), 351–365 (1989)

[MR03] A. Muzalevsky, S. Repin, On two-sided error estimates for approximate solutions of problems in the linear theory of elasticity. Russ. J. Numer. Anal. Math. Model. **18**(1), 65–85 (2003)

[MR08] O. Mali, S. Repin, Estimates of the indeterminacy set for elliptic boundary-value problems with uncertain data. J. Math. Sci. **150**(1), 1869–1874 (2008)

[MR10] O. Mali, S. Repin, Two-sided estimates of the solution set for the reaction-diffusion problem with uncertain data, in *Issue Dedicated to the Jubilee of Prof. R. Glowinski*, ed. by W. Fitzgibbon, Y. Kuznetsov, P. Neittaanmäki, J. Periaux, O. Pironneau. Comput. Methods Appl. Sci., vol. 15 (Springer, New York, 2010), pp. 183–198

[MR11a] O. Mali, S. Repin, Blowup of the energy increment caused by uncertainty of the Poisson' ratio in elasticity problems. Russ. J. Numer. Anal. Math. Model. **26**(4), 413–425 (2011)

[MR11b] A. Mikhaylov, S. Repin, Estimates of deviations from exact solution of the Stokes problem in the vorticity-velocity-pressure formulation. Zap. Nauč. Semin. POMI **397**(42), 73–88 (2011). In Russian

[MRV09] D. Meidner, R. Rannacher, J. Vihharev, Goal-oriented error control of the iterative solution of finite element equations. J. Numer. Math. **17**(2), 143–172 (2009)

[MS09] M.S. Mommer, R. Stevenson, A goal-oriented adaptive finite element method with convergence rates. SIAM J. Numer. Anal. **47**(2), 861–886 (2009)

[Néd80] J.C. Nédélec, Mixed finite elements in R^3. Numer. Math. **35**(3), 315–341 (1980)

[Nev89a] O. Nevanlinna, Remarks on Picard-Lindelöf iteration I. BIT Numer. Math. **29**(2), 328–346 (1989)

[Nev89b] O. Nevanlinna, Remarks on Picard-Lindelöf iteration II. BIT Numer. Math. **29**(3), 535–562 (1989)

[Nic05] S. Nicaise, On Zienkiewicz-Zhu error estimators for Maxwell's equations. C. R. Math. Acad. Sci. Paris **340**(9), 697–702 (2005)

[Nic06] S. Nicaise, A posteriori error estimation of some cell centered finite volume methods for diffusion-convection-reaction problems. SIAM J. Numer. Anal. **44**(3), 949–978 (2006)

[Nit81] J.A. Nitsche, On Korn's second inequality. RAIRO. Anal. Numér. **15**, 237–248 (1981)

[NK81] M. Natori, H. Kawarada, An application of the integrated penalty method to free boundary problems of Laplace equations. Numer. Funct. Anal. Optim. **3**(1), 1–17 (1981)

[NPW12] P. Neff, D. Pauly, K.-J. Witsch, Maxwell meets Korn: a new coercive inequality for tensor fields in $\mathbb{R}^{N\times N}$ with square-integrable exterior derivative. Math. Methods

Appl. Sci. **35**(1), 65–71 (2012)

[NR01] P. Neittaanmäki, S. Repin, A posteriori error estimates for boundary-value problems related to the biharmonic operator. East-West J. Numer. Math. **9**(2), 157–178 (2001)

[NR04] P. Neittaanmäki, S. Repin, *Reliable Methods for Computer Simulation, Error Control and a Posteriori Estimates* (Elsevier, New York, 2004)

[NR09] P. Neittaanmäki, S. Repin, Computable error indicators for approximate solutions of elliptic problems, in *ECCOMAS Multidisciplinary Jubilee Simposium*, ed. by J. Eberhardsteiner et al., Computational Methods in Applied Sciences (Springer, Berlin, 2009), pp. 199–212

[NR10a] P. Neittaanmäki, S. Repin, Guaranteed error bounds for conforming approximations of a Maxwell type problem, in *Applied and Numerical Partial Differential Equations*, ed. by W. Fitzgibbon, Y.A. Kuznetsov, P. Neittaanmäki, J. Periaux, O. Pironneau. Comput. Methods Appl. Sci., vol. 15 (Springer, New York, 2010), pp. 199–211

[NR10b] P. Neittaanmäki, S. Repin, A posteriori error majorants for approximations of the evolutionary Stokes problem. J. Numer. Math. **18**(2), 119–134 (2010)

[NR12] A. Nazarov, S. Repin, Exact constants in Poincare type inequalities for functions with zero mean boundary traces. Technical report, 2012. arXiv:1211.2224

[NRRV10] J.M. Nordbotten, T. Rahman, S. Repin, J. Valdman, A posteriori error estimates for approximate solutions of the Barenblatt-Biot poroelastic model. Comput. Methods Appl. Math. **10**(3), 302–314 (2010)

[NRT08] P. Neittaanmäki, S. Repin, P. Turchyn, New a posteriori error indicators in terms of linear functionals for linear elliptic problems. Russ. J. Numer. Anal. Math. Model. **23**(1), 77–87 (2008)

[NST06] P. Neittaanmäki, J. Sprekels, D. Tiba, *Optimization of Elliptic Systems. Theory and Applications*. Springer Monographs in Mathematics (Springer, New York, 2006)

[NW03] P. Neff, C. Wieners, Comparison of models for finite plasticity: a numerical study. Comput. Vis. Sci. **6**(1), 23–35 (2003)

[OBN$^+$05] J.T. Oden, I. Babushka, F. Nobile, Y. Feng, R. Tempone, Theory and methodology for estimation and control of errors due to modeling, approximation, and uncertainty. Comput. Methods Appl. Mech. Eng. **194**(2–5), 195–204 (2005)

[OC00] M. Olshanskii, E. Chizhonkov, On the best constant in the *infsup* condition for prolonged rectangular domains. Ž. Vyčisl. Mat. Mat. Fiz. **67**(3), 387–396 (2000). In Russian

[Ohl01] M. Ohlberger, A posteriori error estimate for finite volume approximations to singularly perturbed nonlinear convection-diffusion equations. Numer. Math. **87**(4), 737–761 (2001)

[Oña09] E. Oñate, *Structural Analysis with the Finite Element Method—Linear Statics. Volume 1. Basis and Solids*. Lecture Notes on Numerical Methods in Engineering and Sciences (International Center for Numerical Methods in Engineering (CIMNE), Barcelona, 2009). With a foreword by Robert L. Taylor

[OP01] J.T. Oden, S. Prudhomme, Goal-oriented error estimation and adaptivity for the finite element method. Comput. Math. Appl. **41**(5–6), 735–756 (2001)

[OPHK01] J.T. Oden, S. Prudhomme, D. Hammerand, M. Kuczma, Modeling error and adaptivity in nonlinear continuum mechanics. Comput. Methods Appl. Mech. Eng. **190**(49–50), 6663–6684 (2001)

[Ort68] J.M. Ortega, The Newton-Kantorovich theorem. Am. Math. Mon. **75**, 658–660 (1968)

[Ost72] A. Ostrowski, Les estimations des erreurs a posteriori dans les procédés itératifs. C. R. Acad. Sci. Paris Sér. A-B **275**, A275–A278 (1972)

[Pan85] P. Panagiotopoulos, *Inequality Problems in Mechanics and Applications* (Birkhäuser, Boston, 1985)

[Pay07] L.E. Payne, A bound for the optimal constant in an inequality of Ladyzhenskaya and Solonnikov. IMA J. Appl. Math. **72**(5), 563–569 (2007)

[PCL94] A.I. Pehlivanov, G.F. Carey, R.D. Lazarov, Least-squares mixed finite elements for second-order elliptic problems. SIAM J. Numer. Anal. **31**(5), 1368–1377 (1994)

[Pot85] F. Potra, Sharp error bounds for a class of Newton-like methods. Libertas Math. **5**, 71–84 (1985)

[PP80] F.-A. Potra, V. Ptak, Sharp error bounds for Newton's process. Numer. Math. **34**(1), 63–72 (1980)

[PP98] J. Peraire, A.T. Patera, Bounds for linear-functional outputs of coercive partial differential equations: local indicators and adaptive refinement, in *Advances in Adaptive Computational Methods in Mechanics*, ed. by P. Ladevéze, J.T. Oden (Elsevier, New York, 1998), pp. 199–216

[PPB12] G.M. Porta, S. Perotto, F. Ballio, Anisotropic mesh adaptation driven by a recovery-based error estimator for shallow water flow modeling. Int. J. Numer. Methods Fluids **70**(3), 269–299 (2012)

[PR09] D. Pauly, S. Repin, Functional a posteriori error estimates for elliptic problems in exterior domains. J. Math. Sci. (N.Y.) **163**(3), 393–406 (2009)

[PRR11] D. Pauly, S. Repin, T. Rossi, Estimates for deviations from exact solutions of the Cauchy problem for Maxwell's equations. Ann. Acad. Sci. Fenn. Math. **36**(2), 661–676 (2011)

[PS08] D. Peterseim, S. Sauter, The composite mini element-coarse mesh computation of Stokes flows on complicated domains. SIAM J. Numer. Anal. **46**(6), 3181–3206 (2008)

[PS11] D. Peterseim, S. Sauter, Finite element methods for the Stokes problem on complicated domains. Comput. Methods Appl. Mech. Eng. **200**(33–36), 2611–2623 (2011)

[PTVF07] W.H. Press, S.A. Teukolsky, W.T. Vetterling, B.P. Flanner, *Numerical Recipes. The Art of Scientific Computing*, 3rd edn. (Cambridge University Press, Cambridge, 2007)

[PW60] L.E. Payne, H.F. Weinberger, An optimal Poincaré inequality for convex domains. Arch. Ration. Mech. Anal. **5**, 286–292 (1960)

[Qn00] J. Qi-nian, Error estimates of some Newton-type methods for solving nonlinear inverse problems in Hilbert scales. Inverse Probl. **16**, 187–197 (2000)

[Ran00] R. Rannacher, The dual-weighted-residual method for error control and mesh adaptation in finite element methods, in *The Mathematics of Finite Elements and Applications, X*, ed. by J. Whiteman. MAFELAP 1999, Uxbridge (Elsevier, Oxford, 2000), pp. 97–116

[Ran02] R. Rannacher, Adaptive finite element methods for partial differential equations, in *Proceedings of the International Congress of Mathematicians*, vol. III, ed. by L.I. Tatsien. Beijing, 2002 (Higher Ed. Press, Beijing, 2002), pp. 717–726

[Rep94] S. Repin, Numerical analysis of nonsmooth variational problems of perfect plasticity. Russ. J. Numer. Anal. Math. Model. **9**(1), 61–74 (1994)

[Rep97a] S. Repin, A posteriori error estimation for nonlinear variational problems by duality theory. Zap. Nauč. Semin. POMI **243**, 201–214 (1997)

[Rep97b] S. Repin, A posteriori error estimation for variational problems with power growth functionals based on duality theory. Zap. Nauč. Semin. POMI **249**, 244–255 (1997)

[Rep99a] S. Repin, A posteriori estimates for approximate solutions of variational problems with strongly convex functionals. J. Math. Sci. **97**(4), 4311–4328 (1999)

[Rep99b] S.I. Repin, A unified approach to a posteriori error estimation based on duality error majorants. Math. Comput. Simul. **50**(1–4), 305–321 (1999). Modelling'98 (Prague)

[Rep00a] S. Repin, Estimates of deviations from exact solutions of elliptic variational inequalities. Zap. Nauč. Semin. POMI **271**, 188–203 (2000)

[Rep00b] S. Repin, A posteriori error estimation for variational problems with uniformly convex functionals. Math. Comput. **69**(230), 481–500 (2000)

[Rep01a] S. Repin, Estimates for errors in two-dimensional models of elasticity theory. J. Math. Sci. (N.Y.) **106**(3), 3027–3041 (2001)

[Rep01b] S.I. Repin, Two-sided estimates of deviation from exact solutions of uniformly elliptic equations, in *Proceedings of the St. Petersburg Mathematical Society, Volume IX*, ed. by N.N. Uralceva. Amer. Math. Soc. Transl. Ser. 2, vol. 209 (Am. Math. Soc., Providence, 2003), pp. 143–171. In Russian

[Rep02a] S. Repin, Estimates of deviations from exact solutions of initial boundary-value problem for the heat equation. Atti Accad. Naz. Lincei Cl. Sci. Fis. Mat. Natur. Rend. Lincei (9) Mat. Appl. **13**(2), 121–133 (2002)

[Rep02b] S. Repin, A posteriori estimates for the Stokes problem. J. Math. Sci. (N.Y.) **109**(5), 1950–1964 (2002)

[Rep04] S. Repin, Estimates of deviations from exact solutions for some boundary-value problems with incompressibility condition. Algebra Anal. **16**(5), 124–161 (2004)

[Rep07] S. Repin, Functional a posteriori estimates for Maxwell's equation. J. Math. Sci. (N.Y.) **142**(1), 1821–1827 (2007)

[Rep08] S. Repin, *A Posteriori Error Estimates for Partial Differential Equations* (de Gruyter, Berlin, 2008)

[Rep09a] S. Repin, Estimates of deviations from exact solutions of initial boundary-value problems for the wave equation. J. Math. Sci. (N.Y.) **159**(2), 229–240 (2009)

[Rep09b] S. Repin, Estimates of deviations from exact solutions of variational inequalities based upon Payne-Weinberger inequality. J. Math. Sci. (N.Y.) **157**(6), 874–884 (2009)

[Rep12] S. Repin, Computable majorants of constants in the Poincare and Friedrichs inequalities. J. Math. Sci. (N.Y.) **186**(2), 153–166 (2012)

[RF04] S.I. Repin, M.E. Frolov, An estimate for deviations from the exact solution of the Reissner-Mindlin plate problem. Zap. Nauč. Semin. POMI **310**, 145–157 (2004)

[Rhe80] W.C. Rheinboldt, On a theory of mesh-refinement processes. SIAM J. Numer. Anal. **17**(6), 766–778 (1980)

[Rit09] W. Ritz, Über eine neue Methode zur Lösung gewisser Variationsprobleme der mathematischen Physik. J. Reine Angew. Math. **135**, 1–61 (1909)

[Roc70] R.T. Rockafellar, *Convex Analysis* (Princeton University Press, Princeton, 1970)

[Rou97] T. Roubíček, *Relaxation in Optimization Theory and Variational Calculus*. de Gruyter Series in Nonlinear Analysis and Applications, vol. 4 (de Gruyter, Berlin, 1997)

[RR12] S. Repin, T. Rossi, On the application of a posteriori estimates of the functional type to quantitative analysis of inverse problems, in *Numerical Methods for Differential Equations, Optimization, and Technological Problems*, ed. by S. Repin, T. Tiihonen, T. Tuovinen (Springer, Berlin, 2012), pp. 109–120

[RS05] S. Repin, A. Smolianski, Functional-type a posteriori error estimates for mixed finite element methods. Russ. J. Numer. Anal. Math. Model. **20**(4), 365–382 (2005)

[RS06] S. Repin, S. Sauter, Functional a posteriori estimates for the reaction-diffusion problem. C. R. Math. Acad. Sci. Paris **343**(5), 349–354 (2006)

[RS07] S. Repin, R. Stenberg, A posteriori error estimates for the generalized Stokes problem. J. Math. Sci. (N.Y.) **142**(1), 1828–1843 (2007)

[RS08] S. Repin, R. Stenberg, A posteriori estimates for a generalized Stokes problem. Zap. Nauč. Semin. POMI **362**(39), 272–302 (2008)

[RS10a] S. Repin, S. Sauter, Computable estimates of the modeling error related to Kirchhoff-Love plate model. Anal. Appl. **8**(4), 409–428 (2010)

[RS10b] S. Repin, S. Sauter, Estimates of the modeling error for the Kirchhoff-Love plate model. C. R. Math. Acad. Sci. Paris **348**(17–18), 1039–1043 (2010)

[RSS03] S. Repin, S. Sauter, A. Smolianski, A posteriori error estimation for the Dirichlet problem with account of the error in the approximation of boundary conditions. Computing **70**(3), 205–233 (2003)

[RSS04] S. Repin, S. Sauter, A. Smolianski, A posteriori estimation of dimension reduction errors for elliptic problems in thin domains. SIAM J. Numer. Anal. **42**(4), 1435–1451 (2004)

[RSS07] S. Repin, S. Sauter, A. Smolianski, Two-sided a posteriori error estimates for mixed formulations of elliptic problems. SIAM J. Numer. Anal. **45**(3), 928–945 (2007)

[RSS12a] S. Repin, T. Samrowski, S. Sauter, Combined a posteriori modeling-discretization error estimate for elliptic problems with complicated interfaces. M2AN Math. Model. Numer. Anal. **46**(6), 1389–1405 (2012)

[RSS12b] S. Repin, T. Samrowski, S. Sauter, Two-sided estimates of the modeling error for elliptic homogenization problems. Technical report 12, University of Zurich, 2012

[RT91] J.E. Roberts, J.-M. Thomas, Mixed and hybrid methods, in *Handbook of Numerical Analysis*, vol. 2, ed. by I. Elishakoff (North-Holland, Amsterdam, 1991), pp. 523–639

[RT11] S. Repin, S. Tomar, Guaranteed and robust error bounds for nonconforming approximations of elliptic problems. IMA J. Numer. Anal. **31**(2), 597–615 (2011)

[RV08] S. Repin, J. Valdman, Functional a posteriori error estimates for problems with nonlinear boundary conditions. J. Numer. Math. **16**(1), 51–81 (2008)

[RV09] S. Repin, J. Valdman, Functional a posteriori error estimates for incremental models in elasto-plasticity. Cent. Eur. J. Math. **7**(3), 506–519 (2009)

[RV10] R. Rannacher, B. Vexler, Adaptive finite element discretization in PDE-based optimization. GAMM-Mitt. **33**(2), 177–193 (2010)

[RV12] R. Rannacher, J. Vihharev, Balancing discretization and iteration error in finite element a posteriori error analysis, in *Numerical Methods for Differential Equations, Optimization, and Technological Problems*, ed. by S. Repin, T. Tiihonen, T. Tuovinen (Springer, Berlin, 2012), pp. 85–108

[RWW10] R. Rannacher, A. Westenberger, W. Wollner, Adaptive finite element solution of eigenvalue problems: balancing of discretization and iteration error. J. Numer. Math. **18**(4), 303–327 (2010)

[RX96] S. Repin, L.S. Xanthis, A posteriori error estimation for elastoplastic problems based on duality theory. Comput. Methods Appl. Mech. Eng. **138**(1–4), 317–339 (1996)

[RX97] S. Repin, L.S. Xanthis, A posteriori error estimation for nonlinear variational problems. C. R. Acad. Sci., Sér. 1 Math. **324**(10), 1169–1174 (1997)

[Saa03] Y. Saad, *Iterative Methods for Sparse Linear System*, 2nd edn. (Society for Industrial and Applied Mathematics, Philadelphia, 2003)

[Sar82] J. Saranen, On an inequality of Friedrichs. Math. Scand. **51**(2), 310–322 (1982)

[Sch97] G.I. Schueller, A state-of-the-art report on computational stochastic mechanics. Probab. Eng. Mech. **12**(4), 197–321 (1997)

[Sch08] J. Schöberl, A posteriori error estimates for Maxwell equations. Math. Comput. **77**(262), 633–649 (2008)

[SDW+10] J.F. Shepherd, M.W. Dewey, A.C. Woodbury, S.E. Benzley, M.L. Staten, S.J. Owen, Adaptive mesh coarsening for quadrilateral and hexahedral meshes. Finite Elem. Anal. Des. **46**(1–2), 17–32 (2010)

[SF73] G. Strang, G. Fix, *An Analysis of the Finite Element Method* (Prentice Hall, Englewood Cliffs, 1973)

[SG89] A. Samarski, A. Gulin, *Numerical Methods* (Nauka, Moscow, 1989). In Russian

[Sha76] G. Shafer, *A Mathematical Theory of Evidence* (Princeton University Press, Princeton, 1976)

[SMGG12] Y. Sirois, F. McKenty, L. Gravel, F. Guibault, Hybrid mesh adaptation applied to industrial numerical combustion. Int. J. Numer. Methods Fluids **70**(2), 222–245 (2012)

[SO97] E. Stein, S. Ohnimus, Coupled model- and solution-adaptivity in the finite element method. Comput. Methods Appl. Mech. Eng. **150**(1–4), 327–350 (1997)

[SO00] M. Schulz, O. Steinbach, A new a posteriori error estimator in adaptive direct boundary element methods: the Dirichlet problem. Calcolo **37**(2), 79–96 (2000)

[Sob50] S.L. Sobolev, *Some Applications of Functional Analysis in Mathematical Physics* (Izdt. Leningrad. Gos. Univ., Leningrad, 1950). In Russian

[SRO07] E. Stein, M. Rüter, S. Ohnimus, Error-controlled adaptive goal-oriented modeling and finite element approximations in elasticity. Comput. Methods Appl. Mech. Eng.

196(37–40), 3598–3613 (2007)
[SS09] J. Schöberl, R. Stenberg, Multigrid methods for a stabilized Reissner-Mindlin plate formulation. SIAM J. Numer. Anal. **47**(4), 2735–2751 (2009)
[ST87] Y. Stephan, R. Temam, Finite element computation of discontinuous solutions in the perfect plasticity theory, in *Computational Plasticity, Part I, II*, ed. by D.R.J. Owen, E. Hinton, E. Oñate. Barcelona, 1987 (Pineridge, Swansea, 1987), pp. 243–256
[SV08] M. Schmich, B. Vexler, Adaptivity with dynamic meshes for space-time finite element discretizations of parabolic equations. SIAM J. Sci. Comput. **30**(1), 369–393 (2007/2008)
[SW05] Sh. Sun, M.F. Wheeler, $L^2(H^1)$ norm a posteriori error estimation for discontinuous Galerkin approximations of reactive transport problems. J. Sci. Comput. **22/23**, 501–530 (2005)
[SW10] D. Schötzau, T.P. Wihler, A posteriori error estimation for hp-version time-stepping methods for parabolic partial differential equations. Numer. Math. **115**(3), 475–509 (2010)
[SXZ06] R. Schneider, Y. Xu, A. Zhou, An analysis of discontinuous Galerkin methods for elliptic problems. Adv. Comput. Math. **25**(1–3), 259–286 (2006)
[SZ90] L.R. Scott, S. Zhang, Finite element interpolation of non-smooth functions satisfying boundary conditions. Math. Comput. **54**(190), 483–493 (1990)
[Tem79] R. Temam, *Navier-Stokes Equations: Theory and Numerical Analysis* (North-Holland, Amsterdam, 1979)
[TG51] S. Timoshenko, J.N. Goodier, *Theory of Elasticity* (McGraw-Hill, New York, 1951)
[TO76] K. Tsuruta, K. Ohmori, A posteriori error estimation for Volterra integro-differential equations. Mem. Numer. Math. **3**, 33–47 (1976)
[TR09] S. Tomar, S. Repin, Efficient computable error bounds for discontinuous Galerkin approximations of elliptic problems. J. Comput. Appl. Math. **226**(2), 358–369 (2009)
[Tre26] E. Trefftz, Ein Gegenstück zum Ritzschen Verfahren, in *Proc. 2nd. Int. Cong. Appl. Mech. Zurich*, ed. by G. Massing (Orell Füssli Verlag, Zürich, 1926), pp. 131–137
[TV05] D. Tiba, R. Vodak, A general asymptotic model for Lipschitzian curved rods. Adv. Math. Sci. Appl. **15**(1), 137–198 (2005)
[TY45] S. Timoshenko, D.H. Young, *Theory of Structures* (McGraw-Hill, New York, 1945)
[Val09] J. Valdman, Minimization of functional majorant in a posteriori error analysis based on $H(\mathrm{div})$ multigrid-preconditioned CG method. Adv. Numer. Anal. **2009**, 164519 (2009)
[Var62] R.S. Varga, *Matrix Iterative Analysis* (Prentice Hall, Englewood Cliffs, 1962)
[Ver89] R. Verfürth, A posteriori error estimators for the Stokes equations. Numer. Math. **55**(3), 309–325 (1989)
[Ver96] R. Verfürth, *A Review of a Posteriori Error Estimation and Adaptive Mesh-Refinement Techniques* (Wiley, New York, 1996)
[Ver98] R. Verfürth, A posteriori error estimates for nonlinear problems. $L^r(0,T;L^\rho(\Omega))$-error estimates for finite element discretizations of parabolic equations. Math. Comput. **67**(224), 1335–1360 (1998)
[Ver00] V.M. Vergbitskiy, *Numerical Methods. Linear Algebra and Non-linear Equations* (Nauka, Moscow, 2000)
[Ver03] R. Verfürth, A posteriori error estimates for finite element discretizations of the heat equation. Calcolo **40**(3), 195–212 (2003)
[Ver05] R. Verfürth, Robust a posteriori error estimates for nonstationary convection-diffusion equations. SIAM J. Numer. Anal. **43**(4), 1783–1802 (2005)
[Voh07a] M. Vohralík, A posteriori error estimates for finite volume and mixed finite element discretizations of convection-diffusion-reaction equations, in *Paris-Sud Working Group on Modelling and Scientific Computing 2006–2007*, ed. by J.-F. Gerbeau, S. Labbé. ESAIM Proc., vol. 18 (EDP Sci., Les Ulis, 2007), pp. 57–69
[Voh07b] M. Vohralík, A posteriori error estimates for lowest-order mixed finite element discretizations of convection-diffusion-reaction equations. SIAM J. Numer. Anal. **45**(4),

1570–1599 (2007) (electronic)

[VW08] B. Vexler, W. Wollner, Adaptive finite elements for elliptic optimization problems with control constraints. SIAM J. Control Optim. **47**(1), 509–534 (2008)

[Wah95] L.B. Wahlbin, *Superconvergence in Galerkin Finite Element Methods*. Lecture Notes in Mathematics, vol. 1605 (Springer, Berlin, 1995)

[Wan00] J. Wang, Superconvergence analysis of finite element solutions by the least-squares surface fitting on irregular meshes for smooth problems. J. Math. Study **33**(3), 229–243 (2000)

[Woh11] B. Wohlmuth, Variationally consistent discretization schemes and numerical algorithms for contact problems. Acta Numer. **20**, 569–734 (2011)

[WY02] J. Wang, X. Ye, Superconvergence analysis for the Navier-Stokes equations. Appl. Numer. Math. **41**(4), 515–527 (2002)

[Yam80] T. Yamamoto, Error bounds for computed eigenvalues and eigenvectors. Numer. Math. **34**(2), 189–199 (1980)

[Yam82] T. Yamamoto, Error bounds for computed eigenvalues and eigenvectors. II. Numer. Math. **40**(2), 201–206 (1982)

[Yam01] N. Yamamoto, A simple method for error bounds of the eigenvalues of symmetric matrices. Linear Algebra Appl. **324**(1–3), 227–234 (2001)

[YC06] J. Yang, Y. Chen, A unified a posteriori error analysis for discontinuous Galerkin approximations of reactive transport equations. J. Comput. Math. **24**(3), 425–434 (2006)

[Ye11] X. Ye, A posterior error estimate for finite volume methods of the second order elliptic problem. Numer. Methods Partial Differ. Equ. **27**(5), 1165–1178 (2011)

[Zad65] L.A. Zadeh, Fuzzy sets. Inf. Control **8**, 338–353 (1965)

[Zad78] L.A. Zadeh, Fuzzy sets as basis for theory of possibility. Fuzzy Sets Syst. **1**(1), 3–28 (1978)

[ZBZ98] O.C. Zienkiewicz, B. Boroomand, J.Z. Zhu, Recovery procedures in error estimation and adaptivity: adaptivity in linear problems, in *Advances in Adaptive Computational Methods in Mechanics*, ed. by P. Ladeveze, J.T. Oden. Cachan, 1997. Stud. Appl. Mech., vol. 47 (Elsevier, Amsterdam, 1998), pp. 3–23

[Zei86] E. Zeidler, *Nonlinear Functional Analysis and Its Applications. I: Fixed-Point Theorems* (Springer, New York, 1986)

[Zio09] E. Zio, Reliability engineering: old problems and new challenges. Reliab. Eng. Syst. Saf. **94**, 125–141 (2009)

[ZN05] Zh. Zhang, A. Naga, A new finite element gradient recovery method: superconvergence property. SIAM J. Sci. Comput. **26**(4), 1192–1213 (2005)

[Zou10] Q. Zou, Hierarchical error estimates for finite volume approximation solution of elliptic equations. Appl. Numer. Math. **60**(1–2), 142–153 (2010)

[ZSM05] Y. Zhou, R. Shepard, M. Minkoff, Computing eigenvalue bounds for iterative subspace matrix methods. Comput. Phys. Commun. **167**(2), 90–102 (2005)

[ZZ87] O.C. Zienkiewicz, J.Z. Zhu, A simple error estimator and adaptive procedure for practical engineering analysis. Int. J. Numer. Methods Eng. **24**(2), 337–357 (1987)

[ZZ88] J.Z. Zhu, O.C. Zienkiewicz, Adaptive techniques in the finite element method. Commun. Appl. Numer. Methods **4**, 197–204 (1988)

Index

O. Mali et al., *Accuracy Verification Methods*,
Computational Methods in Applied Sciences 32, DOI 10.1007/978-94-007-7581-7,

Printed in the United States
By Bookmasters

Printed in the United States
By Bookmasters